EMS Tracts in Mathematics 28

EMS Tracts in Mathematics

This series includes advanced texts and monographs covering all fields in pure and applied mathematics. *Tracts* will give a reliable introduction and reference to special fields of current research. The books in the series will in most cases be authored monographs, although edited volumes may be published if appropriate. They are addressed to graduate students seeking access to research topics as well as to the experts in the field working at the frontier of research.

For a complete listing see our homepage at www.ems-ph.org.

10 Vladimir Turaev, *Homotopy Quantum Field Theory*
11 Hans Triebel, *Bases in Function Spaces, Sampling, Discrepancy, Numerical Integration*
12 Erich Novak and Henryk Woźniakowski, *Tractability of Multivariate Problems. Volume II: Standard Information for Functionals*
13 Laurent Bessières et al., *Geometrisation of 3-Manifolds*
14 Steffen Börm, *Efficient Numerical Methods for Non-local Operators. $\mathcal{H}^2$-Matrix Compression, Algorithms and Analysis*
15 Ronald Brown, Philip J. Higgins and Rafael Sivera, *Nonabelian Algebraic Topology. Filtered Spaces, Crossed Complexes, Cubical Homotopy Groupoids*
16 Marek Janicki and Peter Pflug, *Separately Analytical Functions*
17 Anders Björn and Jana Björn, *Nonlinear Potential Theory on Metric Spaces*
18 Erich Novak and Henryk Woźniakowski, *Tractability of Multivariate Problems. Volume III: Standard Information for Operators*
19 Bogdan Bojarski, Vladimir Gutlyanskii, Olli Martio and Vladimir Ryazanov, *Infinitesimal Geometry of Quasiconformal and Bi-Lipschitz Mappings in the Plane*
20 Hans Triebel, *Local Function Spaces, Heat and Navier–Stokes Equations*
21 Kaspar Nipp and Daniel Stoffer, *Invariant Manifolds in Discrete and Continuous Dynamical Systems*
22 Patrick Dehornoy with François Digne, Eddy Godelle, Daan Kramer and Jean Michel, *Foundations of Garside Theory*
23 Augusto C. Ponce, *Elliptic PDEs, Measures and Capacities. From the Poisson Equation to Nonlinear Thomas–Fermi Problems*
24 Hans Triebel, *Hybrid Function Spaces, Heat and Navier-Stokes Equations*
25 Yves Cornulier and Pierre de la Harpe, *Metric Geometry of Locally Compact Groups*
26 Vincent Guedj and Ahmed Zeriahi, *Degenerate Complex Monge–Ampère Equations*
27 Nicolas Raymond, *Bound States of the Magnetic Schrödinger Operator*

Antoine Henrot
Michel Pierre

Shape Variation and Optimization

A Geometrical Analysis

European Mathematical Society

Authors:

Antoine Henrot
Institut Elie Cartan, UMR CNRS 7502
Université de Lorraine
Boulevard des Aiguillettes, B.P. 70239
54506 Vandœuvre-lès-Nancy Cedex
France
E-mail: Antoine.Henrot@univ-lorraine.fr

Michel Pierre
Univ Rennes
École Normale Supérieure de Rennes
Institut de Recherche Mathématique de Rennes
Campus de Ker Lann
35170 Bruz
France
E-mail: Michel.Pierre@ens-rennes.fr

2010 Mathematical Subject Classification: 49Q10, 49Q05, 49Q12, 49K20, 49K40, 53A10, 35R35, 35J20, 58E25, 31B15, 65K10, 93B27, 74P20, 74P15, 74G65, 76M30.

Key words: Shape optimization, optimum design, calculus of variations, variations of domains, Hausdorff convergence, continuity with respect to domains, Γ-convergence, shape derivative, geometry of optimal shapes, Laplace-Dirichlet problem, Neumann problem, overdetermined problems, isoperimetric inequality, capacity, potential theory, spectral theory, homogenization.

ISBN 978-3-03719-178-1

The Swiss National Library lists this publication in The Swiss Book, the Swiss national bibliography, and the detailed bibliographic data are available on the Internet at http://www.helveticat.ch.

Contact address:

European Mathematical Society Publishing House
Seminar for Applied Mathematics
ETH-Zentrum SEW A21
CH-8092 Zürich
Switzerland

Email: info@ems-ph.org
Homepage: www.ems-ph.org

Typeset using the authors' TEX files: Alison Durham, Manchester, UK
Printing and binding: Beltz Bad Langensalza GmbH, Bad Langensalza, Germany
∞ Printed on acid free paper
9 8 7 6 5 4 3 2 1

Foreword

This book is essentially the English version of the one published in French by the same authors [188], except for some additions and updates. It originated from graduate courses given in past years at the Universities of Nancy, Besançon, and Rennes in France.

The goal of these graduate courses was to provide an introduction to modern approaches to shape optimization, relying only on undergraduate level as a prerequisite, but reaching actual, current open questions of this very active domain.

The book is written in this same initial spirit. Some specific directions are more particularly developed, but all necessary mathematical tools and proofs are then provided in order to offer a self-contained presentation.

For instance, we provide all necessary knowledge on the classical capacity associated with the energy space H^1. We also develop the particular case of shape optimization associated with the Dirichlet problem for the Laplace operator: this is a simple but typical example that is significant among all the main questions that arise in shape optimization associated with more complex systems of partial differential equations. As apparent from this model example, it is important to understand the behavior of these systems under variations of their underlying domains. This explains why the two keywords "variation" and "optimization" appear in the title of this book.

In the same spirit, we chose to devote one full and extended chapter to the main question of differentiation with respect to shapes. This is a rather difficult but unavoidable topic that can rapidly become technical. We aimed at a mathematically rigorous presentation while being at the same time concerned with providing efficient tools for the actual computations of the shape derivatives (rigorous analysis and efficient calculus are somehow antagonistic in this topic).

We have also described all the various topologies on open subsets of $\mathbb{R}^N$, which are mostly used in the variation of shapes and in continuity properties for the associated PDEs. We tried to provide some kind of "FAQ" on this question.

The last two chapters address different questions. One is about qualitative geometric properties of optimal shapes: We chose to present several explicit examples in order to describe as many different methods as possible. The other one contains an introduction to quite different points of view in shape optimization which are recent and still in progress.

And we thought it was interesting to add a bibliographical footnote each time a new (noncontemporary) mathematician was quoted. Among other sources, we used the excellent book [168] by B. Hauchecorne and D. Suratteau, as well as the rich website http://www-history.mcs.st-and.ac.uk/Indexes/Full_Alph.html.

We would like to thank all the colleagues who helped us with their remarks on the French version and on preliminary versions of this one, in particular Marc Dambrine,

Evans Harrell, Antoine Laurain, Morgan Pierre, Yannick Privat, Takéo Takahashi, Michiel van den Berg and his group in Bristol.

Bruz and Nancy, April, 2017.

Antoine Henrot
Université de Lorraine
Institut Elie Cartan

Michel Pierre
Univ Rennes
École Normale Supérieure de Rennes
Institut de Recherche Mathématique de Rennes

Contents

Chapter 1

Introduction and examples

1.1 Introduction

Optimizing the shape of an object to make it the most efficient or the most resistant or the most streamlined or the lightest or the most noiseless or the most stealthy or the cheapest is clearly a very old task. But the development of computers and the recent explosion of modeling and scientific computing have given this topic new life. Many new and interesting questions have appeared. A mathematical topic was born: *shape optimization* (or *optimum design*).

In this field, we look for maxima or minima of functions, that is, we try to solve and analyze problems of the following kind: Find a solution Ω^* of

$$\Omega^* \in \mathcal{O}, \quad F(\Omega^*) = \min_{\Omega \in \mathcal{O}} F(\Omega), \tag{1.1}$$

where $\mathcal{O}$ is a class of subsets in $\mathbb{R}^N$ and F is a functional defined on $\mathcal{O}$ with values in $\mathbb{R}$. The elements of $\mathcal{O}$ are called *admissible* shapes or domains and F is called a *shape functional*.

The mathematical questions associated with (1.1) include the following:

- Existence of a solution. For that purpose, one needs to study
  - the dependence of the functional $F(\Omega)$ with respect to the domains Ω;
  - the control of the variation of $F(\cdot)$, or related functionals, with respect to the shapes;
  - the compactness properties of the class of admissible shapes.
- Identification of necessary or sufficient optimality conditions. Conditions of first order (the first derivative vanishes or is proportional to the derivative of the constraints) or second order (positivity of the Hessian) are obtained through the tool of the *shape derivative*.
- Geometric and qualitative properties of optimal shapes: connectedness, convexity, symmetries, regularity, or singularities of domains.
- The effective computation of optimal shapes.

This book deals with all of these points, except for questions related specifically to numerical calculations. For these questions, we refer, e.g., to [255] or [165].

In the first chapter, we present several problems of shape optimization. The first ones have an academic flavor whereas later there are some more concrete situations with possible industrial applications. Our intention was to gain insight into some representative problems. Many further examples can be found in the literature, for example in the books quoted below.

The second chapter is devoted to the study of some topologies for subsets of $\mathbb{R}^N$. Indeed, the mathematical proofs of existence for minimization problems most often consist in introducing some *adequate* topology together with compactness and continuity arguments. Thus, Chapter 3 is devoted to the study of continuity with respect to the domain of the solution of a boundary value problem. As shown in the examples below, many shape functionals involve the solution of some partial differential equation set on the domain. We concentrate, in particular, on the model problem of the Laplace equation with Dirichlet boundary conditions. We also recall definitions and properties of the classical capacity associated to the Sobolev space H^1. Chapter 4 presents several approaches to prove existence of a *classical solution* for a shape optimization problem. This kind of problem is *generically ill posed*. Therefore, to prove existence of a solution, one is often led either to restrict the class of admissible domains by imposing geometric restrictions, or else to extend the class of admissible domains by allowing *relaxed solutions*. This last point is explained in Chapter 7. Before that, Chapter 5 is devoted to the notion of the shape derivative, which allows the writing of first- and second-order optimality conditions. Finally, Chapter 6 studies the geometric properties of the solution of some shape optimization problems, such as possible symmetries, convexity, star-shapedness, and connectedness.

Several books on shape optimization have appeared during recent years. Let us mention, in particular, Pironneau 1984 [255], which gives a general overview on the topic; Banichuk 1990 [31] on structural optimization; Sokołowski–Zolésio 1992 [280], a rather general book that describes, among other things, the way to get optimality conditions; Bendsoe 1995 [35], which is more specifically devoted to structural optimization (many books on this topic are in fact written by mechanicians; we refer to [12] for a more complete bibliography); Haslinger–Neittaanmäki 1996 [165] about numerical aspects and, in particular, the finite element approach; Kawohl–Pironneau–Tartar–Zolésio 2000 [207], containing a series of lectures on recent progress on the topic; Tartar 2000 [287], which gathers the fundamental ideas of the author about homogenization; Delfour–Zolésio 2001 [131] with an approach based more on differential geometry and a deep use of the oriented-distance function; Allaire 2002 [12] with an extended view on topological optimization using the homogenization method; Bendsoe–Sigmund 2003, [38] also devoted to topological optimization; Laporte–Le Tallec 2003 [225], which gives an overview of different numerical methods in shape optimization; Haslinger–Mäkinen 2003 [164] with a particular focus on applications; Bucur–Buttazzo 2005 [65], where the authors study deeply the question of existence and the capacitary tools; Allaire 2007 [13], which is also a very good introduction to the topic; Mohammadi–Pironneau 2010 [236] for another overview of numerical methods in shape optimization, but for fluid mechanics; Plotnikov–Sokołowski 2012 [257], more specifically devoted to shape optimization problems for the compressible Navier–Stokes equations; Novotny–Sokołowski 2013 [247], which presents a complete study of the topological derivative in shape optimization; and the recent book [182] on spectral theory aspects. Let us also recall the important works by F. Murat and J. Simon [242], [243] on the shape derivative. Of course, many other books will be quoted at appropriate places in this book.

1.2 Some academic examples

1.2.1 Isoperimetric problems

These questions go back to ancient times. A classical example is the following: Suppose you have a fence with a given length and you want to find the shape of the largest field that can be surrounded by this fence. The Greeks already knew that the answer to this isoperimetric problem was given by a disk. The mathematical translation of this property is an isoperimetric inequality: if Ω is a plane domain with finite area $|\Omega|$ and perimeter $P(\Omega)$, we have

$$|\Omega| \leq \frac{1}{4\pi}|P(\Omega)|^2, \tag{1.2}$$

with equality only for the disk. The name "Queen Dido's problem" is often associated with a variant of this problem: The unknown domain Ω may have a part of its boundary a priori prescribed. Indeed, as mentioned in Virgil's *Aeneid*, Queen Dido, after escaping from Tyre, arrived at the site of the future Carthage. The local chief granted her a field as large as she could encircle with an oxhide. She had the idea of cutting the oxhide into a very thin (and long) strip. Then, she was led to the problem of finding the largest field surrounded by this strip but with the coast as a given part of its boundary. One can prove (e.g., using a first-order optimality condition; see Chapter 5) that the "free boundary", made by the strip, is the arc of a circle.

The analogue of the previous isoperimetric inequalityin any dimension N is given by

$$|\Omega|^{N-1} \leq \frac{1}{N^N V_N} P(\Omega)^N, \tag{1.3}$$

where the perimeter $P(\Omega)$ corresponds to the surface area of the boundary of the N-dimensional domain Ω, $|\Omega|$ is its volume (or Lebesgue[1] measure), and V_N is the

[1]Henri LEBESGUE, 1875–1941, French, taught in Rennes, Poitiers, and Collège de France. He made major

volume of the unit ball. Here, equality holds if and only if Ω is a ball. In fact, one had to wait until the 19th century to see a mathematical proof of (1.2). J. Steiner,[2] (see [283]) gave several subtle proofs assuming the existence of a domain achieving the minimum. His proof was later completed by C. Carathéodory.[3] The isoperimetric inequality in dimension 3 is due to H. A. Schwarz,[4] cf. [273], and the general case (1.3) to E. Schmidt[5] in [271]. Several other proofs appeared afterwards. We refer to [42], [45], or [250]. Let us mention, for example, an elementary proof of (1.2) using Fourier[6] series, in [42]. Most of the historical data that appear here are from [30]. We refer also to the nice survey by R. Osserman [250] and to F. Morgan's book [238] for other references on the isoperimetric inequality.

In the same way, we can consider the problem where we reverse the game with the volume $|\Omega|$ and the perimeter (or boundary surface) $P(\Omega)$:

$$\min\{P(\Omega),\ \Omega \subset \mathbb{R}^N,\ \text{measurable},\ |\Omega| = S_0\}. \tag{1.4}$$

In that context, the solution is still given by the ball. This is an obvious consequence of the isoperimetric inequality (1.2) or (1.3). We note that, whereas these isoperimetric questions can be easily stated and understood, the mathematical proofs are often difficult and some related problems remain open to this day. For example, if we ask for the surface in $\mathbb{R}^3$ of minimal area that contains a doubly connected domain Ω with components of specified volumes v_1 and v_2, the question is still open. It is only recently, in 1995, that J. Hass, M. Hutchings, and R. Schlafly, cf. [166] and [167], solved the particular case $v_1 = v_2$. They proved that the solution is given by two pieces of sphere separated by a disk. Let us remark that this question is closely related to the classical problems of minimal surface and of soap bubbles (see below). Indeed, the surface of a soap film has a minimal area for the volume that it contains, because the bubble tries to minimize its elastic energy, which is proportional to its area. A possible generalization of these questions is presented below: It is the problem of capillary surfaces.

---

contributions to integration theory and Fourier series.

[2]Jacob STEINER, 1796–1863, Swiss, was a professor in Berlin. He was considered the greatest geometer since the Greek, Appollonius. We will see him again in Chapter 6 where Steiner symmetrization appears.

[3]Constantin CARATHÉODORY, 1873–1950, German of Greek origin, made contributions to measure theory, the calculus of variations, and conformal mappings.

[4]Hermann Amandus SCHWARZ, 1843–1921, German, student of Weierstrass, professor at Göttingen and Berlin. He worked on analytic functions, partial differential equations, and potential theory. He is known for the famous Cauchy–Schwarz inequality and for the Schwarz theorem on the equality of second cross derivatives.

[5]Erhard SCHMIDT, 1876–1959, German, professor in Berlin. He worked in functional analysis, Hilbert spaces (he is the author of the famous Schmidt orthogonalization algorithm), and integral equations.

[6]Joseph FOURIER, 1768–1830, French, professor at the École Polytechnique until 1794, Prefect of Isère, then of Rhône. He was elected in 1816 to the French Académie des Sciences, then to the Académie française. He is well known for his mathematical modeling of heat conduction, including his famous dissertation on the analytical theory of heat with the introduction of trigonometric series. Nowadays, he is certainly one of the most cited mathematicians (although he was long neglected by the mathematical community); see [201].

In the previous examples, it is not easy to prove the existence of a minimizer. We refer, for instance, to the paper by F. Almgren and J. Taylor in [16] for the double-bubble problem, who use sophisticated tools from geometric measure theory. Chapter 4 of this book is devoted to this question of existence of a minimizer, especially for shape optimization problems linked to partial differential equations. Finally, let us remark that in the previous problems, we do not really have uniqueness of the solution, but only uniqueness up to displacements.

1.2.2 Minimal surfaces and capillary surfaces

The minimal surface problem is to find the equilibrium surface taken by a soap bubble or film. It is illustrated in many mathematical sections of scientific museums. One has a rigid frame (usually not planar) that is dipped into soapy water. The question is to determine the shape of the soap film fixed to the frame. Mathematically, we consider a curve γ in $\mathbb{R}^3$ and we can prove (by the principle of least energy) that the shape of the soap film corresponds to a surface S of boundary γ with a minimal area. Thus, we must solve the shape optimization problem

$$|S^*| = \min\{|S|,\ S \text{ surface of boundary } \gamma\},$$

where we denote by $|S|$ the area of the surface S. The surface S^* is called the **minimal surface** associated to γ. We will prove, in Chapter 5, that such a surface has the property that its mean curvature vanishes everywhere (outside γ).

The problem of capillary surfaces is a generalization of the previous one. One considers a tank containing some liquid. This liquid adheres to the wall by capillary action. The question is to find the shape of the air–liquid interface (the "free surface" of the liquid). For progress on this topic, we refer to the book by R. Finn [138]. Mathematically, the data consist of a bounded smooth open set D in $\mathbb{R}^3$, corresponding to the tank, and the volume of the liquid V_0. If we denote by Ω the domain occupied by the liquid, we can write the total energy of the system "tank plus liquid", which is the sum of the energy of surface tension, namely,

$$E_1(\Omega) := \text{Area}(\partial\Omega \cap D) + \cos\gamma\,\text{Area}(\partial\Omega \cap \partial D)$$

and, in addition, of the potential energy of gravity, given by

$$E_2(\Omega) := -\int_\Omega K(x)\,dx,$$

where γ is a prescribed angle and K is a bounded function, both being properties of the liquid and the tank wall.

The principle of least energy states that the shape of the optimal Ω is the one that minimizes the total energy. Hence, we are led to the shape optimization problem

$$\min\{E_1(\Omega) + E_2(\Omega),\ \Omega \subset D, |\Omega| = V_0\}.$$

With the tools developed in Chapter 5, one can prove that the solution of this problem has mean curvature that is equal to K everywhere on the free boundary $\partial\Omega \cap D$, which makes a fixed angle θ with the wall of the tank.

1.2.3 Eigenvalue problems

It is possible to consider a shape optimization problem for any functional depending on the shape of a domain. Another example that has brought about many serious works since the early part of the 20th century is the minimization of the eigenvalues of the Laplace[7] operator with different boundary conditions. Let us give some insight into this question, to be recalled in later chapters. For a more complete survey of the question, we refer to [180] or to the books [181], [182] and the references given therein.

1.2.3.1 Eigenvalues of the Laplacian with Dirichlet boundary conditions. The problem of minimizing the eigenvalues of the Laplacian with Dirichlet[8] boundary conditions is a long story. It is written

$$\min\{\lambda_k(\Omega),\ \Omega \subset D,\ |\Omega| = A \text{ (given in } (0,\infty))\}, \tag{1.5}$$

where D is an open subset of $\mathbb{R}^N$ and $\lambda_k(\Omega)$, $k \geq 1$ is the kth eigenvalue of the Laplace operator on Ω with homogeneous Dirichlet boundary conditions. In fact, this question appeared (with $D = \mathbb{R}^N$ and $k = 1$) in the celebrated book of Lord Rayleigh,[9] *The Theory of Sound* (e.g., in the 1894 edition). Thanks to some explicit computation and "physical evidence", Lord Rayleigh conjectured that the disk should minimize λ_1.

It was in the 1920s that Faber[10] [136] and Krahn[11] [216] simultaneously solved Rayleigh's conjecture using a symmetrization technique. This classical proof is given in Theorem 6.1.9. The case of the second eigenvalue of the Dirichlet–Laplacian is an easy corollary as shown by E. Krahn in [217]: The minimum of λ_2 is achieved by the union of two disjoint identical balls. Krahn's remark actually went unnoticed and this result is often attributed to P. Szego, for instance by G. Pólya[12] in [259], and

[7]Pierre Simon, Marquis de LAPLACE, 1749–1827, great French astronomer and mathematician. He contributed deeply to the theory of gravitation, to celestial mechanics, and to mathematical tools including ordinary and partial differential equations. He also made remarkable contributions to probability theory.

[8]Peter Gustav LEJEUNE-DIRICHLET, 1805–1859, German, professor in Breslau, Berlin, and Göttingen Universities. He made contributions in number theory, Fourier series, and to the Dirichlet problem.

[9]John RAYLEIGH, 1842–1919, English, more a physicist than a mathematician. He worked on acoustics, optics, and wave propagation. He received the Nobel Prize in Physics in 1904.

[10]Georg FABER, 1877–1966, German, professor in Munich. The polynomial that allows the development of analytic functions in a plane domains bears his name.

[11]Edgar KRAHN, 1894–1961, Estonian, played an important role in the development of mathematics in Estonia. He worked in differential geometry, probability, gas dynamics, and elasticity.

[12]György PÓLYA, 1887–1985, Hungarian, then naturalized American, taught in Zürich before going to the

has also been rediscovered several times, for instance by Hong in [198]. The proof is also given in Chapter 6.

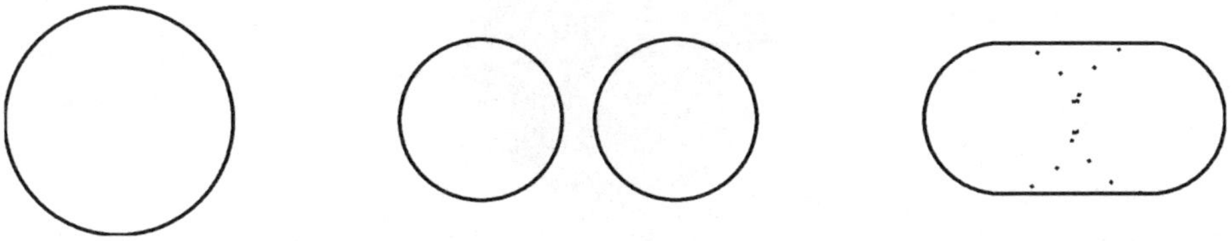

Figure 1.1. The disk minimizes λ_1 (left); two identical disks minimize λ_2 (center); the "stadium" does not minimize λ_2 among convex sets, but almost (right).

Things are much more complicated from the third eigenvalue onwards. The existence of a minimizer for λ_k, $k \geq 3$ (always with a fixed volume) is now proved. A proof was given in 1993 by Buttazzo–Dal Maso [79] in the case when D is bounded. It is obtained as a particular case of their general result for functionals that are nonincreasing with respect to the inclusion (see the discussion in Chapter 4). The question of existence is much more difficult when minimizing among all domains in $\mathbb{R}^N$ (with prescribed Lebesgue measure) because of the lack of compactness. A first proof of existence was obtained for λ_3 in [69]. Then a proof for all $k \geq 3$ was obtained more recently in [64], [231], [72] by different methods including concentration–compactness, shape subsolutions, or surgery. We refer to the end of Chapter 4 and to [182] for a more detailed discussion of this unbounded situation.

The domain that realizes the optimum in $\mathbb{R}^N$ is however not known when $k \geq 3$. For λ_3, it is conjectured that it is the ball if $N = 2$ and the union of three identical balls if $N \geq 4$. In dimension $N = 3$, numerical computations suggest that the optimum is some kind of "pinched ball" (see [22] or [182, Ch. 11]).

When $N = 2$, Wolf and Keller proved in [293] that the disk is a *local* minimizer for λ_3. It is also proved in [69] that the optimum for λ_3 is connected in dimensions $N = 2, 3$; see Corollary 6.4.4 in Chapter 6.

For λ_4 and $N = 2$, it is conjectured that the minimum is achieved by the union of two disks with radii in a ratio of $\sqrt{j_{0,1}/j_{1,1}}$, where $j_{0,1}$ and $j_{1,1}$ are respectively the first positive zeros of the Bessel[13] functions J_0 and J_1; see Figure 1.3.

P. Szego was led to conjecture that any eigenvalue of the Dirichlet–Laplacian in the plane should achieve its minimum for a ball or a union of balls. This conjecture turns out to be false, as is now known. Wolf and Keller [293] noticed that for the 13th eigenvalue, the square is better than any union of disks of the same area! Numerical computations (see [251], [22], [40], Figure 1.3) suggest that, as soon as the 5th

United States. His works are famous in number theory, functional analysis, and statistics.

[13]Friedrich Wilhelm BESSEL, 1784–1846, born in Westphalia (Germany). An astronomer and geodesist, he studied the functions named after him, useful for the three-body problem.

Figure 1.2. The three-dimensional domain that minimizes λ_3 (by courtesy of A. Berger and E. Oudet).

eigenvalue for $N = 2$, the minimum is no longer attained by the disk or by a union of disks. A detailed analysis may be found in the recent paper [39], where it is, in particular, rigorously proved that for $k \geq 5$ in $\mathbb{R}^2$, minimizers are not unions of disks.

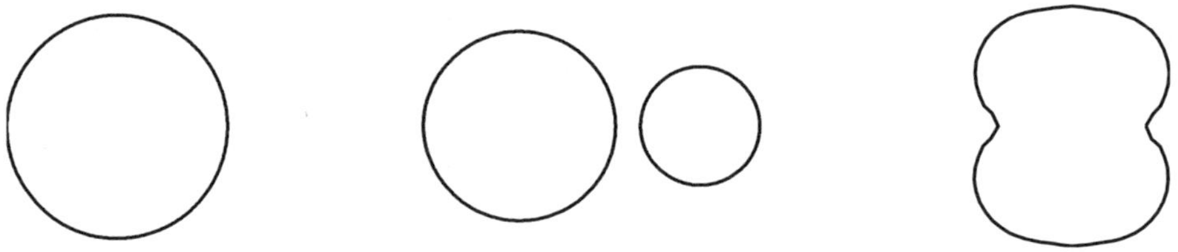

Figure 1.3. The disk probably minimizes λ_3 (left); two disks that probably minimize λ_4 (center); a candidate to be the minimizer of λ_5 (right).

Let us mention here two other noteworthy results in this context. The first one was obtained by M. Ashbaugh and R. Benguria. They were able to solve a long-standing conjecture of Payne–Pólya–Weinberger that the ball realizes the minimum of the ratio $\frac{\lambda_1}{\lambda_2}$ among sets of given volume; cf. [24]. The second one is the so-called **gap problem**. An important quantity in quantum mechanics is the gap between the two first eigenvalues of the Laplacian or, more generally, of a Schrödinger operator: $\gamma := \lambda_2 - \lambda_1$. Some bounds have been obtained for the gap when Ω is assumed to be convex, in terms of the diameter of Ω (see, e.g., [279], [296]), but the optimal bound has been conjectured for a long time (presumed stated for the first time by Michiel Van den Berg in [289]). This problem can be seen as the shape optimization problem for the functional $\lambda_2 - \lambda_1$ among convex sets with a diameter constraint. It has been recently solved by B. Andrews and J. Clutterbuck in [20], where the authors proved that, for any convex domain in $\mathbb{R}^N$,

$$\lambda_2(\Omega) - \lambda_1(\Omega) \geq \frac{3\pi^2}{d^2},$$

where d is the diameter of Ω. This bound is optimal since, for a sequence of rectangles (in 2 dimensions) $\Omega_\varepsilon = (0, \sqrt{d^2 - \varepsilon^2}) \times (0, \varepsilon)$ the inequality is sharp (indeed, we have $\lambda_2(\Omega_\varepsilon) - \lambda_1(\Omega_\varepsilon) \to 3\pi^2/d^2$). In other words, in spite of the convexity constraint, which may help in many situations to prove existence (see Chapter 4), in this shape optimization problem there is no existence of a minimizer.

Currently, the list of open problems for eigenvalues is impressive. We state some of them below, but for other problems and a complete bibliography, we refer to [23], [26], [180], [181], [182] [252], [253], [272], [294], [295].

Open Problem 1. Let D be a fixed bounded open set. According to Buttazzo and Dal Maso's theorem (see [79] or Theorem 4.7.6 and Corollary 4.7.12), problem (1.5) has a solution that is a priori a quasi-open set (see Definition 3.3.38).

Assume $k = 1$. If the constant A is small enough so that D contains a ball of volume A, then this ball is the solution of the problem according to the previous discussion. The interesting case is thus when D does not contain a ball of volume A. In [173], M. Hayouni proved that there always exists a solution that is an open set. In fact, if D is connected, every minimizer is open (see [57]) and rather regular (analytic up to a small singular set in any dimension; see [58]). If D is not connected, it is possible to build pathological examples; see Exercise 4.10. If D does not contain a ball of measure A, it can be proved that the minimizer Ω^* touches the boundary of D; see [181]. A possible proof, assuming regularity of Ω^* is as follows. Using the optimality condition as in Chapter 5, one can prove that the normal derivative of the first eigenfunction has to be constant on the free part of the boundary of Ω^*. Then by Serrin's theorem (see [275] and Theorem 6.1.11), Ω^* must be a ball. Thus the boundary of Ω^* consists of some pieces that lie on the boundary of D as well as free parts. One may wonder whether these free parts could be pieces of a sphere (as for minimal surfaces). It is not in fact the case, as proved in [184]. But the free parts are regular (analytic) in dimension $N = 2$ [58], and, in higher dimensions they are regular outside a set of small Hausdorff dimension. Its optimal dimension is not known (the conjecture is $N-7$; see the discussion in [182]). Finally, one may wonder whether Ω^* is convex when D is convex.

Open Problem 2. Prove that, in the class of polygons with n sides with given area, λ_1 is minimum for the regular polygon. This result is known only for $n = 3$ and $n = 4$ (it is due to Pólya and can be easily proved using Steiner symmetrization), but for $n \geq 5$ this technique is no longer efficient since Steiner symmetrization increases the number of sides!

Open Problem 3. Let $A \in (0, \infty)$, $D = \mathbb{R}^N$, $k = 3$. Prove that the solution of (1.5) is a ball if $N = 2$ and the union of three disjoint identical balls if $N \geq 4$.

Open Problem 4. Prove that λ_4 is minimal in (1.5) for the union of two balls in dimension $N = 2$ and $D = \mathbb{R}^N$.

Open Problem 5. The minimum of λ_2 in (1.5) being achieved by the union of two identical balls in $\mathbb{R}^N$, it is tempting to consider the same problem in the class of connected open sets. In fact, it is easy to prove that the minimum is not achieved in this class. Indeed if one looks at the set Ω_ε obtained by joining two balls with a thin pipe of thickness ε (a dumbbell shape), a simple argument based on the so-called γ-convergence (see Chapter 3) shows that the eigenvalues of Ω_ε converge to the eigenvalues of the two balls when $\varepsilon \to 0$. Therefore, the infimum of λ_2 is the same with or without the connectedness assumption and is achieved by a nonconnected set (two disjoint balls as already said).

Now it is interesting to ask the same question in the class of **convex sets**. It is easy to prove the existence of a solution (see, e.g., [111]), but it is much more difficult to identify it. It was conjectured that the minimizer is the convex hull of two identical balls (the so-called **stadium**; see Figure 1.1), which is the convex set "closest" to the absolute minimum. However, in [184], the authors refute this conjecture. Nevertheless, they provide numerical results and analytic arguments indicating that the optimal domain is close to the stadium. But its regularity is different as proved in [220]: The junction between the flat and strictly convex parts is only $C^{1,1/2}$ while it is $C^{1,1}$ for the stadium. It also remains open to prove that the optimal domain has one or two axes of symmetry.

1.2.3.2 Eigenvalues of the Laplacian with other boundary conditions. Let us denote by $0 = \mu_1 \leq \mu_2 \leq \cdots \leq \mu_n$ the eigenvalues of the Laplacian with Neumann boundary conditions. The minimization problem for the eigenvalues of the Neumann-Laplacian with volume constraint is trivial. Indeed, if one considers a very long rectangle like $(0, L) \times (0, l)$, one can always assume, if L is large enough, that the nth eigenvalue of the rectangle is $\mu_n = \frac{(n-1)^2\pi^2}{L^2}$. Therefore, letting L go to ∞, we see that the infimum of μ_n is always 0. In fact, it is achieved for any open set that is the union of at least n connected components. If we want to consider an interesting problem concerning the eigenvalues of the Neumann-Laplacian, we must look at the *maximization* problem among sets of given volume. For example, G. Szegő in dimension 2 and H. Weinberger in any dimension, proved in [286] and [290] (see also [272]) that the maximum of μ_2 is achieved by the ball. The case of the second nontrivial eigenvalue μ_3 has been (partially) solved by A. Girouard, N. Nadirashvili, and I. Polterovich [152] in 2 dimensions and for simply connected domains. The union of two identical balls is still the optimal domain. More generally, the existence of a convex domain that maximizes the nth eigenvalue μ_n was proved by S. Cox and M. Ross in [112]. For more results and a general survey, we refer to [181] and [34]. It is still an open problem to prove existence for $n \geq 3$ and, of course, to identify the possible maximizer.

Now, let us consider a third type of boundary condition, or Robin boundary condition, like $\frac{\partial u}{\partial n} + \alpha u = 0$ on the boundary of Ω. It can be seen as an interpolation between the Neumann case (for $\alpha = 0$) and the Dirichlet case (for $\alpha = +\infty$). For the first eigenvalue and for $\alpha > 0$ one can prove a result similar to the Rayleigh–Faber–Krahn inequality: the ball minimizes the first eigenvalue of the Laplacian with this boundary condition. This result is due to M. H. Bossel in dimension 2 and has been generalized to higher dimensions by D. Daners; see [48] and [122]. The case of the second eigenvalue for $\alpha > 0$ has been solved by J. Kennedy; see [210]: the union of two identical balls is still the optimal domain. For higher eigenvalues, the situation is more complex: The existence of an optimal domain is not known and, moreover, it has been proved that the optimizers should depend on the parameter α. Let us observe that, on the other hand, the case $\alpha < 0$ is largely open even for the first eigenvalue, where this is now the maximization problem that is relevant. In [141], P. Freitas and D. Krejcirik observe that in 2 dimensions the ball is the maximizer for small values of the negative parameter α but it is not true for large values as asymptotically the ball loses out to spherical shells! It is the only situation known so far for the Laplace operator where the first eigenvalue is not optimized by a ball! We refer to the survey chapter in [182, Ch. 4] for more details on the Robin problem.

Let us also mention another classical eigenvalue problem for the Laplace operator, which is the Steklov[14] problem. In that case, this is the maximization problem that is interesting. Under the volume constraint, the fact that the ball is the extremal domain has been proved by R. Weinstock in 2 dimensions (see [292]) and generalized by F. Brock in higher dimensions; see [59]. A perimeter constraint appears to be more natural from the following viewpoint: the Steklov eigenvalues are the eigenvalues of the Dirichlet-to-Neumann operator, which is an operator defined on the boundary of the domain. In that case, R. Weinstock has proved that the disk is the maximizer in the class of simply connected planar domains. However, this topological assumption cannot be removed, as can be seen from the example of an annulus. Moreover, for general Euclidean domains the question of existence of a maximizer for the first eigenvalue with a perimeter constraint remains open. Nevertheless, it is known that for simply connected planar domains, the kth normalized Steklov eigenvalue is maximized in the limit by a disjoint union of $k - 1$ identical disks for any $k \geq 2$. For more details, we refer to the survey by A. Girouard and I. Polterovich [153] reprinted in [182, Ch. 5].

[14]Vladimir Andreevich STEKLOV, 1864–1926, Russian. He made important contributions to the study of boundary value problems for partial differential equations with applications to electrostatics and hydrodynamics.

1.2.3.3 How to place an obstacle. Let Ω be a fixed bounded domain in $\mathbb{R}^N$ and let ω be a hole for which only the shape is fixed. For example, we can consider a circular hole and, in this case, ω is a small disk.

The problem is to determine the location of the hole ω in the set Ω that maximizes or minimizes the first eigenvalue $\lambda_1(\Omega\backslash\omega)$ of the Laplacian with Dirichlet boundary conditions. Physical intuition suggests that $\lambda_1(\Omega\backslash\omega)$ will be minimum when the hole ω touches the boundary of Ω, while it will be maximum when ω is in the "center" of Ω in some sense. The maximization problem was first studied by J. Hersch in [195]. He proved that, among every annular domain of area A whose exterior and interior boundaries have respective lengths L and L_0 satisfying $L^2 - L_0^2 = 4\pi A$, the one that realizes the maximum first eigenvalue of the Dirichlet–Laplacian is the circular ring with concentric circles. This result has been rediscovered several times (see, e.g., S. Kesavan in [211] or E. Harrell, P. Kröger, and K. Kurata in [163]). In this last paper, several extra results are given when Ω is convex with some hyperplane symmetries. The general case seems to remain open. Let us remark that one can also consider a more general question where the measure of the hole or its perimeter is given but not its shape. See, for example, [194] for a discussion on this kind of problem.

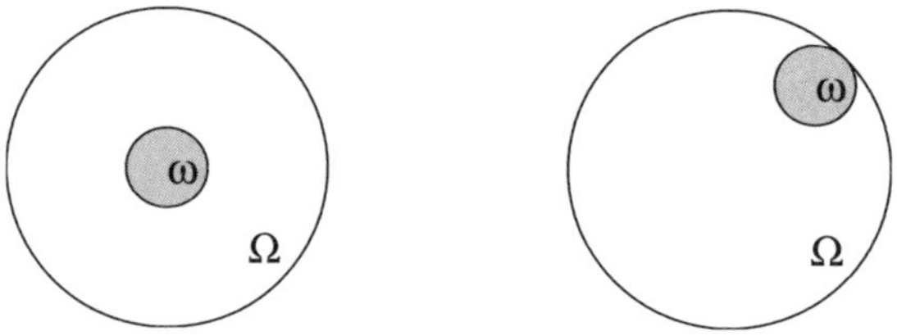

Figure 1.4. The location of the hole that maximizes $\lambda_1(\Omega\backslash\omega)$ (left); one location that minimizes $\lambda_1(\Omega\backslash\omega)$ (right).

1.2.3.4 Eigenvalues of other operators. One can also be interested in similar questions for other operators. For example, some famous conjectures concern the bi-Laplacian operator $\Delta^2 = \Delta(\Delta)$.

The clamped plate. We consider the eigenvalue problem

$$\begin{cases} \Delta^2 u = \Gamma(\Omega) u & \text{in } \Omega, \\ u = 0 & \text{on } \partial\Omega, \\ \frac{\partial u}{\partial n} = 0 & \text{on } \partial\Omega. \end{cases} \tag{1.6}$$

Lord Rayleigh conjectured that the first eigenvalue of this problem should be minimized by the ball. G. Pólya and G. Szegő gave a proof of this conjecture *assuming* that the eigenfunction was nonnegative. Unfortunately, this is not always true for the bi-Laplacian, which does not have a maximum principle. Recently, this conjecture has been proved in dimension 2 by N. S. Nadirashvili in [246] and by M. Ashbaugh and R. Benguria in dimension 3; see [25]. The problem remains open in higher dimensions.

The buckling plate. Here we consider the eigenvalue problem

$$\begin{cases} -\Delta^2 u = \Lambda(\Omega)\Delta u & \text{in } \Omega, \\ u = 0 & \text{on } \partial\Omega, \\ \frac{\partial u}{\partial n} = 0 & \text{on } \partial\Omega. \end{cases} \tag{1.7}$$

This time, it is G. Pólya and G. Szegő who conjectured that the ball should minimize the first eigenvalue of this problem (among sets of given volume). They were able to prove it, assuming that the first eigenfunction was nonnegative, which is not true in general. The problem remains open despite important contributions. More precisely, H. Weinberger and B. Willms found a nice proof assuming the existence and regularity of a minimizing domain; see [181] or [207]. Later, M. Ashbaugh and D. Bucur in [27] proved the existence of a minimizer in the class of simply connected open sets. Now, it remains to prove regularity to fill the gap.

1.3 Some other examples with applications

1.3.1 Electromagnetic shaping of liquid metal

Let us consider a column of liquid metal falling into an electromagnetic field created by an alternating current running through vertical conductors (see Figure 1.5). We are interested in the shape of a horizontal section of the column (we assume that the problem is locally invariant under vertical translations). Therefore, it is still a *free boundary problem*. The unknown shape is an equilibrium for the sum of the different forces: The electromagnetic forces created by the total electromagnetic field, the surface tension forces, and the difference in pressure between the interior and the exterior of the liquid.

The vertical invariance hypothesis allows us to consider a two-dimensional model. We also assume that the frequency of the current is high so that we can use a magnetostatic model involving the effective values of the physical quantities. We refer to [185], [186] for more details and a mathematical study of this problem, and to [108] and [254] for a numerical study and several examples.

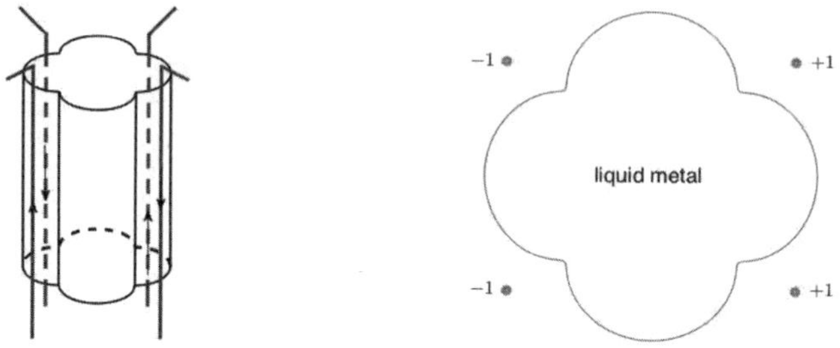

Figure 1.5. Column of liquid metal between linear conductors (left); its horizontal section (right).

Let us denote by Ω the horizontal section of the column, which is the domain occupied by the liquid metal; see Figure 1.5. The total magnetic field $\vec{B}$ is a solution of Maxwell's[15] equations:

$$\begin{cases} \operatorname{rot} \vec{B} = \mu_0 \vec{j_0} & \text{in } \Omega^c, \\ \operatorname{div} \vec{B} = 0 & \text{in } \Omega^c, \\ \vec{B}.\vec{n} = 0 & \text{on } \partial\Omega, \end{cases} \tag{1.8}$$

where $\mu_0 = 4\pi \times 10^{-7}$ is the magnetic permeability of the vacuum, $\vec{j_0} = (0, 0, j_0)$ is the (vertical) vector of current density, and $\vec{n}$ is the exterior unit normal vector to Ω. The relation $\operatorname{div} \vec{B} = 0$ implies that the field $\vec{B}$ derives from a potential, that is, $\vec{B}$ can be written

$$\vec{B} = \begin{pmatrix} \frac{\partial\psi}{\partial y} \\ -\frac{\partial\psi}{\partial x} \\ 0 \end{pmatrix}, \tag{1.9}$$

where ψ is a function defined on Ω^c (the complement of Ω) and bounded at ∞. Of course, ψ is defined only up to a constant. Moreover, the relation $\vec{B}.\vec{n} = 0$ on $\partial\Omega$ shows that ψ is constant on $\partial\Omega$, and since ψ is defined up to a constant, we can assume without loss of generality that $\psi = 0$ on $\partial\Omega$. Therefore, system (1.8) can be written

$$\begin{cases} -\Delta\psi = \mu_0 j_0 & \text{in } \Omega^c, \\ \psi = 0 & \text{on } \partial\Omega, \\ \psi & \text{bounded at } \infty. \end{cases} \tag{1.10}$$

[15]James Clerk MAXWELL, 1831–1879, Scottish physicist and mathematician. He discovered the fundamental equations of electromagnetism.

In this case, the electromagnetic energy is

$$J_1(\Omega) := \frac{1}{2\mu_0}\int_{\Omega^c} |\nabla\psi(x)|^2\,dx - \int_{\Omega^c} j_0\psi(x)\,dx \tag{1.11}$$

and the energy of surface tension is

$$J_2(\Omega) := \tau\int_{\partial\Omega} ds = \tau P(\Omega), \tag{1.12}$$

where $\tau \in [0,\infty)$ is a characteristic of the metal (it also depends on the temperature) and is called the constant of surface tension. The total energy is thus given by

$$J(\Omega) = J_1(\Omega) + J_2(\Omega) = \frac{1}{2\mu_0}\int_{\Omega^c} |\nabla\psi(x)|^2\,dx - \int_{\Omega^c} j_0\psi(x)\,dx + \tau P(\Omega), \tag{1.13}$$

where ψ is a solution of (1.10). Taking into account that the area of the section is prescribed, seeking the equilibria leads to the following shape optimization problem:

$$\min\{J(\Omega);\ \ |\Omega| = S_0\}. \tag{1.14}$$

Remark 1.3.1. This example is representative of many shape optimization problems where the functional to be minimized is defined as

$$J(\Omega) = F(\Omega, \psi_\Omega), \tag{1.15}$$

where ψ_Ω is the solution of a partial differential equation on Ω. More examples follow.

1.3.2 Optimization of a magnet

In some electronic devices (for example, in the injection system of a car), one can find equipment such as depicted in Figure 1.6. It is a magnet facing a toothed wheel. The magnetic field created by the magnet is different depending on whether it is a hollow or a tooth facing the magnet. A small probe located near the magnet measures the jump in the magnetic field. To make the device as reliable as possible, it is convenient to let this jump be as large as possible. For that purpose, one can essentially act on the shape of the magnet.

Let Ω_0 denote the magnet, Γ_0 its boundary, S the probe, and let K^{T} (resp. K^{H}) denote the toothed wheel when a tooth (resp., a hollow) faces the magnet. We denote by Ω^{T} and Ω^{H} the complementary set in $\mathbb{R}^3$ of K^{T} and K^{H}. We introduce φ^{T} (resp., φ^{H}), the scalar magnetic potential in each situation. We compute φ^{T} by solving the exterior problem

$$\begin{cases} -\Delta\varphi^{\mathrm{T}} = L_0 & \text{in } \Omega^{\mathrm{T}}, \\ \varphi^{\mathrm{T}} = 0 & \text{on } \partial\Omega^{\mathrm{T}}, \\ \varphi^{\mathrm{T}}(X) \to 0 & \text{when } |X| \to +\infty, \end{cases} \tag{1.16}$$

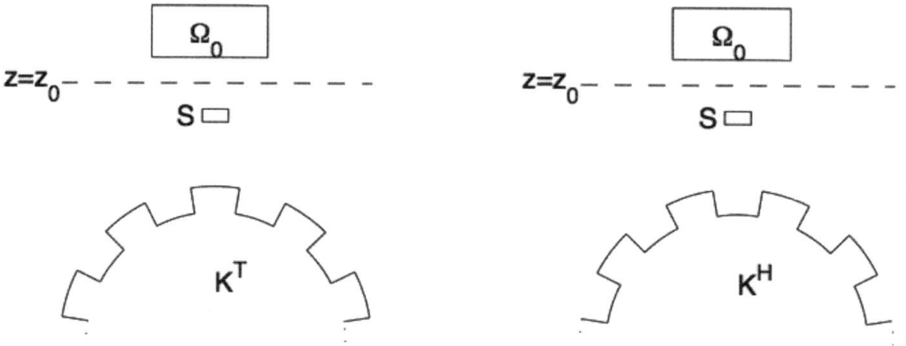

Figure 1.6. The magnet facing a tooth (left), and facing a hollow (right).

where L_0 is the linear form on $H_0^1(\Omega^{\mathrm{T}})$ defined by

$$\langle L_0, w\rangle := \int_{\Gamma_0} \mathbf{M_0}.\mathbf{n} w \, d\sigma. \tag{1.17}$$

(Here $\mathbf{M_0}$ is the magnetization vector, $\mathbf{n}$ is the exterior normal vector to Γ_0.) The problem is to find the shape of the magnet that maximizes the functional

$$J(\Omega_0) = \frac{1}{2}\int_S |\nabla\varphi_\omega^{\mathrm{T}}(X)|^2 - |\nabla\varphi_\omega^{\mathrm{H}}(X)|^2 \, dX. \tag{1.18}$$

This problem was studied during an industrial PhD; we refer to [193] for more details and results.

1.3.3 Image segmentation

The aim here is to "segment" an image. This means finding curves bounding areas where the image has a homogeneous (or almost constant) grey level. This technique is used in spatial and medical imaging.

Mathematically, an image, say black and white, is given by a bounded open set D in $\mathbb{R}^2$ (for example, a square) and a function $g(x, y)$ defined on D with values in $[0, 1]$. This function is the "grey level": To each point (actually, to each pixel) is associated a number between 0 and 1. The paler the image is, the closer this number is to 0 and conversely for darker points with 1. We want to determine a set $K \subset D$, where K is a compact set (typically a union of curves) and a function $u : D \mapsto [0,1]$ satisfying the following criteria:

- u is close to the original image g;
- u is smooth and has small variations in each connected component of $D\backslash K$ (for example, it could be constant);
- the total length of K should remain small (otherwise, the image would segment into many small parts, which would not be convenient for its interpretation).

A model proposed by Mumford and Shah in 1985 consists in minimizing the functional

$$J(K,u) := a\int_D (u-g)^2\,dx + b\int_{D\backslash K} |\nabla u(x)|^2\,dx + c\,\text{“Length”}(K), \tag{1.19}$$

among closed sets $K \subset D$ and functions $u \in H^1(D\backslash K)$. In (1.19), each integral clearly corresponds to one of the criteria defined previously, and the positive constants a, b, c give more weight to one or another criterion. The word “Length” has to be made precise. It depends, of course, on the choice of the class of admissible sets K. It is usual to replace the length by the one-dimensional Hausdorff[16] measure (see Chapter 3). This functional has given rise to many works in recent years. For a starting point and an exhaustive bibliography, we refer, for example, to [8], [46], [237] and [124], [125], [127], [226]. See also the book by G. David [126] for regularity questions and [51], [53], [91] for several numerical approaches.

1.3.4 Identification of cracks or defects

An important issue in nondestructive control of structures or materials is the detection of cracks or defects inside the structure. Generally, the interior of the material is out of reach and we have to identify the defects from observations made on the boundary of the domain.

A possible model uses the conduction properties (thermal or electrical) of the material. Let us denote the material by Ω. If γ is the (unknown) crack, one can impose some flux f on the boundary of Ω. Then the temperature (or the potential) u_γ is the solution of

$$\begin{cases} \Delta u_\gamma = 0 & \text{in } \Omega\backslash\gamma, \\ \dfrac{\partial u_\gamma}{\partial n} = 0 & \text{on } \gamma, \\ \dfrac{\partial u_\gamma}{\partial n} = f & \text{on } \partial\Omega. \end{cases} \tag{1.20}$$

[16]Félix HAUSDORFF, 1868–1942, German, taught in Leipzig, Greisswald, and Bonn. He is famous for his works on topology and metric spaces.

Now, we measure $u_\gamma = g$ on some part of the boundary (or on the whole boundary) and, using this measure, we try to recover the defect. It is possible to consider this problem as a shape optimization problem (in fact, it is a geometric inverse problem). Indeed, to find γ, one can try to minimize the least-square functional

$$J(\gamma) = \int_{\partial\Omega} (u_\gamma - g)^2 \, dx.$$

Mathematically, several questions are interesting:

- Identifiability, which is the question of uniqueness. The problem is to determine whether for different data f and g, it would be possible to find a unique crack γ.
- Identification, which is the question of existence of such a crack. What is the class of functions g that corresponds to such a situation? For other kinds of problems, what are the data and the measures that we need in order to identify the unknown?
- Sensitivity with respect to the data is a crucial question. In inverse problems, one often encounters very bad behavior with respect to small perturbations on data or measures. Therefore, the numerical approach is generally tough.

One can also imagine other kinds of problems or models. For example, it is classical to try to detect oil fields thanks to electrostatic properties of the subsoil. In this case, the geometric unknown is no longer a crack, but one has to identify several layers of the subsoil characterized by their conductivity. As an introduction to these questions, we refer, for example, to [9], [21], [47], [61], [144], [146].

1.3.5 Reinforcement or insulation problems

Let us consider a conductive body Ω. We wish to put some insulator around its boundary (or part of it) in order to improve its behavior. Let us denote by $\varepsilon h(x)$, $x \in \partial\Omega$ the thickness of the insulator at a point x of the boundary; see Figure 1.7. We can introduce

$$\Gamma_\varepsilon(h) = \{x + \delta h(x) n(x),\ x \in \partial\Omega,\ 0 < \delta < \varepsilon, \},$$

where $n(x)$ is the unit normal vector at point x, and we set $\Omega_\varepsilon(h) := \overline{\Omega} \cup \Gamma_\varepsilon(h)$.

If we are interested in a thermal insulation problem we can, for example, assume that the body Ω has conductivity equal to 1, while the insulating layer has weak conductivity ε (we can take the same parameter as in the definition of the layer without loss of generality, as it just means to normalize the function h).

Now, we can consider several optimization problems. For example, given a heat source f, we compute the temperature u inside $\Omega_\varepsilon(h)$, which is the solution of the

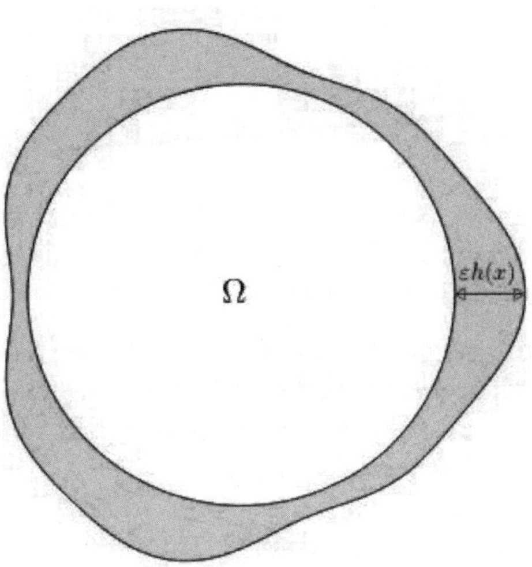

Figure 1.7. Reinforcement.

following problem, where χ_Ω denotes the characteristic function of Ω:

$$\begin{cases} -\operatorname{div}\left[(\chi_\Omega + \varepsilon(1-\chi_\Omega))\,\nabla u\right] = f & \text{in } \Omega_\varepsilon(h), \\ u = 0 & \text{on } \partial\Omega_\varepsilon(h). \end{cases} \tag{1.21}$$

Then, we must choose the thickness h that maximizes the total heat

$$J(h) = \int_{\Omega_\varepsilon(h)} f(x)u(x)\,dx,$$

with u the solution of (1.21). If we look for a model that is more independent of the possible heat sources, it is convenient to take into account the eigenvalues of the corresponding elliptic problem. Let us denote by $\lambda(\Omega, \varepsilon, h)$ the first eigenvalue of the problem

$$\begin{cases} -\operatorname{div}\left[(\chi_\Omega + \varepsilon(1-\chi_\Omega))\,\nabla \phi\right] = \lambda(\Omega, \varepsilon, h)\phi & \text{in } \Omega_\varepsilon(h), \\ \phi = 0 & \text{on } \partial\Omega_\varepsilon(h). \end{cases} \tag{1.22}$$

This eigenvalue determines the rate with which $\Omega_\varepsilon(h)$ dissipates heat, and it gives a very good measure of the insulating quality. So, we are led to look for h that *minimizes* $\lambda(\Omega, \varepsilon, h)$. Let us note that, in the literature, it is especially the limit case $\varepsilon \to 0$ that has been studied. In particular, A. Friedman in [143] showed that

$\lambda(\Omega, \varepsilon, h)$ converges, when $\varepsilon \to 0$, to the first eigenvalue of the Robin problem:

$$\begin{cases} -\Delta\phi = \xi\phi & \text{in } \Omega, \\ \phi + h\dfrac{\partial\phi}{\partial n} = 0 & \text{on } \partial\Omega. \end{cases} \tag{1.23}$$

We refer to [110] for a study of the corresponding minimization problem and to [77] for the minimization problem related to (1.21). One can also read [1], [268], [52], [68], and the references therein for other reinforcement problems.

It is also possible to consider the same kind of questions in acoustics. For example, one can look for an insulating layer on the wall of a room that will improve its acoustic properties. This is an important question for a concert hall and, in this case, one wants to obtain very good reproduction of sound in some specific part of the hall. Another problem concerns vehicles or homes near a noise source (heavy traffic for example). In this case, the goal is to minimize the noise. Thus, criteria can be disparate, and the mathematical tools can also vary. One can choose to work with the time variable, for example with the wave equation, or in the frequency variable, for example with the Helmholtz [17] equation. To give a specific example, let us consider the problem of a car. The noise source f can be its engine and has compact support C. The interior of the car is another region D. We want to find the best insulating layer ω that we can put on the ceiling of the car (of course, we can assume that the volume of this insulating layer is given). In the time variable, a simple model consists in solving the wave equation with an absorbing condition on the insulating part:

$$\begin{cases} \frac{\partial^2 u}{\partial t^2} - \Delta u = f & \text{in } (0,T)\times\Omega, \\ u(0,x) = 0,\ \frac{\partial u}{\partial t}(0,x) = 0 & \text{in } \Omega, \\ \frac{\partial u}{\partial n} + \alpha u = 0 & \text{on } \partial\omega \text{ (absorbing condition)}, \\ \frac{\partial u}{\partial n} = 0 & \text{on } \partial\Omega\backslash\partial\omega \text{ (reflecting condition)}. \end{cases} \tag{1.24}$$

Then, we can look for a domain ω that minimizes the functional

$$J(\omega) := \int_0^T \int_D |u(t,x)|^2 \, dt \, dx.$$

An advantage of working in frequency variables could be to make the problem more independent of the noise source and of the time. We could also try to counteract specific (unpleasant) frequencies. A possible problem is to choose ω in such a way that there are no frequencies in some given interval. Here we recover problems very similar to those encountered in Section 1.2.3.

[17]Hermann von HELMHOLTZ, German, 1821–1894. A philosopher and a physicist, he worked on the propagation of acoustic waves.

1.3.6 Composite materials and structural optimization

In structural mechanics or in thermal sciences, we often encounter the following situation: We have two (or more) materials with different mechanical or thermal properties. One of the materials might be air or a vacuum. We would like to produce a composite material with the best possible properties. For example, if the two materials have thermal conductivities α and β, we denote by A (resp., $\Omega\backslash A$) the part of the body occupied by some material of conductivity α (resp., β). Then the conductivity of the whole domain Ω is

$$a(x) = \alpha\chi_A(x) + \beta(1 - \chi_A(x)), \tag{1.25}$$

where χ_A denotes the characteristic function of A. To compute the temperature u inside Ω (with a heat source f), we have to solve the equation

$$\begin{cases} -\operatorname{div}(a(x)\nabla u) = f & \text{in } \Omega, \\ u = 0 & \text{on } \partial\Omega. \end{cases} \tag{1.26}$$

This may lead to minimizing a functional of the form

$$J(A) := \int_A g(x, u(x))\, dx + \int_{\Omega\backslash A} h(x, u(x))\, dx. \tag{1.27}$$

This problem will be studied in Chapter 7. Generally, it has no classical solution, and we have to use a relaxed form and homogenization techniques to get a well-posed problem.

We have a similar situation in structural mechanics and much work has been done in that domain (see the references at the beginning of this chapter). A classical model problem is the following. We try to build a structure with maximal rigidity or minimal weight (an electrical mast for example). This structure has to be contained in some design region Ω in $\mathbb{R}^N$ that is subject to particular boundary conditions. For example, one can assume that some part of the boundary $\partial\Omega_N$ is subject to an external surface force f, while on the other part $\partial\Omega_D$, we impose no displacement. The unknown structure ω is an elastic material, represented by a subset of Ω obtained by removing some holes from the whole domain. Generally, on the boundary of these holes, we consider a boundary condition of zero traction. Now, if we assume that the elastic behavior of ω is characterized by an isotropic tensor A, we can write the elasticity system

$$\begin{cases} \sigma = Ae(u),\ e(u) = (\nabla u + \nabla^t u)/2,\ \operatorname{div}\sigma = 0 & \text{in } \omega, \\ u = 0 & \text{on } \partial\Omega_D, \\ \sigma \cdot n = f & \text{on } \partial\Omega_L, \\ \sigma \cdot n = 0 & \text{on } \partial\omega\backslash\partial\Omega. \end{cases} \tag{1.28}$$

A classical quantity that gives a good measure of the rigidity of the structure is the *compliance* defined by

$$c(\omega) = \int_{\partial\Omega_L} f \cdot u = \int_\omega Ae(u) \cdot e(u) = \int_\omega A^{-1}\sigma \cdot \sigma. \tag{1.29}$$

Now the problem consists in maximizing $c(\omega)$ with a given volume of material (or minimizing the volume with $c(\omega)$ fixed). Figure 1.8, kindly provided by G. Allaire, illustrates this kind of problem. One of the main difficulties is that the topology of the

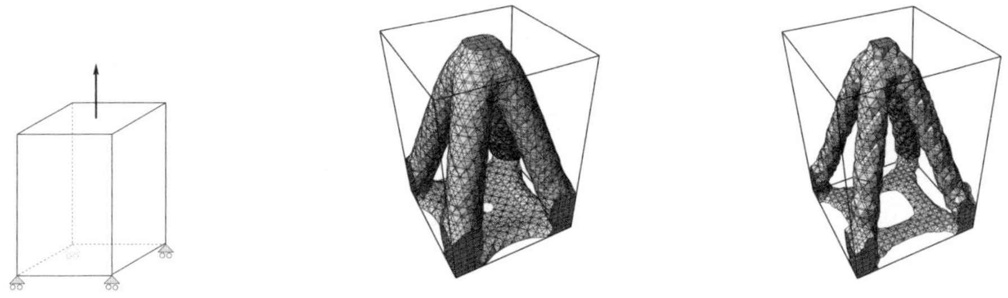

Figure 1.8. Tetrapody: boundary conditions (left), relaxed optimal shape (middle), penalized optimal shape (right).

desired structure is a priori unknown. Therefore, the usual methods, which consist in continuously moving the boundary (and based on the use of the domain derivative presented in Chapter 5), become inefficient. Moreover, in general, there are no classical solutions, that is, no optimal structure ω^*. Heuristically, one can guess this situation by checking that we can always improve a given structure by replacing a big hole with many small holes, increasing rigidity. This is the reason why the methods coming from homogenization are so powerful in that context. The first mathematical works in that domain are those by Murat and Tartar [244], Cherkaev and Kohn [97], Cherkaev and Lurie [98], Kohn and Strang [215]. Several important works later proved the efficiency of these methods for practical problems; see, for example, [14], [15], [36]. We refer to [12], [13], and the references therein for more details.

1.3.7 Examples in aeronautics or fluid mechanics

From an historical point of view, aeronautics is certainly one of the first industrial domains that has been concerned with shape optimization. Among the classical problems that have inspired much research, let us mention

- optimization of a wing profile or an airfoil to improve the drag (see below) or the lift;
- attempts to make airplanes less noisy (in particular, supersonic airplanes);
- research on stealth planes (or submarines);
- optimization of the blades of a fan or a pump.

In the penultimate case, one wants to make the plane as invisible as possible to military surveillance radar. For that purpose, it is possible to play not only with the plane's shape, but also with the width or the composition of the paint: We recover questions similar to the reinforcement or insulation problems presented in Section 1.3.5.

As an example, let us consider an explicit model for the problem of optimization of a wing profile. The reader will find more details in [255], [236]. Let us introduce the drag $\mathcal{T}$ of a wing A as the quantity

$$\mathcal{T} = \int_A \left[\mu(\nabla u + \nabla u^T) - \frac{2\mu}{3}\operatorname{div} u\right] n - \int_A Pn, \tag{1.30}$$

where u is the velocity of the fluid, μ its viscosity, and P its pressure. Velocity and pressure can be computed, for example, through the solution of the Navier[18]–Stokes[19] system (compressible or incompressible) in the complement of the domain occupied by the plane. In some cases, the Euler[20] system is a better model. Then, we are interested in minimizing $\mathcal{T}u_\infty$, where u_∞ is the velocity of the wing in the fluid. This is a tough problem, both from a theoretical viewpoint (due to the intrinsic difficulties of the nonlinear Navier–Stokes system for high Reynolds[21] number) and from a numerical viewpoint (tridimensional nonlinear system, unbounded domain, etc.). Of course, while improving the drag, good lift must be preserved! Thus, this problem can be seen as a multiobjective optimization problem or, sometimes, as a competition problem where one can use the methods of game theory. This field of research is still very active, in particular from a numerical viewpoint (one-shot methods, incomplete sensitivities, stochastic perturbations to study robustness, etc.). As well as the book by Pironneau [255] already quoted, we refer to [199], [235], [236].

[18]Claude NAVIER, 1785–1836, taught at the École Polytechnique and at the École des Ponts and Chaussées; he worked in fluid mechanics.

[19]George STOKES, 1819–1903, British physicist and mathematician, with several important contributions to partial differential equations in fluid mechanics.

[20]Leonhard EULER, 1707–1783, Swiss, was a professor in Saint Petersburg, then in Berlin, then again in Saint Petersburg. His name is everywhere in mathematics due to his huge work in almost all domains and with exceptional creativity.

[21]Osborne REYNOLDS, 1842–1912, British engineer, known for his works in hydrodynamics, hydraulics, and lubrication theory.

Chapter 2

Topologies on domains of $\mathbb{R}^N$

2.1 Why do we need a topology?

In Chapter 1, we saw several examples of problems like (1.1), namely, of the type

$$\min_{\Omega \in \mathcal{O}} F(\Omega), \tag{2.1}$$

where $\mathcal{O}$ is a class of subsets or "domains" in $\mathbb{R}^N$ and $F : \mathcal{O} \to \mathbb{R}$ is a "shape functional". One of the first tasks of the mathematician is

(1) to check that $m = \inf\{F(\Omega),\ \Omega \in \mathcal{O}\}$ is finite ($m > -\infty$);

(2) (to try) to prove that this inf is achieved (and is therefore a minimum): In this case, we get an *existence* theorem.

The first step is generally easy. For the existence of a minimizer, a systematic method consists in working with a minimizing sequence Ω_n, namely, a sequence of domains Ω_n in $\mathcal{O}$ such that

$$\lim_{n\to\infty} F(\Omega_n) = m. \tag{2.2}$$

Such a sequence always exists if $\mathcal{O}$ is not empty, by the definition of the infimum. Then we have to prove that Ω_n "converges" (in a sense to be made precise) to a domain $\Omega^\star$ in $\mathcal{O}$, *and* that $F(\Omega^\star) \le m$.

So, we see that it is important and natural to define a topology on the class of admissible domains. Then the existence of the limit $\Omega^\star$ will come from a *compactness* property of the sequence Ω_n for this topology. Finally, the inequality "$F(\Omega^\star) \le m$" corresponds to the lower semicontinuity of the functional F "along" this sequence Ω_n. Thus, the proof of existence of a minimizer comes from a theorem like "A lower semicontinuous functional on a compact set achieves its minimum". We can summarize this general approach by the following proposition.

Proposition 2.1.1. *Let us assume that $\mathcal{O}$ is endowed with a topology τ such that*

(1) $F : \mathcal{O} \mapsto \mathbb{R}$ is lower semicontinuous for the sequences, that is,

$$\Omega_n \xrightarrow{\tau} \Omega \implies F(\Omega) \le \liminf_{n\to\infty} F(\Omega_n);$$

(2) any sequence F-bounded is sequentially compact, that is,

$$\sup_n |F(\Omega_n)| < +\infty \implies \exists\, (\Omega_{n_k})_{k\geq 1}, \exists\, \Omega \in \mathcal{O},\ \Omega_{n_k} \stackrel{\tau}{\longrightarrow} \Omega.$$

Then if F is bounded from below, there exists $\Omega^\star \in \mathcal{O}$ such that

$$F(\Omega^\star) = \min\{F(\Omega),\ \Omega \in \mathcal{O}\}.$$

The proof of this proposition is essentially contained in the above discussion. Let us remark that we need sequential properties only.

Remark 2.1.2. As we have seen in different examples presented in the previous chapter, the functional $\Omega \mapsto F(\Omega)$ is often of the kind $F(\Omega) = J(\Omega, u_\Omega)$, where u_Ω is the solution of a partial differential equation (PDE) set on Ω (or a system of such equations). So, property (1) of the previous proposition needs a "good" continuous dependence of the solution of the PDE u_Ω with respect to Ω. This subject is studied in depth in Chapter 3 in the case of elliptic operators.

2.2 Different topologies on domains

2.2.1 Introduction

A feature of shape optimization problems is the freedom we have to choose the topology on the family of domains to prove existence of a minimizer. This freedom is at the same time a difficulty and a richness. Most often, there is no "natural" topology. This is in fact a rather general situation for optimization problems,[1] but the situation becomes more complicated in shape optimization since there is no "canonical" topology on the class of subsets in $\mathbb{R}^N$, or even on the class of open sets in $\mathbb{R}^N$.

As the reason for introducing a topology is to apply Proposition 2.1.1, we are faced with two contradictory demands. Indeed, if we wish the family of admissible domains to be compact, then we should choose a topology as coarse as possible (meaning that it contains a few open sets). Conversely, if we wish the functional we minimize to be continuous or lower semicontinuous (l.s.c.), we should choose a topology as fine as possible. The art of analysis is to play between these two antagonistic demands.

[1]When we are faced with an optimization problem like $\min\{J(a); a \in A\}$, it is not a priori a question of topology. The introduction of a topology on A is a mathematical framework to prove the existence of a minimizer thanks to a powerful tool of analysis: compactness.

We present here in detail three topologies that are very useful for families of domains "without too much regularity". The first one is even defined for all measurable sets in $\mathbb{R}^N$; the other two are defined for families of *open sets* in $\mathbb{R}^N$. Depending on the context, however, we will sometimes work with other more adapted topologies. Some comments will be made on that point at the end of this chapter.

2.2.2 Convergence of characteristic functions

A first natural idea is to make a "correspondence" between the sets (which are Lebesgue measurable) E in $\mathbb{R}^N$ and their characteristic function χ_E (by definition this function equals 1 on E and 0 outside). This function belongs to the functional space $L^\infty(\mathbb{R}^N)$, and also to any space $L^p(\mathbb{R}^N)$, $1 \leq p \leq \infty$, as soon as E is bounded (or, more generally, has finite Lebesgue measure). Then we can use any of the classical topologies on these functional spaces and apply it to characteristic functions.

Of course, χ_E is only defined almost everywhere. So, this representation of domains does not distinguish between an open set and the same open set minus some compact set of zero measure. Now, it is known (see also the next chapter) that the solution of the Dirichlet problem for the Laplace operator on an open set is changed by such a modification. This point of view will lead to serious restrictions for some domain functionals. Nevertheless, it is interesting to use it in some cases.

The topology given by characteristic functions has an interesting property that will be developed in Section 2.2.6 on compactness; see in particular Proposition 2.2.28. It is the following: for any sequence of measurable sets E_n, the corresponding sequence of characteristic functions χ_{E_n} is *weakly-$*$ relatively compact in $L^\infty(\mathbb{R}^N)$*, which means that there exist $\chi \in L^\infty(\mathbb{R}^N)$ and a subsequence $(n_k)_{k\geq 0}$ such that

$$\forall \psi \in L^1(\mathbb{R}^N), \quad \lim_{k\to\infty} \int_{\mathbb{R}^N} \chi_{E_{n_k}} \psi = \int_{\mathbb{R}^N} \chi \psi. \tag{2.3}$$

Unfortunately, the limit χ is not, in general, a characteristic function (see Exercise 2.1). We know only that it takes values between 0 and 1 (see Proposition 2.2.28). Nevertheless, the limit will be a characteristic function if the convergence is "strong" in the sense that it takes place in L^p_{loc} for some $p \in [1, \infty)$. Indeed, it is then possible to extract a subsequence that converges almost everywhere. Therefore, in the limit χ takes only the values 0, 1 and it coincides with the characteristic function of the set where it takes the value 1.

It is interesting to notice that, conversely, the weak-$*$ limit is a characteristic function only if the convergence is strong. More precisely, we have

Proposition 2.2.1. *If $(E_n)_{n\geq 1}$ and E are measurable sets in $\mathbb{R}^N$ such that χ_{E_n} weakly-$*$ converges in $L^\infty(\mathbb{R}^N)$ (in the sense of (2.3)) to χ_E, then $\chi_{E_n} \longrightarrow \chi_E$ in $L^p_{\mathrm{loc}}(\mathbb{R}^N)$ for any $p < +\infty$ and almost everywhere.*

Proof. By assumption, we have

$$\forall \psi \in L^1(\mathbb{R}^N), \quad \lim_{n\to\infty} \int_{\mathbb{R}^N} (\chi_{E_n} - \chi_E)\psi(x)\,dx = 0. \tag{2.4}$$

Let us denote by B_R the ball of center 0 and radius R and by E^c the complement of E. Taking $\psi = \chi_{B_R}\chi_{E^c}$ in (2.4) yields

$$0 = \lim_{n\to\infty} \int_{\mathbb{R}^N} \chi_{E_n}\chi_{B_R}\chi_{E^c}(x)\,dx = \lim_{n\to\infty} |B_R \cap (E_n \setminus E)|,$$

where $|\cdot|$ denotes the Lebesgue measure in $\mathbb{R}^N$. Now, if we take $\psi = \chi_{B_R}$ in (2.4), we get

$$0 = \lim_{n\to\infty} \int_{B_R} (\chi_{E_n} - \chi_E)(x)\,dx = \lim_{n\to\infty} \big\{|B_R \cap (E_n \setminus E)| - |B_R \cap (E \setminus E_n)|\big\}.$$

So, we also have $|B_R \cap (E \setminus E_n)| \to 0$. Now

$$\int_{B_R} |(\chi_{E_n} - \chi_E)(x)|^p\,dx = |B_R \cap (E \setminus E_n)| + |B_R \cap (E_n \setminus E)|, \tag{2.5}$$

which proves the proposition. □

Remark 2.2.2. In general, the convergence does not take place in the whole $L^p(\mathbb{R}^N)$ as shown by the following example. Take E_n to be the ball of center x_n and radius 1, where x_n is a sequence going to ∞. Then, $\chi_{E_n} \rightharpoonup 0$ weakly-$*$ in $L^\infty(\mathbb{R}^N)$ (0 is the characteristic function of the empty set), but $\|\chi_{E_n}\|_{L^p}$ does not converge to 0 since it is a nonzero constant.

Following the previous remarks, it is quite natural to choose as a (first) definition of convergence of the sequence E_n, the one that appears in Proposition 2.2.1, namely:

Definition 2.2.3. Let $(E_n)_{n\geq 1}$ and E be measurable sets in $\mathbb{R}^N$. We say that E_n converges in the sense of characteristic functions to E when n goes to ∞ if

$$\chi_{E_n} \longrightarrow \chi_E \quad \text{in } L^p_{\text{loc}}(\mathbb{R}^N), \quad \forall p \in [1,\infty). \tag{2.6}$$

Remark 2.2.4. In (2.6), one can choose any $p < \infty$ because $|\chi_{E_n} - \chi_E|$ takes only the values 0 and 1, so, for any finite p, $|\chi_{E_n} - \chi_E|^p = |\chi_{E_n} - \chi_E|$. Therefore $\|\chi_{E_n} - \chi_E\|_{L^p} = \|\chi_{E_n} - \chi_E\|_{L^1}^{1/p}$. On the contrary, the case $p = \infty$ has no interest since $\|\chi_{E_n} - \chi_E\|_{L^\infty} = 1$ as soon as E_n and E are different on a set of positive measure.

Remark 2.2.5. It is noteworthy that, if the sets E_n stay in a fixed set B with finite measure, this convergence coincides with the one given by the metric δ defined on measurable sets of $\mathbb{R}^N$ (modulo equality almost everywhere) as follows: For E_1, E_2 measurable in $\mathbb{R}^N$,

$$\delta(E_1, E_2) := \arctan\big(\operatorname{meas}(E_1 \Delta E_2)\big), \tag{2.7}$$

where Δ denotes the symmetric difference of the sets E_1, E_2 namely,

$$E_1 \Delta E_2 = (E_1 \backslash E_2) \cup (E_2 \backslash E_1).$$

The equivalence of these two convergences immediately follows from the identity (2.5) where we replace B_R by B. Note that the arctan function could be replaced in (2.7) by any increasing function since we are dealing with sets E_1, E_2 included in a set B with finite measure. But (2.7) also defines a good metric if B is replaced by any unbounded subset of $\mathbb{R}^N$ (see Exercise 2.10).

Let us recall that the quotient space of measurable sets in $\mathbb{R}^N$ modulo equality almost everywhere, endowed with this metric is complete. Moreover, applying Baire's[2] theorem to this space gives powerful results on uniform integrability of functions (see Exercise 2.10).

Remark 2.2.6 (Other kinds of weak convergence on characteristic functions). Let us assume that the sets $(E_n)_{n\geq 1}$ are contained in a set B of finite measure. Then the functions χ_{E_n} are also bounded in $L^p(R^N)$ for any $p \in [1, \infty)$. For $p \in (1, \infty)$, by reflexivity of L^p, one can extract χ_{E_n} a subsequence that converges weakly in $L^{p'}(\mathbb{R}^N)$, where p' is conjugate to p (and $1/p + 1/p' = 1$). This means that there exists $\chi \in L^p(\mathbb{R}^N)$ such that

$$\forall\, \psi \in L^{p'}(\mathbb{R}^N), \quad \lim_{n\to\infty} \int_{\mathbb{R}^N} \chi_{E_n} \psi = \int_{\mathbb{R}^N} \chi \psi. \tag{2.8}$$

This weak convergence is often denoted by

$$\chi_{E_n} \overset{\sigma(L^p, L^{p'})}{\longrightarrow} \chi_E,$$

where $\sigma(L^p, L^{p'})$ is classical notation for the associated weak topology. Here, the first argument L^p of $\sigma(\cdot,\cdot)$ indicates the space to which the functions belong and the second argument $L^{p'}$ denotes the space of test functions (we refer, e.g., to [54]). This notation can also be applied to $p = 1$, $p' = \infty$ while the weak-$*$ L^∞ topology can be denoted similarly, as $\sigma(L^\infty, L^1)$.

Of course, the functions χ obtained in (2.8) and (2.3) coincide, since the integrals $\int \chi\psi$ are the same in both cases for any $\psi \in L^1 \cap L^{p'}$.

[2] René BAIRE, 1874–1932, French, taught first in "collèges", then in the Universities of Montpellier and Dijon. He worked on irrational numbers, real functions, and set theory.

Let us remark that, in the statement of Proposition 2.2.1, the assumption on weak-$*$ L^∞ convergence (or $\sigma(L^\infty, L^1)$) can be replaced by convergence in $\sigma(L^p, L^{p'})$ for any p in $[1, \infty)$. For $p = 1$, the assumption would nevertheless be stronger since it would imply convergence in $L^p(\mathbb{R}^N)$ for any finite integer p and imply conservation of mass at ∞ (replace χ_{B_R} by the constant function equal to 1 in the proof). See also Exercise 2.10.

2.2.3 Hausdorff convergence of open sets

2.2.3.1 Definitions. In this section, we restrict ourselves to *open* sets and we will always assume that they are "confined": There exists a *fixed* (large) compact set B in $\mathbb{R}^N$ that contains all the open sets we consider. We denote by $\mathcal{K}_B$ the family of nonempty compact sets included in B. Let us begin with the definition of the Hausdorff distance on $\mathcal{K}_B$. We denote by $d(\cdot,\cdot)$ the Euclidean[3] distance in $\mathbb{R}^N$.

Definition 2.2.7. Let K_1 and K_2 be in $\mathcal{K}_B$. We set

$$\begin{aligned} \forall x \in B, \quad d(x, K_1) &:= \inf_{y \in K_1} d(x, y), \\ \rho(K_1, K_2) &:= \sup_{x \in K_1} d(x, K_2), \\ d^{\mathrm{H}}(K_1, K_2) &:= \max(\rho(K_1, K_2), \rho(K_2, K_1)). \end{aligned} \tag{2.9}$$

In fact, since K_1, K_2 are (not empty) compact sets, the inf and sup are achieved. It is an elementary exercise to prove that d^{H} is a distance on $\mathcal{K}_B$ known as the *Hausdorff distance*. It can be proved that $\mathcal{K}_B$, endowed with d^{H}, is complete and even compact (see Theorem 2.2.25).

We can now define convergence in the sense of Hausdorff for open sets in B.

Definition 2.2.8. Let $(\Omega_n)_{n\geq 1}$ and Ω be open sets included in B. We say that the sequence Ω_n converges in the sense of Hausdorff to Ω if

$$d^{\mathrm{H}}(B\backslash\Omega_n, B\backslash\Omega) \longrightarrow 0 \quad \text{when } n \to \infty. \tag{2.10}$$

We will denote this convergence by $\Omega_n \xrightarrow{\mathrm{H}} \Omega$.

Remark 2.2.9. The terminology "convergence in the sense of Hausdorff" will be used both for compact sets in B and for open sets in B even though the notions are different. Usually the difference is clear from the context and the notation is then kept simple.

---

[3]EUCLID, about 330–275 BC, Greek mathematician whose 13 books composing *The Elements* had a huge influence on the development of mathematics.

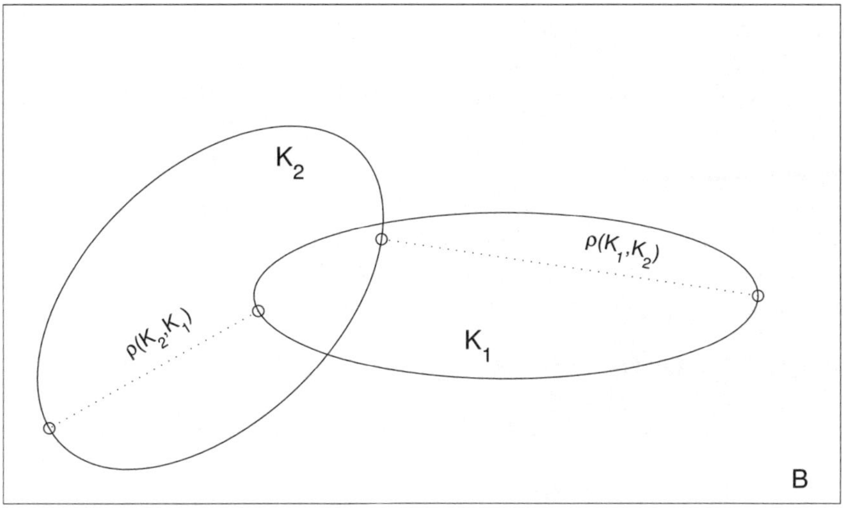

Figure 2.1. Hausdorff distance: $d^{\mathrm{H}}(K_1, K_2) = \max(\rho(K_1, K_2), \rho(K_2, K_1))$.

Remark 2.2.10. Hausdorff convergence is associated with the metric defined on the class $\mathcal{O}_B$ of open sets of $\mathbb{R}^N$ included in B by

$$\forall\, \Omega_1, \Omega_2 \in \mathcal{O}_B, \quad d_{\mathrm{H}}(\Omega_1, \Omega_2) := d^{\mathrm{H}}(B\backslash\Omega_1, B\backslash\Omega_2). \tag{2.11}$$

Let us remark that $B\backslash\Omega$ is indeed a nonempty compact set if $\Omega \in \mathcal{O}_B$: It contains ∂B.

Remark 2.2.11. The definition of this metric does not in fact depend on B. More precisely, if $\hat{B}$ is another compact set containing the open sets Ω_1, Ω_2, we have

$$d^{\mathrm{H}}(B\backslash\Omega_1, B\backslash\Omega_2) = d^{\mathrm{H}}(\hat{B}\backslash\Omega_1, \hat{B}\backslash\Omega_2).$$

This can be proved thanks to the following elementary lemma, which is interesting in its own right.

Lemma 2.2.12. *If $\Omega_2\backslash\Omega_1$ is not empty (i.e., Ω_2 is not included in Ω_1), then*

$$\rho(B\backslash\Omega_1, B\backslash\Omega_2) = \rho(\Omega_2\backslash\Omega_1, \partial\Omega_2) \quad [= \rho(\overline{\Omega_2\backslash\Omega_1}, \partial\Omega_2)]. \tag{2.12}$$

Proof. For any x in B,

$$d(x, B\backslash\Omega_2) = \begin{cases} 0 & \text{if } x \in B\backslash\Omega_2, \\ d(x, \partial\Omega_2) & \text{if } x \in \Omega_2. \end{cases}$$

So, if $\Omega_2\backslash\Omega_1$ is not empty,

$$\rho(B\backslash\Omega_1, B\backslash\Omega_2) = \sup_{x\in B\backslash\Omega_1} d(x, B\backslash\Omega_2) = \sup_{x\in\Omega_2\backslash\Omega_1} d(x, \partial\Omega_2),$$

and the lemma follows. □

If $\Omega_2\backslash\Omega_1$ is empty, then of course $\rho(B\backslash\Omega_1, B\backslash\Omega_2) = 0$. Setting $\rho(\emptyset, E) := 0$ for all E, we can state the following corollary, which shows the independence of the distance d_{H} with respect to B, for open bounded sets, as claimed above.

Corollary 2.2.13. *Let Ω_1,Ω_2 be two open bounded sets in $\mathbb{R}^N$. Then*

$$d_{\mathrm{H}}(\Omega_1, \Omega_2) = \max\{\rho(\Omega_2\backslash\Omega_1, \partial\Omega_2), \rho(\Omega_1\backslash\Omega_2, \partial\Omega_1)\}. \tag{2.13}$$

Remark 2.2.14. If $\Omega_1 = \emptyset$, then $d_{\mathrm{H}}(\emptyset, \Omega_2)$ is the radius of a largest ball included in Ω_2.

2.2.3.2 Properties, examples, and counterexamples. The Hausdorff distance is described in some detail in [132], [169], [267]. This section lists several results about d_{H} that are useful in shape optimization. The proofs are in general elementary so we will just sketch them. Let us begin with the *Hausdorff convergence of compact sets*:

(1) A nonincreasing sequence of nonempty compact sets converges to its intersection. If K_n is such a sequence, let $K := \cap_n K_n$ and let $x_n \in K_n$ be such that $d(x_n, K) = \rho(K_n, K)$. Then there exists a subsequence of x_n converging to some $x_\infty \in K$. The inequality $|d(x_\infty, K) - d(x_n, K)| \leq d(x_\infty, x_n)$ implies $\lim_{n\to\infty} d(x_n, K) = d(x_\infty, K) = 0$. Therefore, $\rho(K_n, K)$ tends to 0 and of course, $\rho(K, K_n) = 0$.

(2) A nondecreasing sequence of nonempty compact sets contained in B converges to the closure of their union.
If K_n is such a sequence, let $K := \overline{\cup_n K_n}$ and let $x_n \in K$ be such that $d(x_n, K_n) = \rho(K, K_n)$. Then there exists a subsequence of x_n converging to some $x_\infty \in K$. There exists also a subsequence n_p and $x^p \in K_{n_p}$ with $x_\infty = \lim_{p\to\infty} x^p$. Thus $\lim_{p\to\infty} d(x_\infty, K_{n_p}) = 0$. It follows that the nonincreasing sequence $d(x_\infty, K_n)$ converges to 0. Finally, $\lim \rho(K, K_n) = \lim d(x_n, K_n) = \lim d(x_\infty, K_n) = 0$.

(3) If K_n converges to K in the sense of Hausdorff,

$$\begin{aligned} K = \cap_n \left(\overline{\cup_{p\geq n} K_p}\right) &= \{x \in B;\ \exists\, x_{n_p} \in K_{n_p}, x_{n_p} \overset{p\to\infty}{\longrightarrow} x\} \\ &= \{x \in B;\ \exists\, x_n \in K_n, x_n \overset{n\to\infty}{\longrightarrow} x\}. \end{aligned} \tag{2.14}$$

According to property (1) above, $F_n := \overline{\cup_{p \geq n} K_p}$ converges to $F := \cap_n F_n$; on the other hand, $d^{\mathrm{H}}(K_n, F_n) = \rho(F_n, K_n) \leq \sup_{p\geq n} \rho(K_p, K_n) \to 0$ since K_n is a Cauchy[4] sequence. It follows that $F = K$, which implies the first equality above. The two other equalities are easy.

(4) **Inclusion.** Inclusion is preserved by Hausdorff convergence.
The mapping $(K_1, K_2) \in \mathcal{K}_B \to \rho(K_1, K_2)$ is continuous, even Lipschitz continuous for d^{H}. Since $\left[(K_1 \subset K_2) \Leftrightarrow \rho(K_1, K_2) = 0\right]$, $K_1^n \subset K_2^n$ implies $K_1 \subset K_2$ if K_i^n converge to K_i in the sense of Hausdorff for $i = 1, 2$.

(5) Here is another useful definition of the Hausdorff distance. For $\alpha > 0$ and K compact, let us define $K^\alpha = \{x \in \mathbb{R}^N; d(x, K) \leq \alpha\}$:

$$d^{\mathrm{H}}(K_1, K_2) = \inf\{\alpha > 0;\ K_2 \subset K_1^\alpha \text{ and } K_1 \subset K_2^\alpha\}. \tag{2.15}$$

This is an immediate consequence of the equivalence

$$(K_2 \subset K_1^\alpha) \Leftrightarrow \sup_{x \in K_2} d(x, K_1) \leq \alpha.$$

(6) **Functions $d(\cdot, K_n)$.** We have the equivalence

$$K_n \xrightarrow{d^{\mathrm{H}}} K \Leftrightarrow d(\cdot, K_n) \xrightarrow{L^\infty(B)} d(\cdot, K).$$

This will be proved later (see Proposition 2.2.27).

Let us now collect some properties of **Hausdorff convergence of open sets** which will be used in the sequel. The first three properties can be deduced from similar properties of compact sets just by taking the complement.

(1) A nondecreasing sequence of open sets included in B converges in the sense of Hausdorff to its union.

(2) A nonincreasing sequence of open sets converges to the interior of the intersection of all the open sets.

(3) **Inclusion.** The inclusion is stable for Hausdorff convergence, that is,

$$\left.\begin{array}{l} \Omega_n^1 \xrightarrow{\mathrm{H}} \Omega^1, \\ \Omega_n^2 \xrightarrow{\mathrm{H}} \Omega^2, \\ \Omega_n^1 \subset \Omega_n^2 \text{ for all } n, \end{array}\right\} \Longrightarrow \Omega^1 \subset \Omega^2. \tag{2.16}$$

[4] Augustin-Louis CAUCHY, 1789–1857, was one of the major mathematicians of the 19th century. He did significant work in analysis: functions of one complex variable, differential equations, notion of continuity, integrals, and also in group theory, determinants, quadratic forms, and the mathematical study of elasticity.

Remark 2.2.15. Nevertheless, it is *not true* that

$$(F \text{ compact} \subset \Omega_n \text{ open sets}, \ \Omega_n \xrightarrow{H} \Omega \text{ open set}) \text{ implies } F \subset \Omega.$$

Example: $\{0\} \subset (-1/n, 1/n) \xrightarrow{H} \emptyset$.

(4) **Proposition 2.2.16.** *Let Ω_n be a sequence of open sets which converges in the Hausdorff topology to the open set Ω and let $x \in \partial\Omega$. Then there exists a sequence of points x_n, with $x_n \in \partial\Omega_n$, that converges to x.*

Let $x \in \partial\Omega$ and let us assume, for the purpose of contradiction, that its distance to $\partial\Omega_n$ does not converge to 0. Then there exists a closed ball $\overline{B}(x,\eta), \eta > 0$ whose intersection with a subsequence $\partial\Omega_{n_k}$ is empty. By connectedness of $\overline{B}(x,\eta)$, it is contained either in Ω_{n_k} or in its complement. By stability of inclusion for Hausdorff convergence of open sets and for closed sets, we have $B(x,\eta) \subset \Omega$ or $\overline{B}(x,\eta) \subset \Omega^c$, which contradicts x being on the boundary of Ω.
Let us remark that, according to (2.14), Proposition 2.2.16 shows that, if $\partial\Omega_n$ Hausdorff-converges to a compact set K, then $\partial\Omega \subset K$. The inclusion is, in general, strict as shown by the previous example $\Omega_n = (-1/n, 1/n)$ where $\partial\Omega = \emptyset, K = \{0\}$.

(5) **Intersection.** Finite intersection is stable for Hausdorff convergence:

$$\left.\begin{aligned} \Omega_n^1 &\xrightarrow{H} \Omega^1, \\ \Omega_n^2 &\xrightarrow{H} \Omega^2, \end{aligned}\right\} \Longrightarrow \Omega_n^1 \cap \Omega_n^2 \xrightarrow{H} \Omega^1 \cap \Omega^2. \tag{2.17}$$

We use $d_H(\Omega_n^1 \cap \Omega_n^2, \Omega^1 \cap \Omega^2) \le \max\{d_H(\Omega_n^1, \Omega^1), d_H(\Omega_n^2, \Omega^2)\}$.

(6) **Union.** For the union, only inclusion holds:

$$\left.\begin{aligned} \Omega_n^1 &\xrightarrow{H} \Omega^1, \\ \Omega_n^2 &\xrightarrow{H} \Omega^2, \\ \Omega_n^1 \cup \Omega_n^2 &\xrightarrow{H} \Omega, \end{aligned}\right\} \Longrightarrow \Omega^1 \cup \Omega^2 \subset \Omega. \tag{2.18}$$

This follows immediately from the previous point (3). The inclusion can, however, be strict as shown by the one-dimensional example

$$\Omega_n^1 = (0,1) \backslash \bigcup_{k=1}^{n-1} \left\{\frac{k}{n}\right\}, \quad \Omega_n^2 = \bigcup_{k=1}^{n-1} \left(\frac{k}{n} - \frac{1}{2^n}, \frac{k}{n} + \frac{1}{2^n}\right).$$

We have $\Omega_n^1 \xrightarrow{H} \emptyset$ and $\Omega_n^2 \xrightarrow{H} \emptyset$ while $\Omega_n^1 \cup \Omega_n^2 = (0,1) \xrightarrow{H} (0,1)$. It is even possible that $\Omega_n^1 \cup \Omega_n^2$ does not converge. For example, let us construct from the above sequence, the new sequence

$$\widehat{\Omega}_n^1 := \Omega_n^1, \quad \widehat{\Omega}_n^2 := \begin{cases} \Omega_n^1 & \text{if } n \text{ even,} \\ \Omega_n^2 & \text{if } n \text{ odd.} \end{cases}$$

Once more, $\widehat{\Omega}_n^1, \widehat{\Omega}_n^2 \overset{H}{\longrightarrow} \emptyset$, but $\widehat{\Omega}_n^1 \cup \widehat{\Omega}_n^2$ has the two accumulation points $\emptyset$ and $(0, 1)$.

(7) **Proposition 2.2.17.** *If the sequence of open sets Ω_n converges to Ω and if K is a compact set contained in Ω, then K is contained in Ω_n for n large enough.*

Since $0 < \inf_{x\in K} d(x, B\backslash\Omega)$ and $d(\cdot, B\backslash\Omega) \le d(\cdot, B\backslash\Omega_n)+d^{\mathrm{H}}(B\backslash\Omega_n, B\backslash\Omega)$, this yields $\inf_{x\in K} d(x, B\backslash\Omega_n) > 0$ for n large enough.

Remark 2.2.18. We will, in particular, use this property in the following way. If φ is a function with compact support in Ω, then φ has its support contained in Ω_n for n large enough.

(8) **Convexity.** Convexity is preserved by Hausdorff convergence of open sets:

$$\Big((\Omega_n)_{n\ge1} \text{ convex}, \Omega_n \overset{H}{\longrightarrow} \Omega\Big) \Rightarrow \Omega \text{ convex.} \tag{2.19}$$

According to the previous property, if $x, y \in \Omega$, then $x, y \in \Omega_n$ for n large; but then the segment $(x, y) \subset \Omega_n$ and this is preserved at the limit since inclusion is preserved. On the contrary, the convex hull is not preserved by Hausdorff convergence. Let us consider again the example of the open set Ω_n^1 of point (6) above. The sequence Ω_n^1 converges to the empty set, but the sequence of convex hulls converges to the interval $(0, 1)$. Nevertheless, the convex hull is a continuous mapping for Hausdorff convergence of compact sets (see Exercise 2.5).

(9) **Connectedness.** Hausdorff convergence does not preserve the connectedness of open sets. We can find a counterexample only in dimensions larger than or equal to 2 because a sequence of open intervals which converges in the sense of Hausdorff can converge only to an interval. As a counterexample, one can choose in $\mathbb{R}^2$,

$$\Omega_n = (0, 2) \times (0, 1)\backslash\{1\} \times [\tfrac{1}{n}, 1] \text{ or } \Omega_n = B(0, 2)\backslash\{e^{2ik\pi/n}, 0 \le k < n\},$$

which converge respectively to $(0, 1) \times (0, 1), \cup(1, 2) \times (0, 1)$, and the ball of radius 2 minus the unit circle (see Figure 2.2).
In contrast, it can be proved that Hausdorff convergence preserves connectedness for *compact sets*. Since this property is important for the sequel (see the following proposition and Šverak's theorem 3.4.14), we point out this result and give a complete proof.

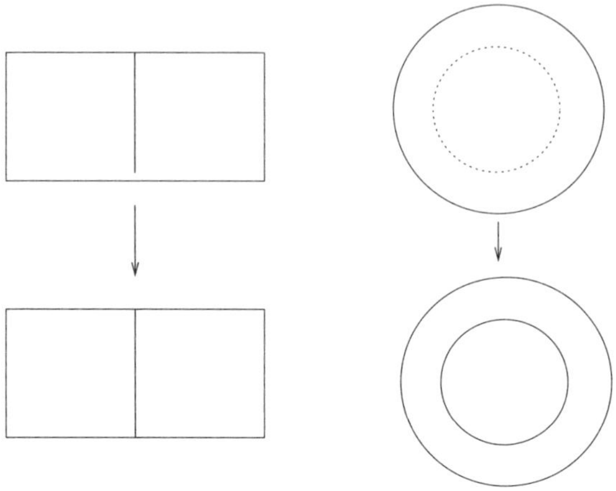

Figure 2.2. Hausdorff distance does not preserve connectedness for open sets.

Proposition 2.2.19. *Let K_n be a sequence of connected compact sets which converges in the sense of Hausdorff to a compact set K. Then K is connected. More generally, if K_n has at most p connected components with $p \geq 1$ independent of n, then K has the same property.*

Remark 2.2.20. This result will be used for open sets: Let Ω_n be a sequence of open sets whose complement in B has at most p connected components. Let us assume that the sequence Ω_n Hausdorff-converges to Ω. Then the complement of Ω has also at most p connected components.

Proof. Let us assume, for the purpose of contradiction, that K has (strictly) more than p connected components. It can be written as the disjoint union of $p+1$ nonempty closed sets $F^i, i = 1, \ldots, p+1$. Indeed, since K is not connected, it is the union of two nonempty closed sets; if $p \geq 2$, one of these closed sets is not connected and is itself the union of two disjoint closed sets. By induction, we find that K is the disjoint union of $p+1$ closed sets.
Let us now define

$$\eta := \min\{|x_i - x_j|;\ x_i \in F_i,\ x_j \in F_j,\ 1 \leq i < j \leq p+1\}.$$

Since the sets F^i are compact and disjoint, we have $\eta > 0$. Then the $p+1$ open sets $\omega_i = \{x \in B;\ d(x, F^i) < \eta/2\}$ are disjoint by construction. Now, for n large enough, $d^{\mathrm{H}}(K_n, K) < \eta/2$; so, $K_n \subset \cup_i \omega_i$ and the sets ω_i define a covering of K_n by $p+1$ disjoint open sets. This implies that K_n has at least $p+1$ connected components as soon as $K_n \cap \omega_i \neq \emptyset$ for any $i = 1, \ldots, p+1$. It is certainly the case for n large enough because, for any i,

$$\sup_{x\in F^i} d(x, K_n) \le \sup_{x\in K} d(x, K_n) < \eta/2.$$

We have a contradiction with the fact that K_n has at most p connected components. □

(10) **Volume.** Hausdorff convergence does not preserve volume. In fact, for open sets, the volume is a lower semicontinuous function, as will be proved in Proposition 2.2.23. This is another property that shows a common phenomenon for Hausdorff convergence of open sets, namely, *collapsing at the limit*. As we have already seen in some examples, the limit can be strictly "smaller" than the sets of the sequence. Some other examples will be shown below.

(11) **Perimeter.** The perimeter of a general open set will be defined later (see Chapter 4). It corresponds to the measure of the lateral boundary ($P(\Omega) = \int_{\partial\Omega} d\sigma(x)$) for regular open sets. Let us show that it is a function that is neither lower semicontinuous (l.s.c.), nor upper semicontinuous (u.s.c.) for Hausdorff convergence. The example of a sequence of domains that look like stamps with smaller and smaller teeth (cf. Figure 2.3), shows that the perimeter of the limit domain can be strictly smaller than the limit inf of the perimeters of the open sets in the sequence. For an example in the reverse sense, let us fix an annulus $\Omega = B(0,1)\backslash\overline{B(0,R)}$, $R < 1$. Let us consider a grid of size $\frac{1}{n}$ laid on the central hole for which we keep only the nodes located in the hole, say $x_1, x_2, \dots, x_p$. We set $\Omega_n = B(0,1)\backslash\left(\bigcup_{i=1}^{p}\{x_i\}\right)$ and it can be proved that $\Omega_n \xrightarrow{H} \Omega$, but $P(\Omega_n) = 2\pi < P(\Omega) = 2\pi(1+R)$.

(12) **Diameter.** The diameter is not continuous for Hausdorff convergence of open sets; it is only lower semicontinuous.

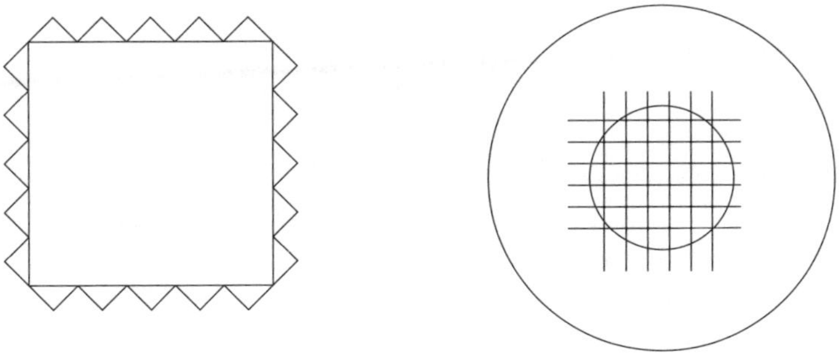

Figure 2.3. The perimeter is not continuous with respect to Hausdorff distance.

2.2.4 Compact convergence

This last notion of convergence is perhaps less natural a priori, but it will be useful when dealing with the question of continuity of the solution of a partial differential equation with respect to domain variations; see [224], [208]. Thus, it is an interesting notion in our situation (we refer to the remark at the end of the first section).

Definition 2.2.21. Let $(\Omega_n)_{n\in N}$ and Ω be open sets in $\mathbb{R}^N$. We say that Ω_n converges in the sense of compact sets to Ω (denoted $\Omega_n \xrightarrow{K} \Omega$) if

(i) for all K compact $\subset \Omega$, we have $K \subset \Omega_n$ for n large enough;

(ii) for all L compact $\subset \overline{\Omega}^c$, we have $L \subset \overline{\Omega}_n^c$ for n large enough.

An (important) drawback of this notion of convergence is the nonuniqueness of the limit. Indeed, it is clear that if the sequence Ω_n converges in the sense of compact sets to Ω, then it also converges to any open set ω such that $\omega \subset \Omega$ and $\overline{\omega} = \overline{\Omega}$. In fact, the topology associated to this convergence is not separated (= not a Hausdorff topology). It is defined on the family $\mathcal{O}$ of open sets in $\mathbb{R}^N$ from the following basis of open sets $\mathcal{V}_{K,L}$. To any pair of compact sets K, L in $\mathbb{R}^N$, let us associate the family of open sets in $\mathbb{R}^N$,

$$\mathcal{V}_{K,L} := \{\omega \in \mathcal{O};\ K \subset \omega,\ L \subset \overline{\omega}^c\}.$$

It can be easily checked that $\{\mathcal{V}_{K,L}; K, L \text{ compact sets in } \mathcal{O}\}$ defines a basis of open sets on the class $\mathcal{O}$ of open sets in $\mathbb{R}^N$ (stability by finite intersection) and that the associated topology generates convergence in the sense of compact sets (see Exercise 2.8). But this topology is not separated: It is possible to exhibit two distinct open sets Ω_1 and Ω_2 in $\mathbb{R}^N$ such that any $\mathcal{V}_{K,L}$ containing Ω_1 intersects any $\mathcal{V}_{\hat{K},\hat{L}}$ containing Ω_2. For example, let us choose for Ω_1 the unit ball and

$$\Omega_2 := B(0,1) \bigcap \bigcup_{k=1}^{\infty} B(x_k, r_k), \tag{2.20}$$

where x_k is a *dense* sequence of points in Ω_1 and where $r_k > 0$ with $\sum_{k\geq 1} r_k^N < 1$ in such a way that Ω_2 is distinct from Ω_1. We have $\Omega_2 \subset \Omega_1$ and $\overline{\Omega}_1 = \overline{\Omega}_2$, and the ball $B_n := B(0, 1 + 1/n)$ belongs to any neighborhood of one or the other set as soon as n is large enough. Hence, B_n converges to both open sets Ω_1 and Ω_2 in the sense of compact sets when n goes to ∞. Moreover, since the choice of the points x_k and the radius r_k is infinite, the sequence B_n has in fact an infinite number of possible limits! Nevertheless, it is true that, except for $B(0,1)$, the other possible limits are rather pathological.

Finally, this situation is similar to the convergence in the sense of characteristic functions. We have uniqueness of the limit only if we work modulo a convenient equivalence relation, namely,

$$\Omega_1 \simeq \Omega_2 \iff \overline{\Omega}_1 = \overline{\Omega}_2. \tag{2.21}$$

Proposition 2.2.22. *The quotient topology on the quotient space of $\mathcal{O}$ by the equivalence relation* (2.21) *is a Hausdorff (separated) topology (see Exercise 2.8).*

There exist several other notions of convergence for open sets that can be easily described: convergence of boundaries, convergence in the sense of Kuratowski,[5] convergence in the topological sense, but we will not use them here. Nevertheless, a very important notion of convergence will be introduced in the next chapter: γ-convergence. It is much less explicit from a geometrical point of view, but will play an important role in the question of continuity of solutions of PDEs set on variable domains. Finally, for regular open sets, it can be useful also to introduce convergence in the sense of C^k-diffeomorphisms of $\mathbb{R}^N$, with $k \geq 1$. We will say that $\Theta_n(\Omega)$ converges to a regular open set Ω if Θ_n converges to the identity for the norm of C^k.

2.2.5 Links between the different notions of convergence

As shown by the following examples, the three notions of convergence we considered are not comparable: None of them is stronger than any other. However, we will point out interesting relations between them (see also Exercises 2.6 and 2.7 at the end of this chapter).

First counterexample. Let Ω be the unit disk in $\mathbb{R}^2$ centered at the origin without the interval $[0,1] \times \{0\}$ and $\Omega_n := B(0, 1 + 1/n)$. Then, it is easy to check that $\Omega_n \xrightarrow{K} \Omega$ while Ω_n converges in the sense of Hausdorff to $B(0,1)$ and therefore not to Ω. In fact, the nonuniqueness of the limit in the sense of compact sets is used here. As explained above, Ω_n converges indeed to any open set contained in $B(0,1)$ the closure of which is the closed unit ball $\overline{B}(0,1)$. For example, it also converges to the set Ω_2 defined in (2.20). Since this set Ω_2 does not have the same measure as $B(0,1)$, it shows that it is also not the limit of Ω_n in the sense of characteristic functions.

Second counterexample (see [255]). In the plane, let us consider $F = [0,3]^2$ and set (see Figure 2.4)

$$\Omega_n := \{(x,y) \in F,\ 0 < x < 3,\ 0 < y < 2 + \sin(nx)\}, \quad K_n = \Omega_n^c, \tag{2.22}$$

$$\Omega := (0, 3 \times (0,1)), \quad K = \Omega^c. \tag{2.23}$$

Then Ω_n converges to Ω in the sense of Hausdorff, since

[5] Kazimierz KURATOWSKI, 1896–1980, Polish, taught in Lwów, then Warsaw. He worked on real functions, topology, set theory, and graphs.

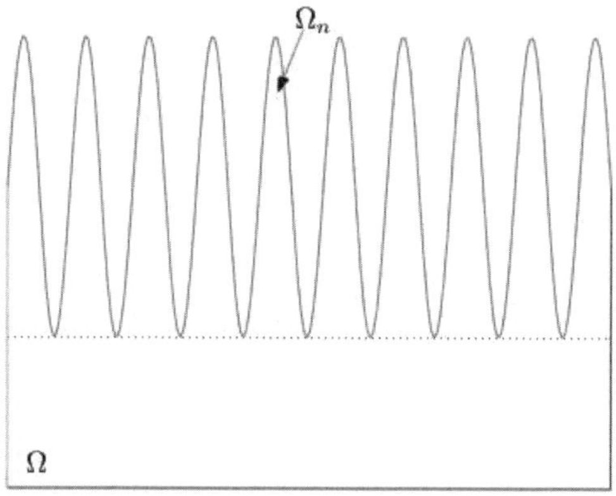

Figure 2.4. Ω_n converges to Ω in the sense of Hausdorff.

- $\rho(K_n, K) = 0$ because $K_n \subset K$;
- for all $x \in K$, we have $d(x, K_n) \leq \pi/n$ and then $\rho(K, K_n) \to 0$.

However Ω_n does not converge to Ω in the sense of compact sets, since any compact set of the kind $[a,b] \times [c,d]$ with $0 < a < b < 3$, $1 < c < d < 3$ is included in $\overline{\Omega}^c$ but cannot be included in any $\overline{\Omega_n}^c$ for n large. Similarly, it is clear that Ω_n does not converge to Ω in the sense of characteristic functions, since

$$\int_F |\chi_{\Omega_n} - \chi_\Omega| \, dx = \int_F \chi_{\Omega_n \setminus \Omega}(x) \, dx = \int_0^3 \int_1^{2+\sin(nx)} dy \, dx = 3 + \frac{1 - \cos(3n)}{n},$$

which does not go to 0 when n goes to ∞.

In fact, it can be proved that Ω_n converges neither in the sense of compact sets, nor in the sense of characteristic functions (let us note that, according to Exercise 2.6, otherwise the limit ω should satisfy $\overline{\omega} = \overline{\Omega}$).

Third counterexample. In $\mathbb{R}$, let us set

$$\Omega_n := \bigcup_{k=0}^{2^n-1} \left(\frac{k}{2^n}, \frac{k+1}{2^n} \right) = [0,1] \setminus \bigcup_{k=0}^{2^n} \left\{ \frac{k}{2^n} \right\}. \tag{2.24}$$

By taking the complement, it is easy to see that $K_n := [0,1] \setminus \Omega_n = \bigcup_{k=0}^{2^n} \{\frac{k}{2^n}\}$ converges in the Hausdorff sense to the interval $[0,1]$. This implies that Ω_n converges in the sense of Hausdorff to the empty set. Conversely, it is clear that Ω_n does not converge in the sense of compact sets to the empty set (nor to any other open set). But since $\chi_{\Omega_n} = \chi_{(0,1)}$ a.e., we have convergence of Ω_n to $(0,1)$ in the sense of characteristic functions.

According to Exercise 2.1, the sets Ω_n can also be chosen such that the characteristic functions only weakly converge.

The previous examples show that there are no complete implications among the three notions of convergence we have defined. Nevertheless, we can mention four strong connections among them:

- We saw in Proposition 2.2.17, that convergence in the sense of Hausdorff implies "half" of the convergence in the sense of compact sets.
- The following proposition shows a similar property. Hausdorff convergence implies "half" of the convergence in the sense of characteristic functions.
- Exercise 2.6 proves that if Ω_n converges in the sense of compact sets to Ω and in the sense of Hausdorff to $\widetilde{\Omega}$, then $\Omega \subset \widetilde{\Omega} \subset \overline{\Omega}$. In particular $\overline{\Omega} = \overline{\widetilde{\Omega}}$. Thus, $\widetilde{\Omega} = \Omega$ if Ω coincides with the interior of $\overline{\Omega}$, namely, if Ω is "saturated" for the natural equivalence relation (2.21) defined for convergence in the sense of compact sets. Such a set Ω (which also verifies $\partial\Omega = \partial\overline{\Omega}$) is said to be a Carathéodory open set.
- The second part of Exercise 2.6 emphasizes an analogous link between convergence in the sense of compact sets and convergence in the sense of characteristic functions. See also Exercise 2.7.

Proposition 2.2.23. *Let Ω_n and Ω be open sets included in some fixed set B. If $\Omega_n \xrightarrow{H} \Omega$ then*

(i) $|\Omega \backslash \Omega_n| \to 0$*;*

(ii) $\chi_\Omega \le \liminf \chi_{\Omega_n}$ *a.e.;*

(iii) if, moreover, $\chi_{\Omega_n} \xrightarrow{\sigma(L^\infty, L^1)} \chi$*, then* $\chi_\Omega \le \chi$*.*

Remark 2.2.24. Since

$$\begin{aligned} \|\chi_{\Omega_n} - \chi_\Omega\|_{L^1} &= \int_B |\chi_{\Omega_n}(x) - \chi_\Omega(x)|\, dx \\ &= \int_{\Omega_n \backslash \Omega} \chi_{\Omega_n}(x)\, dx + \int_{\Omega \backslash \Omega_n} \chi_\Omega(x)\, dx \\ &= |\Omega_n \backslash \Omega| + |\Omega \backslash \Omega_n|, \end{aligned}$$

it is clear that, also in this case, convergence in the sense of Hausdorff implies "half" of the convergence in the sense of characteristic functions.

Proof of Proposition 2.2.23. As usual, let us define $K = B\backslash\Omega$ and $K_n = B\backslash\Omega_n$. Let us fix a sequence of positive numbers ε_n decreasing to 0 and such that $\varepsilon_n \geq \rho(K_n, K)$. Since $d^{\mathrm{H}}(K_n, K) \longrightarrow 0$, we have

$$\chi_{\Omega\backslash\Omega_n} = \chi_{K_n\backslash K} \leq \chi_{\{x\in B;\ 0<d(x,K)<\varepsilon_n\}}.$$

But, since $\{x \in B;\ 0 < d(x, K) < \varepsilon_n\}$ decreases to the empty set, the classical lemma of Beppo Levi[6] implies that

$$\lim_{n\to\infty} \int_B \chi_{\Omega\backslash\Omega_n}(x)\,dx = 0,$$

and therefore $|\Omega\backslash\Omega_n| \longrightarrow 0$. Since, moreover,

$$\chi_\Omega = \chi_{\Omega\backslash\Omega_n} + \chi_{\Omega\cap\Omega_n} \leq \chi_{\Omega\backslash\Omega_n} + \chi_{\Omega_n}, \tag{2.25}$$

and $\chi_{\Omega\backslash\Omega_n} \longrightarrow 0$ a.e. (according to the first part of the proof), this yields

$$\chi_\Omega \leq \liminf_{n\to\infty} \chi_{\Omega_n} \text{ a.e.,} \tag{2.26}$$

and, in the case of (iii), using also Lebesgue dominated convergence,

$$\int_B \chi_\Omega \psi \leq \lim_{n\to\infty} \int_B \chi_{\Omega_n}\psi = \int_B \chi\psi, \tag{2.27}$$

which gives the desired result in (iii). □

2.2.6 Compactness results

As we saw in Proposition 2.1.1, it is important to be able to extract a convergent subsequence from a sequence Ω_n of open sets. In other words, we would like to work with a topology with good compactness properties. As is shown below, this is the case for the Hausdorff topology.

2.2.6.1 Convergence in the sense of Hausdorff

Theorem 2.2.25. *Let K_n be a sequence of compact sets contained in a fixed compact set B. Then there exist a compact set K contained in B and a subsequence K_{n_k} that converges in the sense of Hausdorff to K as $k \longrightarrow \infty$.*

Corollary 2.2.26. *Let Ω_n be a sequence of open sets contained in some fixed compact set B. Then there exist an open set Ω contained in B and a subsequence Ω_{n_k} that converges in the sense of Hausdorff to Ω when $k \longrightarrow \infty$. In other words, $\{\Omega;\ \Omega \subset B\}$ is compact for the Hausdorff metric d_{H}.*

[6]Beppo LEVI, 1875–1961, Italian, in Argentina after 1939. He worked on the theory of integrable functions and also in quantum mechanics.

For any compact set K in $\mathbb{R}^N$, we will denote by $d_K(x)$ the (continuous) function "Euclidean distance to K", defined by $d_K(x) = d(x, K) = \min\{d(x, y),\ y \in K\}$. The following proposition gives an equivalent definition of the Hausdorff distance.

Proposition 2.2.27. *If K_1 and K_2 are two compact sets, then*

$$d^{\mathrm{H}}(K_1, K_2) = \|d_{K_1} - d_{K_2}\|_{L^\infty(\mathbb{R}^N)} = \|d_{K_1} - d_{K_2}\|_{L^\infty(K_1 \cup K_2)}. \tag{2.28}$$

In particular,

$$K_n \xrightarrow{\mathrm{H}} K \iff d_{K_n} - d_K \text{ converges uniformly to } 0 \text{ in } \mathbb{R}^N.$$

Proof. Let us define $\sigma(K_2, K_1) := \|d_{K_1} - d_{K_2}\|_{L^\infty(K_1 \cup K_2)}$. For any $x \in K_2$,

$$d(x, K_1) = |d(x, K_1) - d(x, K_2)| \le \sigma(K_2, K_1).$$

Therefore, taking the maximum over $x \in K_2$, we have $\rho(K_2, K_1) \le \sigma(K_2, K_1)$. Since the right-hand side is symmetric in K_1 and K_2, reversing the role of K_1 and K_2 yields

$$d^{\mathrm{H}}(K_1, K_2) \le \sigma(K_2, K_1).$$

Conversely, for $x \in \mathbb{R}^N$ fixed, let us denote by k_1 and k_2 the points in K_1 and K_2 respectively such that

$$d(x, K_1) = d(x, k_1), \quad d(x, K_2) = d(x, k_2).$$

For any $y \in K_1$, we have $d(x, y) \le d(x, k_2) + d(k_2, y)$, so by taking the minimum for $y \in K_1$, we get

$$d(x, K_1) \le d(x, k_2) + d(k_2, K_1) = d(x, K_2) + d(k_2, K_1).$$

This implies

$$d(x, K_1) - d(x, K_2) \le d(k_2, K_1) \le \rho(K_2, K_1) \le d^{\mathrm{H}}(K_1, K_2).$$

By symmetry, it can be deduced that

$$\forall\, x \in \mathbb{R}^N, \quad |d(x, K_1) - d(x, K_2)| \le d^{\mathrm{H}}(K_1, K_2),$$

and then, taking the upper bound for $x \in \mathbb{R}^N$ gives

$$\sigma(K_2, K_1) \le \|d_{K_1} - d_{K_2}\|_{L^\infty(\mathbb{R}^N)} \le d^{\mathrm{H}}(K_1, K_2),$$

with which Proposition 2.2.27 follows. □

Proof of Theorem 2.2.25. Let us consider the sequence d_{K_n} in the vector space of continuous functions $\mathcal{C}(B)$ endowed with the uniform norm.

- The sequence d_{K_n} is uniformly bounded (e.g., by the diameter of B).
- The sequence d_{K_n} is equicontinuous, because

$$|d_{K_n}(x) - d_{K_n}(y)| = |d(x, K_n) - d(y, K_n)| \leq d(x, y).$$

Therefore, according to the Arzelà–Ascoli[7] theorem, the family d_{K_n} is relatively compact in $\mathcal{C}(B)$. Hence a subsequence, still denoted by d_{K_n}, converges uniformly on B to a continuous function f. Let us introduce $K = \{x \in B;\ f(x) = 0\}$, which is a compact subset of B. The end of the proof consists in proving that $f = d_K$, and the theorem follows, thanks to Proposition 2.2.27.

First of all, since $|d_{K_n}(x) - d_{K_n}(y)| \leq d(x, y)$, by passing to the limit we get $|f(x) - f(y)| \leq d(x, y)$ and, in particular, for any $y \in K$, $f(x) = |f(x)| \leq d(x, y)$, which implies that

$$f(x) \leq d_K(x) \tag{2.29}$$

by taking the minimum on the right-hand side.

Now, let x be fixed in $\mathbb{R}^N$, and, for any integer n, introduce a point $x_n \in K_n$ such that $d(x, K_n) = d(x, x_n)$. Since (x_n) is a sequence in B, a subsequence x_{n_k} can be extracted which converges to a point $y \in B$. So

$$f(x) = \lim d_{K_{n_k}}(x) = \lim d(x, x_{n_k}) = d(x, y).$$

But, since $f(y) = \lim d_{K_{n_k}}(y) \leq \lim d(y, x_{n_k}) = 0$, by definition of K, it can be deduced that $y \in K$. Therefore $f(x) = d(x, y) \geq d_K(x)$, and the result follows using (2.29). □

Proof of Corollary 2.2.26. Let $K_n := B \setminus \Omega_n$. By Theorem 2.2.25, there exist a compact $K \subset B$ and a subsequence of K_n converging to K in the sense of Hausdorff. Since $\partial B \subset K_n$, we still have $\partial B \subset K$ at the limit so that $\Omega := B \setminus K = \text{Interior}(B) \setminus K$ is open and is the Hausdorff limit of a subsequence of Ω_n by Definition 2.2.8. Whence the corollary. □

2.2.6.2 Convergence in the sense of characteristic functions. As mentioned earlier, convergence in the sense of characteristic functions has the following compactness property:

[7]Giulio ASCOLI, 1843–1896, Italian, worked in functional analysis. Cesare ARZELÀ, 1847–1912, Italian, is also the author of pioneering works in functional analysis.

Proposition 2.2.28. *Let E_n be a sequence of measurable sets in $\mathbb{R}^N$. Then a subsequence can be extracted from the sequence χ_{E_n} that converges weak-$*$ in $L^\infty(\mathbb{R}^N)$) (i.e., in $\sigma(L^\infty, L^1)$) to a function $\chi \in L^\infty(\mathbb{R}^N)$. Moreover, the function χ satisfies $0 \le \chi \le 1$ a.e.*

This result follows from the Banach[8]–Alaoglu[9] weak-$*$ compactness theorem, which we recall next (see, e.g., [54]). It then suffices to apply it to $X = L^1(\mathbb{R}^N)$ and $X' = L^\infty(\mathbb{R}^N)$.

Lemma 2.2.29. *The unit ball of the dual space X' of a Banach space X is compact for the weak-$*$ topology $\sigma(X', X)$. Moreover, if X is separable, then the unit ball is sequentially compact.*

Remark 2.2.30 (About $0 \le \chi \le 1$). Positivity is preserved by passing to the limit in $\sigma(L^\infty, L^1)$. Indeed, if χ is the limit of a sequence $\chi_n \ge 0$, then for any $\psi \in L^1$ with $\psi \le 0$, we have $\int \chi_n \psi \le 0$ and in the limit, $\int \chi\psi \le 0$. This implies that $\chi \ge 0$ a.e. Applying this remark to $1 - \chi_n$, it follows that for all n, $\chi_n \le 1 \Rightarrow \chi \le 1$.

If one considers a family of sets with uniformly bounded measure (this is obviously the case when the sets E_n are all contained in a fixed ball B), the sequence χ_{E_n} is then bounded in every space $L^p(\mathbb{R}^N)$. Since $L^p(\mathbb{R}^N)$ is reflexive for $1 < p < \infty$, the sequence χ_{E_n} is also weakly relatively compact in L^p. Since $\mathcal{C}_0^\infty \subset L^{p'} \cap L^1$, the subsequence introduced in Proposition 2.2.28 converges weakly in L^p to the same limit χ. As already mentioned, this limit χ is generally no longer a characteristic function (see Exercise 2.1). To show that it takes values between 0 and 1, it is possible to resort to Mazur's[10] theorem, which states that if a sequence converges weakly in a Banach space X (i.e., in $\sigma(X, X')$), then there exists a sequence of convex combinations which strongly converges to the same limit (see [54]). In our case, since every function χ_{E_n} takes values only between 0 and 1, the same is true of any convex combination of the χ_{E_n}. Now, from any strongly convergent sequence in $L^p(\mathbb{R}^N)$, a subsequence that converges almost everywhere can be extracted. Then, we have $0 \le \chi(x) \le 1$, a.e. x.

In the case where the limit function χ is in fact a characteristic function, we have already seen that the convergence is strong (see Proposition 2.2.1).

We will see later (in Chapter 4) that if the sets E_n have a perimeter that is uniformly bounded, then the sequence χ_{E_n} is compact in the strong topology.

---

[8]Stefan BANACH, 1892–1945, Polish, was from the famous Lwów school. He made a huge contribution to functional analysis, in particular by introducing the spaces named after him.

[9]Leonidas ALAOGLU, 1914–1981, was born to Greek parents in Canada. His doctoral thesis, under the direction of Lawrence M. Graves, is the source of the Banach–Alaoglu theorem

[10]Stanislaw MAZUR, 1905–1981, Polish, taught in Lwów, Łódź, and Krakow. He worked in functional analysis.

2.2.6.3 Convergence in the sense of compact sets. The situation is not so good here. Let us explain why with an example.

Consider the case, in dimension 2 to keep matters simple, where Ω is defined to be the epigraph of a continuous, positive function f on $[a, b]$ with values in $\mathbb{R}$, viz.

$$\Omega = \{(x, y) \in \mathbb{R}^2,\ a < x < b,\ 0 < y < f(x)\}.$$

Assume that the open sets Ω_n are defined in the same way with continuous, positive functions f_n on $[a, b]$. Then we have

Proposition 2.2.31. *The following two implications hold:*

(i) $\left[\Omega_n \xrightarrow{K} \Omega\right]$ *implies that f_n uniformly converges to f on every compact subset of (a, b).*

(ii) If f_n uniformly converges to f on $[a, b]$, then $\left[\Omega_n \xrightarrow{K} \Omega\right]$.

Remark 2.2.32. Equivalence does not hold in (i), as shown by simple examples such as $f(x) = 1$ and $f_n(x) = 1 + x^n$ on the interval $(0, 1)$. On the other hand, $\Omega_n \xrightarrow{K} \Omega$ does not imply uniform convergence, as shown by the example $f(x) = 2$, $f_n(x) = 2 - x^n$.

This proposition shows that, unlike in the case of convergence in the sense of Hausdorff, compactness is far from being immediate in this case. Indeed, even if we assume the functions f_n to be uniformly bounded, it is well known (by the Arzelà–Ascoli theorem), that an extra assumption is needed, namely, the equicontinuity of the sequence, to be able to extract a subsequence that will be uniformly convergent (even on any compact subset). Equicontinuity can be obtained if we assume some control on the slopes. This would be the case, for example, if the functions f_n are assumed to be uniformly Lipschitz[11] (viz. with a common Lipschitz constant).

Proof of Proposition 2.2.31. For any positive number ε introduce

$$K_\varepsilon = \{(x, y) \in \mathbb{R}^2,\ a + \varepsilon \le x \le b - \varepsilon,\ \varepsilon \le y \le f(x) - \varepsilon\}.$$

Now assume that $\Omega_n \xrightarrow{K} \Omega$. By definition, $K_\varepsilon \subset \Omega_n$ for n large enough, and consequently

$$\forall\, x \in [a + \varepsilon, b - \varepsilon], \quad f_n(x) \ge f(x) - \varepsilon.$$

Similarly, with the exterior compact set $L_\varepsilon := \{a \le x \le b; f(x) + \varepsilon \le y \le \varepsilon^{-1}\}$, we have $L_\varepsilon \subset \overline{\Omega}^c$ and then $L_\varepsilon \subset \overline{\Omega}_n^c$ for n large enough. In particular,

$$\forall\, x \in [a, b], \quad f_n(x) \le f(x) + \varepsilon.$$

[11]Rudolph LIPSCHITZ, 1832–1903, German, taught in Berlin, Breslau, and Bonn. He is known for his work on differential equations and Riemannian geometry.

This proves (i). Assertion (ii) is also easy: if K is any compact set contained in Ω, we have $K \subset K_\varepsilon$ for some $\varepsilon > 0$, and the uniform convergence of f_n to f implies $K_\varepsilon \subset \Omega_n$ for n large. We follow the same idea with a compact set L contained in the exterior of Ω: it can be seen that $L \cap [a,b]\times[0,\infty)$ is contained in L_ε for some ε since $f > 0$ (otherwise, L could cross the x-axis), and then by the uniform convergence of f_n, we have $L_\varepsilon \subset \overline{\Omega}_n^c$ for n large enough. Since $\overline{\Omega}_n \subset [a,b]\times[0,\infty)$ for the remaining part of L, the inclusion in $\overline{\Omega}_n^c$ follows. □

Exercise 2.9 generalizes the previous example, where the open sets are defined as the level sets of continuous functions.

2.3 Sequence of sets with bounded perimeter

In this section, we will see that, for a sequence of open sets with bounded perimeter, the corresponding sequence of characteristic functions is relatively compact *in the strong topology*. This is very useful for shape optimization problems because the perimeter often arises in the functional we try to minimize. In particular, this will be the case when the functional contains the energy of surface tension (see some examples in the first chapter). The perimeter can also correspond to some particular "cost" for the shape we try to optimize (instead of the volume, which would be the natural cost in some other problems).

First of all, we need to define the perimeter of any measurable set and not only of regular open sets. For that purpose, we follow the theory of generalized perimeter introduced by de Giorgi.[12]

2.3.1 Definition of the perimeter, properties

Let D be any open set in $\mathbb{R}^N$. Let us denote by $\mathcal{D}(D;\mathbb{R}^N)$ the vector space of functions from D into $\mathbb{R}^N$ that are C^∞ with compact support, that is,

$$\varphi = (\varphi_1, \varphi_2, \dots, \varphi_N) \in \mathcal{D}(D;\mathbb{R}^N) \quad \text{if} \quad \varphi_i \in \mathcal{D}(D) \text{ for any } i = 1, \dots, N.$$

The space $\mathcal{D}(D;\mathbb{R}^N)$ is endowed with the norm

$$\|\varphi\|_\infty := \sup_{x\in D} \left[\left(\sum_{i=1}^N \varphi_i(x)^2\right)\right]^{1/2} = \sup_{x\in D} |\varphi(x)|.$$

[12]Ennio DE GIORGI, 1928–1996, Italian mathematician. He spent his entire career at the Scuola Normale Superiore di Pisa, where he trained many students. He made important contributions to the calculus of variations (minimal surfaces and Γ-convergence, which he introduced in 1975), geometric measure theory, and partial differential equations (in particular, regularity theory).

Definition 2.3.1. Let Ω be a measurable set in D. The perimeter of Ω relative to D (or simply perimeter if $D = \mathbb{R}^N$), is the number defined by

$$P_D(\Omega) = \sup\Big\{\int_\Omega \operatorname{div}(\varphi)\,dx;\ \varphi \in \mathcal{D}(D;\mathbb{R}^N),\ \|\varphi\|_\infty \le 1\Big\}.$$

If $D = \mathbb{R}^N$, we simply denote the perimeter by $P_{\mathbb{R}^N}(\Omega) = P(\Omega)$.

Remark 2.3.2. It is also possible to write $\int_\Omega \operatorname{div}(\varphi)\,dx$ in the following way (which will be useful later):

$$\begin{aligned}\int_\Omega \operatorname{div}(\varphi)\,dx &= \int_D \chi_\Omega\left(\sum_{i=1}^N \frac{\partial\varphi_i}{\partial x_i}\right)dx = \left\langle \chi_\Omega, \sum_{i=1}^N \frac{\partial\varphi_i}{\partial x_i}\right\rangle_{\mathcal{D}'\times\mathcal{D}}\\ &= -\sum_{i=1}^N \left\langle \frac{\partial\chi_\Omega}{\partial x_i}, \varphi_i\right\rangle_{\mathcal{D}'\times\mathcal{D}} = -\langle \nabla\chi_\Omega, \varphi\rangle_{\mathcal{D}'(D;\mathbb{R}^N)\times\mathcal{D}(D;\mathbb{R}^N)},\end{aligned}$$

where $\mathcal{D}'(D;\mathbb{R}^N) = \mathcal{D}'(D)^N$ is the space of distributions $(T_1,\dots,T_N)$ with, for all $i = 1,\dots,N$, $T_i \in \mathcal{D}'(D)$.

If the set Ω is regular, the previous definition coincides with the usual definition of perimeter (or " lateral surface") of the boundary of Ω in D.

Proposition 2.3.3. *If Ω is a bounded open set of class C^1, then $P_D(\Omega) = \int_{\partial\Omega\cap D} d\sigma$ (where $d\sigma$ is the elementary surface element on $\partial\Omega$).*

We refer to page 212 of Chapter 5 for the definition and the main properties of an open set of class C^1.

Proof of Proposition 2.3.3. If $\varphi \in \mathcal{D}(D;\mathbb{R}^N)$, then Green's classical formula yields

$$\int_\Omega \operatorname{div}(\varphi)\,dx = \int_\Omega \sum_{i=1}^N \frac{\partial\varphi_i}{\partial x_i}\,dx = \int_{\partial\Omega} \varphi_i n_i\,d\sigma = \int_{\partial\Omega\cap D} \varphi.\vec{n}\,d\sigma.$$

Now

$$\varphi(x).\vec{n}(x) \le |\varphi(x)| \le \|\varphi\|_\infty \le 1,$$

and so

$$\int_\Omega \operatorname{div}(\varphi)\,dx \le \int_{\partial\Omega\cap D} d\sigma \quad\text{and}\quad P_D(\Omega) \le \int_{\partial\Omega\cap D} d\sigma.$$

To obtain the reverse inequality, we would like to exhibit a function φ such that $\varphi = \vec{n}$ on $\partial\Omega$ or, at least, to approximate $\vec{n}$ with a sequence of such functions $\varphi^q \in \mathcal{D}(D;\mathbb{R}^N)$. This can be done in the following way. Let $\epsilon > 0$ be given, and let $D_\epsilon \subset D$ be a compact set such that $\int_{\partial\Omega\cap D} d\sigma - \int_{\partial\Omega\cap D_\epsilon} d\sigma \le \epsilon$. By the compactness property, there exists a family of C^1- diffeomorphisms $\psi_i, i = 1, \dots, p$ defining $\partial\Omega \cap D_\epsilon$ as explained in the definition of C^1-regularity on page 212. Without loss of generality, one can assume that the domains of definition $\mathcal{O}_i$ of the functions ψ_i are all contained in D. Using the associated partition of unity (ξ_i) such that $\sum_i \xi_i \equiv 1$ on a neighborhood of $\partial\Omega \cap D_\epsilon$, one obtains an extension of $\vec{n}$ according to formula (5.44). Therefore

$$\varphi(x) = \sum_{i=1}^{p} \xi_i(x)\vec{n}\left(\psi_i \circ \pi_i \circ \psi_i^{-1}(x)\right),$$

where π_i is a convenient orthogonal projection and φ is continuous. So, if $x \in \partial\Omega \cap D_\epsilon$, it holds $\varphi(x) = \sum_1^p \xi_i(x)\vec{n}(x) = \vec{n}(x)$. Using a regularization process, setting $\varphi^q := \varphi * \rho_q$ (where ρ_q is a mollifier), it is possible to construct a sequence of functions in $\mathcal{D}(D;\mathbb{R}^N)$ which converges uniformly to φ and for which we have

$$\int_\Omega \operatorname{div}(\varphi^q)\, dx = \int_{\partial\Omega\cap D} \varphi^q.\vec{n}(\sigma)\, d\sigma \quad \longrightarrow \quad \int_{\partial\Omega\cap D} \varphi.\vec{n}\, d\sigma \geq \int_{\partial\Omega\cap D} d\sigma - 2\epsilon,$$

and the result follows. □

Let us now give another interpretation of the perimeter in terms of Radon[13] measures. First of all, let us recall that if $f \in L^1(D;\mathbb{R}^N)$, the norm $\|f\|_1 = \int_D |f(x)|\, dx$, where $|\cdot|$ denotes the Euclidean norm, can be defined by duality as

$$\|f\|_1 = \sup\{\textstyle\int_D f(x).\varphi(x)\, dx;\ \varphi \in \mathcal{D}(D;\mathbb{R}^N),\ \|\varphi\|_\infty \leq 1\}.$$

More generally, the previous formula gives the total variation of f when it is a Radon measure. We recall that a Radon measure on D with values in $\mathbb{R}^N$ is a continuous linear form on the vector space $C_0^0(D;\mathbb{R}^N)$ of continuous functions with values in $\mathbb{R}^N$ and with compact support in D (in other words, a Radon measure is a distribution of order 0). More precisely, we can state

Proposition 2.3.4. *Let $\mu = (\mu_1, \mu_2, \dots, \mu_N) \in \mathcal{D}'(D;\mathbb{R}^N)$ (i.e., $\mu_i \in \mathcal{D}'(D)$). Then, μ is a Radon measure with finite total mass on $\mathbb{R}^N$ if and only if*

$$\|\mu\|_1 = \sup\{\langle \mu, \varphi\rangle_{\mathcal{D}'\times\mathcal{D}};\ \varphi \in \mathcal{D}(D;\mathbb{R}^N),\ \|\varphi\|_\infty \leq 1\} < +\infty.$$

For the proof, we refer to [274]. We will denote by $\mathcal{M}_b(D)$ the real Radon measures on D with finite total mass and we introduce

$$\mathcal{M}_b(D;\mathbb{R}^N) = \{\mu = (\mu_1, \mu_2, \dots, \mu_N) \quad \text{with } \mu_i \in \mathcal{M}_b(D)\}$$

[13] Johann RADON, 1887–1956, Austrian mathematician. He taught in Vienna, but also in several German universities. His main work is related to measure theory and integration.

endowed with the norm $\|.\|_1$ defined above. The important point is that if $\mu \in \mathcal{D}'(D, \mathbb{R}^N)$ is a distribution such that

$$\forall \varphi \in \mathcal{D}(D, \mathbb{R}^N), \quad |\langle \mu, \varphi \rangle| \leq C\|\varphi\|_\infty,$$

then μ is a Radon measure with finite total mass. This is a definition by duality. In the same spirit, if $\mu \in \mathcal{D}'(D, \mathbb{R}^N)$ satisfies

$$\forall \varphi \in \mathcal{D}(D, \mathbb{R}^N), \quad |\langle \mu, \varphi \rangle| \leq C\|\varphi\|_2,$$

then μ is a function in $L^2(D, \mathbb{R}^N)$. More generally, if a distribution is continuous for some norm on $\mathcal{D}(\mathbb{R}^N)$, then it can generally be identified with an element of the dual space of the completion of $\mathcal{D}(\mathbb{R}^N)$ for this norm.

Corollary 2.3.5. *Let $\Omega \subset D$ be a measurable set. Then*

$$P_D(\Omega) < +\infty \iff \nabla \chi_\Omega \textit{ is a measure of finite total mass}$$

with

$$P_D(\Omega) = \|\nabla \chi_\Omega\|_1.$$

It is the total variation of the gradient (in the sense of distributions) of the characteristic function.

Remark 2.3.6. There are some useful properties of the perimeter. For example, for any open set $D \subset \mathbb{R}^N$,

$$\forall\, A, B \text{ measurable}, \quad P_D(A \cup B) + P_D(A \cap B) \leq P_D(A) + P_D(B).$$

We refer to [154] for the proof. Is uses the fact that $\chi_{A \cup B} = \chi_A + \chi_B - \chi_A \chi_B$ and then, *formally*,

$$\begin{aligned} \nabla \chi_{A \cup B} &= (1 - \chi_B)\nabla \chi_A + (1 - \chi_A)\nabla \chi_B, \\ \Rightarrow \quad |\nabla \chi_{A \cup B}| &\leq (1 - \chi_B)|\nabla \chi_A| + (1 - \chi_A)|\nabla \chi_B| \\ &\leq |\nabla \chi_A| + \nabla \chi_B| - |\nabla \chi_{A \cap B}|. \end{aligned}$$

Of course, these expressions are valid only for regular approximations f_A^n, f_B^n of χ_A, χ_B. A useful lemma, needed to pass to the limit, claims that it is possible to choose these approximations in such a way that $\|\nabla f_A^n\|_1$ converges to $\|\nabla \chi_A\|_1$ (and similarly for B).

2.3.2 Continuity and compactness

Let us first recall some continuity results for the volume and the perimeter, with respect to convergence of characteristic functions. Some of these results have been stated previously.

Proposition 2.3.7. *Let Ω_n and Ω be measurable and bounded sets in $\mathbb{R}^N$.*

(1) If $\chi_{\Omega_n} \to \chi_\Omega$ in $L^1_{\mathrm{loc}}(D)$, then $|\Omega| \leq \liminf |\Omega_n|$,
and $P_D(\Omega) \leq \liminf P_D(\Omega_n)$.

(2) If $\chi_{\Omega_n} \to \chi_\Omega$ in $L^1(D)$, then $|\Omega| = \lim |\Omega_n|$,
and $P_D(\Omega) \leq \liminf P_D(\Omega_n)$.

Thus, the perimeter is lower semicontinuous (l.s.c.) for the strong convergence of characteristic functions.

Proof. For the volume, (1) simply follows from Fatou's[14] lemma: The assumption implies that $\chi_{\Omega_{n_k}} \to \chi_\Omega$ a.e. as $k \to \infty$ for some subsequence $(n_k)_{k \geq 1}$. Therefore,

$$\int_{\mathbb{R}^N} \chi_\Omega = |\omega| \leq \liminf \int_{\mathbb{R}^N} \chi_{\Omega_{n_k}} = |\Omega_{n_k}|.$$

Point (2) is obvious for the volume.

For the perimeter, if $\varphi \in \mathcal{D}(D; \mathbb{R}^N)$, we have

$$\begin{aligned}\int_\Omega \operatorname{div}(\varphi)\,dx &= \lim_n \int_{\Omega_n} \operatorname{div}(\varphi)\,dx \\ &\leq \liminf_n \sup\{\textstyle\int_{\Omega_n} \operatorname{div}(\varphi)\,dx;\ \|\varphi\|_\infty \leq 1\} \\ &= \liminf_n P_D(\Omega_n).\end{aligned}$$

The inequality follows by taking the supremum over φ on the left-hand side. □

Remark 2.3.8. If $\chi_{\Omega_n} \rightharpoonup \chi_\Omega$ in L^1 (weak convergence), it was proved in Proposition 2.2.1 that $\chi_{\Omega_n} \to \chi_\Omega$ in $L^1_{\mathrm{loc}}(\mathbb{R}^N)$. Therefore, point (1) in Proposition 2.3.7 above applies.

Remark 2.3.9. If $\chi_{\Omega_n} \to \chi_\Omega$ in $L^1_{\mathrm{loc}}(\mathbb{R}^N)$, the inequality can be strict for the volume in (1): As an example, consider the case of a ball going to ∞.

Let us now state some important compactness properties. Recall that $\mathcal{M}_b(D)$ is the dual space of $\mathcal{C}_0(D)$ endowed with the uniform norm, where $\mathcal{C}_0(D)$ is the linear space of continuous functions on D that tend to 0 at the boundary of D, in the sense that

$$\forall\, \varepsilon > 0, \quad \exists\, K_\varepsilon \text{ compact } \subset D \text{ such that } |u| < \varepsilon \text{ outside } K_\varepsilon.$$

This is a separable space. Therefore, the unit ball of $\mathcal{M}_b(D)$ is sequentially compact in $\mathcal{M}_b(D)$ for the weak-$*$ topology (also denoted $\sigma(\mathcal{M}_b(D), \mathcal{C}_0(D))$. More precisely,

[14]Pierre FATOU, 1878–1929, French mathematician and astronomer. He spent his whole working life at the Observatoire de Paris and contributed to integration theory, Taylor series, and functions of complex variables.

Proposition 2.3.10. *If μ_n is a sequence of Radon measures on D such that $\|\mu_n\|_1 \leq C$, then there exist a subsequence μ_{n_k} and $\mu \in \mathcal{M}_b(D)$ such that $\mu_{n_k} \overset{*}{\rightharpoonup} \mu$ in the topology $\sigma(\mathcal{M}_b(D), C_0(D))$.*

This proposition has an important application for sequences of sets with bounded perimeter:

Theorem 2.3.11. *Let Ω_n be a sequence of measurable subsets of an open set D in $\mathbb{R}^N$. Assume that*

$$|\Omega_n| + P_D(\Omega_n) \leq C \quad \textit{independently of } n.$$

Then there exist a measurable set $\Omega \subset D$ and a subsequence Ω_{n_k} such that

$$\chi_{\Omega_{n_k}} \longrightarrow \chi_\Omega \quad \textit{in } L^1_{\mathrm{loc}}(D),$$

and

$$\nabla \chi_{\Omega_{n_k}} \overset{*}{\rightharpoonup} \nabla \chi_\Omega \quad \textit{in } \sigma(\mathcal{M}_b(D)^N, C_0(D)^N).$$

Moreover, if D is of finite measure, the convergence of $\chi_{\Omega_{n_k}}$ to χ_Ω takes place in $L^1(D)$.

For the proof, we will use the following result.

Proposition 2.3.12. *Let D be an open set of $\mathbb{R}^N$ and let f_n be a sequence of functions in $L^1(D)$ such that ∇f_n is an N-tuple of Radon measures with finite total mass (i.e., $\nabla f_n \in \mathcal{M}_b(D)^N$) and*

$$\|f_n\|_{L^1} + \|\nabla f_n\|_1 \leq C.$$

Then there exist $f \in L^1(D)$, with $\nabla f \in \mathcal{M}_b(D)^N$, and a subsequence f_{n_k} such that $f_{n_k} \longrightarrow f$ in $L^1_{\mathrm{loc}}(D)$ and $\nabla f_{n_k} \overset{*}{\rightharpoonup} \nabla f$ in the weak-$*$ topology $\sigma(\mathcal{M}_b(D)^N, C_0(D)^N)$.

Remark 2.3.13. Let us denote by $\mathrm{BV}(D) = \{f \in L^1(D);\ \nabla f \in \mathcal{M}_b(D)^N\}$ the set of functions with bounded variation in D. The previous proposition asserts that the embedding of $\mathrm{BV}(D)$ into $L^1_{\mathrm{loc}}(D)$ is compact.

Remark 2.3.14. In general, the convergence does not take place in L^1, but only in L^1_{loc}. For example, if $f \in \mathcal{D}(\mathbb{R}^N)$ and $f_n(x) = f(x+n)$, then the sequence f_n satisfies the assumptions of the proposition and $f_n \longrightarrow 0$ in $L^1_{\mathrm{loc}}(\mathbb{R}^N)$, but not in $L^1(\mathbb{R}^N)$. In the same way (but it is more difficult to see), even if D is bounded, but with an irregular boundary, compactness does not occur in general in L^1. On the other hand, if D is bounded with a Lipschitz boundary (see Definition 2.4.5), it can be proved that the embedding of $\mathrm{BV}(D)$ into $L^1(D)$ is compact.

Proof of Proposition 2.3.12 (see also [298]). According to Proposition 2.3.10, there exist a subsequence f_{n_k} and $f \in \mathcal{M}_b(D)$, $\mu \in \mathcal{M}_b(D)^N$ such that

$$f_{n_k} \overset{*}{\rightharpoonup} f \quad \text{(in the topology } \sigma(\mathcal{M}_b(D), C_0(D)))$$

and

$$\nabla f_{n_k} \overset{*}{\rightharpoonup} \mu \quad \text{(in the topology } \sigma(\mathcal{M}_b(D)^N, C_0(D)^N)).$$

Moreover, since $\mathcal{D}(D) \hookrightarrow C_0(D)$, the convergence also takes place in the sense of distributions, and the derivative being continuous in $\mathcal{D}'(D)$ this proves $\mu = \nabla f$.

It remains to prove the convergence of f_{n_k} to f in $L^1_{\text{loc}}(D)$. For that purpose, we use a classical criterion for compactness in L^1_{loc}, namely, uniform continuity of translations (see [54]). We have

$$f_n(x+h) - f_n(x) = \int_0^1 \nabla f_n(x+th).h\,dt,$$

and therefore on $D_\varepsilon = \{x \in D / d(x, \partial D) > \varepsilon\}$, for any h with $|h| < \varepsilon$,

$$\int_{D_\varepsilon} |f_n(x+h) - f_n(x)|\,dx \le \int_0^1 dt\,|h| \int_{D_\varepsilon} |\nabla f_n(x+th)|\,dx \le \|\nabla f_n\|_1 |h|,$$

which implies, according to the estimate given by the assumption, that

$$\lim_{h\to 0} \{\sup_n \|f_n(x+h) - f_n(x)\|_{L^1(D_\varepsilon)}\} = 0.$$

This is the expected compactness criterion. □

Proof of Theorem 2.3.11. Let us apply Proposition 2.3.12 with $f_n = \chi_{\Omega_n}$. This gives the existence of $f \in L^1(D)$ such that $\nabla f \in \mathcal{M}_b(D)^N$, with

$$\chi_{\Omega_{n_k}} \xrightarrow{L^1_{\text{loc}}} f \quad \text{and} \quad \nabla \chi_{\Omega_{n_k}} \overset{*}{\rightharpoonup} \nabla f.$$

But this implies that f is a characteristic function, since by convergence almost everywhere of a subsequence, we have $f(1-f) = 0$ a.e. in D. Therefore, if we set $\Omega := \{x \in D; f(x) = 1\}$, we have $f = \chi_\Omega$. Moreover, since $\chi_{\Omega_{n_k}} = 0$ outside D, convergence takes place in $L^1_{\text{loc}}(\mathbb{R}^N)$ and hence, if D is of finite measure, using the dominated convergence theorem, we obtain convergence in $L^1(D)$. □

2.4 Sequences of uniformly regular open sets

The geometric results we are going to present here are historically among the first that turned out to have significant applications to existence questions in shape optimization. The idea is that we often expect optimal shapes to be regular. Therefore, it does not seem too restrictive to a priori require some regularity constraints on the set of admissible shapes. And as is shown here, such uniform regularity assumptions will ensure good compactness and convergence properties.

We are going to study the class of uniformly Lipschitz domains. As we will see, it is equivalent to considering domains satisfying a uniform cone condition. In shape optimization problems, it will be sometimes more convenient to use this point of view, due to Denise Chenais; see [95], [96] (see also [6] for the following definition).

Definition 2.4.1. Let y be a point in $\mathbb{R}^N$, ξ a unit vector, and ε a positive real number. Let $C(y,\xi,\varepsilon)$ be the cone of vertex y (without its vertex), of direction ξ and dimension ε, defined by

$$C(y,\xi,\varepsilon) = \{z \in \mathbb{R}^N,\ (z-y,\xi) \geq \cos(\varepsilon)|z-y| \text{ and } 0 < |z-y| < \varepsilon\}.$$

An open set Ω is said to have the ε-cone property if

$$\forall\, x \in \partial\Omega,\ \exists\, \xi_x \text{ unit vector such that } \forall\, y \in \overline{\Omega} \cap B(x,\varepsilon),\ C(y,\xi_x,\varepsilon) \subset \Omega.$$

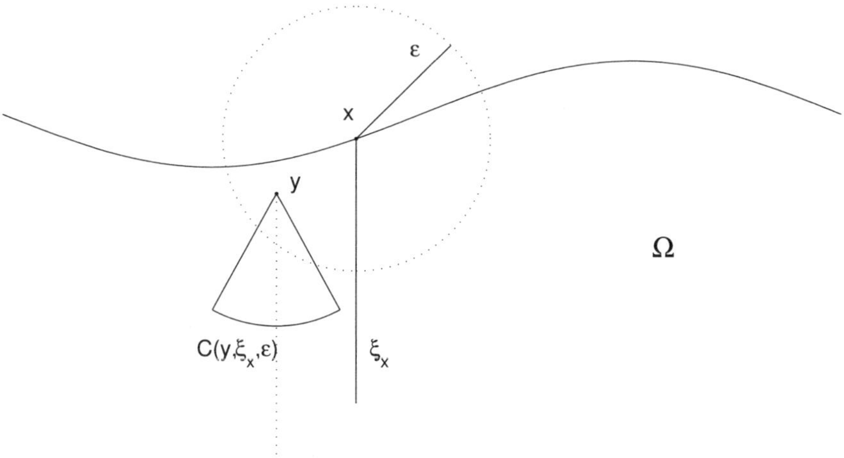

Figure 2.5. The ε-cone property.

Remark 2.4.2. It is important that the choice of the direction ξ_x be uniform with respect to *every* point y in the ball $B(x,\varepsilon)$. For instance, the following open sets *do not* satisfy this property:

- $\mathbb{R}^N\backslash 0$, $\mathbb{R}^N\backslash\{(x_1, x_2, \ldots, x_{N-1}, 0)\}$;
- the union of two exterior tangent balls or the domain between two interior tangent balls;
- any open set in $\mathbb{R}^2$ whose boundary has a cusp;
- $\{(x, y) \in \mathbb{R}^2;\ xy > 0\}$.

On the other hand, convex open sets, or more generally open sets that are star shaped with respect to all the points of a ball, have the ϵ-cone property; see below.

Definition 2.4.3. An open set Ω is said to be *star shaped* with respect to a point $x_0 \in \Omega$ if $y \in \Omega$ implies $[x_0, y] \subset \Omega$.

We say that it is *star shaped with respect to an open ball* $B \subset \Omega$ if it is star shaped with respect to every point in B. In this case,

$$\forall\, y \in \overline{\Omega},\ \forall\, z \in B, \quad \{ty + (1-t)z,\ 0 \le t < 1\} \subset \Omega. \tag{2.30}$$

Indeed, if $y_n \in \Omega$ converges to y, the point z_n such that $ty + (1-t)z = ty_n + (1-t)z_n$, so $z_n = z + t(y - y_n)/(1-t)$, belongs to B for n large enough.

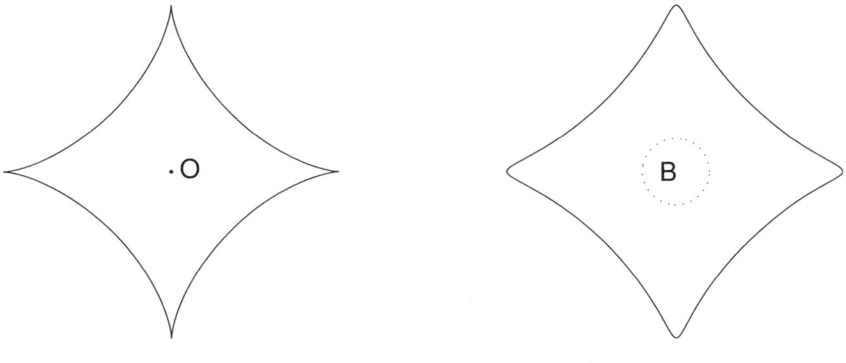

Figure 2.6. A star-shaped open set (left), an open set star shaped with respect to a ball (right).

Proposition 2.4.4. *A bounded convex open set has the ε-cone property. More generally, this is the case for a bounded open set star shaped with respect to a ball.*

Proof. A convex open set is clearly star shaped with respect to any ball that it contains. So, it suffices to prove the second assertion.

Let Ω be star shaped with respect to $B(x_0, r) \subset \Omega$. For $x \in \partial\Omega$ and $y \in B(x, r/2) \cap \overline{\Omega}$, introduce the point $z = z(y) = y + x_0 - x \in B(x_0, r/2)$. We have $B(z, r/2) \subset B(x_0, r)$ and thus the set

$$\mathcal{C} := \{ty + (1-t)\hat{z};\ 0 \le t < 1,\ \hat{z} \in B(z, r/2)\}$$

is, according to (2.30), contained in Ω. This set defines a portion of a cone with vertex y, direction $\xi_x = (z-y)/|z-y| = (x_0-x)/|x_0-x|$ and "half-angle" θ such that $\sin\theta = r/(2|x_0-x|)$. In particular, it contains $C(y,\xi_x,\varepsilon)$ where ε is defined by

$$\varepsilon := \min\{L, a\sin(r/(2m))\}, \quad L := d(x_0,\partial\Omega)/2, \quad m := \sup_{x\in\partial\Omega} |x_0-x|.$$

□

In fact, this ε-cone property for an open set Ω is completely equivalent to the (uniform) Lipschitz regularity of its boundary. This is not completely obvious, and we present here a proof mainly due to D. Chenais [95], [96]. Let us first recall the definition of an open set with Lipschitz boundary.

Definition 2.4.5. An open set Ω in $\mathbb{R}^N$ is said to have a Lipschitz boundary, if for some L, a, $r \in (0,\infty)$, for any $x_0 \in \partial\Omega$, there exist an orthogonal coordinate system with origin at $x_0 = 0$, a cylinder $K = K' \times (-a,a)$ centered at the origin, with K' open ball in $\mathbb{R}^{(N-1)}$ of radius r, and a function $\varphi : K' \to (-a,a)$, L-Lipschitz continuous with $\varphi(0) = 0$, and

$$\begin{aligned} \partial\Omega \cap K &= \{(x', \varphi(x'); x' \in K')\}, \\ \Omega \cap K &= \{(x', x_N); x' \in K', x_N > \varphi(x')\}. \end{aligned}$$

Remark 2.4.6. This definition means that $\partial\Omega$ is, in a neighborhood of each of its points, the graph of a Lipschitz function *and Ω is only on one side of its boundary.* This is a different notion than saying that $\partial\Omega$ is a Lipschitz manifold. Such a manifold is defined as follows: For any $x \in \partial\Omega$, there exists a neighborhood V of x and a mapping $\psi : V \to \mathbb{R}^N$ such that ψ is one to one, ψ and ψ^{-1} are Lipschitz continuous, and $\Omega \cap V = \{x \in V, \psi_N(x) > 0\}$, where we have set $\psi = (\psi_1, \dots, \psi_N)$.

It is easy to see that Definition 2.4.5 implies this one by setting

$$\psi(x) := (x_1, \dots, x_{N-1}, x_N - \varphi(x')).$$

On the other hand, the converse is false: There is a counterexample in [156] where the boundary is even not the graph of a continuous function. Let us notice that, as soon as ψ is C^1 in the previous definition, then, according to the implicit function theorem, it is always possible to construct φ of class C^1, which satisfies the properties of Definition 2.4.5.

Theorem 2.4.7. *An open set Ω with a bounded boundary has the ε-cone property if and only if it has a Lipschitz boundary.*

Proof. Let us assume first that Ω has a Lipschitz boundary. For $x_0 \in \partial\Omega$, we introduce a local coordinate system, as in Definition 2.4.5, where $x_0 = 0$,

$$K = K' \times (-a,a), \quad \varphi : K' \to (-a,a),$$

with the same notation. Let us choose as the direction of the cone in $x_0 = 0$ the "vertical" direction: $\xi := e_N$ (the Nth vector of the orthogonal basis), and set $\varepsilon = \min(a/2, r/2, \arctan(1/L))$.

Now fix any point $y \in \overline{\Omega} \cap B(0, \varepsilon)$. We have to prove that the cone $C(y, \xi, \varepsilon)$ is contained in Ω. First of all, thanks to the choice of ε, for any $z \in C(y, \xi, \varepsilon)$, we have $z \in K$, since, if $z = (z', z_N)$,

$$\max(|z'|, |z_N|) \le |z| = |z - x_0| \le |z - y| + |y - x_0| < 2\varepsilon \le \min(a, r).$$

Moreover, since

$$z_N - y_N = (z - y, \xi) > |z - y| \cos\varepsilon,$$

by definition of the cone, it follows that

$$(z_N - y_N)^2 > \cos^2\varepsilon(|z' - y'|^2 + |z_N - y_N|^2)$$

and therefore

$$z_N - y_N = |z_N - y_N| > \frac{1}{\tan\varepsilon}|z' - y'|.$$

This implies that

$$z_N - \varphi(z') = z_N - y_N + y_N - \varphi(y') + \varphi(y') - \varphi(z') > \frac{1}{\tan\varepsilon}|z' - y'| - L|z' - y'| \ge 0,$$

which proves that $z \in \Omega$. Now, using a classical compactness argument ($\partial\Omega$ is compact because it is bounded), one can choose a single number $\varepsilon > 0$ that works for any boundary point.

Conversely, let us now assume that Ω has the ε_0-cone property for some number $\varepsilon_0 > 0$. Choose ε such that $2\varepsilon < \varepsilon_0$, $\tan^2\varepsilon \le 1$. Let $x_0 \in \partial\Omega$ be fixed and let ξ be the unit vector in the direction of the cone associated with x_0. We still work in a local coordinate system with $x_0 = 0$, $e_N = \xi$ chosen such that (see above)

$$C(y, \xi, \varepsilon) = \{z \in B(y, \varepsilon),\ z_N - y_N > \tfrac{1}{\tan\varepsilon}|z' - y'|\}.$$

Let $K' := \{y' \in \mathbb{R}^{N-1};\ |y'| < \varepsilon \tan\varepsilon\}$ and let K be the cylinder $K' \times (-\varepsilon, \varepsilon)$. Our aim is to define the boundary of Ω in K as the graph of a Lipschitz function. Let us remark that $K \subset B(0, 2\varepsilon)$.

First, we check that $\partial\Omega$ cannot intersect the basis of the cylinder K. For that purpose, we want to show that $C(0, -\xi, 2\varepsilon) \subset \Omega^{\mathrm{c}}$. Indeed, let $z \in C(0, -\xi, 2\varepsilon)$; if we had $z \in \Omega$, by the 2ε-cone property, this would imply, since $z \in B(x, 2\varepsilon)$, that $C(z, \xi, 2\varepsilon) \subset \Omega$. Since, however, the point 0 belongs both to the cone and to Ω^{c}, we get a contradiction and therefore z does not belong to Ω.

Because of the inclusion $C(0, \xi, 2\varepsilon) \subset \Omega$ (see Figure 2.7), $\partial\Omega$ is "constrained" to go out through the lateral boundary of the cylinder K. Then let us define the function $\varphi : K' \to \mathbb{R}$ by

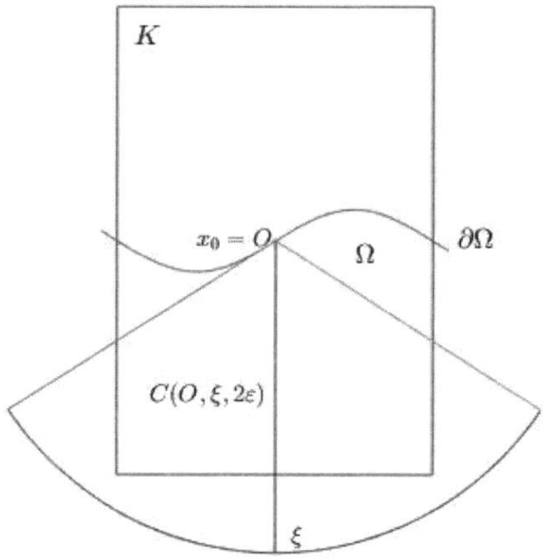

Figure 2.7. Proof of equivalence between ε-cone and uniformly Lipschitz.

$$\varphi(y') = \inf\{y_N \in [-\varepsilon, \varepsilon];\ (y', y_N) \in \Omega\}$$

(the above set is not empty because $(y', \varepsilon) \in \Omega$ for any $y' \in K'$). It is easy to check that, for any $y' \in K'$,

- $-\varepsilon < \varphi(y') < \varepsilon$;
- $(y', \varphi(y')) \in \partial\Omega$;
- $(y', y_N) \in \Omega \cap K \Longrightarrow \varphi(y') \leq y_N$ by definition, and then $\varphi(y') < y_N$ according to the line above;
- $\varphi(y') < y_N < \varepsilon \Longrightarrow y = (y', y_N) \in \Omega$: Indeed, by the 2ε-cone property applied to $(y', \varphi(y')) \in B(0, 2\varepsilon)$, the point $(y', y_N) \in C((y', \varphi(y')), \xi, 2\varepsilon)$, which is included into Ω.

Thus, we have

$$\Omega \cap K = \{(y', y_N) \in K;\ y_N > \varphi(y')\}.$$

In the same way, it can be proved that

$$\partial\Omega \cap K = \{(y', \varphi(y'),\ y' \in K'\},$$

where it remains only to check the inclusion $\subset$. Now if $(y', \eta) \in \partial\Omega \cap K$, according to the fourth point above, we have $\eta \leq \varphi(y')$; but it is impossible for $\eta < \varphi(y')$, since in that case, $(y', \varphi(y')) \in C((y', \eta), \xi, \varepsilon) \subset \Omega$, which is a contradiction.

It remains to prove that φ is (uniformly) Lipschitz continuous. Let $y', z' \in K'$. Since $(z', \varphi(z')) \in \partial\Omega$ and therefore is not in Ω, $(z', \varphi(z'))$ cannot belong to the cone $C((y', \varphi(y')), \xi, \varepsilon)$, which is contained in Ω. So

$$\varphi(z') - \varphi(y') \le \frac{1}{\tan\varepsilon}|z' - y'|,$$

and the same inequality holds in the reverse sense, by exchanging the roles of y', z'. Therefore

$$|\varphi(z') - \varphi(y')| \le \frac{1}{\tan\varepsilon}|z' - y'|,$$

which proves that φ is Lipschitz continuous with Lipschitz constant $\frac{1}{\tan\varepsilon}$. □

Remark 2.4.8 (Uniformity of geometric constants). It is a consequence of the above proof that, if Ω has the ε-cone property, its boundary is Lipschitz in the sense of Definition 2.4.5 with constants L, a, r which only depend on ε. Conversely, if Ω has a Lipschitz boundary with constants L, a, r, which are uniform (this is the case in the theorem), then it has the ϵ-cone property with some ϵ depending only on L, a, r.

Remark 2.4.9. If Ω is an open set with Lipschitz boundary, then $\overline{\Omega}^{c}$, the exterior of Ω, has the same property because it has the same boundary. Therefore, a consequence of the theorem is the following: If Ω is an open set with a bounded boundary that has the ε-cone property, then its exterior $\overline{\Omega}^{c}$ has the same property.

In this section, we consider a given ball D and a real number $\varepsilon > 0$, both fixed. Let us introduce the class of open sets,

$$\mathcal{O}_\varepsilon = \{\Omega \text{ open set}, \ \Omega \subset D, \ \Omega \text{ has the } \varepsilon\text{-cone property}\}. \tag{2.31}$$

The following result shows the good compactness properties of the class $\mathcal{O}_\varepsilon$ for the different topologies defined in this chapter. Moreover, the convergence behaves very well.

Theorem 2.4.10. *Let Ω_n be a sequence of open sets in the class $\mathcal{O}_\varepsilon$. Then there exist an open set $\Omega \in \mathcal{O}_\varepsilon$ and a subsequence Ω_{n_k} that converges to Ω in the sense of Hausdorff, in the sense of characteristic functions and in the sense of compacts. Moreover, $\overline{\Omega}_{n_k}$ and $\partial\Omega_{n_k}$ converge in the sense of Hausdorff respectively to $\overline{\Omega}$ and $\partial\Omega$.*

Proof. According to the results proved in the previous sections, we already know that there exist a subsequence Ω_{n_k}, an open set Ω, and $\chi \in L^\infty(D)$ such that Ω_{n_k} converges in the sense of Hausdorff to Ω, and $\chi_{\Omega_{n_k}} \rightharpoonup \chi$ in L^∞ weak-$*$ with $\chi_\Omega \le \chi \le 1$.

Let us introduce now, as usual, the complementary compact sets $F_n = \overline{D}\backslash\Omega_n$ and $F = \overline{D}\backslash\Omega$. Without loss of generality (possibly by enlarging D), let us assume that $\rho(\partial B, \partial\Omega) \ge \varepsilon$. Let us first prove that F has the ε-cone property.

Let $x \in \partial\Omega$. From Proposition 2.2.16, we know that there exists a sequence of points $x_n \in \partial\Omega_n$ converging to x. For each n, let us denote by ξ_n the direction of the cone associated to x_n. By compactness of the unit sphere, up to another subsequence, one can assume that ξ_n converges to a unit vector ξ. Then let $y \in B(x,\varepsilon) \cap F$; by the definition of Hausdorff convergence, there exists a sequence $y_n \in F_n$ converging to y. Since $|y_n - x_n| \to |y - x| < \varepsilon$, we have $|y_n - x_n| < \varepsilon$ for n large enough. Now we apply the ε-cone property to F_n at y_n to conclude that

$$C(y_n, \xi_n, \varepsilon) \subset F_n.$$

It is easy to check that $\overline{C}(y_n, \xi_n, \varepsilon)$ converges, both in the sense of Hausdorff and in the sense of characteristic functions, to $\overline{C}(y, \xi, \varepsilon)$. Since it has already been proved that inclusion is preserved by Hausdorff convergence of compact sets, we have

$$C(y, \xi, \varepsilon) \subset \overline{C}(y, \xi, \varepsilon) \subset F,$$

which proves that F (and hence Ω) has the ε-cone property.

Let us now study more precisely f and prove that $f = 0$ a.e. on F. This fact, together with $f = 1$ a.e. on Ω, will prove that $f = \chi_\Omega$, and then we can conclude, since $\chi_{\Omega_{n_k}} \to \chi_\Omega$ thanks to Proposition 2.2.1.

Let $x \in \partial\Omega$ and $y \in B(x,\varepsilon) \cap F$. Let us fix $\phi \in L^1(D)$ and let us introduce the two cones $C_n = C(y_n, \xi_n, \varepsilon)$ and $C = C(y, \xi, \varepsilon)$. Since $C_n \subset F_n$, applying the ε-cone property to F_n for n large enough yields

$$\int_{C_n} \phi = \int_{C_n} \chi_{F_n}\phi = \int_{C_n} \chi_D \phi - \int_{C_n} \chi_{\Omega_n}\phi.$$

But χ_{C_n} converges (strongly) to χ_C and $\chi_{\Omega_{n_k}} \to f$ in L^∞ weak-$*$. So we can pass to the limit when n_k goes to ∞ in the above equality and we get

$$\int_C \phi = \int_C \chi_D \phi - \int_C f\phi = \int_C \phi - \int_C f\phi,$$

which means that for all $\phi \in L^1(B)$, $\int_C f\phi = 0$, implying that $f = 0$ a.e. on C, for any $x \in \partial\Omega$ and for any cone $C(y, \xi, \varepsilon)$. Consequently, $f = 0$ a.e. on $\{y \in F / d(y, \partial\Omega) < \varepsilon\}$.

Let us now consider the other points y: We still fix a sequence $y_n \in F_n$ that converges to y, and then

- either there exists a subsequence y_{n_k} such that $d(y_{n_k}, \partial\Omega_{n_k}) \geq \varepsilon$ (i.e., $B(y_{n_k}, \varepsilon) \subset F_{n_k}$) and then the previous argument can be repeated as well, by replacing C_n everywhere by $B(y_{n_k}, \varepsilon)$ to conclude that $f = 0$ a.e. on $B(y, \varepsilon)$;

- or else $d(y_n, \partial\Omega_n) < \varepsilon$ for any $n \geq n_0$ and then by the ε-cone property, there exists $C(y_n, \xi_n, \varepsilon) \subset F_n$ and $C(y_n, \xi_n, \varepsilon) \to C(y, \xi, \varepsilon)$ (up to a subsequence). Then we repeat the same proof, replacing C_n by $C_n \cap \overline{B}$ (since $C \cap \overline{B} \subset \overline{B}\backslash\Omega$) to prove that $f = 0$ a.e. in a neighborhood of y. It can be deduced that $f = 0$ a.e. on $\{y \in F / d(y, \partial\Omega) \geq \varepsilon\}$.

For convergence in the sense of compacts, we just have to prove the property for compact sets outside Ω, thanks to Proposition 2.2.17. Let L be such a compact set; without loss of generality, we may assume that L has a nonempty interior. For the purpose of contradiction, let us assume that there exists a subsequence Ω_{n_k} such that $L \cap \overline{\Omega}_{n_k} \neq \emptyset$. Then we are faced with the following alternatives:

- either (for some subsequence) $L \subset \Omega_{n_k}$, but then $|\Omega_{n_k} \backslash \Omega| \geq |L| > 0$, and this contradicts convergence in the sense of characteristic functions;

- or else $L \cap \partial\Omega_{n_k} \neq \emptyset$. Then let us consider a sequence of points $x_{n_k} \in L \cap \partial\Omega_{n_k}$. It is always possible to assume that (up to a subsequence) the sequence converges to a point $x \in L$. By the ε-cone property applied to each x_{n_k}, there exists a cone $C(x_{n_k}, \xi_k, \varepsilon)$ contained in Ω_{n_k}. As above, one can assume moreover that $C(x_{n_k}, \xi_k, \varepsilon)$ converges in the sense of Hausdorff to $C(x, \xi, \varepsilon)$, which, by continuity of inclusion under Hausdorff convergence, is a cone contained in Ω. This, however, contradicts the fact that $x \in L \subset \overline{\Omega}^c$.

□

2.5 Exercises

Exercise 2.1. In $\mathbb{R}$, let us consider the following sequence of open sets:

$$\Omega_n := \bigcup_{k=0}^{2^{n-1}-1} \left(\frac{2k}{2^n}, \frac{2k+1}{2^n} \right).$$

Prove that the sequence $f_n = \chi_{\Omega_n}$ converges weakly in $L^p(0,1)$, for any $1 \leq p \leq \infty$, to the constant function equal to $1/2$. What is the limit of Ω_n in the sense of Hausdorff and in the sense of compact sets?

Exercise 2.2. Is the volume function continuous for the three kinds of convergence seen in this chapter? Consider the same question for the diameter.

Exercise 2.3. Let us assume that a sequence of open sets Ω_n converges in the sense of compact sets to an open set Ω. Does it imply that

$$\forall\, x \in \partial\Omega, \quad d(x, \partial\Omega_n) \to 0?$$

Consider the same question for convergence in the sense of characteristic functions and for Hausdorff convergence.

Exercise 2.4. This exercise is inspired by [197].

(1) Give an example to show that Hausdorff convergence of a sequence of open sets Ω_n (resp., of closed sets F_n) to an open set Ω (resp., a closed set F) does not imply, in general, convergence in the sense of Hausdorff of the boundary of Ω_n to the boundary of Ω.

(2) Let K_n be a sequence of compact sets which converges in the sense of Hausdorff to a compact K. Prove that ∂K_n has at least 1 accumulation point (in the sense of Hausdorff convergence), and that any accumulation point F satisfies

$$\partial K \subset F \subset K.$$

(3) Let K be a compact set. For any positive number δ, let us denote by $B_\delta(K)$ the set of points in $\mathbb{R}^N$ whose distance to K is less than δ. Let us now introduce the functions

$$g_K(x,\delta) := d(x, \mathbb{R}^N \backslash B_\delta(K)), \qquad g_K(\delta) := \sup_{x \in \partial K} g_K(x,\delta).$$

Prove the following properties:

(A) $g_K(0) = 0$;

(B) g_K is a nonincreasing function;

(C) $g_K(\delta) \geq \delta$;

(D) $g_K(\delta)$ is equal to the Hausdorff distance of ∂K to the boundary of the set $B_\delta(K)$.

(4) Let $\mathcal{K}$ be a family of compact sets in $\mathbb{R}^N$. We define the function $g_{\mathcal{K}}$ by

$$g_{\mathcal{K}} = \sup_{K \in \mathcal{K}} g_K(\delta).$$

Prove that $g_{\mathcal{K}}$ is continuous on the right at 0 if and only if, for any sequence of compact sets K_n in $\mathcal{K}$, we have

$$K_n \xrightarrow{\text{H}} K \implies \partial K_n \xrightarrow{\text{H}} \partial K.$$

Exercise 2.5. Prove that the convex hull of a compact set is compact.

Let K_n be a sequence of compact sets converging in the sense of Hausdorff to the compact set K. Prove that the convex hull of K_n converges in the sense of Hausdorff to the convex hull of K.

Exercise 2.6. Let Ω_n be a sequence of open sets which converges in the sense of compact sets to an open set Ω and in the sense of Hausdorff to $\widetilde{\Omega}$. Prove that $\Omega \subset \widetilde{\Omega} \subset \overline{\Omega}$. Deduce from the previous fact that, if Ω_n is a sequence of open sets, contained in some fixed ball, that converges in the sense of compact sets to a Carathéodory open set Ω (this means $\partial\Omega = \partial\overline{\Omega}$), then Ω_n converges in the sense of Hausdorff to Ω.

In the same way, prove that, if Ω_n is a sequence of open sets, contained in some fixed ball, that converges in the sense of compact sets to an open set Ω and in the sense of characteristic functions to $\widetilde{\Omega}$, then $\Omega \subset \overline{\widetilde{\Omega}}$ and $\widetilde{\Omega} \subset \overline{\Omega}$.

Exercise 2.7. Prove that, if the sequence Ω_n converges in the sense of compact sets to Ω and if the boundary of Ω has zero measure, then the sequence Ω_n also converges in the sense of characteristic functions.

Exercise 2.8. Let us denote by $\mathcal{O}$ the class of open sets in $\mathbb{R}^N$. For each pair of compact sets (K, L) in $\mathbb{R}^N$, we introduce

$$\mathcal{V}_{K,L} := \{\omega \in \mathcal{O};\ K \subset \omega,\ L \subset \overline{\omega}^c\}.$$

(1) Prove that the family

$$\{\mathcal{V}_{K,L};\ K, L \text{ compact sets in } \mathbb{R}^N\},$$

is stable under finite intersection and therefore is a basis of open sets on $\mathcal{O}$. We denote by τ the topology generated by this basis.

(2) Prove that the convergence of a sequence of open sets for this topology coincides with the convergence in the sense of compact sets.

(3) For a given open set Ω in $\mathbb{R}^N$, determine the intersection of all $\mathcal{V}_{K,L}$ containing Ω. Prove that the topology τ is not separated(= not a Hausdorff topology).

(4) Let Ω and $\hat{\Omega}$ be two open sets that are limits of the same sequence of open sets for the topology τ. Prove that these two sets have the same closure.

(5) Let Ω_1, Ω_2 be two open sets such that $\overline{\Omega_1} \neq \overline{\Omega_2}$. Prove that there exist two τ-open sets $\mathcal{V}_1, \mathcal{V}_2$ such that

$$\Omega_1 \in \mathcal{V}_1, \quad \Omega_2 \in \mathcal{V}_2, \quad \mathcal{V}_1 \cap \mathcal{V}_2 = \emptyset.$$

(6) Let $\mathcal{O}_Q$ be the quotient space of $\mathcal{O}$ by the equivalence relation defined by

$$\Omega_1 \simeq \Omega_2 \iff \overline{\Omega_1} = \overline{\Omega_2}.$$

Prove that $\mathcal{O}_Q$ endowed with the quotient topology of τ is separated.

Exercise 2.9. Let $f_n, f : \mathbb{R}^N \to \mathbb{R}$ be continuous and let us consider the open sets contained in some fixed closed B and defined by

$$\Omega_n = \{x \in B,\ f_n(x) > 0\} \quad \text{and} \quad \Omega = \{x \in B,\ f(x) > 0\}.$$

We assume that the open sets Ω_n are not empty. The aim of this exercise is to prove the following compactness result:

Proposition 2.5.1. *Let us assume that the functions f_n are of class C^2 on B with, moreover,*

(i) for all $x \in B$, for all $n \in \mathbb{N}$, for all i, j, $\left|\frac{\partial f_n}{\partial x_i}(x)\right| \leq M$ *and* $\left|\frac{\partial^2 f_n}{\partial x_i \partial x_j}(x)\right| \leq M$;

(ii) for all $x \in B$, for all $n \in \mathbb{N}$, $|f_n(x)| + |\nabla f_n(x)| \geq \alpha$.

Then there exists a subsequence of Ω_n which converges in the sense of compact sets to an open set $\Omega \subset B$.

(1) Prove first the following lemma: Assume that $\partial\Omega = \{x \in \mathbb{R}^N / f(x) = 0\}$. Then, if f_n converges uniformly to f on B, we have $\Omega_n \xrightarrow{K} \Omega$.

(2) Prove the proposition.

Exercise 2.10. Let us denote by Σ the quotient space of the set of (Lebesgue)-measurable subsets in R^N by the equivalence relation:

$$E_1 \sim E_2 \iff \chi_{E_1} = \chi_{E_2} \text{ a.e.}$$

(1) Prove that

$$\delta(E_1, E_2) := \arctan\left(\text{meas}(E_1 \Delta E_2)\right)$$

defines a distance on Σ.

(2) Let $(E_n)_{n\geq 1}$, E be given measurable sets in $\mathbb{R}^N$. Prove that the three following properties are equivalent:

$$\begin{cases} \delta(E_n, E) \to 0, \\ \chi_{\Omega_n} - \chi_E \overset{\sigma(L^1(\mathbb{R}^N), L^\infty(\mathbb{R}^N))}{\longrightarrow} 0, \\ \chi_{E_n} - \chi_E \overset{L^1(\mathbb{R}^N)}{\longrightarrow} 0. \end{cases} \tag{2.32}$$

(3) Prove that Σ endowed with this distance is a complete space.

(4) Let $(f_n)_{n\geq 1}$ be a sequence of integrable functions from $\mathbb{R}^N$ into $\mathbb{R}$ such that, for any measurable set E in $\mathbb{R}^N$,

$$\lim_{n\to\infty} \int_E f_n \text{ exists.} \tag{2.33}$$

Prove (local uniform integrability of the sequence f_n)

$$\sup_n \int_E |f_n| \to 0 \text{ when } |E| \to 0. \tag{2.34}$$

Hint: Use the Baire property of (Σ, δ).

(5) If the sequence f_n weakly converges in $L^1(\mathbb{R}^N)$, then property (2.33) is satisfied. Prove that (2.34) remains satisfied if f_n is a relatively weakly compact sequence in $L^1(\mathbb{R}^N)$.

(6) Prove that all these properties remain true if we replace $\mathbb{R}^N$ equipped with the Lebesgue measure by any σ-finite measure space.

Exercise 2.11. Using the notation of Section 2.4, prove that if y_n converges to y in $\mathbb{R}^N$ and ξ_n converges to ξ on the unit sphere of $\mathbb{R}^N$, then $C(y_n, \xi_n, \varepsilon)$ converges, both in the sense of Hausdorff and in the sense of characteristic functions to $C(y, \xi, \varepsilon)$.

Exercise 2.12. Let Ω_n be a sequence of open sets having the ε-cone property, that converges to Ω in one of the three possibilities of this chapter. Prove that

$$\overline{\Omega_n} \xrightarrow{\text{H}} \overline{\Omega} \quad \text{and} \quad \partial\Omega_n \xrightarrow{\text{H}} \partial\Omega.$$

Chapter 3

Continuity with respect to domains

As we saw in the previous chapter, the existence of optimal shapes requires some continuity or at least lower semicontinuity of the functional to be minimized. This also implies having continuity of the solution of the associated partial differential equation with respect to the variations of the domain in an adequate topology.

In this chapter, we analyze in detail the continuity (or the noncontinuity) of the mapping $\Omega \longrightarrow u_\Omega \in H_0^1(D)$, where u_Ω is the solution of the Dirichlet problem (see Proposition 3.1.20) on a variable open subset Ω of a fixed open set D. In particular, we look at the case when the family of open sets Ω is equipped with the topology of Hausdorff convergence.

A complete analysis requires using the concept of *capacity*. In fact, here it will simply be the classical "electrostatic" capacity associated with the energy norm of the space H^1. We will recall all necessary definitions and give complete proofs of the relevant properties (which are spread out in the literature). We will also recall all that is necessary to know on the spaces H^1, H_0^1, H^{-1}, requiring only a limited knowledge on derivatives in the sense of distributions.

We analyze the precise and "simple" example of the Dirichlet problem for the Laplacian operator and we aim to give all necessary details. Indeed, although done on a specific example, our approach is quite general and carries over to more sophisticated operators. In fact, we show how it may even be extended to other second-order operators like the elasticity operator with variable coefficients or the Laplacian operator with Neumann[1] boundary conditions. We also give some hints on how to extend it to the bi-Laplacian operator, which is the simplest example among fourth-order operators. As mentioned at the end of the chapter, our approach may also be applied to nonlinear operators like the so-called p-Laplacian. It is then necessary to work with *nonlinear capacities* associated with the Sobolev[2] space $W^{1,p}$ (and $W^{m,p}$ for higher-order operators). These capacities will not be studied here: We deliberately choose to concentrate on the H^1 (or $W^{1,2}$) capacity to keep the presentation simple and also because it is a very significant model for the use of capacities in these continuity questions for the solutions of partial differential equations. For more elaborate presentations, we refer to [19], [178], [209], [224], [232], [170], [4], [230], [177], [140].

---

[1]Carl Gottfried NEUMANN, 1832–1925, German, was a professor in Halle, Basel, Tübingen, and Leipzig. He contributed to potential theory and differential equations.

[2]Sergeï SOBOLEV, 1908–1989, Russian, worked most of the time in Moscow but also in Novosibirsk. He is well known for his work on partial differential equations and for the functional spaces named after him.

3.1 The Dirichlet problem

3.1.1 The space H_0^1 and its dual space H^{-1}

We fix D an open subset of $\mathbb{R}^N$ and we denote by $\mathcal{C}_0^\infty(D)$ the space of infinitely differentiable functions from $\mathbb{R}^N$ into $\mathbb{R}$ with compact support in D. We define the norm $||\cdot||_{H^1}$ by

$$\forall\, u \in \mathcal{C}_0^\infty(D), \quad ||u||_{H^1}^2 := \int_D u^2 + |\nabla u|^2, \tag{3.1}$$

where $|\cdot|$ denotes the Euclidean norm in $\mathbb{R}^N$, which is computed here for the values of the gradient $\nabla u(x) = (\frac{\partial u(x)}{\partial x_1}, \dots, \frac{\partial u(x)}{\partial x_N})$.

Definition 3.1.1. We define

$$H^1(D) := \{u \in L^2(D);\ \nabla u \in L^2(D)^N\},$$

equipped with the norm $||\cdot||_{H^1}$ defined as in (3.1), but where ∇u is to be understood in the sense of distributions.

We denote by $H_0^1(D)$ the closure of $\mathcal{C}_0^\infty(D)$ in $H^1(D)$.

The space $H^1(D)$ may be isometrically identified with a subspace of $L^2(D) \times L^2(D)^N$ by the one-to-one mapping

$$\mathcal{I} : u \in H^1(D) \longrightarrow (u, \nabla u) \in L^2(D) \times L^2(D)^N.$$

The image of $\mathcal{I}$ is a closed subspace by the continuity of differentiation in the sense of distributions. Thus it is a separable Hilbert[3] space as $L^2(D)$. Consequently, this same separability property holds for $H^1(D)$ as well as for its closed subspace $H_0^1(D)$.

Remark 3.1.2. It follows from its definition that $H_0^1(D)$ is also a closed subspace of $H_0^1(\widehat{D})$ for all open sets $\widehat{D}$ containing D. Indeed, we may notice that, in Definition 3.1.1, the convergence of ∇u_n holds also in $L^2(\mathbb{R}^N)$ and its limit is also equal to ∇u, computed in the sense of distributions *in the whole space* $\mathbb{R}^N$. Thus for $u \in H_0^1(D)$, we will not make any difference between u considered as being defined from D into $\mathbb{R}$ and u extended by zero to the whole space $\mathbb{R}^N$ (which is therefore in $H_0^1(\mathbb{R}^N)$).

Remark 3.1.3. There are two particular cases.

---

[3]David HILBERT, 1862–1943, German, got involved in almost all domains of mathematics. He left us a tremendous body of work, together with an exceptionally fruitful research program with the series of 23 problems that he presented at the Paris Conference of the International Congress of Mathematicians in 1900.

Case 1: $D = \mathbb{R}^N$**.** Then

$$H_0^1(\mathbb{R}^N) = H^1(\mathbb{R}^N) = \{u \in L^2(\mathbb{R}^N);\ \nabla u \in L^2(\mathbb{R}^N)^N\}.$$

Indeed if $u \in H^1(\mathbb{R}^N)$, we may approximate it in H^1, by truncation and regularization, by the functions $u_n := \rho_n * (\varphi_n u)$ of C_0^∞ where

- ρ_n is a regularizing sequence converging to the Dirac[4] mass at the origin, for instance,

$$\left.\begin{aligned} &\rho_n(x) := n^N \rho_0(nx) \text{ with } \rho_0 \in C_0^\infty(B(0,1)),\ 0 \le \rho_0 \le 1,\\ &\rho_0 = 1 \text{ on a neighborhood of the origin, } \textstyle\int \rho_0 = 1; \end{aligned}\right\} \tag{3.2}$$

- $\varphi_n(x) := \rho_0(x/n)$. Thus φ_n converges to 1 almost everywhere and $\varphi_n u$ converges to u in L^2. Moreover $\nabla(\varphi_n u) = u\nabla\varphi_n + \varphi_n \nabla u$ converges in $(L^2)^N$ to ∇u since $\nabla\varphi_n = \nabla\rho_0(\cdot/n)/n$ converges a.e. to 0 and remains bounded so that $u\nabla\varphi_n$ converges to 0 in L^2. By convolution, $(u_n, \nabla u_n)$ converges in $L^2(D) \times L^2(D)^N$ to $(u, \nabla u)$.

Case 2: $D = (0,1)$**.** As $H_0^1(D)$ is a subspace of $H^1(\mathbb{R})$, let us emphasize some one-dimensional properties of $H^1(\mathbb{R})$. First, $H^1(\mathbb{R})$ is embedded into $L^\infty(\mathbb{R})$ and even into the space of $1/2$-Hölder[5]-continuous functions: To see it, we write that for all $u \in C_0^\infty(\mathbb{R})$ and all $x, y \in \mathbb{R}$,

$$|u(y) - u(x)| = \left|\int_x^y u'(t)\,dt\right| \le \left\{\int_{\mathbb{R}} u'^2(t)\,dt\right\}^{1/2} |y - x|^{1/2}, \tag{3.3}$$

$$\begin{aligned} |u^2(y) - u^2(x)| &= \left|\int_x^y 2u(t)u'(t)\,dt\right| \\ &\le 2\left\{\int_{\mathbb{R}} u^2(t)\,dt\right\}^{1/2} \left\{\int_{\mathbb{R}} u'^2(t)\,dt\right\}^{1/2}. \end{aligned} \tag{3.4}$$

Choosing x close to $-\infty$ and taking the supremum in $y \in \mathbb{R}$ in this second inequality, we obtain

$$||u||^2_{L^\infty(\mathbb{R})} \le 2||u||_{L^2(\mathbb{R})}||u'||_{L^2(\mathbb{R})}. \tag{3.5}$$

These two inequalities propagate by density to all $u \in H^1(\mathbb{R}) = H_0^1(\mathbb{R})$. In particular, any function u of $H^1(\mathbb{R})$ has a continuous representative $\tilde{u}$ and this representative is unique since two continuous functions that are equal a.e. are equal everywhere.

[4]Paul DIRAC, 1902–1984, English of Swiss origin, physicist and mathematician, contributed to functional analysis and mathematical physics.

[5]Ludwig Otto HÖLDER, 1859–1937, German, was Weierstrass' student at the University of Tübingen. He is well known for the inequality and the functions spaces named after him, but he also left us important contributions in group theory and mathematical logic.

In fact, the subspace $H_0^1(0,1)$ may be explicitly described as

$$H_0^1(0,1) = \{u \in H^1(0,1);\ \tilde{u}(0) = \tilde{u}(1) = 0\}. \tag{3.6}$$

Inclusion from the left to the right follows from the definition of $H_0^1(\Omega)$ and from the embeddings mentioned above. Conversely, for u in the space on the right of this equality, we first check that its extension by 0 to $\mathbb{R}$ is in $H^1(\mathbb{R})$: Indeed, a simple computation shows that its distribution derivative is equal to the extension by 0 of the derivative of u in $(0,1)$ (the continuity of the extension at 0 and 1 implies that there is no contribution of the distribution derivative at these two points). Thus $u' \in L^2(\mathbb{R})$ and $u \in H^1(\mathbb{R})$.

Let $n \geq 3$, $\lambda_n = n/(n-2)$, and $u_n(x) := u\big(1/2 + \lambda_n(x - 1/2)\big)$: then u_n is compactly supported in $[1/n, 1-1/n]$ and converges in H^1 to u. We then introduce $v_n := u_n * \rho_{2n}$, where ρ_n is the regularizing sequence introduced in (3.2) so that $v_n \in C_0^\infty(0,1)$ and converges in H^1 to u.

If now Ω is a bounded open set in dimension 1, we deduce

Proposition 3.1.4.

$$H_0^1(\Omega) = \{u \in H_0^1(D);\ \tilde{u} = 0 \ on\ D\backslash\Omega\}. \tag{3.7}$$

Proof. Let $\mathcal{H}$ be the space on the right-hand side of this inequality. The inclusion $H_0^1(\Omega) \subset \mathcal{H}$ follows easily from the definition of $H_0^1(\Omega)$ since H^1-convergence implies *uniform convergence* in dimension 1. For the other inclusion, we use that Ω is the union of disjoint intervals I_p, $p \in \mathcal{P} \subset \mathbb{N}$. The continuous representative $\tilde{u}$ vanishes on $D\backslash\Omega$. The characterization (3.6) tells us that if $u \in \mathcal{H}$, then its restriction to I_p belongs to $H_0^1(I_p)$. Therefore it is the limit in this space of a sequence $u_n^p \in C_0^\infty(I_p)$ such that

$$\|u_n^p - u\|_{H_0^1(I_p)} \leq 2^{-n}2^{-p}.$$

We then check that the sequence of functions $\sum_{p \leq n} u_n^p$, which belong to $C_0^\infty(\Omega)$, converges in $H_0^1(D)$ to u as $n \to \infty$. □

In dimension $N \geq 2$, functions in H_0^1 are not necessarily continuous. Therefore it is more difficult to find a characterization of $H_0^1(D)$ similar to (3.7). Nevertheless, a similar description in terms of quasi-continuous functions may be given (see Theorem 3.3.42).

Definition 3.1.5. Let $\mathcal{D}'(D)$ denote the space of distributions on D. We introduce

$$H^{-1}(D) := \Big\{T \in \mathcal{D}'(D);\ T = f_0 + \textstyle\sum_{1 \leq i \leq N} \frac{\partial f_i}{\partial x_i},\ f_0, \dots, f_N \in L^2(D)\Big\}, \tag{3.8}$$

equipped with the norm

$$\|T\|_{H^{-1}(D)}^2 := \inf\Big\{ \textstyle\sum_{0 \leq i \leq N} \|f_i\|_{L^2}^2 \text{ for all choices of } f_i \Big\}.$$

By Riesz's[6] theorem (see, e.g., [54, Thm. V.5] or [196, Thm. 3.1]), for all L in the dual space $(H_0^1)'(D)$ of $H_0^1(D)$, there exists a unique $u \in H_0^1(D)$ such that

$$\forall\, v \in H_0^1(D), \quad L(v) = \int_D uv + \nabla u \nabla v. \tag{3.9}$$

The restriction of L to $\mathcal{C}_0^\infty(D)$ satisfies

$$\forall\, v \in \mathcal{C}_0^\infty(D), \quad L(v) = \langle u - \Delta u, v\rangle_{\mathcal{D}' \times \mathcal{C}_0^\infty}, \tag{3.10}$$

where $\Delta = \sum_{1\le i\le N} \partial^2/\partial x_i^2$ is the Laplacian operator. Thus the restriction $T := L_{|\mathcal{C}_0^\infty(D)}$ of L to $\mathcal{C}_0^\infty(D)$ is a distribution in D that belongs to $H^{-1}(D)$ since $T = u - \Delta u = u - \sum_i \partial_{x_i}(\partial_{x_i} u)$ where $u, \partial_{x_i} u \in L^2$ (cf. 3.8). More precisely , we have

Proposition 3.1.6. *The mapping $L \to T := L_{|\mathcal{C}_0^\infty(D)}$ is an isometric isomorphism from $\left(H_0^1(D)\right)'$ onto $H^{-1}(D)$. Moreover, the operator $I - \Delta$ is also an isometry from $H_0^1(D)$ onto $H^{-1}(D)$. Thus for all $f \in H^{-1}(D)$, there exists a unique solution of*

$$u \in H_0^1(D),\ u - \Delta u = f \text{ in } \mathcal{D}'(D), \tag{3.11}$$

and we have $||u||_{H_0^1(D)} = ||f||_{H^{-1}(D)}$.

Proof. Let $L \in (H_0^1)'(D)$; we already know (see (3.9), (3.10)) that $T := L_{|\mathcal{C}_0^\infty(D)}$ is a distribution belonging to $H^{-1}(D)$ with $T = u - \sum_i \partial_{x_i}(\partial_{x_i} u)$, $u \in H_0^1(D)$, where $u = \mathcal{R}(L)$, $\mathcal{R}$ being the Riesz isometry of $\left(H_0^1(D)\right)'$ onto $H_0^1(D)$. Using the definition of the H^{-1}-norm, we also have

$$||L||^2_{(H_0^1)'} = ||u||^2_{H_0^1} = \int_D u^2 + |\nabla u|^2 \ge ||T||^2_{H^{-1}}.$$

Conversely, let $T \in H^{-1}(D)$; it can be written $T = f_0 + \sum_{1\le i\le N} \partial_{x_i} f_i$. Thus for all $v \in \mathcal{C}_0^\infty(D)$,

$$\begin{aligned} |\langle T, v\rangle_{\mathcal{D}'\times\mathcal{C}_0^\infty}| &= \left| \int_D f_0 v - \sum_{1\le i\le N} f_i \frac{\partial v}{\partial x_i} \right| \\ &\le \left(\sum_{0\le i\le N} ||f_i||^2_{L^2} \right)^{1/2} \cdot \left(||v||^2_{L^2} + ||\nabla v||^2_{L^2} \right)^{1/2}. \end{aligned}$$

[6] Frigyes RIESZ, 1880–1956, Hungarian. His main contributions are in functional analysis, where several famous results are named after him. His brother Marcel, 1880–1969, is also known for his work on series and on L^p-spaces.

Hence T extends by density, and in a unique way, to a linear continuous form L on $H_0^1(D)$: Moreover, we have the estimate

$$||L||^2_{(H_0^1)'} \leq \sum_{0\leq i\leq N} ||f_i||^2_{L^2},$$

and, passing to the infimum over all choices of the f_i,

$$||L||_{(H_0^1)'} \leq ||T||_{H^{-1}}.$$

This proves that $L \longrightarrow T = L_{|C_0^\infty(D)}$ is an isometry. By composition, $I - \Delta$ is also an isometry (note that $(I - \Delta)(u) = \mathcal{R}(u)_{|C_0^\infty(D)}$). □

We deduce from Proposition 2.3.12 the classical Rellich[7] theorem.

Theorem 3.1.7. *Let D be an open subset of $\mathbb{R}^N$ and let $(u_n)_{n\geq 0}$ be a bounded sequence in $H^1(D)$. Then we may pass to a subsequence of u_n that converges strongly in $L^2_{\mathrm{loc}}(D)$ and a.e.*

If D is bounded, the embedding of $H_0^1(D)$ into $L^2(D)$ is compact.

Proof. Let u_n be a bounded sequence in $H^1(D)$. By Proposition 2.3.12, there exists $(u_{n_k})_{k\geq 1}$ that converges at least in $L^1_{\mathrm{loc}}(D)$. Let us set $U_{(p,q)} = (u_{n_p} - u_{n_q})^2$. Then $U_{(p,q)}$ and $\nabla U_{(p,q)}$ are bounded in L^1. Thus $U_{(p,q)}$ is also relatively compact in L^1_{loc}. We deduce that u_{n_k} also converges in L^2_{loc} as $k \longrightarrow \infty$. Up to a subsequence, we may assume that a.e. convergence holds.

If now u_n is a bounded sequence in $H_0^1(D)$ with D bounded, we may introduce an open ball B containing the compact set $\overline{D}$ and we may see u_n as a bounded sequence in $H_0^1(B)$. Thus, by the previous point, we may pass to a subsequence that converges in $L^2(D)$. □

Remark. Note that, even for bounded D, the embedding of $H^1(D)$ into $L^2(D)$ is not necessarily compact. On the other hand, it is the case if D has a Lipschitz boundary (see, e.g., [54]). Note also that the embedding of $H_0^1(D)$ into $L^2(D)$ is compact as soon as D is of finite measure.

3.1.2 Lip $\circ H^1 \subset H^1$

We will recall here the stability of H^1 under the action of nonlinear functions, not necessarily *Lipschitz continuous*.

[7]Franz RELLICH, 1906–1955, was director of the famous Göttingen Mathematical Institute after the Second World War. He is known for several results on compactness of operators and their spectral properties in connection with perturbation theory.

It is simple to see that, for D an open subset of $\mathbb{R}^N$, if $u \in H^1(D)$ and $G \in C^1(\mathbb{R})$ with G' bounded and $G(0) = 0$, then $G(u) \in H^1(D)$ and $\nabla G(u) = G'(u)\nabla u$. This extends to Lipschitz continuous functions, with an a priori difficulty in giving a meaning to this formula. Indeed, a Lipschitz continuous function is differentiable "only" a.e. Thus if E is the negligible set where G' does not necessarily exist, $G'(u(x))$ is not a priori defined on $\{x \in D;\ u(x) \in E\}$, which may be not negligible at all (think of the case when $E = \{0\}$ and when u vanishes on an open set). In fact we have the following lemma where derivatives have to be understood in the sense of distributions and where we use the classical space

$$W^{1,1}(D) := \big\{u \in L^1(D);\ \nabla u \in \big(L^1(D)\big)^N\big\}. \tag{3.12}$$

Lemma 3.1.8. *Let I be a bounded open subset of $\mathbb{R}$ and let $u \in W^{1,1}(I)$. Then for all negligible set $A \subset \mathbb{R}$,*

$$u' = 0 \text{ a.e. on } \{x \in I;\ u(x) \in A\}. \tag{3.13}$$

If $G : \mathbb{R} \to \mathbb{R}$ is Lipschitz continuous, then

$$G(u) \in W^{1,1}(I), \text{ and } G(u)' = G'(u)u' \text{ a.e. on } I(*). \tag{3.14}$$

Remark 3.1.9. $(*)$ If G is Lipschitz continuous, it is differentiable a.e., $G' \in L^\infty(\mathbb{R})$, and $G(\hat r) - G(r) = \int_r^{\hat r} G'(t)\,dt$ for all $r, \hat r \in \mathbb{R}$ (see [266], [135]). Since G' is defined only a.e., formula (3.14) is a priori ambiguous: In fact, if A denotes the negligible set where G' may not exist, since $u' = 0$ a.e. on $A_G := \{x \in I,\ u(x) \in A\}$, we naturally set $G'(u)u' := 0$ on A_G.

Remark 3.1.10. If $u \in W^{1,1}(a,b)$, it follows from Lemma 3.1.8 that $u' = 0$ a.e. on the set $\{x \in (a,b);\ u(x) = 0\}$. In fact, if $u \in W^{1,1}(D)$, we also have $\nabla u = 0$ a.e. on the set $\{x \in D;\ u(x) = 0\}$. To see it, we apply Lemma 3.1.8 to the partial derivatives $\partial_{x_i} u$ and we use Fubini's[8] theorem as follows. Let us denote $x = (x_1, x')$. The function $\xi \to \partial_{x_1} u(\xi, x')$ is integrable a.e. x' so that a.e. x', $\partial_{x_1} u(\cdot, x') = 0$ a.e. on the set $\{\xi;\ u(\xi, x') = 0\}$; applying Fubini's theorem, we deduce that $\partial_{x_1} u = 0$ a.e. on the set $\{x \in D;\ u = 0\}$. We do the same for the other partial derivatives.

Proof of Lemma 3.1.8. The starting point is the following: if $G \in C^1(\mathbb{R})$ and if $u \in W^{1,1}(I)$, then

$$G(u) \in W^{1,1}(I), \text{ and a.e. } G(u)' = G'(u)u'. \tag{3.15}$$

[8]Guido FUBINI, 1879–1943, of Venetian origin, former student of the Scuola Normale Superiore di Pisa; he was then professor in Pisa, Catania, Turin, and Princeton. His contributions are particularly in functional analysis and integration theory.

Indeed, this is well known when moreover $u \in C^1(\mathbb{R})$. If not, we approximate u by convolution in I. We pass to the limit as follows: if u_n denotes the approximating sequence, then $G(u_n)$, $G'(u_n)$ converge to $G(u)$, $G'(u)$ uniformly on compact subsets of I, u_n' converges in $L^1_{\mathrm{loc}}(I)$ to u', and $G(u_n)'$ converges to $G(u)'$ in the sense of distributions. Thus, we deduce at the limit that $G(u)' = G'(u)u' \in L^1(I)$ and consequently $G(u) \in W^{1,1}(I)$.

Let us now set $G(r) = \int_0^r p(s)\,ds$, where first $p = \chi_\omega$, with ω an open subset of I. This function p is the increasing limit of continuous functions p_n and the relation (3.15) holds for $G_n(r) = \int_0^r p_n(s)\,ds$. It is preserved at the limit by monotonicity, that is,

$$\text{a.e. } G(u)' = p(u)u'. \tag{3.16}$$

If $p = \chi_A$, where A is a Lebesgue-measurable subset of $\mathbb{R}$, there exists a nonincreasing sequence of open sets ω_n containing A whose measure tends to the measure of A. The relation (3.16) with p replaced by χ_{ω_n} is again preserved at the limit by monotonicity with $p = \chi_{A'}$, where $A' = \cap\omega_n$ and with $G(r) = \int_0^r p(s)\,ds$.

If A is of zero measure (and so is A'), G is identically equal to zero and the relation then says that $\chi_{A'}(u)u' = 0$ a.e., that is, $u' = 0$ a.e. on $\{x \in I;\ u(x) \in A'\} \supset \{x \in I;\ u(x) \in A\}$, whence (3.13).

In the case where the set A is of positive and finite measure, by (3.13), $\chi_{A'\setminus A}(u)u' = 0$ a.e. so that (3.16) holds at the same time for $p = \chi_{A'}$ and for $p = \chi_A$, since the antiderivative G is the same in both cases.

By linearity, (3.16) may be extended to simple functions p. Next, if p is bounded measurable, it is the uniform limit of simple functions. Again, we may pass to the limit in (3.16) and by (3.13), we may even replace p by any function that is a.e. equal to p, whence (3.14). □

Proposition 3.1.11. *Let $u \in H^1(D)$ and let $G : \mathbb{R} \to \mathbb{R}$ be Lipschitz continuous with $G(0) = 0$. Then $G(u) \in H^1(D)$ and we have $\nabla G(u) = G'(u)\nabla u$. If $u \in H_0^1(D)$, then $G(u) \in H_0^1(D)$.*

Proof. For the first part, let us first remark that $G(u) \in L^2$ since we have the pointwise inequality $|G(u)| \le \|G'\|_{L^\infty}|u|$. Next we apply Lemma 3.1.8 to each partial function $\xi \to u(x_1, \dots, x_{i-1}, \xi, x_{i+1}, \dots, x_N)$. Using Fubini's theorem as in Remark 3.1.10, we deduce $\partial_{x_i} G(u) \in L^1_{\mathrm{loc}}$ and $\nabla G(u) = G'(u)\nabla u$ (with the same meaning as in Remark 3.1.9). From this, we deduce that $\nabla G(u) \in L^2$ since $G' \in L^\infty$ and $\nabla u \in L^2$. Whence the first part.

For the H_0^1 part, any $u \in H_0^1$ is the limit in H^1 of a sequence of functions $u_n \in C_0^\infty(D)$; since $G(0) = 0$, then $G(u_n)$ is compactly supported in D and, by the first part, $G(u_n) \in H^1(D)$, that is to say, $G(u_n) \in H_0^1(D)$ (we may use a convolution to prove it). Since G is Lipschitz continuous, $G(u_n)$ converges in L^2 to $G(u)$. On the other hand, $\nabla G(u_n) = G'(u_n)\nabla u_n$ is bounded in L^2. Thus $G(u_n)$ is bounded in

$H_0^1(D)$ and we can pass to a weakly converging subsequence. Consequently, its limit $G(u)$ also belongs to $H_0^1(D)$ (since a strongly closed subspace is also weakly closed; see [54]). □

We will apply Proposition 3.1.11 to the particular Lipschitz functions $G(r) = r^+$ $(= \max\{r, 0\})$, $G(r) = r^-$ $(= \max\{-r, 0\})$, and $G(r) = |r|$.

Here and later, we will simply write

$$[u \geq 0] := \{x;\ u(x) \geq 0\}, \quad [u \geq v] := \{x;\ u(x) \geq v(x)\}, \ldots, \tag{3.17}$$

where the variable x belongs to different domains that will be clear from the context.

Corollary 3.1.12. *Let $u \in H^1(D)$ (resp., $H_0^1(D)$). Then, u^+, u^-, $|u| \in H^1(D)$ (resp., $H_0^1(D)$) and we have*

$$\nabla u^+ = \nabla u \chi_{[u>0]}, \qquad \nabla |u| = \operatorname{sign}(u) \nabla u. \tag{3.18}$$

Also let $v \in H^1(D)$ (resp., $H_0^1(D)$). Then $\sup(u, v)$, $\inf(u, v) \in H^1(D)$ *(resp. $H_0^1(D)$) and we have*

$$\nabla \sup(u, v) = \nabla u \chi_{[u \geq v]} + \nabla v \chi_{[u \leq v]}. \tag{3.19}$$

Moreover the mappings $u \to u^+$, u^-, $|u|$, $(u, v) \to$ $\sup(u, v)$, $\inf(u, v)$ *are continuous from H^1 into itself, equipped with its strong topology or with its weak topology.*

Proof of Corollary 3.1.12. The first part immediately follows from Proposition 3.1.11.

For the second one, if u_n converges to u in H^1, we see, thanks to the identity $\nabla |u_n| = \operatorname{sign}(u_n) \nabla u_n$, that $\nabla |u_n|$ converges to $\nabla |u|$ in L^2. The continuity of $u \to |u|$ follows. For the rest, we remember that

$$\sup(u, v) = \frac{u + v + |u - v|}{2}, \qquad \inf(u, v) = \frac{u + v - |u - v|}{2}.$$

If now u_n converges weakly to u in H^1, then $|u_n|$ is bounded in H^1. Thus, up to a subsequence, $|u_n|$ converges weakly in H^1 and also a.e. (see Theorem 3.1.7). Thus, the limit is necessarily equal to $|u|$ and the whole sequence converges. □

Notation. Given a space H of real-valued functions, we will throughout denote by H^+ the subset of functions of H with nonnegative values.

Corollary 3.1.13. *The space $C_0^\infty(D)^+$ is dense in $H_0^1(D)^+$.*

Proof. If $u \in H_0^1(D)^+$, it may be approximated by functions $u_n \in C_0^\infty(D)$. By Corollary 3.1.12, u_n^+ converges in H_0^1 to $u^+ = u$. Obviously, u_n^+ does not generally belong to $C_0^\infty(D)$. But it has a compact support so that, by convolution with a regularizing sequence, it may be approximated in H^1 by nonnegative functions of $C_0^\infty(D)$, whence the result. □

Corollary 3.1.14. *Let $v \in H^1(\mathbb{R}^N)$ such that there exists $w \in H^1_0(D)$ with $|v| \leq w$ a.e. Then $v \in H^1_0(D)$.*

Proof. It is sufficient to prove $v^+ \in H^1_0(D)$ (similarly we will have $v^- \in H^1_0(D)$). Let $w_n \in C_0^\infty(D)^+$ converge to w in H^1: Then $\inf\{w_n, v^+\}$ converges to $\inf\{w, v^+\} = v^+$. Since the function $\inf\{w_n, v^+\}$ is compactly supported in D, after convolution with an adequate regularizing sequence, we obtain a sequence in $C_0^\infty(D)$ that converges to v^+. □

Proposition 3.1.15. *Let $u \in H^1(\mathbb{R}^N)$ be continuous (which precisely means that u has a continuous representative still denoted by u) and such that $u = 0$ everywhere outside the open set D. Then $u \in H^1_0(D)$.*

Proof. We may assume $u \geq 0$. The assumption implies that, for $\epsilon > 0$, $(u - \epsilon)^+$ is compactly supported in D. By convolution with a regularizing sequence, we can approximate $(u - \epsilon)^+$ in H^1 by functions in $C_0^\infty(D)$. Since $(u - \epsilon)^+$ converges to u in H^1 as $\epsilon \to 0$, we deduce that u can itself be approximated in H^1 by functions in C_0^∞. □

Remark 3.1.16. The same result holds if u is only assumed to be continuous on a neighborhood of the boundary of D.

Note that the "converse" is false: A function in $H^1_0(D)$ that is continuous on $\mathbb{R}^N$ is not necessarily equal to zero everywhere outside D (take for instance D = a disk without its center in $\mathbb{R}^2$).

We will see later that a function in $H^1_0(D)$ vanishes *"quasi-everywhere"* — but only quasi-everywhere — outside D.

3.1.3 The Poincaré inequality

Let us end this review on the space $H^1_0(\Omega)$ with its classical equivalent norm when Ω is bounded, and the well-known Poincaré[9] inequality.

Proposition 3.1.17. *Assume D is bounded in one direction. Then there exists $C > 0$ depending only on D such that*

$$\forall\, u \in H^1_0(D), \quad \int_D u^2 \leq C^2 \int_D |\nabla u|^2. \tag{3.20}$$

[9](Jules) Henri POINCARÉ, 1854–1912, was born in Nancy, France. He was at the same time a mathematician, physicist, astronomer, and philosopher; his work was exceptionally innovative and covered an impressive variety of domains: complex analysis, differential and algebraic equations, topology, number theory, probability, relativity and quantum physics, and the n-body problem.

Corollary 3.1.18. *Assume D is bounded in one direction. Then*

$$u \longrightarrow \left\{ \int_D |\nabla u|^2 \right\}^{1/2}$$

is a Hilbert norm that is equivalent to the H^1-norm (3.1) *on $H_0^1(D)$.*

This corollary is straightforward.

Proof of Proposition 3.1.17. Assume for instance that

$$D \subset [-a, a] \times \mathbb{R}^{N-1}.$$

For all $u \in C_0^\infty(D)$ and $x = (x_1, x') \in \mathbb{R} \times \mathbb{R}^{N-1}$,

$$\begin{aligned} u^2(x_1, x') &= \int_{-\infty}^{x_1} 2u(\xi, x')u_{x_1}(\xi, x')\, d\xi \\ &\leq 2 \left\{ \int_{\mathbb{R}} u^2(\xi, x')\, d\xi \right\}^{1/2} \left\{ \int_{\mathbb{R}} u_{x_1}^2(\xi, x')\, d\xi \right\}^{1/2}. \\ \int_{-\infty}^{+\infty} u^2(x_1, x')\, dx_1 &= \int_{-a}^{+a} u^2(x_1, x')\, dx_1 \\ &\leq 4a \left\{ \int_{\mathbb{R}} u^2(\xi, x')\, d\xi \right\}^{1/2} \left\{ \int_{\mathbb{R}} u_{x_1}^2(\xi, x')\, d\xi \right\}^{1/2}. \end{aligned}$$

We integrate this inequality with respect to x' on $\mathbb{R}^{N-1}$ and we apply the Schwarz inequality to obtain, after simplifying,

$$\int_D u^2 \leq (4a)^2 \int_D u_{x_1}^2 \leq (4a)^2 \int_D |\nabla u|^2.$$

□

Remark 3.1.19 (About the best constant in (3.20)[10]). When the Poincaré inequality (3.20) holds for an open set D, we have

$$0 < C_p := \inf \left\{ \textstyle\int_D |\nabla u|^2;\ u \in H_0^1(D),\ \int_D u^2 = 1 \right\}. \tag{3.21}$$

We may wonder whether this infimum is reached: It is the case if D is bounded since (see Theorem 3.1.7) the embedding of $H_0^1(D)$ into $L^2(D)$ is compact. Then from a minimizing sequence we may pass to a subsequence u_n that converges strongly in L^2 and weakly in H_0^1 (by weak sequential compactness of the unit ball in the Hilbert

[10]And a first thought about the existence of minima in variational problems on spaces of functions.

space $H_0^1(D)$); at the limit, the constraint $\int_D u^2 = 1$ is preserved and we use that $u \to \int_D |\nabla u|^2$ is l.s.c. for the weak-convergence in H_0^1.

In fact, as soon as the minimum is reached in (3.21), a minimizing function is necessarily an eigenfunction of the Laplace operator associated with its smallest eigenvalue λ_1, that is to say, a solution of

$$u \in H_0^1(D), \quad -\Delta u = \lambda_1 u \text{ in } D. \tag{3.22}$$

To prove it, we argue that the function

$$t \in \mathbb{R} \to \int_D |\nabla(u+tv)|^2 \Big/ \int_D (u+tv)^2$$

reaches a minimum at $t = 0$ so that its derivative (which does exist) vanishes. Elementary computations lead to

$$\int_D \nabla u \nabla v - \lambda_1 \int_D uv = 0, \text{ with } \lambda_1 = \int_D |\nabla u|^2,$$

whence (3.22) since v is arbitrary in $\mathcal{C}_0^\infty(D)$.

Note that $\lambda_1 = \int_D |\nabla u|^2 > 0$: Indeed if not, ∇u is identically equal to zero and u is constant on connected components of D. But since $u \in H_0^1(D)$, this implies $u \equiv 0$ (see Exercise 3.1) which contradicts $\int_D u^2 = 1$.

To show that it is the smallest eigenvalue, we consider another eigenfunction $\hat{u} \in H_0^1(D)$, that is,

$$-\Delta\hat{u} = \lambda\hat{u} \text{ in the sense that } \forall\, v \in H_0^1(D), \int_D \nabla\hat{u}\nabla v = \lambda \int_D \hat{u}v,$$

and we apply the Poincaré inequality (where $C_p = \lambda_1$), that is,

$$\lambda \int_D \hat{u}^2 = \int |\nabla\hat{u}|^2 \geq \lambda_1 \int_D \hat{u}^2.$$

Whence $\lambda \geq \lambda_1$.

In fact the minimum is not always met in (3.21) even if $C_p > 0$! For instance, it is not reached if D is the strip $\mathbb{R} \times [0,1]$ in $\mathbb{R}^2$ while the Poincaré inequality holds by Proposition 3.1.17.

One may prove it by a "Pohozaev multiplicator" technique, which consists here in multiplying by xu_x the identity (3.22) (which, as we have seen, holds with a nonzero function u if the minimum is reached). We then have, after integrating on

$D_R := [-R, R] \times [0, 1]$,

$$\lambda \int_{D_R} x u u_x = - \int_{D_R} x u_x u_{xx} - x u_x u_{xy},$$

$$\lambda \int_{D_R} x \partial_x (u^2/2) = - \int_{D_R} x \partial_x (u_x^2/2) - \int_{-R}^{R} dx [x u_x u_y]_{y=0}^{y=1} + \int_{D_R} x u_{xy} u_y.$$

We integrate by parts the first two integrals, then the last one as well, and we use that $u_{xy} u_y = \partial_x (u_y^2/2)$. A careful analysis proves that all the integrated terms tend to 0 as $R \to \infty$, thanks to the fact that the involved function belongs to $H^1(D)$. We are left with

$$-\frac{\lambda}{2} \int_D u^2 = \frac{1}{2} \int_D u_x^2 - u_y^2.$$

Multiplying (3.22) by u, we also have

$$\lambda \int_D u^2 = \int_D u_x^2 + u_y^2.$$

We deduce $0 = \int_D u_x^2$, which is a contradiction.

3.1.4 The Dirichlet problem for the Laplacian

Proposition 3.1.20. *Let Ω be a bounded open subset of $\mathbb{R}^N$ and let $f \in H^{-1}(\Omega)$. Then there exists a unique solution $u = u_\Omega^f$ to the problem*

$$u \in H_0^1(\Omega), \quad \langle f, v \rangle_{H^{-1}(\Omega) \times H_0^1(\Omega)} = \int_\Omega \nabla u \nabla v \quad \forall\, v \in H_0^1(\Omega). \tag{3.23}$$

Moreover, u is the unique solution of

$$u \in H_0^1(\Omega), \quad -\Delta u = f \text{ in } \mathcal{D}'(\Omega), \tag{3.24}$$

and also of the minimization problem

$$J(u) = \min \left\{ J(v); v \in H_0^1(\Omega) \right\},$$

where $J(v) = \frac{1}{2} \int_\Omega |\nabla v|^2 - \langle f, v \rangle_{H^{-1}(\Omega) \times H_0^1(\Omega)}$. Finally we have

$$J(u) = -\frac{1}{2} \int_\Omega |\nabla u|^2 = -\frac{1}{2} \langle f, u \rangle_{H^{-1}(\Omega) \times H_0^1(\Omega)}. \tag{3.25}$$

Proof. Since Ω is bounded, by the Poincaré inequality and its Corollary 3.1.18, $[v \to \int_\Omega |\nabla v|^2]$ is a norm that is equivalent to the H^1-norm on $H_0^1(\Omega)$. By the Riesz theorem (3.9) (replacing T by f), we deduce the existence of a unique $u \in H_0^1(\Omega)$ such that

$$\forall\, v \in H_0^1(\Omega), \quad \langle f, v\rangle_{H^{-1}(\Omega)\times H_0^1(\Omega)} = \int_\Omega \nabla u \nabla v,$$

which is exactly problem (3.23). Note that, in particular,

$$\forall\, v \in C_0^\infty(\Omega), \quad \langle f, v\rangle_{H^{-1}(\Omega)\times H_0^1}(\Omega) = \langle -\Delta u, v\rangle_{\mathcal{D}'\times C_0^\infty}.$$

This is the expression of (3.24), which is equivalent to (3.23) by density of $C_0^\infty(\Omega)$ in $H_0^1(D)$.

We denote this solution by u_Ω^f.

For all $v \in H_0^1(\Omega)$, we have

$$J(u_\Omega^f + v) - J(u_\Omega^f) = \int_\Omega \nabla u_\Omega^f \nabla v - \langle f, v\rangle + \frac{1}{2}\int_\Omega |\nabla v|^2 = \frac{1}{2}\int_\Omega |\nabla v|^2,$$

which is strictly positive for all nonzero $v \in H_0^1(\Omega)$. Whence the second point.

For the third one, we choose $v = u_\Omega^f$ in (3.23) to obtain

$$\int_\Omega |\nabla u_\Omega^f|^2 = \langle f, u_\Omega^f\rangle_{H^{-1}\times H_0^1}.$$

□

Remark 3.1.21. When Ω is the empty set, the Dirichlet problem is not interesting. However, to keep a uniform treatment, we will as usual set $u_\Omega^f = 0$ when $\Omega = \emptyset$. This is compatible with the maximum principle that we state now.

Proposition 3.1.22. *The mapping* $f \in H^{-1}(D) \to u_\Omega^f$ *is nondecreasing. Moreover, if* $f \geq 0$ *we have*

$$\Omega_1 \subset \Omega_2 \;\Rightarrow\; u_{\Omega_1}^f \leq u_{\Omega_2}^f.$$

Proof. For the first point, by linearity, it is sufficient to prove $(f \leq 0) \Rightarrow (u_\Omega^f \leq 0)$. Let us denote u_Ω^f by u. We apply (3.23) with $v = u^+$, which belongs to $H_0^1(\Omega)$ (see Corollary 3.1.12). We obtain

$$\int_\Omega \nabla u \cdot \nabla u^+ = \langle f, u^+\rangle \leq 0.$$

Since the first integral is equal to $\int_\Omega |\nabla u^+|^2$ (see Corollary 3.1.12), we deduce $u^+ = 0$.

For the second point, let us denote $u_{\Omega_i}^f$ by u_i, $i = 1, 2$. For all $v \in H_0^1(\Omega_1)$, we have $\int_D \nabla(u_1 - u_2)\nabla v = 0$. But $u_2 \geq 0$ and therefore $(u_1 - u_2)^+ \leq u_1^+$. This proves (see Corollary 3.1.14) that $(u_1 - u_2)^+ \in H_0^1(\Omega_1)$. Choosing $v = (u_1 - u_2)^+$ in the above relation, we deduce $v = 0$, that is to say, $u_1 \leq u_2$. □

3.2 Continuity for the Dirichlet problem

3.2.1 Problem statement

Let D be a bounded open subset of $\mathbb{R}^N$ that is meant to contain all variable open subsets that will be considered. Also let $f \in H^{-1}(D)$.

For all open subsets $\Omega \subset D$, we will denote the solution of the Dirichlet problem associated with Ω and f by u_Ω^f or sometimes only u_Ω if it is not ambiguous.

Let us recall that, since $H_0^1(\Omega)$ is a subspace of $H_0^1(D)$, the distribution $f \in H^{-1}(D)$ acts as well on $H_0^1(\Omega)$ functions and its restriction to Ω belongs to $H^{-1}(\Omega)$. Moreover, we have

$$\forall\, v \in H_0^1(\Omega), \quad \langle f, v\rangle_{H^{-1}(D)\times H_0^1(D)} = \langle f, v\rangle_{H^{-1}(\Omega)\times H_0^1(\Omega)}. \tag{3.26}$$

Most of the time, this will be denoted by

$$\langle f, v\rangle_{H^{-1}\times H_0^1}.$$

The continuity problem. We are given a sequence of open subsets $\Omega_n \subset D$ converging in some sense to an open subset Ω. When is it possible to claim that $u_{\Omega_n}^f$ "converges" to u_Ω^f?

We will see that most of the "classical" topologies on the family of open subsets, like those considered in the previous chapter, *do not imply this convergence*. Thus the right question is rather, under which extra conditions on the sequence $(\Omega_n)_n$ does continuity hold? Or also, is it possible to define a specific convergence on the sequence $(\Omega_n)_n$ that will exactly imply convergence of the $u_{\Omega_n}^f$?

One could also ask, under which condition on f does continuity hold? But as we will easily check in Section 3.2.3, the answer is completely independent of f: The situation is "good" for any f as soon as it is good for $f \equiv 1$.

3.2.2 Some easy facts

Let us give a first fundamental estimate.

Proposition 3.2.1. *There exists C depending only on D such that, for all open subsets $\Omega \subset D$, the solution of* (3.23) *satisfies*

$$\|u_\Omega^f\|_{H_0^1(D)} \le C\|f\|_{H^{-1}(D)}. \tag{3.27}$$

Proof. Let $u = u_\Omega^f$. From (3.25), we deduce

$$\int_\Omega |\nabla u|^2 \le \|f\|_{H^{-1}}\|u\|_{H_0^1(D)}.$$

By Poincaré's inequality in D, there exists $k > 0$ *depending only on* D such that

$$k\|u\|^2_{H^1_0(D)} \le \int_D |\nabla u|^2 = \int_\Omega |\nabla u|^2 \le \|f\|_{H^{-1}} \|u\|_{H^1_0(D)}.$$

Estimate (3.27) follows. □

Corollary 3.2.2. *Let* $(\Omega_n)_n$ *be* any *sequence of open subsets of* D*. Then there exists a subsequence* $k \longrightarrow u^f_{\Omega_{n_k}}$ *and* $u^\star \in H^1_0(D)$ *such that* $u^f_{\Omega_{n_k}}$ *converges weakly in* $H^1_0(D)$ *to* $u^\star$ *as* $k \longrightarrow +\infty$*. Moreover, if there exists an open subset* $\Omega \subset D$ *such that* $u^\star = u^f_\Omega$*, then the convergence is strong in* $H^1_0(D)$*.*

Proof. The first part is a direct consequence of the uniform bound proved in (3.27) and of the weak compactness of sequences of the closed unit ball of the Hilbert space $H^1_0(D)$. The last part is deduced by passing to the (weak) limit in the property

$$\int_D |\nabla u^f_{\Omega_n}|^2 = \langle f, u^f_{\Omega_n} \rangle_{H^{-1}\times H^1_0},$$

coupled with the same property for u^f_Ω. Indeed, we then have

$$\lim_{n\to+\infty} \int_D |\nabla u^f_{\Omega_n}|^2 = \langle f, u^f_\Omega \rangle_{H^{-1}\times H^1_0} = \int_D |\nabla u^f_\Omega|^2,$$

which implies that the convergence in $L^2(D)^N$ of $\nabla u^f_{\Omega_n}$ is strong. □

Obviously we would like that this limit $u^\star$ be exactly u^f_Ω, where Ω is the limit in some sense of the Ω_n. A main step in this direction can be made in the frequent situation where

$$\begin{cases} \text{there exists an open subset } \Omega \subset D \text{ such that,} \\ \text{for all compact subsets } K \subset \Omega, \\ \text{there exists } n_0 \in \mathbb{N} \text{ such that, for all } n \ge n_0,\ K \subset \Omega_n. \end{cases} \tag{3.28}$$

We know that this is the case for Hausdorff convergence and for convergence in the sense of compact sets (see Chapter 2). With this property, we then have

Proposition 3.2.3. *Assume* (3.28) *holds. Then there exist* $u^\star \in H^1_0(D)$ *and a subsequence* $u^f_{\Omega_{n_k}}$ *which converges weakly in* $H^1_0(D)$ *to* $u^\star$ *satisfying*

$$\forall v \in H^1_0(\Omega), \quad \int_\Omega \nabla u^\star \cdot \nabla v = \langle f, v \rangle_{H^{-1}(\Omega)\times H^1_0(\Omega)}. \tag{3.29}$$

Proof. Let us denote $u_n = u^f_{\Omega_n}$. We already know about the subsequence and $u^\star$. Now let φ be fixed in $C_0^\infty(\Omega)$. Thanks to the assumption (3.28), we have $\varphi \in C_0^\infty(\Omega_n)$ for n large enough, and therefore

$$\int_{\Omega_n} \nabla u_n \cdot \nabla \varphi = \langle f, \varphi \rangle_{H^{-1}\times H_0^1}, \tag{3.30}$$

or by extending to the whole set D,

$$\int_D \nabla u_n \cdot \nabla \varphi = \langle f, \varphi \rangle_{H^{-1}\times H_0^1}.$$

Thus after passing to the weak limit in $H_0^1(D)$, we obtain

$$\int_\Omega \nabla u^\star \cdot \nabla \varphi = \int_D \nabla u^\star \cdot \nabla \varphi = \langle f, \varphi \rangle_{H^{-1}\times H_0^1}. \tag{3.31}$$

Since this equality (3.31) is valid for all $\varphi \in C_0^\infty(\Omega)$, we deduce (3.29) by density of $C_0^\infty(\Omega)$ in $H_0^1(\Omega)$. □

Thus, to be able to claim that u_{Ω_n} converges to u_Ω, one needs only to answer the following question:

$$\text{When can one say that } u^\star \in H_0^1(\Omega)? \tag{3.32}$$

We will see that it is not always the case. However, if it is the case, we can state the following.

Proposition 3.2.4. *Assume (with the notation and assumptions of Proposition 3.2.3) that $u^\star \in H_0^1(\Omega)$. Then*

- $u^\star = u^f_\Omega$;
- *the whole sequence $u^f_{\Omega_n}$ converges to u^f_Ω;*
- *the convergence is strong in $H_0^1(D)$.*

Proof. The first point is easily deduced from (3.29) in Proposition 3.2.3 and from the uniqueness of the solution of the Dirichlet problem in Ω. This same uniqueness ensures that the whole sequence converges. The last point may be deduced from Corollary 3.2.2. □

3.2.3 The situation does not depend on f

It is important to know that the fact of whether the situation is "good" or not depends on the open subsets Ω_n, Ω but not on f. Indeed the following result, due to V. Šverak (1992) in [285], shows that it is enough to know what happens in the particular situation where $f \equiv 1$, this for *any* sequence of open subsets.

Theorem 3.2.5. *Let Ω_n, Ω be open subsets of D. If $u^1_{\Omega_n}$ converges in $L^2(D)$ to u^1_Ω, then $u^f_{\Omega_n}$ converges to u^f_Ω strongly in $H^1_0(D)$ for all $f \in H^{-1}(D)$.*

Proof. The main argument is in fact the maximum principle (cf. Proposition 3.1.22).

Let us denote $u^f_{\Omega_n} = u^f_n$, $u^f_\Omega = u^f$. Note first that, by Corollary 3.2.2, the convergence of u^1_n to u^1 holds strongly in $H^1_0(D)$.

We first prove the result when $f \in L^\infty(D)$. As $-M \leq f \leq M$, the maximum principle says

$$-Mu^1_n \leq u^f_n \leq Mu^1_n, \text{ a.e.} \tag{3.33}$$

(Mu^1_n is obviously the solution of the Dirichlet problem with right-hand side equal to M.) But it is known that, up to a subsequence, u^f_n converges weakly in $H^1_0(D)$ to some function $u^\star$. It is sufficient to prove that $u^\star = u^f$ and, by uniqueness of the limit, we will have convergence of the whole sequence strongly in $H^1_0(D)$.

At the limit, we keep the inequality

$$-Mu^1 \leq u^\star \leq Mu^1 \text{ a.e.} \tag{3.34}$$

Since $u^1 \in H^1_0(\Omega)$, so is $u^\star$ by Corollary 3.1.14.

Now let $\varphi \in C^\infty_0(D)$, $\varphi \geq 0$. Since $u^1 > 0$ on Ω (strong maximum principle: see, e.g., [151]; see also the remark below), $\varphi \leq pu^1$ for p large enough. Then $\varphi_n := \inf\{\varphi, pu^1_n\}$ belongs to $H^1_0(\Omega_n)$ and converges to φ strongly in $H^1_0(D)$ by Corollary 3.1.12. We have

$$\forall p, \quad \int_D \nabla u^f_n \nabla \varphi_n = \int_D f\varphi_n, \quad \text{which implies} \int_D \nabla u^\star \nabla \varphi = \int_D f\varphi.$$

This shows that $u^\star = u^f$. We then conclude by density of $L^\infty(D)$ in $H^{-1}(D)$ (this is easy to check) and thanks to the uniformity with respect to the open subsets $\Omega \subset D$ of the norm of the mapping $f \in H^{-1}(D) \to u^f_\Omega \in H^1_0(D)$ (estimation (3.27)). Thus if f_p is a sequence in $L^\infty(D)$ that converges strongly to f in $H^{-1}(D)$, with the same notation as above, we have

$$\begin{aligned}\|u^f_n - u^f\|_{H^1_0(D)} &\leq \|u^f_n - u^{f_p}_n\|_{H^1_0(D)} + \|u^{f_p}_n - u^{f_p}\|_{H^1_0(D)} + \|u^{f_p} - u^f\|_{H^1_0(D)} \\ &\leq 2C\|f_p - f\|_{H^{-1}(D)} + \|u^{f_p}_n - u^{f_p}\|_{H^1_0(D)},\end{aligned}$$

and the expected result follows from the first part of the proof which said that, for p fixed, $u_n^{f_p}$ converges to u^{f_p}. □

Remark. One can find a different proof of $u^* = u^f$ in Chapter 4 (see Proposition 4.7.2). Obviously, one can assume that the data f_n move as well:

Corollary 3.2.6. *Under the assumptions of Theorem 3.2.5, if f_n converges to f in $H^{-1}(D)$, then $u_{\Omega_n}^{f_n}$ converges to u_Ω^f in $H_0^1(D)$.*

Proof. We simply write

$$\|u_{\Omega_n}^{f_n} - u_\Omega^f\| \le \|u_{\Omega_n}^{f_n} - u_{\Omega_n}^f\| + \|u_{\Omega_n}^f - u_\Omega^f\| \le C\|f_n - f\|_{H^{-1}} + \|u_{\Omega_n}^f - u_\Omega^f\|.$$

□

We will now indicate the simplest situations where it is possible to give a positive answer to question (3.32). In what follows, we denote the solution of the Dirichlet problem (3.23), (3.24) on Ω_n (resp., Ω) by $u_{\Omega_n}^f$ (resp., u_Ω^f) or u_n (resp., u) if not ambiguous.

3.2.4 Nondecreasing sequences

Theorem 3.2.7. *Let $(\Omega_n)_n$ be a nondecreasing sequence of open subsets of D and $\Omega = \bigcup_n \Omega_n$. Then $u_n = u_{\Omega_n}^f$ converges to $u = u_\Omega^f$ in $H_0^1(D)$. It is more generally the case if $(\Omega_n)_n$ is a sequence of open subsets included in the open set Ω and converging in the sense of Hausdorff to Ω.*

Remark. This convergence result is particularly useful for the numerical approximation of the Dirichlet problem when Ω is approximated from inside by polygonal open subsets Ω_n. If Ω_n increases to Ω or if the boundary of the Ω_n converges in an adequate sense (for instance, in the sense of Hausdorff) to the boundary of Ω, then the solution follows.

Proof of Theorem 3.2.7. We already know that a nondecreasing sequence of open sets converges in the sense of Hausdorff to its union. Thus it is sufficient to prove the second part of the theorem.

We know that (3.28) holds for Hausdorff convergence of open sets. Then let $u^\star$ be a weak limit of a subsequence of u_n according to Proposition 3.2.3 (still denoted by u_n). We know that it is sufficient to prove that $u^\star \in H_0^1(\Omega)$. Since $u_n \in H_0^1(\Omega_n)$, there exists $w_n \in C_0^\infty(\Omega_n)$ such that

$$\|u_n - w_n\|_{H_0^1(D)} \le 1/n.$$

We deduce that w_n also weakly converges to $u^\star$ in $H_0^1(D)$. Since $\Omega_n \subset \Omega$, it follows that $w_n \in C_0^\infty(\Omega)$. Since w_n converges weakly to $u^\star$ in $H^1(D)$, we deduce $u^\star \in H_0^1(\Omega)$. □

The result of Proposition 3.2.7 will be improved later in terms of Newtonian capacity (see Proposition 3.4.2): It is indeed sufficient that the exceeding part $\Omega_n \backslash \Omega$ be small.

Also interesting is the following corollary, which shows a pointwise lower semi-continuity of the mapping $\Omega \rightarrow u_{\Omega}^{f}$ for Hausdorff convergence.

Corollary 3.2.8. *Assume $f \geq 0$. If the open sets Ω_n converge in the sense of Hausdorff to the open set Ω, then*

$$\text{for a.e. } x \in D, \quad u_{\Omega}^{f}(x) \leq \liminf u_{\Omega_n}^{f}(x).$$

Proof. By Proposition 3.1.22, $u_{\Omega \cap \Omega_n}^{f} \leq u_{\Omega_n}^{f}$. Since Hausdorff convergence is stable with respect to intersections (see Chapter 2), then $\Omega \cap \Omega_n$ converges in the sense of Hausdorff to Ω. By Theorem 3.2.7 above, $u_{\Omega \cap \Omega_n}^{f}$ converges to u_{Ω}^{f}. Then we pass to the limit in the previous inequality. □

Remark 3.2.9. It is also well known that a nonincreasing sequence of open sets converges in the sense of Hausdorff to the interior of the intersection of these open sets. But as we will see later, the solution of the Dirichlet problem does not necessarily "follow".

3.2.5 The dimension-1 case

Theorem 3.2.10. *Assume D is a bounded interval in $\mathbb{R}$. Let Ω_n be a sequence of open subsets of D which converges in the sense of Hausdorff to some open subset $\Omega \subset D$. Then u_{Ω_n} converges to u_{Ω} in $H_0^1(D)$.*

Proof. In dimension 1, the H^1-norm is stronger than Hölder norms. By Ascoli's theorem, we may assume that the subsequence described in Proposition 3.2.3 also converges uniformly in D to $u^{\star}$ (or more precisely to its continuous representative $\tilde{u}^{\star}$). If $x \in D \backslash \Omega$, there exists a sequence $x_n \in D \backslash \Omega_n$ converging to x. By uniform convergence, $u_n(x_n)$ converges to $\tilde{u}^{\star}(x)$ (recall that we denote $u_n := u_{\Omega_n}$). But $u_n(x_n) = 0$ by (3.7). Thus $\tilde{u}^{\star}(x) = 0$. Therefore $\tilde{u}^{\star} = 0$ on $D \backslash \Omega$. Thanks to (3.7), one may deduce that $u^{\star} \in H_0^1(\Omega)$. As before, we know that this is sufficient to guarantee the convergence of the whole sequence u_n to u_{Ω}. □

This continuity result, when coupled with the compactness of *any* sequence of bounded subsets of a given bounded set, leads to a situation that is completely understood in dimension 1. Obviously, it mainly relies on the compact embedding of H^1 into L^{∞}. This is false as soon as the dimension is equal to 2 or more. And the continuity result above is false as well. In other words, it is indeed possible that the limit $u^{\star}$ of Proposition 3.2.3 does not belong to $H_0^1(\Omega)$, where Ω is the Hausdorff limit of the sequence $(\Omega_n)_n$ (which always exists up to a subsequence). The goal of next subsection is to provide counterexamples in this direction.

3.2.6 Counterexamples to continuity in dimension 2

Let $(x_n)_n$ be a dense subsequence in the unit disk D of $\mathbb{R}^2$. Let $\Gamma_n := \{x_1, x_2, \dots, x_n\}$ and $\Omega_n := D \backslash \Gamma_n$. Then Ω_n converges in the sense of Hausdorff to the empty set $\emptyset$, but $u^f_{\Omega_n}$ does not converge to $u^f_\emptyset \equiv 0$. Indeed, $H^1_0(\Omega_n) = H^1_0(D)$ so that $u^f_{\Omega_n} \equiv u^f_D$. This comes from that the fact — which will be developed later in this book — that a point is of zero *capacity* in dimension 2. More precisely, when n is given, one can find a sequence $(v^p)_p \in C^\infty_0(D)$ such that

$$v^p \to 0 \text{ in } H^1_0(D), \quad v^p = 1 \text{ on a neighborhood of } \Gamma_n, \quad 0 \le v^p \le 1$$

(we refer to Section 3.3 for the definition of such a sequence). Thus any $v \in C^\infty_0(D)$ is the limit in $H^1_0(D)$ of $v(1 - v^p) \in C^\infty_0(\Omega_n)$, whence the equality $H^1_0(\Omega_n) = H^1_0(D)$ claimed above.

One can even make "holes" around any point x_n in such a way that the sequence $u^f_{\Omega_n}$ does not converge to 0 yet. More precisely, let $\Omega_n := D \backslash \cup_{1 \le k \le n} \overline{B}(x_k, r_k)$, where r_n is a sequence of positive real numbers such that

$$\sum_{n \ge 1} \frac{-1}{\log(r_n)} < \eta, \qquad \pi \sum_{n \ge 1} r_n^2 < 1. \tag{3.35}$$

The sequence of open sets $(\Omega_n)_n$ decreases to $E = D \backslash \cup_{n \ge 1} \overline{B}(x_n, r_n)$, which is of positive measure by the second condition on the sequence r_n. This sequence $(\Omega_n)_n$ still converges in the sense of Hausdorff to the empty set. But again, $u^1_{\Omega_n}$ *does not converge to* $u^1_\emptyset \equiv 0$.

Indeed, let $\psi \in C^\infty_0(D)^+$ and let us consider the function

$$\forall x \in D, \quad \varphi(x) := \psi(x) \left[1 + \sum_{n \ge 1} \alpha_n \log(|x - x_n|) \right]^+,$$

where $\alpha_n = -1/\log(r_n)$. This function is well defined thanks to the first condition on the sequence r_n: Indeed, since each function $[x \to \log|x - x_n|]$ belongs to $L^1(D)$, the series $\sum_{n \ge 1} \alpha_n \log(|x - x_n|)$ converges in $L^1(D)$ and its norm is bounded above by $k\eta$, where $k = \|\log(\cdot)\|_{L^1(2D)}$.

The function φ vanishes on the union of the balls $\overline{B}(x_n, r_n)$, that is to say outside E, and it is also compactly supported in each Ω_n. It is not identically zero if η is chosen small enough and if the support of ψ is large enough. Moreover, it belongs to $H^1(D)$ (to see that, one can directly compute its H^1-norm or notice that it is superharmonic and bounded in D, and then use Exercise 3.5). Therefore it belongs to $H_0^1(\Omega_n)$. In particular, we have

$$\int_{\Omega_n} \nabla u_{\Omega_n}^1 \nabla \varphi = \int_{\Omega_n} \varphi.$$

At the limit, if $u^\star$ is the weak limit of a subsequence $u_{\Omega_n}^1$,

$$\int_D \nabla u^\star \nabla \varphi = \int_E \varphi.$$

Since E is of positive measure and since φ is nonnegative and not identically zero on E, this proves that $u^\star$ is not the zero function.

In fact, when Ω_n contains holes whose number tends to ∞ with n, continuity generally fails. *It may even happen that* $u_{\Omega_n}^f$ *converges to the solution of a quite different problem that is not governed by the pure Laplace operator.* Let us for instance consider the classical situation in homogenization theory where the open sets Ω_n are obtain by taking off from D a large number of small holes that are uniformly spread over D. The Hausdorff limit of these Ω_n is the empty set. But the limit of $u_n = u_{\Omega_n}^f$ depends on the size of these holes. As a first intuitive approach, we may state the following rule:

- if the holes are "small", then u_n converges to u_D;
- if the holes are large, then u_n converges to 0;
- there exists a critical size of the holes such that the weak limit $u^\star$ of u_n according to Proposition 3.2.3 is the solution *of a new partial differential equation on* D.

Let us for instance consider the following two-dimensional example due to F. Murat and D. Cioranescu [105] (see also [213]). Then, the following can be proved (we refer to Chapters 4 and 7 for more details).

Let $D = (0,1)^2$ and, for $0 < i, j < n$, let $x_{ij} = \left(\frac{i}{n}, \frac{j}{n}\right)$, and also let $\Omega_n = D \setminus \bigcup_{0<i,j<n} B(x_{i,j}, r_n)$. By Proposition 3.2.3, there exists a subsequence of u_n that converges weakly in $H_0^1(D)$ to some $u^\star$ and we can here characterize $u^\star$ according to the size of the holes:

Proposition 3.2.11. *Let* $u_n = u_{\Omega_n}^f$ *and* $u^\star$ *be as above. Then*

- *if* $\frac{\log r_n}{n^2} \xrightarrow[n\to+\infty]{} -\infty$, *then* $u^\star = u$ *(and* $u_n \to u$ *in* $H_0^1(D)$*);*
- *if* $\frac{\log r_n}{n^2} \xrightarrow[n\to+\infty]{} 0$, *then* $u^\star = 0$ *(and* $u_n \to 0$ *in* $H_0^1(D)$*);*

and the critical case is

- *if* $\frac{\log r_n}{n^2} \xrightarrow[n\to+\infty]{} -d < 0$, *then* $u^\star$ *is the solution of*

$$u^\star \in H_0^1(\Omega) \quad \textit{and} \quad -\Delta u^\star + \frac{2\pi}{d} u^\star = f.$$

Proof. A proof of this result is proposed in Exercise 3.8.

Remark 3.2.12. Here the Hausdorff limit of the open sets Ω_n is the empty set again. This is a sign of a genuine difficulty and of a new phenomenon (now well known in homogenization theory). We learned in previous discussions that difficulties will mainly arise from the size of $\Omega_n \backslash \Omega$ where Ω is the Hausdorff limit of the Ω_n (if it was empty for n large, we know by Proposition 3.2.7 that everything would be all right). Here this difference is as big as possible since it is equal to the whole set Ω_n. However, if the holes are large enough, in other words if the Ω_n are not too large, we will nevertheless be in a favorable situation where indeed $u^\star \in H_0^1(\Omega)$, which means $u^\star = 0$ and the convergence is strong. In all other cases, the Hausdorff limit of Ω_n is not significant.

3.2.7 Sequence of uniformly Lipschitz open sets

The counterexample of the previous subsection shows that continuity with respect to Hausdorff convergence does not hold for the Dirichlet problem. Thus it is necessary to add extra conditions on the variable domains to expect good convergence of the solutions. We now state a classical sufficient condition, in terms of *uniform regularity of the* Ω_n which in particular avoids the pathology of the previous example. For this we recall the notation (2.31) introduced in Chapter 2:

$$\mathcal{O}_\varepsilon = \{\Omega \text{ open}, \ \Omega \subset D, \ \Omega \text{ satisfies the } \varepsilon\text{-cone condition}\}.$$

Theorem 3.2.13 (Chenais). *Let* $(\Omega_n)_n$ *be a sequence of open sets of the class* $\mathcal{O}_\varepsilon$ *converging in the sense of Hausdorff to the open set* Ω. *Then* u_n *converges to* $u = u_\Omega$.

Remark 3.2.14. In connection with the remark following Theorem 3.2.7, note that this result is also useful for the numerical approximation of u_Ω^f where Ω is approximated by polygonal open sets from inside or partially from outside as well. If they stay in the class $\mathcal{O}_\epsilon$, it is sufficient to check their Hausdorff convergence or even their convergence in the sense of compact sets or of the characteristic functions. We indeed have

Corollary 3.2.15. *Let* $(\Omega_n)_n$ *be a sequence of open sets from the class* $\mathcal{O}_\varepsilon$ *converging to the open set* $\Omega \in \mathcal{O}_\epsilon$ *in the sense of compact sets or of the characteristic functions. Then* u_n *converges to* $u = u_\Omega$.

Proof of Corollary 3.2.15. By Theorem 2.4.10, there exists a subsequence $(\Omega_{n_k})_k$ converging in the three topologies previously defined to an open set $\widetilde{\Omega} \in \mathcal{O}_\epsilon$. It is then sufficient to verify that $\Omega = \widetilde{\Omega}$ and we finish by using Theorem 3.2.13. To obtain the equality, we check that two open sets of $\mathcal{O}_\epsilon$ which are equal a.e. are indeed equal and that two open sets of $\mathcal{O}_\epsilon$ which are compact limits of the same sequence of open subsets are equal (since they have the same closure). □

Proof of Theorem 3.2.13. By Theorem 2.4.10, Ω still belongs to the class $\mathcal{O}_\varepsilon$ and convergence also holds in the sense of characteristic functions. Let $u^\star$ be a limit of u_n according to Proposition 3.2.3. By passing to the a.e. limit in $0 = u_n(\chi_D - \chi_{\Omega_n})$ a.e., we obtain that $u = 0$ a.e. on $D \backslash \Omega$. We then finish thanks to Proposition 3.2.4 and to the following proposition, which guarantees that $u^\star \in H_0^1(\Omega)$. □

Proposition 3.2.16. *Let Ω be a bounded open set with a Lipschitz boundary. Then*

$$\begin{aligned} H_0^1(\Omega) &= \{u \in H^1(\mathbb{R}^N);\ u = 0 \text{ a.e. on } \Omega^c\} \\ &= \{u \in H^1(\mathbb{R}^N);\ u = 0 \text{ a.e. on } \overline{\Omega}^c\}. \end{aligned} \tag{3.36}$$

Proof. Let us denote by V_1 the second space and by V_2 the third one. Obviously $V_1 \subset V_2$. On the other hand, for $u \in H_0^1(\Omega)$, there exists a sequence $u_n \in C_0^\infty(\Omega)$ converging to u in $H^1(\mathbb{R}^N)$. Up to a subsequence, we may assume that the convergence holds a.e. This provides the inclusion $H_0^1(\Omega) \subset V_1$.

Let us now prove that $V_2 \subset H_0^1(\Omega)$ (the proof is very much like the proof of [54, Prop. IX.18]).

Let $(\mathcal{V}_i)_{i=1,\dots,p}$ be a covering of the boundary of Ω by neighborhoods associated with its Lipschitz property according to Definition 2.4.5. Let $\zeta_i \in C_0^\infty(\mathcal{V}_i)$ be a partition of unity associated with $\cup_i \mathcal{V}_i$. It is sufficient to prove that $u_i = \zeta_i u \in H_0^1(\mathcal{V}_i)$. For each i (see Definition 2.4.5), we may assume, up to a translation and up to an orthogonal change of coordinates, that $\mathcal{V}_i = \mathcal{B} \times (-a, a)$, where $\mathcal{B}$ is an open ball centered at 0 in $\mathbb{R}^{N-1}$ and that there exists $\varphi : \mathcal{B} \to (-a, a)$ Lipschitz continuous with $\varphi(0) = 0$ and

$$\partial\Omega \cap \mathcal{V}_i = \{(x', \varphi(x'));\ x' \in \mathcal{B}\}, \quad \Omega \cap \mathcal{V}_i = \{(x', x_N);\ x' \in \mathcal{B},\ x_N > \varphi(x')\}.$$

Let us set $(X', X_N) = (x', x_N - \varphi(x'))$ and

$$U(X', X_N) = u_i(x', x_N) = u_i(X', X_N + \varphi(X')).$$

Assuming all functions are of class C^1, one has

$$\nabla_{X'} U = \nabla_{x'} u + \partial_{x_N} u_i \varphi', \quad \partial_{X_N} U = \partial_{x_N} u_i, \quad \nabla_{x'} u = \nabla_{X'} U - \partial_{x_N} u_i \varphi'.$$

We deduce that $[U \in H^1(\mathbb{R}^N)]$ is equivalent to $[u_i \in H^1(\mathbb{R}^N)]$ with equivalence of the norms: One may for instance first do the above computation for u, φ, U of class

C^1, then make the usual changes of variable and see that the H^1-norms of u and U are equivalent with a constant that depends only on the L^∞-norm of φ'. Next we approximate u in H^1 by a function of $\mathcal{C}_0^\infty(\mathcal{V}_i)$ and we approximate φ by convolution. The equivalence of the norms is kept at the limit.

Let us introduce $U_n = \rho_n * U$, where ρ_n is a regularizing sequence with support in $\{x \in \mathbb{R}^N;\ 1/2n < x_N < 1/n\}$ (so that ρ_n "pushes up" the support of U). Indeed,

$$\operatorname{Support} U_n \subset \operatorname{Support} U + \operatorname{Support} \rho_n \subset [x_N \geq 1/2n] \subset [x_N > 0],$$

which means that $U_n \in \mathcal{C}_0^\infty([x_N > 0])$. On the other hand, U_n converges to U in $H^1(\mathbb{R}^N)$. Thus $U \in H_0^1([x_N > 0])$. It follows that $u_i \in H_0^1(\mathcal{V}_i)$: To check it, one may consider $u_i^n(x', x_N) = U_n(x', x_N - \varphi(x'))$ which converges to u_i in H^1. It is compactly supported in $\mathcal{V}_i$ and is therefore itself in $H_0^1(\mathcal{V}_i)$ since, by convolution with a regularizing sequence, one may approximate it in H^1 by functions in $\mathcal{C}_0^\infty(\mathcal{V}_i)$. □

In fact, it turns out that the two main ingredients to show that $u^\star \in H_0^1(\Omega)$ in the proof of Theorem 3.2.13 are the above proposition and the fact that Ω_n converges to Ω in the sense of Hausdorff and of the characteristic functions. Thus, in the same way, one can prove

Proposition 3.2.17. *Let $(\Omega_n)_n$ be a sequence of open sets converging in the sense of Hausdorff and of the characteristic functions to an open set Ω satisfying the first equality of* (3.36). *Then u_n converges to $u = u_\Omega$.*

Remark 3.2.18. It follows from this proposition that, if Ω_n is a sequence of open sets whose measure tends to 0, then u_n converges to $0 = u_\emptyset$, since Ω_n converges also in the sense of Hausdorff to the empty set. Obviously this may also be proved by noticing that

$$\|u_n\|_{H_0^1(\Omega_n)} \leq C(D)\|f\|_{H^{-1}(\Omega_n)} \to 0.$$

Remark 3.2.19. Equalities (3.36) are valid when Ω is regular enough: We will see later (see Theorem 3.4.6) that Ω is then called *stable*. Sufficient conditions are given in [140], [155], [284]. Note that the first equality alone corresponds to a weaker property (see Exercise 3.9) and is more easily satisfied. However, it already requires that the open set be without a crack. Several other extra regularity conditions may be added. Most of the time they require the use of the capacity associated with the space $H_0^1(D)$, which is absolutely necessary to provide more refined results on this question of continuity with respect to domains.

3.3 Capacity associated with the H^1-norm

3.3.1 Definitions and first properties

In a classical way, we first define the capacity of compact sets, then of open sets, and finally of any set.

Definition 3.3.1. Given any compact subset of $\mathbb{R}^N$, we set

$$\mathrm{cap}(K) = \inf\left\{\|v\|^2_{H^1(\mathbb{R}^N)};\ v \in C_0^\infty(\mathbb{R}^N),\ v \geq 1 \text{ on } K\right\} < +\infty. \tag{3.37}$$

Definition 3.3.2. Given ω an open subset of $\mathbb{R}^N$, we set

$$\mathrm{cap}(\omega) := \sup\{\mathrm{cap}(K);\ K \text{ compact},\ K \subset \omega\}. \tag{3.38}$$

To justify the next definition, let us first check the following property.

Lemma 3.3.3. *For any compact subset $K \subset \mathbb{R}^N$,*

$$\mathrm{cap}(K) = \inf\{\mathrm{cap}(\omega);\ \omega \textit{ open},\ K \subset \omega\}.$$

Proof. By definition, $\mathrm{cap}(K) \leq \inf\{\mathrm{cap}(\omega);\ \omega \text{ open},\ K \subset \omega\}$. Since the mapping $K \to \mathrm{cap}(K)$ is monotone nondecreasing with respect to inclusion, if $\omega \subset K_1$ with ω open and K_1 compact, one has $\mathrm{cap}(\omega) \leq \mathrm{cap}(K_1)$. Now let $\epsilon > 0$ and $v \in C_0^\infty(\mathbb{R}^N)$ with $v \geq \chi_K$, $\|v\|^2_{H^1} \leq (1+\epsilon)\,\mathrm{cap}(K)$. Let us consider the open set $\omega = [(1+\epsilon)v > 1]$ and the compact set $K_1 = [(1+\epsilon)v \geq 1]$. One has

$$K \subset \omega \subset K_1, \quad \mathrm{cap}(\omega) \leq \mathrm{cap}(K_1) \leq (1+\epsilon)^2\|v\|^2_{H^1} \leq (1+\epsilon)^2[\mathrm{cap}(K) + \epsilon].$$

Lemma 3.3.3 follows. □

Thus we may set the following general definition.

Definition 3.3.4. Given $E \subset \mathbb{R}^N$, we set

$$\mathrm{cap}(E) := \inf\{\mathrm{cap}(\omega);\ \omega \text{ open},\ E \subset \omega\}. \tag{3.39}$$

In fact, we have the following property that provides an alternative and direct definition of the capacity of any subset of $\mathbb{R}^N$ (we agree that the next infimum is $+\infty$ if the family of functions v involved is empty):

Proposition 3.3.5. *For any $E \subset \mathbb{R}^N$,*

$$\mathrm{cap}(E) = \inf\left\{\|v\|^2_{H^1(\mathbb{R}^N)};\ v \geq 1 \textit{ a.e. on a neighborhood of } E\right\}.$$

Thus, $\mathrm{cap}(E) = 0$ *is equivalent to the existence of a sequence v_n converging to* 0 *in $H^1(\mathbb{R}^N)$ and greater than or equal to* 1 *on a neighborhood of E.*

Remark 3.3.6. Obviously, any set of zero capacity is also of zero Lebesgue measure since, up to a subsequence, we may assume that the sequence v_n of the previous proposition converges a.e. to 0.

Remark 3.3.7. One may be willing to define the capacity of any set E, like for open sets, by

$$\text{cap}_\star(E) = \sup\{\text{cap}(K);\ K \text{ compact}, K \subset E\}.$$

But this definition of "interior" capacity is different from the one we gave above.

In fact, a set E that satisfies $\text{cap}(E) = \text{cap}_\star(E)$ is called **capacitable**. G. Choquet ([100], [101]) gave a rather precise description of capacitable sets. For instance, for the capacity defined here, any Borelian set is capacitable. This may be proved by using the properties stated in Proposition 3.3.9 and the general results of G. Choquet in [100], [101].

Remark 3.3.8. Let us emphasize that the definition of the capacity of a compact set involves "regular" test functions, but this is not the case for the capacity of any set, even not of an open set. Indeed, if a regular function is greater than or equal to 1 on a set, so it is on its closure. Thus working only with regular test functions would not make the difference between the capacity of a set and of its closure, which would not be right.

Note also that the weaker constraint "$v \geq 1$ a.e. on E" in Proposition 3.3.5 would not be sufficient. Indeed, this would not for instance allow the "thin" compact sets of zero measure and of positive capacity to be "seen".

Proof of Proposition 3.3.5. Let us denote by $\widetilde{\text{cap}}(E)$ the right-hand side of Proposition 3.3.5. Assume first that $E = K$ is compact. If v_n is a minimizing sequence in the definition of $\text{cap}(K)$, the sequence $(1 + 1/n)v_n$ is also minimizing and $(1 + 1/n)v_n$ is greater than 1 *on a neighborhood of* K: Thus $\widetilde{\text{cap}}(K) \leq \text{cap}(K)$. On the other hand, let v_n be a minimizing sequence in the definition of $\widetilde{\text{cap}}(K)$. By regularization and truncation, one may approximate it in H^1, through the same method as in Remark 3.1.3 (Case 1), by functions in $C_0^\infty(\mathbb{R}^N)$ that are greater than 1 on a neighborhood of K. This shows $\text{cap}(K) \leq \widetilde{\text{cap}}(K)$.

Now let $E = \omega$ be an open set. By the previous analysis, for all compact sets $K \subset \omega$, $\text{cap}(K) = \widetilde{\text{cap}}(K) \leq \widetilde{\text{cap}}(\omega)$ so that $\text{cap}(\omega) \leq \widetilde{\text{cap}}(\omega)$. The reverse inequality is trivial if $\text{cap}(\omega) = +\infty$. Assume $\text{cap}(\omega) < +\infty$. Let K_n be a nondecreasing sequence of compact subsets of ω such that $\text{cap}(K_n)$ increases to $\text{cap}(\omega)$. Up to enlarging the K_n, it can be assumed that their union is equal to ω. Let v_n be a sequence of functions in $C_0^\infty(\mathbb{R}^N)$ such that

$$\|v_n\|_{H^1}^2 \leq \text{cap}(K_n) + 1/n \leq \text{cap}(\omega) + 1/n, \quad v_n \geq 1 \text{ on } K_n.$$

Up to a subsequence, it may be assumed that v_n converges weakly in $H^1(\mathbb{R}^N)$ and a.e. to a function v (cf. [54]). In particular, $v \geq 1$ a.e. on ω. This implies $\widetilde{\text{cap}}(\omega) \leq \|v\|_{H^1}^2 \leq \liminf \|v_n\|_{H^1}^2 \leq \lim \text{cap}(K_n) = \text{cap}(\omega)$.

Finally, for any set E and any open set ω with $E \subset \omega$, we immediately have $\widetilde{\text{cap}}(E) \leq \widetilde{\text{cap}}(\omega) = \text{cap}(\omega)$ and therefore $\widetilde{\text{cap}}(E) \leq \text{cap}(E)$. The reverse inequality is trivial if $\widetilde{\text{cap}}(E) = +\infty$. If not, there exists a minimizing sequence v_n in the definition of $\widetilde{\text{cap}}(E)$ and a neighborhood ω_n of E such that

$$v_n \geq 1 \text{ a.e. on } \omega_n \supset E, \quad \|v_n\|_{H^1}^2 \geq \text{cap}(\omega_n) \geq \text{cap}(E).$$

The inequality $\widetilde{\text{cap}}(E) \geq \text{cap}(E)$ follows. □

This capacity is a way to measure the "thinness" of sets: It is a real-valued function defined on the whole family of subsets of $\mathbb{R}^N$ that is very much like a measure, except for some main properties. We summarize next those properties of $\text{cap}(\cdot)$ that guarantee that it is a "good capacity":

Proposition 3.3.9. *(1)* $A \subset B \Longrightarrow \text{cap}(A) \leq \text{cap}(B)$.

(2) Let K_n be a sequence of compact sets with $K_{n+1} \subset K_n$ for all $n \geq 0$ and let $K := \cap_n K_n$. Then $\text{cap}(K) = \lim_{n\to\infty} \text{cap}(K_n)$.

(3) Let E_n be a sequence of sets with $E_n \subset E_{n+1}$ for all $n \geq 0$ and let $E := \cup_n E_n$. Then $\text{cap}(E) = \lim_{n\to\infty} \text{cap}(E_n)$.

(4) (Strong subadditivity) Given any subsets A, $B \subset \mathbb{R}^N$, we have

$$\text{cap}(A \cup B) + \text{cap}(A \cap B) \leq \text{cap}(A) + \text{cap}(B). \tag{3.40}$$

Remark 3.3.10. It follows from these properties that, if A_n is any sequence of sets, then

$$\text{cap}(\cup_n A_n) \leq \sum_n \text{cap}(A_n). \tag{3.41}$$

To obtain this, we may apply claims (3) and (4) of the above proposition to the nondecreasing sequence $E_n = \cup_{1 \leq p \leq n} A_p$. In particular, if the A_n are of zero capacity, then $\text{cap}(\cup_n A_n) = 0$. As for measures, this property will be repeatedly used.

Proof of Proposition 3.3.9. The first claim is easily deduced from the same property, which we know is valid for compact sets.

Next let K_n, K be as in (2). By monotonicity, $\text{cap}(K) \leq \lim \text{cap}(K_n)$. If ω is an open set containing K, for n large enough $K_n \subset \omega$. We deduce $\lim \text{cap}(K_n) \leq \text{cap}(\omega)$. Together with Lemma 3.3.3, this proves the reverse inequality.

Now let E_n, E be as in (3). By monotonicity, $\lim \mathrm{cap}(E_n) \le \mathrm{cap}(E)$. The reverse inequality is trivial if $\lim \mathrm{cap}(E_n) = +\infty$. It is much more delicate in general and will require using claim (4). Note however that it is easy if $E_n = \omega_n$ is open for all n. Indeed, if K is a compact set included in the open set $\omega = \cup_n \omega_n$, then for n large enough, $K \subset \omega_n$. Thus, $\mathrm{cap}(K) \le \mathrm{cap}(\omega_n)$ and by passing to the supremum over the family of compact sets K included in ω, we obtain $\mathrm{cap}(\omega) \le \lim_n \mathrm{cap}(\omega_n)$.

For general E_n, let us assume by contradiction that there exists $\epsilon > 0$ such that $\lim_n \mathrm{cap}(E_n) + \epsilon < \mathrm{cap}(E)$. Let us set $\rho_n = 1 - 1/n$. We are going to prove by induction that there exists a nondecreasing sequence of open sets ω_n such that

$$E_n \subset \omega_n \quad \text{and} \quad \mathrm{cap}(\omega_n) \le \mathrm{cap}(E_n) + \varepsilon \rho_n. \tag{3.42}$$

Then by passing to the limit, we will have, with $\omega = \cup_n \omega_n$,

$$E \subset \omega, \quad \text{and} \quad \mathrm{cap}(E) \le \mathrm{cap}(\omega) = \lim_n \mathrm{cap}(\omega_n) \le \lim_n \mathrm{cap}(E_n) + \epsilon < \mathrm{cap}(E),$$

which is a contradiction.

Property (3.42) is valid for $n = 1$ with some open set ω_1 by definition of $\mathrm{cap}(E_1)$; assume that the property holds for n. By definition of $\mathrm{cap}(E_{n+1})$, one can find an open set $\hat{\omega}$ containing E_{n+1} such that

$$\mathrm{cap}(E_{n+1}) \le \mathrm{cap}(\hat{\omega}) \le \mathrm{cap}(E_{n+1}) + \varepsilon(\rho_{n+1} - \rho_n). \tag{3.43}$$

Let us then set $\omega_{n+1} = \omega_n \cup \hat{\omega}$ and let us use formula (3.40):

$$\mathrm{cap}(\omega_{n+1}) \le \mathrm{cap}(\omega_n) + \mathrm{cap}(\hat{\omega}) - \mathrm{cap}(\omega_n \cap \hat{\omega}).$$

But $E_n \subset \omega_n \cap \hat{\omega}$ so that by using monotonicity, the induction hypothesis, and formula (3.43), we have

$$\begin{aligned} \mathrm{cap}(\omega_{n+1}) &\le \mathrm{cap}(E_n) + \varepsilon\rho_n + \mathrm{cap}(E_{n+1}) + \varepsilon(\rho_{n+1} - \rho_n) - \mathrm{cap}(E_n) \\ &= \mathrm{cap}(E_{n+1}) + \varepsilon\rho_{n+1}. \end{aligned}$$

We will now prove claim (4) with the help of Proposition 3.3.5. Let φ_1 and φ_2 be two functions in $H^1(\mathbb{R}^N)$ such that $\varphi_1 \ge 1$ on a neighborhood of A and $\varphi_2 \ge 1$ on a neighborhood of B. We set $\psi = \sup(\varphi_1, \varphi_2)$ and $\phi = \inf(\varphi_1, \varphi_2)$. The two functions ψ and ϕ are in $H^1(\mathbb{R}^N)$ and their first derivatives are given by (see Corollary 3.1.12)

$$\frac{\partial \psi}{\partial x_i} = \frac{\partial \varphi_2}{\partial x_i} \chi_{[\varphi_2 \ge \varphi_1]} + \frac{\partial \varphi_1}{\partial x_i} \chi_{[\varphi_1 > \varphi_2]} \quad \text{a.e.,} \tag{3.44}$$

$$\frac{\partial \phi}{\partial x_i} = \frac{\partial \varphi_1}{\partial x_i} \chi_{[\varphi_2 \ge \varphi_1]} + \frac{\partial \varphi_2}{\partial x_i} \chi_{[\varphi_1 > \varphi_2]} \quad \text{a.e.} \tag{3.45}$$

We deduce

$$\int_{\mathbb{R}^N} \psi^2 + |\nabla\psi|^2 + \int_{\mathbb{R}^N} \phi^2 + |\nabla\phi|^2 = \int_{\mathbb{R}^N} \varphi_1^2 + |\nabla\varphi_1|^2 + \int_{\mathbb{R}^N} \varphi_2^2 + |\nabla\varphi_2|^2. \quad (3.46)$$

As $\psi \geq 1$ on a neighborhood of $A \cup B$ and $\phi \geq 1$ on a neighborhood of $A \cap B$, it follows that

$$\operatorname{cap}(A \cup B) + \operatorname{cap}(A \cap B) \leq \|\varphi_1\|_{H^1}^2 + \|\varphi_2\|_{H^1}^2.$$

Since this is true for all test functions φ_1 and φ_2 associated with the definitions of $\operatorname{cap}(A)$ and $\operatorname{cap}(B)$ according to Proposition 3.3.5, then claim (4) follows. □

3.3.2 Relative capacity and capacitary potential

In the whole of this subsection, we denote by D a bounded open subset of $\mathbb{R}^N$. We are going to define the "capacity relative to D" associated with the space $H_0^1(D)$ in the same way as we did for the "global" capacity in $\mathbb{R}^N$ in the previous subsection. We will denote it by $\operatorname{cap}_D(\cdot)$ or $\operatorname{cap}(\cdot, D)$.

Definition 3.3.11. Given K a compact subset of D, we define

$$\operatorname{cap}_D(K) = \inf\left\{ \int_D |\nabla v|^2;\ v \in C_0^\infty(D),\ v \geq 1 \text{ on } K\right\} < +\infty. \quad (3.47)$$

Given ω an open subset of D, we set

$$\operatorname{cap}_D(\omega) := \sup\{\operatorname{cap}_D(K); K \text{ compact}, K \subset \omega\}. \quad (3.48)$$

If $E \subset D$, we set

$$\operatorname{cap}_D(E) := \inf\{\operatorname{cap}_D(\omega); \omega \text{ open set}, E \subset \omega\}. \quad (3.49)$$

In order to guarantee the consistency of this definition, we easily check (like we did for the capacity in $\mathbb{R}^N$) that, for any compact set $K \subset D$, $\operatorname{cap}_D(K) = \inf\{\operatorname{cap}_D(\omega);\ \omega \text{ open set}, K \subset \omega\}$. And we also prove

Proposition 3.3.12. *For any $E \subset D$,*

$$\operatorname{cap}_D(E) = \inf\left\{ \int_D |\nabla v|^2;\ v \in H_0^1(D),\ v \geq 1 \text{ a.e. on a neighborhood of } E\right\}.$$

Remark 3.3.13. The proof of this proposition is exactly the same as the proof of Proposition 3.3.5 except for the use of the following lemma, which ensures localization in D for compact sets.

Lemma 3.3.14. *Let K be a compact subset of D and let $v \in H_0^1(D)$ be greater than or equal to 1 on a neighborhood of K. Then for all $\epsilon > 0$, there exists $v_\epsilon \in C_0^\infty(D)$ such that*

$$\|v - v_\epsilon\|_{H_0^1(D)} \leq \epsilon, \quad v_\epsilon \geq 1 \text{ on a neighborhood of } K.$$

Proof. Let $\zeta \in C_0^\infty(D)$ with $\zeta = 1$ on a neighborhood of K. Let us write $v = v\zeta + v(1-\zeta)$. If ρ_p is a regularizing sequence, for p large enough, $(v\zeta) * \rho_p$ is in $C_0^\infty(D)$ and greater than 1 on a neighborhood of K. Moreover, this sequence converges to $v\zeta$ in $H_0^1(D)$ as $p \to \infty$. By definition of $H_0^1(D)$, there exists $z_p \in C_0^\infty(D)$ converging to v in $H_0^1(D)$. But $z_p(1-\zeta)$ is still in $C_0^\infty(D)$, vanishes on a neighborhood of K, and converges to $v(1-\zeta)$ as $p \to \infty$. It follows that, for p large enough, $v^p := (v\zeta) * \rho_p + z_p(1-\zeta)$ is in $C_0^\infty(D)$, is greater than 1 on a neighborhood of K, and converges to v in $H_0^1(D)$ when $p \to \infty$. Thus we may choose $v_\epsilon := v^p$ with p large enough. □

Similarly to the proof of Proposition 3.3.9, one proves that this $H_0^1(D)$-capacity inherits all properties that make it a "good capacity":

Proposition 3.3.15. *The mapping $E \subset D \to \mathrm{cap}_D(E)$ has the same properties as the mapping $E \subset \mathbb{R}^N \to \mathrm{cap}(E)$ as stated in Proposition 3.3.9.*

Remark 3.3.16. Since $v \to \int_D |\nabla v|^2$ defines a norm that is equivalent to the H^1-norm on $H_0^1(D)$ (by the Poincaré inequality; see Corollary 3.1.18), there exists a constant $C = C(D) > 0$ such that $\mathrm{cap}(E) \leq C\,\mathrm{cap}_D(E)$. The reverse inequality is wrong: For instance, if E has some parts closed to the boundary, it is possible that there does not exist any $v \in H_0^1(D)$ greater than 1 on a neighborhood of E, in which case $\mathrm{cap}_D(E) = +\infty$. This is in fact the case for $E = D$ while $\mathrm{cap}(D) < +\infty$ since D is bounded.

Thus the capacity $\mathrm{cap}_D(\cdot)$ does depend on D. However, "locally" inside D, it depends only on the H^1-norm. In particular, the property of being of zero capacity depends on the H^1-norm and on the dimension. More precisely, we have the following.

Proposition 3.3.17. *Let D_1, D_2 be bounded open subsets of $\mathbb{R}^N$ and for $i = 1, 2$, $D_i^\varepsilon := \{x \in D_i;\ d(x, \partial D_i) \geq \varepsilon\}$. Then there exist positive constants C_0, C_1, C_2 depending only on ε, D_1, D_2 such that, for all $A \subset D_1^\varepsilon$, $B \subset D_1^\varepsilon \cap D_2^\varepsilon$,*

$$\mathrm{cap}_{D_1}(A) \leq C_0\,\mathrm{cap}(A), \quad \mathrm{cap}_{D_1}(B) \leq C_1\,\mathrm{cap}_{D_2}(B) \leq C_2\,\mathrm{cap}_{D_1}(B).$$

If, moreover, $E \subset D_1 \cap D_2$, then

$$(\mathrm{cap}_{D_1}(E) = 0) \Leftrightarrow (\mathrm{cap}_{D_2}(E) = 0) \Leftrightarrow (\mathrm{cap}(E) = 0).$$

Proof of Proposition 3.3.17. Let us fix a function $\psi \in C_0^\infty(D_1)$ equal to 1 on a neighborhood of D_1^ε. Let $v \in H^1(\mathbb{R}^N)$ be a test function for the definition of $\mathrm{cap}(A)$ according to Proposition 3.3.5. Then ψv is a test function for the definition of $\mathrm{cap}_{D_1}(A)$ and we have

$$\mathrm{cap}_{D_1}(A) \leq \int |\nabla(\psi v)|^2 \leq C(\psi) \int |\nabla v|^2 + v^2. \tag{3.50}$$

We use the Poincaré inequality in D_1 to bound $\int v^2$ from above by $C\int |\nabla v|^2$. Taking the infimum over all choices of v, we deduce the first inequality of Proposition 3.3.17.

For the second one, we fix $\psi \in C_0^\infty(D_1 \cap D_2)$ equal to 1 on a neighborhood of $D_1^\varepsilon \cap D_2^\varepsilon$. Then we introduce a function v associated with the definition of $\mathrm{cap}_{D_2}(B)$ according to Proposition 3.3.12 and we argue as we just did.

The third inequality may be obtained by interchanging the roles of D_1 and D_2.

Now let $E \subset D_1 \cap D_2$. Let K_n be a sequence of compact sets increasing to $D_1 \cap D_2$. By the above inequalities and Remark 3.3.16, for all n we have

$$(\mathrm{cap}_{D_1}(E \cap K_n) = 0) \Leftrightarrow (\mathrm{cap}_{D_2}(E \cap K_n) = 0) \Leftrightarrow (\mathrm{cap}(E \cap K_n) = 0).$$

The same equivalence for E itself follows thanks to the continuity of all capacities with respect to nondecreasing limits (see Propositions 3.3.9 and 3.3.15). □

Remark 3.3.18. If being of zero capacity does not depend on D, it is not the same for the property that $\mathrm{cap}(\omega_n)$ converges to 0, even for a decreasing limit of open sets ω_n. Indeed (see Exercise 3.7), one can find open sets $D_1 \subset D_2$ and a sequence of decreasing open sets $\omega_n \subset D_1$ such that

$$\lim \mathrm{cap}_{D_2}(\omega_n) = 0,\ \forall\, n,\ \mathrm{cap}_{D_1}(\omega_n) = +\infty.$$

A useful tool to compute the relative capacity of a set is to compute the energy of its *capacitary potential* that we define now. For this, given $A \subset D$, we introduce

$$\Gamma_A := \big\{v \in H_0^1(D);\ \exists\, v_n \xrightarrow{H_0^1} v,\ v_n \geq 1 \text{ a.e. on a neighborhood of } A\big\}.$$

The set Γ_A is obviously a closed convex subset of $H_0^1(D)$.

Definition 3.3.19. We assume that Γ_A is not empty. The **capacitary potential** of A is, by definition, the projection u_A of 0 onto Γ_A for the Hilbert norm $v \to \{\int_D |\nabla v|^2\}^{1/2}$ of $H_0^1(D)$, which is characterized by

$$\int_D |\nabla u_A|^2 = \inf\big\{\textstyle\int_D |\nabla v|^2;\ v \in \Gamma_A\big\}.$$

Remark 3.3.20. Note that if v_n is a minimizing sequence in the latter definition, then v_n converges strongly in $H_0^1(D)$ to u_A. Indeed, up to a subsequence, v_n converges weakly in H^1 to a function $v_\infty \in H_0^1(D)$. The set Γ_A is closed for weak convergence of $H_0^1(D)$ since it is closed for strong convergence and convex (see [54]). Therefore $v_\infty \in \Gamma_A$. We also have

$$\int_D |\nabla v_\infty|^2 \leq \lim \int_D |\nabla v_n|^2 = \int_D |\nabla u_A|^2.$$

By uniqueness of the projection, $v_\infty = u_A$ and since the norms are preserved, the convergence is strong.

Let us now state some properties that turn out to be very useful for the computations of u_A and of $\mathrm{cap}_D(A)$.

Theorem 3.3.21. *Let $A \subset D$. Then*

(1) *if Γ_A is not empty,* $\mathrm{cap}(A) = \int_D |\nabla u_A|^2$; $\mathrm{cap}(A) = +\infty$ *otherwise;*

(2) $-\Delta u_A \geq 0$ *on D,* $\mathrm{Support}(\Delta u_A) \subset \overline{A}$ *(and therefore u_A is harmonic in $D \backslash \overline{A}$);*

(3) $u_A \geq \chi_A$ *a.e.,* $0 \leq u_A \leq 1$ *a.e.;*

(4) *if Γ_A is not empty,* $\mathrm{cap}(A) = -\int_D \Delta u_A$;

(5) $\big(U \in \Gamma_A,\ -\Delta U \geq 0\big) \Rightarrow (u_A \leq U$ *a.e.*$)$.

Proof. Claim (1) is a direct consequence of Proposition 3.3.12. For claim (2), we use that for all $\varphi \in C_0^\infty(D)$, $\varphi \geq 0$, $t > 0$, we have $(u_A + t\varphi) \in \Gamma_A$ so that

$$\int_D |\nabla(u_A + t\varphi)|^2 \geq \int_D |\nabla u_A|^2. \tag{3.51}$$

We deduce (after dividing by t and letting t tend to 0)

$$\int_D \nabla u_A \nabla \varphi \geq 0,$$

or also

$$\langle -\Delta u_A, \varphi \rangle_{\mathcal{D}' \times C_0^\infty} \geq 0.$$

Thus $-\Delta u_A$ is a nonnegative distribution and therefore a nonnegative Radon measure. If now $\varphi \in C_0^\infty(D)$ is compactly supported in $D \backslash \overline{A}$ (and of any sign), we have $u_A + t\varphi \in \Gamma_A$ for all $t \in \mathbb{R}$: Thus we can again use (3.51) and obtain

$$\langle -\Delta u_A, \varphi \rangle_{\mathcal{D}' \times C_0^\infty} = 0,$$

which proves that the support of the measure Δu_A is included in $\overline{A}$.

For claim (3), we already know that $u_A \geq \chi_A$ a.e. (since it is the a.e. limit of functions that are a.e. greater than or equal to 1 on A). To show that $u_A \leq 1$, we notice that $\inf(u_A, 1) \in \Gamma_A$ since, by Corollary 3.1.12, $\inf(u_A, 1) = \lim \inf(u_n, 1)$, where u_n is greater that 1 a.e. on a neighborhood of A and converges to u_A in $H_0^1(D)$. Again by Corollary 3.1.12,

$$\int_D |\nabla[\inf(u_A, 1)]|^2 = \int_D |\nabla u_A|^2 \chi_{[u_A < 1]} \leq \int_D |\nabla u_A|^2.$$

By uniqueness of the projection of 0 onto Γ_A, we deduce that $\inf(u_A, 1) = u_A$ so that $u_A \leq 1$. We prove in the same way that $u_A \geq 0$ by considering the function $(u_A)^+ = \sup(u_A, 0)$.

For claim (4), let us first introduce $u_n \in \mathcal{C}_0^\infty(D)$ converging in $H_0^1(D)$ to u_A. Since $u_A \leq 1$, then $\min\{u_n, 1\}$ converges to u_A as well. Given a regularizing sequence $(\rho_p)_p$, one can choose a sequence of integers p_n so that $\hat{u}_n := \rho_{p_n} * [\min\{u_n, 1\}]$ converges to u_A in $H_0^1(D)$ (and $\hat{u}_n \in \mathcal{C}_0^\infty(D)$). We then have

$$\begin{aligned}\operatorname{cap}_D(A) = \int_D |\nabla u_A|^2 = \lim \int_D \nabla u_A \nabla \hat{u}_n = \langle \hat{u}_n, -\Delta u_A \rangle_{\mathcal{C}_0^\infty \times \mathcal{D}'} = \cdots \\ = \int_D \hat{u}_n d(-\Delta u_A) \leq \int_D -\Delta u_A.\end{aligned}$$

Let us show the reverse inequality, first for the compact sets, then for the open sets. If $A = K$ is compact, u_K is the limit in H^1 of a sequence of functions $v_n \in \mathcal{C}_0^\infty(D)$, $v_n \geq \chi_K$ (see Remark 3.3.20). Thus

$$\int_D -\Delta u_K \leq \int_D v_n d(-\Delta u_K) = \langle v_n, -\Delta u_K \rangle_{\mathcal{C} \times \mathcal{D}'} = \int_D \nabla v_n \nabla u_K,$$

and the latter integral converges to $\int_D |\nabla u_K|^2$. We deduce $\int_D -\Delta u_K \leq \operatorname{cap}_D(K)$ and also the equality thanks to the first step.

If now K_n is a sequence of compacts sets that increases to ω, we check (like in Remark 3.3.20) that u_{K_n} converges in H^1 to u_ω and therefore $-\Delta u_{K_n}$ converges in the sense of distributions to $-\Delta u_\omega$. Since the mass of the measures Δu_{K_n} is bounded, convergence holds in the sense of measures and

$$\int_D -\Delta u_\omega \leq \liminf \int_D -\Delta u_{K_n} = \lim \operatorname{cap}_D(K_n) = \operatorname{cap}_D(\omega),$$

whence the equality since the reverse inequality was already proved.

Similarly, if ω_n is a nonincreasing sequence of open sets such that $\operatorname{cap}_D(\omega_n)$ converges to $\operatorname{cap}_D(A)$, we check that u_{ω_n} converges in H^1 to u_A (see Remark 3.3.20). Arguing as above, we deduce the inequality $\int_D -\Delta u_A \leq \operatorname{cap}_D(A)$ (and therefore the equality thanks to the first step).

Finally let $U \in \Gamma_A$ with $-\Delta U \geq 0$. This means $\int_D \nabla U \nabla v \geq 0$ for all $v \in \mathcal{C}_0^\infty(D)^+$. This is also valid for all $v \in H_0^1(D)^+$ by density (see Proposition 3.1.13). Let us set $V := \inf(u_A, U)$. We check with the help of Corollary 3.1.12 that $V \in \Gamma_A$. Since u_A is the projection of 0 onto Γ_A, we have $\int_D \nabla u_A \nabla(u_A - V) \leq 0$. We then write

$$\int_D |\nabla(u_A - V)|^2 = \int_D \nabla u_A \nabla(u_A - V) - \int_D \nabla V \nabla(u_A - V) \leq - \int_D \nabla V \nabla(u_A - V).$$

By Corollary 3.1.12, the latter integral is equal to

$$\int_{[u_A>U]} \nabla U \nabla (u_A - U)^+ = \int_D \nabla U \nabla (u_A - U)^+ \geq 0.$$

It follows that $u_A - V = 0$, that is, $u_A \leq U$. □

Remark 3.3.22. The last claim of Theorem 3.3.21 says that u_A is the smallest potential U (i.e., $U \in H^1_0(D)$, $-\Delta U \geq 0$) that is greater than or equal to 1 on A in the sense of the definition of Γ_A. This pointwise minimization property is intensively used in potential theory. Note that it implies

$$(E_1 \subset E_2) \Rightarrow (u_{E_1} \leq u_{E_2} \text{ a.e.}). \tag{3.52}$$

3.3.3 How to compute capacities of sets: Some examples

Computations of capacities of sets are made by using the capacitary potentials introduced in Definition 3.3.19, together with their properties described in Theorem 3.3.21, and with the help of Proposition 3.3.23.

Let us start with an example in dimension 1 with $D = (a, b)$ and $K = [x_0, x_1] \subset D$. Then

$$u_K(x) = \begin{cases} \dfrac{x-a}{x_0-a}, & x \in [a, x_0], \\ 1 & x \in [x_0, x_1], \\ \dfrac{b-x}{b-x_1} & x \in [x_1, b]. \end{cases} \tag{3.53}$$

Indeed, it is the only function of $H^1_0(a, b)$ (and therefore continuous) that is harmonic (i.e., affine) outside K and equal to 1 on K. We deduce

$$\mathrm{cap}_D(K) = \frac{1}{x_0 - a} + \frac{1}{b - x_1} > 0.$$

Note that u_K is harmonic outside ∂K and equal to 1 on ∂K. Thus

$$u_K = u_{\partial K} \quad \text{and} \quad \mathrm{cap}_D(K) = \mathrm{cap}(\{x_0, x_1\}) = \mathrm{cap}_D(\partial K). \tag{3.54}$$

This property is general and not particular to dimension 1:

Proposition 3.3.23. *For any compact subset K of a bounded open set D, the following holds:*

$$u_K = u_{\partial K} \quad \textit{and} \quad \mathrm{cap}_D(K) = \mathrm{cap}_D(\partial K).$$

This proposition will be proved later, after Proposition 3.3.37, with the help of the tools of *quasi-continuity*. We use it in the computations below.

Taking $x_0 = x_1$ in (3.54), we obtain that the capacity of a point is (strictly) positive. Thus, *only the empty set is of zero capacity in dimension* 1.

As soon as the dimension is at least 2, the capacity of a point is zero. Let us first compute the capacity of a small ball. Let us choose $D = B_R$ = the ball of radius R around the origin. Let $K_n = B^n$ be the ball of radius $1/n$ around the origin. To find $u_n = u_{B^n}$, we solve

$$\begin{cases} \Delta u_n = 0 & \text{in } B_R \backslash B^n, \\ u_n = 0 & \text{on } \partial B_R, \\ u_n = 1 & \text{on } B^n. \end{cases} \tag{3.55}$$

In dimension 2, the solution is $u_n = -\log(r/R)/\log(nR)$ (for $r = |x|$) on $B_R \backslash B^n$. We have $|\nabla u_n| = 1/r \log(nR)$ and the capacity of B^n is given as

$$\operatorname{cap}_{B_R}(B^n) = \int_{B_R} |\nabla u_n|^2 = \int_0^{2\pi} \int_{1/n}^R \frac{1}{r^2(\log(nR))^2} r\, dr = \frac{2\pi}{\log(nR)}. \tag{3.56}$$

Since $\operatorname{cap}_{B_R}(\{0\}) \leq \lim_{n\to\infty} \operatorname{cap}_{B_R}(B^n)$, we deduce that the capacity of $\{0\}$ is equal to zero.

In dimension $N \geq 3$, the solution of (3.55) is given by

$$u_n = c_n \big[(r/R)^{2-N} - 1\big], \quad c_n = \big[(nR)^{N-2} - 1\big]^{-1},$$

and we obtain

$$\operatorname{cap}_{B_R}(B^n) = N(N-2)\omega_N c_n R^{N-2}, \tag{3.57}$$

where ω_N denotes the volume of the unit ball in $\mathbb{R}^N$. For large n, we therefore have $\operatorname{cap}_{B_R}(B^n) = O(n^{2-N})$ and $\operatorname{cap}(\{0\}) = 0$.

Remark 3.3.24. The values of $b(t) := \operatorname{cap}_{B_{2t}}(B_t) = \operatorname{cap}(B_t, B_{2t})$ (that will be used later) are therefore given by

$$b(t) = 2\pi/\log 2 \text{ if } N = 2, \quad b(t) = b_N t^{N-2} \text{ with } b_N = N(N-2)\omega_N/(1 - 2^{2-N}).$$

Note that this capacity does not depend on t in dimension $N = 2$ and is like t^{N-2} when $N \geq 3$. More generally, we have for any set $E \subset \mathbb{R}^N$

$$\operatorname{cap}(E \cap B_t, B_{2t}) = (2t)^{N-2} \operatorname{cap}(\frac{E}{2t} \cap B_{1/2}, B_1).$$

About the capacity of a continuous arc in dimension 2. In dimension 2 the capacity of a segment is strictly positive and so is the capacity of a continuous arc. Let us find bounds from below that depend only on their size.

Let us denote by D_R the open disk of radius R around the origin. Let us consider a square K centered at the origin with sides of length $2a$ and included in D_1. By

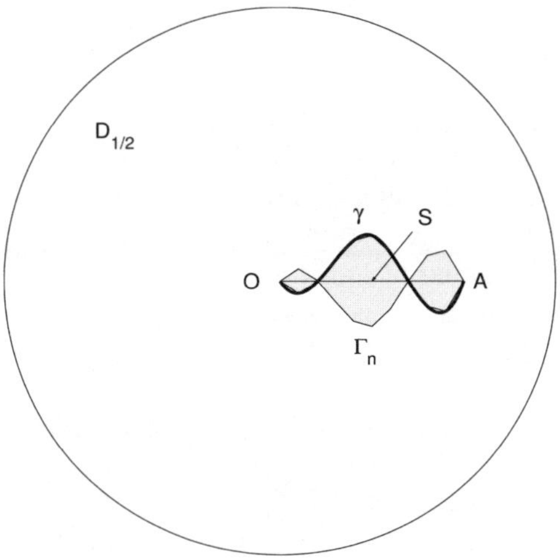

Figure 3.1. Computation of the capacity of a segment.

Proposition 3.3.23, the capacity of the boundary of the square is equal to the capacity of the square itself and is therefore bounded from below by the capacity of the disk of radius a centered at the origin. By (3.56), we have

$$\mathrm{cap}_{D_1}(\partial K) \geq -2\pi/\log a, \quad \text{for any } a \in (0, 1/\sqrt{2}).$$

Thanks to rotation invariance, each side of the square has the same capacity. By subadditivity (see Proposition 3.3.9), it follows that the capacity of a side is bounded from below by $\mathrm{cap}_{D_1}(\partial K)/4 \geq -\pi/2\log a$.

Let us denote by S the segment joining the origin to the point $(a, 0)$ with $a \in (0, 1/2]$. We then have $\mathrm{cap}_{D_1}(S) \geq -k/\log a$ with $k > 0$ independent of a. Indeed, S is the side of a square centered at $(a/2, a/2)$ and included in the disk D with the same center and of radius $b = 1 + \sqrt{2}/4$ that itself contains D_1. Thus

$$\mathrm{cap}_{D_1}(S) \geq \mathrm{cap}_D(S) \geq -\pi/2\log(a/2b),$$

where this last inequality uses a homothety of ratio $1/b$ (see Remark 3.3.24) and the latter result. And there exists $k > 0$ independent of $a \in (0, 1/2]$ such that $-\pi/2\log(a/2b) \geq -k/\log a$.

Now let γ be a continuous arc included in $D_{1/2}$ — that is to say, the image of a continuous mapping from $[0, 1]$ into $D_{1/2}$ — and joining the origin to the point $A = (a, 0)$, $a \leq 1/2$. Let us show that we still have

$$\mathrm{cap}_{D_1}(\gamma) \geq -k/\log a. \tag{3.58}$$

Let us denote by γ^ϵ the set of points in D_1 whose distance to γ is less than or equal to ϵ. Since γ^ϵ is a sequence of compact sets that decreases to γ as ϵ decreases to 0, it follows that $\lim_{\epsilon\to 0} \mathrm{cap}_{D_1}(\gamma^\epsilon) = \mathrm{cap}_{D_1}(\gamma)$.

Fix ϵ small enough so that $\gamma^\epsilon \subset D_1$. We can approximate γ by a sequence of arcs γ_n that are piecewise linear and that join the origin to $(a, 0)$ and in such a way that they are included in γ^ϵ for n large enough. Thus they satisfy $\mathrm{cap}_{D_1}(\gamma^\epsilon) \geq \mathrm{cap}_{D_1}(\gamma_n)$. Up to taking off a finite number of loops, we may assume that γ_n does not have any double points. Let $\widetilde{\gamma_n}$ denote the symmetric image of γ_n with respect to the axis $X = \{(x, 0),\ x \in \mathbb{R}\}$ and $\Gamma_n = \gamma_n \cup \widetilde{\gamma_n}$. By subadditivity, we have

$$\mathrm{cap}_{D_1}(\Gamma_n) \leq \mathrm{cap}_{D_1}(\gamma_n) + \mathrm{cap}_{D_1}(\widetilde{\gamma_n}) = 2\,\mathrm{cap}_{D_1}(\gamma_n).$$

Let C be the polygonal domain "inside Γ_n", that is to say, the union of Γ_n and of the *bounded* connected components of $\mathbb{R}^N \backslash \Gamma_n$. We easily check that it is a compact set whose boundary is included in Γ_n and that contains the segment $S = [O, A]$.[11] We obtain, by using Proposition 3.3.23 and the monotonicity of the capacity,

$$\mathrm{cap}_{D_1}(\gamma^\epsilon) \geq \mathrm{cap}_{D_1}(\Gamma_n) \geq \mathrm{cap}_{D_1}(\partial C) = \mathrm{cap}_{D_1}(C) \geq \mathrm{cap}_{D_1}(S) \geq -k/\log a.$$

This yields (3.58).

By invariance of the capacity with respect to any homothety in dimension 2 (see Remark 3.3.24), it follows that, for any continuous arc γ included in D_t and joining the origin to a point of ∂D_t,

$$\mathrm{cap}_{D_{2t}}(\gamma) = \mathrm{cap}(\gamma, D_{2t}) \geq k_2 = k/\log 2. \tag{3.59}$$

We deduce the following corollary which will be used later.

Corollary 3.3.25. *Let $K \subset \mathbb{R}^2$ be compact, connected, and containing at least 2 points whose distance is equal to a. Then for any $x \in K$,*

$$\forall\, t \in (0, a/2), \quad \mathrm{cap}(K \cap B(x, t), B(x, 2t)) \geq k_2,$$

where k_2 is a "universal" constant.

Indeed, there exists a sequence of connected open sets $\omega_n \supset K$ such that

$$\mathrm{cap}(\omega_n \cap B(x, t), B(x, 2t)) \to \mathrm{cap}(K \cap B(x, t), B(x, 2t)).$$

These open sets ω_n are also arc connected and therefore contain a continuous arc γ joining x to a point of K whose distance is at least $a/2$. But for all $t \in (0, a/2]$,

[11] Note that if a point of S did not belong to C, one could join it by a piecewise linear arc to 2 points of ∂D_1 symmetric with respect to the axis X, and it would cut Γ_n into 2 unconnected pieces.

there exists a point of this arc on $\partial B(x,t)$ (we choose the first of them). We then apply (3.59) to obtain

$$\mathrm{cap}(\omega_n \cap B(x,t), B(x,2t)) \geq \mathrm{cap}(\gamma \cap B(x,t), B(x,2t)) \geq k_2.$$

□

Remark 3.3.26. It can be proved that a compact subset of a segment that is not of zero (linear) measure is of positive capacity in $\mathbb{R}^2$ (see Exercise 3.3 where estimates from above may also be found).

In dimension 3, the capacity of a segment is equal to zero as well as the capacity of a regular arc (see exercises). More generally, the following rule holds:

- If $E \subset \mathbb{R}^N$ is contained in an $(N-2)$-dimensional manifold, then $\mathrm{cap}(E) = 0$;
- if $E \subset \mathbb{R}^N$ contains a piece of a regular $(N-1)$-dimensional manifold, then $\mathrm{cap}(E) > 0$;
- If E is a regular d-dimensional manifold, then

$$[\mathrm{cap}(E) = 0] \Leftrightarrow [d \leq N-2].$$

We refer for instance to [4] for a proof of the above results and for more general computations. Let us also mention the explicit computations of capacities made in [224] for quite a lot of geometric bodies in $\mathbb{R}^2$ and $\mathbb{R}^3$.

3.3.4 Quasi-continuity and quasi-open sets

In dimension 1, H^1 functions are continuous (see (3.3)) or, more precisely, they have a continuous representative. This is false as soon as the dimension is greater than or equal to 2. The correct property is then the so-called *quasi-continuity* and the main theorem of this section is Theorem 3.3.29 below. Let us first define the following.

Definition 3.3.27. Given $D \subset \mathbb{R}^N$, we will say that a property holds **quasi-everywhere (q.e.)** in D if it holds everywhere on D except on a subset of zero capacity.

Definition 3.3.28. A function $f : \mathbb{R}^N \to \mathbb{R}$ is said to be **quasi-continuous** if there exists a nonincreasing sequence of open sets $\omega_n \subset \mathbb{R}^N$ such that

(i) $\lim_{n\to+\infty} \mathrm{cap}(\omega_n) = 0$;

(ii) the restriction of f to the complement $\omega_n^c := \mathbb{R}^N \backslash \omega_n$ of ω_n is continuous.

A continuous function is obviously quasi-continuous. Note that, in dimension 1, since even a point has a positive capacity, quasi-continuity is equivalent to continuity (in the latter definition, ω_n is necessarily empty for n large enough). Note also that this property is weaker than the property of being continuous quasi-everywhere.

Theorem 3.3.29. *Any $f \in H^1(\mathbb{R}^N)$ has a quasi-continuous representative $\tilde{f}$ and it is unique (modulo quasi-everywhere equality).*

Uniqueness of the quasi-continuous representative is a consequence of the following lemma.

Lemma 3.3.30. *Let $f : \mathbb{R}^N \to \mathbb{R}$ be quasi-continuous. Then*

$$(f \geq 0 \text{ a.e.}) \Longrightarrow (f \geq 0 \text{ q.e.}).$$

Proof. Let us show that the set $A = [f < 0]$ is of zero capacity. There exists a sequence of open sets ω_n whose capacity tends to 0 and such that the restriction of f to ω_n^c is continuous. As a consequence, $A \cup \omega_n$ is open. Let v_n be a sequence of functions converging to 0 in H^1 and greater than 1 a.e. on ω_n. Since the Lebesgue measure of A is zero, $v_n \geq 1$ a.e. on the open set $A \cup \omega_n$ as well. Whence $\operatorname{cap}(A \cup \omega_n) \leq \|v_n\|_{H^1}^2$, which proves $\operatorname{cap}(A) = 0$. □

Proof of Theorem 3.3.29. Let $f_p \in C_0^\infty(\mathbb{R}^N)$, converging to f in H^1 and a.e. We extract a subsequence $(f_{p_m})_m$ such that $\sum_{m=1}^\infty 2^{2m} \|f_{p_{m+1}} - f_{p_m}\|_{H^1}^2 < +\infty$. Let us then set

$$\Omega_m := [|f_{p_{m+1}} - f_{p_m}| > 2^{-m}], \quad \omega_n := \cup_{m \geq n} \Omega_m.$$

Since $2^m |f_{p_{m+1}} - f_{p_m}|$ is an H^1 function greater than 1 on the open set Ω_m, it follows that

$$\operatorname{cap}(\Omega_m) \leq 2^{2m} \|\, |f_{p_{m+1}} - f_{p_m}|\, \|_{H^1}^2 \leq 2^{2m} \|f_{p_{m+1}} - f_{p_m}\|_{H^1}^2,$$
$$\operatorname{cap}(\omega_n) \leq \sum_{m \geq n} 2^{2m} \|f_{p_{m+1}} - f_{p_m}\|_{H^1}^2.$$

Thus ω_n is a nonincreasing sequence of open sets whose capacity tends to 0 as n tends to ∞. For $x \in \omega_n$,

$$\forall\, m \geq n, \quad |f_{p_{m+1}}(x) - f_{p_m}(x)| \leq 2^{-m}.$$

Thus, for fixed n, the restriction of f_{p_m} to ω_n^c converges uniformly as $m \to \infty$. The restriction to ω_n^c of the limit $\tilde{f}$ is therefore continuous, this for all n. We extend $\tilde{f}$ by 0 on $\cap_n \omega_n$. Since f_p converges a.e. to f, we may deduce that $\tilde{f}$ is a representative of f. And $\tilde{f}$ is quasi-continuous by construction. □

Notation. From now on, for each $f \in H^1(\mathbb{R}^N)$, we will denote by $\tilde{f}$ its quasi-continuous representative.

Remark 3.3.31. This definition applies in particular to functions in $H_0^1(D)$, for any open subset D of $\mathbb{R}^N$. Note that in this case, we may choose the functions f_p of the latter proof in $C_0^\infty(D)$ and we get the more precise following result (see Remark 3.3.18): Let $f \in H_0^1(D)$; then there exists a sequence of open sets $\omega_n \subset D$ such that $\mathrm{cap}_D(\omega_n) \to 0$ (i.e., $u_{\omega_n} \to 0$ in $H_0^1(D)$) and the restriction of f to ω_n^c is continuous. We will sometimes say that f is $H_0^1(D)$-quasi-continuous.

Remark 3.3.32. We see in the above proof that the functions f_{p_m} converge quasi-everywhere to $\tilde{f}$. In fact, the same approach leads to the following proposition, which turns out to be very useful when dealing with quasi-everywhere inequalities.

Proposition 3.3.33. *Let f_p converge to f in $H^1(\mathbb{R}^N)$. Then there exists a subsequence $(f_{p_m})_m$ such that $\tilde{f}_{p_m}$ converges quasi-everywhere to $\tilde{f}$ as m tends to ∞.*

Proof. We argue exactly as in the proof of Theorem 3.3.29 by extracting a subsequence such that $\sum_{m=1}^{\infty} 2^{2m} \|f_{p_{m+1}} - f_{p_m}\|_{H^1}^2 < +\infty$. We then set

$$\Omega_m := [|\tilde{f}_{p_{m+1}} - \tilde{f}_{p_m}| > 2^{-m}], \quad \omega_n := \cup_{m \geq n} \Omega_m.$$

The rest of the proof is exactly the same, assuming that one starts with the same initial estimate for the capacity of Ω_m, namely,

$$\mathrm{cap}(\Omega_m) \leq 2^{2m} \|f_{p_{m+1}} - f_{p_m}\|_{H^1}^2.$$

This was obvious from the definitions when it was known that Ω_m was open. It is not the case here anymore but this nevertheless follows from the next lemma. □

Lemma 3.3.34. *Let $z \in H^1(\mathbb{R}^N)$ and $E := [|\tilde{z}| > 1]$. Then*

$$\mathrm{cap}(E) \leq \|z\|_{H^1}^2.$$

Proof. Since $\tilde{z}$ is quasi-continuous, there exists a sequence of open sets ω_n whose capacity tends to 0 and such that $E \cup \omega_n$ is open. Denote by v_n a sequence of functions converging to 0 in H^1 and greater than 1 a.e. on ω_n. Thus $|z| + v_n$ is greater than 1 a.e. on the open set $E \cup \omega_n$ and we have

$$\mathrm{cap}(E) \leq \mathrm{cap}(E \cup \omega_n) \leq \| \, |z| + v_n \|_{H^1}^2.$$

The inequality of Lemma 3.3.34 follows. □

Integration by parts with respect to $H^{-1}(D)$ measures of finite energy can be made for $H_0^1(D)$ functions by using their quasi-continuous representatives. For instance, the following holds.

Proposition 3.3.35. *Let $u, v \in H_0^1(D)$ with $-\Delta u \geq 0$. Then, $-\Delta u$ is a measure which does not charge sets of zero capacity and we have*

$$\int_D \nabla u \nabla v = \int_D \tilde{v}\, d(-\Delta u).$$

Proof. Let us start with a remark. Since the distribution $-\Delta u$ is nonnegative, it is a Borelian measure that is finite on compacts sets (a so-called Radon measure). If $w \in C_0^\infty(D)$, one can write

$$\int_D w\, d(-\Delta u) = \langle w, -\Delta u\rangle_{C_0^\infty \times \mathcal{D}'} = \int_D \nabla w \nabla u.$$

If $w \in H_0^1(D)$ is continuous and compactly supported, one may approximate it by convolution in H^1 and uniformly by functions in $C_0^\infty(D)$ and pass to the limit in the latter identity to obtain

$$\int_D w\, d(-\Delta u) = \int_D \nabla w \nabla u \leq \|w\|_{H^1} \|u\|_{H^1}. \tag{3.60}$$

Let $K \subset D$ and $w_n \in C_0^\infty(D)$, $w_n \geq 0$, $w_n \geq 1$ on K and such that $\mathrm{cap}_D(K) = \lim \int_D |\nabla w_n|^2$. We have

$$\int_K d(-\Delta u) \leq \int_D w_n\, d(-\Delta u) \leq \|w_n\|_{H^1} \|u\|_{H^1},$$

and at the limit,

$$\int_K -\Delta u \leq \|u\|_{H^1} \{\mathrm{cap}(K)\}^{1/2}.$$

This estimate carries over to open sets and to all measurable sets by definition of the capacity and by the usual properties of measures. Thus $[\mathrm{cap}(E) = 0] \Rightarrow [-\Delta u(E) = 0]$.

Since $v \in H_0^1(D)$, there exists a sequence $v_n \in C_0^\infty(D)^+$ converging to v in H^1. According to (3.60), one may write

$$\int_D v_n\, d(-\Delta u) = \int_D \nabla v_n \nabla u, \qquad \int_D |v_p - v_q|\, d(-\Delta u) = \int_D \nabla u \nabla |v_p - v_q|.$$

By Proposition 3.3.33, it may be assumed, up to a subsequence, that v_n converges quasi-everywhere to $\tilde{v}$, and therefore $(-\Delta u)$-a.e. as well. The latter inequality shows also that v_n is a Cauchy sequence in $L^1(-\Delta u)$. It is therefore converging in this L^1-space and its limit is necessarily equal to $\tilde{v}$. Thus $\tilde{v} \in L^1(-\Delta u)$ and the identity of Proposition 3.3.35 follows by passing to the limit in the corresponding identity for v_n. □

Remark 3.3.36. It follows from the previous analysis that

$$v \in H_0^1(D) \to \tilde{v} \in L^1(-\Delta u) \text{ is continuous.}$$

One can also prove that, if $u \in H_0^1(D)$ is such that Δu is a measure (not necessarily positive), then this measure does not charge sets of zero capacity (see [157]). On the other hand, the identity of Proposition 3.3.35 is not valid in general without positivity, even if Δu is in $L^1(D)$ (see [55]).

We are now going to state a new characterization of the convex cones Γ_A in terms of quasi-continuous representatives (see the definition of Γ_A just before Definition 3.3.19).

Proposition 3.3.37. *For any $A \subset D$ such that Γ_A is not empty,*

$$\Gamma_A = \{v \in H_0^1(D);\ \tilde{v} \geq 1 \text{ q.e. on } A\}.$$

Proof. If $v \in \Gamma_A$, it is the limit in $H_0^1(D)$ of v_n where $v_n \geq 1$ a.e. on an open neighborhood Ω_n of A. But by Lemma 3.3.30 applied with $f = \tilde{v_n} - 1$, we also have $\tilde{v_n} \geq 1$ q.e. on Ω_n and therefore on A. Since there exists a subsequence of $\tilde{v_n}$ which converges q.e. to $\tilde{v}$, it follows that $\tilde{v} \geq 1$ q.e. on A.

Conversely, let $v \in H_0^1(D)$ with $\tilde{v} \geq 1$ q.e. on A. The goal is to prove that $v \in \Gamma_A$. Since Γ_A is closed in H^1, it is sufficient to prove that, for any positive integer p, $\sup(v, -p) \in \Gamma_A$. Without loss of generality, we may therefore assume that v is bounded below by $-p$.

Set $\Omega_n := [(1 + 1/n)\tilde{v} > 1]$. Then Ω_n contains A and by Remark 3.3.31, there exists a sequence of open sets $\omega_n \subset D$ with $\Omega_n \cup \omega_n$ open and $u_{\omega_n} \to 0$ in $H_0^1(D)$. But $(1 + 1/n)\tilde{v} + (p + 1)u_{\omega_n} \geq 1$ a.e. on $\Omega_n \cup \omega_n$ — which is a neighborhood of A — and converges to v in $H_0^1(D)$ as n tends to ∞. □

We now have the tools to prove Proposition 3.3.23.

Proof of Proposition 3.3.23. Let us denote $w := u_K - u_{\partial K}$. Using Propositions 3.3.35 and 3.3.21, one may write

$$\int_D |\nabla w^+|^2 = \int_D \nabla w^+ \nabla w = \int_D \tilde{w}^+ d(-\Delta w) \leq \int_D \tilde{w}^+ d(-\Delta u_K). \tag{3.61}$$

By Proposition 3.3.21, the support of Δu_K is in K. On the other hand, by Proposition 3.3.37, $u_K = 1$ q.e. on K, $u_{\partial K} = 1$ q.e. on ∂K: In particular, $u_K = 1$ a.e. in the interior of K and therefore Δu_K vanishes in this interior. Thus Δu_K is in fact supported in ∂K. Since $\tilde{w} = \widetilde{u_K} - \widetilde{u_{\partial K}} = 0$ q.e. on ∂K, it follows from (3.61) that $\int_D |\nabla w^+|^2 = 0$, which is to say, $u_K \leq u_{\partial K}$, whence the equality of both since the other inequality is trivial by monotonicity. □

It is now time to give a definition for the following notion that we have implicitly used several times.

Definition 3.3.38. A subset Ω of D is said to be **quasi-open** if there exists a nonincreasing sequence of open sets ω_n such that

$$\lim_{n\to+\infty} \operatorname{cap}(\omega_n) = 0,$$
$$\forall\, n,\ \Omega \cup \omega_n \text{ is open.}$$

Remark 3.3.39. Open sets are obviously quasi-open sets. One also gets easy examples of quasi-open sets by perturbing an open set by (or on) a set of zero capacity. But since a quasi-open set is in fact defined only up to a set of zero capacity, we do not create new objects in this way. On the other hand, we may construct quasi-open sets that are even not equal a.e. to open sets (see Exercise 3.6).

In the latter definition, since $\cup_n(\Omega \cap \omega_n) = \Omega \cup (\cap_n \omega_n)$ and since $\cap_n \omega_n$ is of zero capacity, it follows that a quasi-open set is equal q.e. to a so-called $\mathcal{G}_\delta$, that is, the denumerable intersection of open sets, which is in particular a Borelian set in $\mathbb{R}^N$. As it is natural to consider that a quasi-open set is defined only quasi-everywhere, it will always be possible to consider that it is a $\mathcal{G}_\delta$.

Proposition 3.3.40. *A denumerable union of quasi-open sets is a quasi-open set.*

Proof. Let $(\Omega^p)_{p\geq 1}$ be a sequence of quasi-open sets and, for each p, $(\omega_n^p)_{n\geq 1}$ a sequence of open sets associated with Ω^p according to Definition 3.3.38 and where we moreover assume that $\operatorname{cap}(\omega_n^p) \leq 2^{-(n+p)}$. Set

$$\widetilde{\omega}_n := \cup_{p\geq 1}\omega_n^p, \qquad \Omega := \cup_{p\geq 1}\Omega^p.$$

Then $\Omega \cup \widetilde{\omega}_n = \cup_{p\geq 1}\{\Omega^p \cup \omega_n^p\}$ is an open set and we have

$$\operatorname{cap}(\widetilde{\omega}_n) \leq \sum_{p\geq 1} 2^{-(n+p)} = 2^{-n}.$$

This proves that Ω is quasi-open. □

Proposition 3.3.41. *Let $f : \mathbb{R}^N \to \mathbb{R}$ be quasi-continuous and $\alpha \in \mathbb{R}$. Then $[f > \alpha]$ is quasi-open. In particular, if $u \in H^1(\mathbb{R}^N)$, then $[\tilde{u} > \alpha]$ is quasi-open.*

Proof. If f is quasi-continuous, there exists a sequence of open sets ω_n whose capacity tends to 0 and such that the restriction of f to ω_n^c is continuous. In particular, $[f > \alpha]$ is relatively open in ω_n^c, which means that there exists an open set Ω_n of D such that $[f > \alpha] = \Omega_n \cap (\omega_n^c)$ or also $[f > \alpha] \cup \omega_n = \Omega_n \cup \omega_n$ and this set is indeed open. □

3.3.5 A new definition of $H_0^1(\Omega)$

The last results of this section present a very useful characterization of the space $H_0^1(\Omega)$ when Ω is open. When the boundary of Ω is regular, it may prove that it coincides with the space of functions in $H^1(\mathbb{R}^N)$ that vanish a.e. outside Ω (see, e.g., Proposition 3.2.16). This is not true for any open set as we can see on simple examples. A good point of view is contained in the next theorem. This new characterization will in fact provide later the right definition of $H_0^1(\Omega)$ for any subset Ω of $\mathbb{R}^N$ without any specific structure.

Theorem 3.3.42. *Let D be an open subset of $\mathbb{R}^N$ and let Ω be an open subset of D. Then*

$$(u \in H_0^1(\Omega)) \iff \big(u \in H_0^1(D) \text{ and } \tilde{u} = 0 \text{ q.e. on } D\backslash\Omega\big). \tag{3.62}$$

Proof. The implication "$\Longrightarrow$" is a consequence of Proposition 3.3.33: If $u \in H_0^1(\Omega)$, there exists by definition a sequence of functions $u_n \in C_0^\infty(\Omega)$ that converges to u in $H_0^1(\Omega) \hookrightarrow H_0^1(D)$. By Proposition 3.3.33, one can extract a subsequence that converges q.e. in D to $\tilde{u}$. Since all functions u_n vanish on $D\backslash\Omega$, then $\tilde{u}$ vanishes q.e. on $D\backslash\Omega$.

For the reverse implication, let $u \in H_0^1(D)$ such that $\tilde{u}$ vanishes q.e. on $D\backslash\Omega$, that is to say (working with its extension by 0 to $\mathbb{R}^N\backslash D$), that vanishes q.e. on Ω^c. The question is to approximate u in H^1 by functions in $C_0^\infty(\Omega)$. Using $u = u^+ - u^-$, one may assume $u \geq 0$ without loss of generality. Since $\inf(u, p)$ converges to u in H^1 as p tends to ∞, one may also assume that u is bounded. Finally, if $\zeta \in C_0^\infty(B(0,2))$ with $0 \leq \zeta \leq 1$, $\zeta \equiv 1$ on $B(0,1)$, and if $\zeta_n(x) = \zeta(x/n)$, one checks that $u\zeta_n$ converges to u in H^1. Thus we may also assume that u is identically zero outside a large ball B.

The obvious starting idea is to regularize u by convolution. But, it is also necessary to "reduce" its support inside Ω. For this, we instead regularize $(u-\delta)^+$ which for $\delta > 0$ vanishes in some sense, and at least q.e., near the boundary of Ω.

More precisely, let us introduce $(\omega_n)_n$, the nonincreasing sequence of open sets associated with the quasi-continuity of $\tilde{u}$. Up to increasing ω_n, one can assume that it contains the zero-capacity subset of Ω^c where $\tilde{u}$ may not be equal to zero. Thus

$$\tilde{u} = 0 \text{ on } \big(\Omega \cup \omega_n\big)^c.$$

For $\delta > 0$, $V_n := [\tilde{u} < \delta] \cup \omega_n$ is an open neighborhood of Ω^c whose complement is included in B and is therefore compact (since closed). On this open set, we have

$$(u-\delta)^+(1-u_{\omega_n}) = 0 \text{ a.e.},$$

and this function is therefore equal to zero a.e. outside the compact set V_n^c of Ω. We then consider the function $\rho_p * \big[(u-\delta)^+(1-u_{\omega_n})\big]$, where ρ_p is a regularizing sequence. For p large enough, this function is in $C_0^\infty(\Omega)$. To end the proof, it is sufficient to prove that it does converge in $H_0^1(D)$ to u as successively $p \to +\infty$, $\delta \to 0$, $n \to +\infty$. The main point is that both functions $(u-\delta)^+$ and $(1-u_{\omega_n})$ belong to $H_0^1(D)$ and are *uniformly bounded*, so that their product does converge in H^1 as $\delta \to 0$ and when $n \to +\infty$. □

According to the characterization of $H_0^1(\Omega)$ given by Proposition 3.3.42 for an open set Ω, we are led to extend this definition to $H_0^1(A)$ where now *A is any subset of the open subset D of $\mathbb{R}^N$*.

Definition 3.3.43. Given $A \subset D$, we define

$$H_0^1(A) := \{u \in H_0^1(D);\ \tilde{u} = 0 \text{ q.e. in } D \backslash A\}.$$

By Proposition 3.3.33, this is a closed subspace of $H_0^1(D)$. It is separable as a subset of the separable set $H^1(\mathbb{R}^N)$. In fact, it can always be defined through a quasi-open set (unique), as stated in the following proposition together with some density results that will turn out to be useful later.

Proposition 3.3.44. *Given any $A \subset D$, there exists $\Omega \subset D$ quasi-open and unique such that $H_0^1(A) = H_0^1(\Omega)$. Moreover, $H_0^1(\Omega)$ is dense in $L^2(\Omega)$ and $\{u_\Omega^f;\ f \in L^\infty(D)\}$ is dense in $H_0^1(\Omega)$.*

Proof. Let u_n be a dense sequence in $H_0^1(A)$ and let us introduce

$$\Omega := \cup_n [\tilde{u}_n \neq 0].$$

Then Ω is quasi-open as a denumerable union of quasi-open sets and

$$[\tilde{u}_n \neq 0] \subset A \text{ q.e.} \Rightarrow \Omega \subset A \text{ q.e.} \Rightarrow H_0^1(\Omega) \subset H_0^1(A).$$

Conversely, by the density property, any $u \in H_0^1(A)$ is the limit in H^1 and q.e. of a subsequence $(u_{n_k})_k$. We then may write

$$[\tilde{u} \neq 0] \subset \cup_k [\tilde{u}_{n_k} \neq 0] \subset \Omega \text{ q.e.},$$

so that $H_0^1(A) \subset H_0^1(\Omega)$.

For the uniqueness, let us prove the following more general result:

$$[\Omega_1, \Omega_2 \subset D \text{ quasi-open}, H_0^1(\Omega_1) \subset H_0^1(\Omega_2)] \Rightarrow [\Omega_1 \subset \Omega_2, \text{ q.e.}]. \tag{3.63}$$

Indeed, assume $\text{cap}_D(\Omega_1 \backslash \Omega_2) > 0$. Let ω_n be a nonincreasing sequence of open sets whose capacity tends to 0 with $\Omega_1 \cup \omega_n$ being open sets. We denote by u_{ω_n} the capacitary potential of ω_n in D. For n large enough, $\text{cap}([u_{\omega_n} < 1] \cap (\Omega_1 \backslash \Omega_2)) > 0$. If K_p is a sequence of compact sets that increases to the open set $\Omega_1 \cup \omega_n$ and if

u_{K_p} denotes the capacitary potential of K_p in $H_0^1(\Omega_1 \cup \omega_n)$, we check that, for p large enough, $u_{K_p}(1 - u_{\omega_n})$ does not belong to $H_0^1(\Omega_2)$ while it belongs to $H_0^1(\Omega_1)$. Whence a contradiction.

For the density of $H_0^1(\Omega)$ in $L^2(\Omega)$, let us prove that

$$\big[f \in L^2(\Omega),\ \forall\, v \in H_0^1(\Omega),\ \textstyle\int_\Omega fv = 0\big] \Rightarrow [f \equiv 0]. \tag{3.64}$$

Let ω_n be a sequence of open sets whose capacity tends to 0 and such that $\Omega \cup \omega_n$ is open for all n. There exists $f_n \in C_0^\infty(\Omega \cup \omega_n)$ such that $\|f - f_n\|_{L^2(\Omega\cup\omega_n)} \le 1/n$. As $f_n(1 - u_{\omega_n}) \in H_0^1(\Omega)$, we have $\int_\Omega f f_n(1 - u_{\omega_n}) = 0$. Passing to the limit in n, we obtain $\int_\Omega f^2 = 0$. Whence (3.64).

Finally let us denote $W = \{u_\Omega^f;\ f \in L^\infty(D)\}$. Let $u \in H_0^1(\Omega)$ and $g = -\Delta u$ computed in the sense of distributions in D; thus $g \in H^{-1}(D)$ and one has $u = u_\Omega^g$ since for all $v \in H_0^1(\Omega) \subset H_0^1(D)$,

$$\int_\Omega \nabla u_\Omega^g \nabla v = \langle g, v\rangle_{H_0^1 \times H^{-1}} = \int_D \nabla u \nabla v.$$

There exists a sequence $g_p \in L^\infty(D)$ converging to g in $H^{-1}(D)$. But $u_\Omega^{g_p}$ converges in $H_0^1(\Omega)$ to $u_\Omega^g = u$. Whence the density of W in $H_0^1(\Omega)$. □

Remark 3.3.45. Here is a consequence of (3.63): If Ω is a quasi-open set of positive capacity, then $H_0^1(\Omega) \neq \{0\}$. In particular $|\Omega| > 0$.

Definition 3.3.43 of $H_0^1(A)$ allows us to extend the notion of the Dirichlet problem to any set $A \subset \mathbb{R}^N$ (this is done in Proposition 4.5.1 in the next chapter).

3.4 Back to the Dirichlet problem

We will use the following lemma several times.

Lemma 3.4.1. *Let Ω_n be a sequence of open sets in $\mathbb{R}^N$ and let $u_n \in H_0^1(\Omega_n)$ converge weakly in $H^1(\mathbb{R}^N)$ to $u^\star$. Set $\widetilde{\Omega}_p := \cup_{n\ge p}\Omega_n$, $E = \cap_p \widetilde{\Omega}_p$. Then $\tilde{u}^\star = 0$ q.e. outside E. If moreover there exists a set A such that $\lim_{n\to\infty} \operatorname{cap}(\Omega_n \backslash A) = 0$, then $\tilde{u}^\star = 0$ q.e. outside A.*

Proof. By the Mazur lemma (see [54, Thm. III.1]), there exists a convex combination $v_p = \sum_k \alpha_p^k u_{n_{k,p}}$, $\alpha_p^k \in (0, 1)$, $\sum_k \alpha_p^k = 1$ such that v_p converges strongly in H^1 to $u^\star$. Since $\tilde{u}_n = 0$ q.e. outside Ω_n, $\tilde{v}_p = 0$ q.e. outside $\widetilde{\Omega}_p$. But by Proposition 3.3.33, up to a subsequence, one may assume that $\tilde{v}_p$ converges q.e. to $\tilde{u}^\star$. It follows that $\tilde{u}^\star = 0$ q.e. outside $\widetilde{\Omega}_p$ for all p and therefore $\tilde{u}^\star = 0$ q.e. outside $E = \cap_p \widetilde{\Omega}_p$.

If $\lim_{n\to\infty} \operatorname{cap}(\Omega_n\backslash A) = 0$, up to a subsequence, we may assume $\operatorname{cap}(\Omega_n\backslash A) \le 2^{-n}$ and therefore $\operatorname{cap}(\widetilde{\Omega}_p\backslash A) \le 2^{-p+1}$. Thus $\operatorname{cap}(E\backslash A) = 0$ since $\operatorname{cap}(E\backslash A) \le \operatorname{cap}(\widetilde{\Omega}_p\backslash A)$ for all p. Therefore $\tilde{u}^\star$ vanishes also q.e. outside A. □

3.4.1 Local perturbation

Let us first give an extension of Proposition 3.2.7, where it was assumed that the variable open sets were included in the limit open set. In fact, if the excess of the variable sets over the limit set is not too large, in the sense of capacity, then the same convergence result holds.

Proposition 3.4.2. *Let Ω_n be a sequence of open subsets of D converging to an open set Ω in the sense of Hausdorff and such that*

$$\lim_{n\to+\infty} \operatorname{cap}(\Omega_n\backslash\Omega) = 0.$$

Then $u^f_{\Omega_n}$ converges to u^f_Ω.

Proof. It is sufficient to answer question (3.32) for the weak H^1-limit $u^\star$ of a subsequence of $u_n = u^f_{\Omega_n}$. According to Proposition 3.3.42, we have to show that

$$\tilde{u}^\star = 0 \text{ q.e. outside } \Omega$$

and this is a consequence of the above Lemma 3.4.1. □

Remark 3.4.3. This proposition is useful when one wants to slightly perturb an open set Ω around its boundary. We for instance have the following corollary (see [179] for an application).

Corollary 3.4.4. *Let Ω be an open set and let $S \subset \partial\Omega$ be a compact set of zero capacity. Assume*

$$\Omega_n\backslash\Omega \subset \bigcup_{x\in S} B(x,\epsilon_n), \qquad \Omega\backslash\Omega_n \subset \bigcup_{x\in\partial\Omega} B(x,\epsilon_n),$$

where $\epsilon_n > 0$ is a sequence that decreases to 0. Then $u^f_{\Omega_n}$ converges to u^f_Ω.

Proof. We check that Ω_n converges in the sense of Hausdorff to Ω. On the other hand, the capacity of the compact set $\bigcup_{x\in S} \overline{B}(x,\epsilon_n)$ decreases to the capacity of the compact set S, which is equal to 0. Thus we may apply Proposition 3.4.2. □

Remark 3.4.5. Note that the hypothesis of the latter corollary is not symmetric with respect to $\Omega_n\backslash\Omega$ and $\Omega\backslash\Omega_n$. The control has to be mainly made on the excess of Ω_n over Ω.

3.4.2 Compact convergence and stable open sets

Up to now, we have generally not made any regularity assumption on the open sets we were dealing with. One may obtain more precise convergence results when assuming more a priori regularity on the limit open set Ω or on the Ω_n themselves, in particular when the Ω_n converge to Ω from its exterior.

Let us first assume that the open sets Ω_n *converge to* Ω *in the sense of compact sets*. We already know that the sequence $u_n = u^1_{\Omega_n}$ converges weakly in $H^1_0(D)$ to a function $u^\star$ and that it is sufficient to show (see question (3.32) and Proposition 3.2.5) that $u^\star$ belongs to $H^1_0(\Omega)$ to be able to conclude (strong) convergence of $u^f_{\Omega_n}$ to u^f_Ω for all f.

By Lemma 3.4.1, we know that $\tilde{u}^\star = 0$ q.e. outside $E = \cap_p\big(\cup_{n\ge p}\Omega_n\big)$.

If L is a compact set included in the exterior of Ω (that is to say, in $\overline{\Omega}^{\mathrm{c}}$), by definition of convergence in the sense of compact sets, we will have $L \subset \overline{\Omega}_n^{\mathrm{c}}$ for n large enough and therefore $L \subset E^{\mathrm{c}}$ and $\tilde{u}^\star = 0$ q.e. on L. Since this is true for all compact sets included in the exterior of Ω, we in fact have

$$\tilde{u}^\star \equiv 0 \quad \text{q.e. on } \overline{\Omega}^{\mathrm{c}}. \tag{3.65}$$

The question we asked in (3.32) leads to the following new question in the case of convergence in the sense of compact sets:

$$\text{When can one claim that } \big[\tilde{u}^\star \equiv 0 \text{ q.e. on } \overline{\Omega}^{\mathrm{c}}\big] \Longrightarrow \big[u^\star \in H^1_0(\Omega)\big]? \tag{3.66}$$

or, according to Theorem 3.3.42, when does the following implication hold:

$$(\tilde{u}^\star \equiv 0 \text{ q.e. on } \overline{\Omega}^{\mathrm{c}}) \Longrightarrow (\tilde{u}^\star \equiv 0 \text{ q.e. on } \Omega^{\mathrm{c}})? \tag{3.67}$$

Let us first state, without proof, a proposition that characterizes those open sets for which (3.66) or (3.67) holds. It is due to M. Keldyš; see [208], [224] or [176] and see also [65] for a proof of equivalence between (i) and (iii).

Theorem 3.4.6. *Let Ω be an open subset of D. Then the following four properties are equivalent:*

(i) Any function $v \in H^1(\mathbb{R}^N)$ that vanishes quasi-everywhere on $\overline{\Omega}^{\mathrm{c}}$ vanishes quasi-everywhere on Ω^{c} as well and therefore belongs to $H^1_0(\Omega)$.

(ii) For any open set ω, $\mathrm{cap}(\omega\backslash\overline{\Omega}) = \mathrm{cap}(\omega\backslash\Omega)$.

(iii) For any $x \in \mathbb{R}^N$ and $r > 0$, $\mathrm{cap}(B(x,r)\backslash\overline{\Omega}) = \mathrm{cap}(B(x,r)\backslash\Omega)$.

(iv) $\displaystyle\liminf_{r\to 0} \frac{\mathrm{cap}(B(x,r)\backslash\overline{\Omega})}{\mathrm{cap}(B(x,r)\backslash\Omega)} > 0$ *q.e.* $x \in \partial\Omega$.

*If one of these properties holds, we will say that the open set Ω is **stable**.*

Let us summarize the previous analysis in the following result.

Theorem 3.4.7. *Let Ω_n be a sequence of open subsets of a bounded open set D that converges in the sense of compact sets to an open set Ω. Assume that Ω is stable. Then $u^f_{\Omega_n}$ converges to u^f_{Ω}.*

Remark 3.4.8. Concerning property (i), note that

$$(\tilde{v} = 0 \text{ q.e. on } \overline{\Omega}^{\mathrm{c}}) \Leftrightarrow (v = 0 \text{ a.e. on } \overline{\Omega}^{\mathrm{c}})$$

since $\tilde{v}$ is quasi-continuous and $\overline{\Omega}^{\mathrm{c}}$ is open. A stable open set therefore also satisfies the following property:

$$u = 0 \text{ a.e. on } \Omega^{\mathrm{c}} \Rightarrow u \in H^1_0(\Omega).$$

This property is however weaker (see Exercise 3.9).

Remark 3.4.9. As a fundamental example of stable open sets, let us mention the open sets *with a Lipschitz boundary*: this is stated and proved in Proposition 3.2.16. To prove it, we could also have directly used the uniform cone condition (see Definition 2.4.1 and Theorem 2.4.7). Indeed, this cone condition, which is valid for $\overline{\Omega}^{\mathrm{c}}$ when valid for Ω, implies the following: For all $x \in \partial\Omega$, there exist $\epsilon > 0$ and two unit vectors ξ_1, ξ_2 such that the cones $C(x, \xi_1, \epsilon)$, $C(x, \xi_2, \epsilon)$ satisfy

$$C_1 := C(x, \xi_1, \epsilon) \subset \overline{\Omega}^{\mathrm{c}},$$
$$C(x, \xi_2, \epsilon) \subset \Omega \quad \text{or} \quad \Omega^{\mathrm{c}} \subset [C(x, \xi_2, \epsilon)]^{\mathrm{c}}.$$

But $[C(x, \xi_2, \epsilon)]^{\mathrm{c}} \cap B(x, \epsilon)$ is also a cone centered at x and of direction $-\xi_2$. Thus, $\Omega^{\mathrm{c}} \cap B(x, \epsilon)$ is contained, for some $\eta > 0$, in the cone

$$C_2 := \{z \in B(x, \epsilon);\ (z - x, -\xi_2) \geq \cos(\eta)|z - x],\ 0 < [z - x] \leq \epsilon\}.$$

If for $r \in (0, \epsilon)$, $i = 1, 2$, we set $C_i^r := B(x, r) \cap C_i$, it follows by monotonicity that

$$\frac{\operatorname{cap}(B(x, r)\backslash\overline{\Omega})}{\operatorname{cap}(B(x, r)\backslash\Omega)} \geq \frac{\operatorname{cap}(C_1^r)}{\operatorname{cap}(C_2^r)} = \frac{\operatorname{cap}(C_1^\epsilon)}{\operatorname{cap}(C_2^\epsilon)}, \tag{3.68}$$

this last equality being obtained by (r/ϵ)-scaling. Whence condition (iv) of Theorem 3.4.6.

3.4.3 Capacitary-type constraints

One may find that the constraints of the ε-cone recalled in the previous subsection are a bit too strong. For instance, they would not allow us to deal with open sets containing cracks, which appear in several applications of shape optimization (such as inverse problems to detects cracks in various materials). One way to weaken these constraints is to introduce capacitary-type constraints for the points of the boundary.

Here we follow the approach in [75] and [65] which itself uses estimates from [177]. We denote by D a ball in $\mathbb{R}^N$.

Definition 3.4.10. Let $\alpha, r \in (0, \infty)$. We will say that an open subset $\Omega \subset D$ satisfies the (α, r)-property of capacity density if

$$\forall\, x \in \partial\Omega, \qquad \frac{\operatorname{cap}\left(\Omega^c \cap B(x,r), B(x,2r)\right)}{\operatorname{cap}\left(B(x,r), B(x,2r)\right)} \ge \alpha > 0.$$

For given $\alpha \in (0,1)$, $r_0 > 0$, we will denote by $\mathcal{O}_{\alpha,r_0}$ the family of open subsets of B satisfying the (α, r)-property of capacity density for all $r \in (0, r_0)$.

This property is obviously weaker than the ε-cone condition. More generally, it is satisfied if, for any boundary point x, there exists a cone of size independent of x included in Ω^c. It is also implied by the uniform so-called *corkscrew property* (see [177] and Exercise 3.10).

This boundary regularity implies Hölder continuity up to the boundary of solutions u_Ω^f to the Dirichlet problem when f is regular enough. This is stated now.

Lemma 3.4.11 (Cf. [177]). *Let $\Omega \in \mathcal{O}_{\alpha,r_0}$ and $f \in L^p(D)$, $p > N/2$. Then $u = u_\Omega^f \in C(\overline{D})$, $u = 0$ everywhere outside $\overline{\Omega}$ and there exist $M > 0$, $\delta \in (0,1)$ such that*

$$\forall\, x, y \in \overline{D}, \quad |u(x) - u(y)| \le M|x-y|^\delta,$$

where M, δ depend only on α, r_0, D, $\|f\|_{L^p(D)}$.

Proof. It is known (see, e.g., [151]) that $w := u_D^f$ is Hölder continuous on $\overline{D}$ with constant and exponent depending only on D, $\|f\|_{L^p(D)}$. But, $h = w - u$ is harmonic on Ω and $h - w \in H_0^1(\Omega)$. But, since $\Omega \in \mathcal{O}_{\alpha,r_0}$, it is also regular in the sense of Wiener[12] (see Definition 3.69 below) and this implies (see, e.g., [177, Thm. 6.27])

$$\forall\, x \in \partial\Omega, \qquad \lim_{y \to x, y \in \Omega} h(y) = w(y).$$

[12]Norbert WIENER, 1894–1964, was born in the USA to a Russian father. He worked on lots of mathematical topics like Brownian motion, normed vector spaces, harmonic analysis, Tauberian theorems, and we also owe the introduction of cybernetics to him.

In particular, u has a continuous representative on $\overline{\Omega}$ that vanishes on $\partial\Omega$. It follows that we can extend it as a continuous function on $\overline{D}$ that is equal to zero on $\overline{\Omega}^{\mathrm{c}}$.

Using the Hölder continuity of w on $\overline{D}$ and therefore on $\partial\Omega$, by [177, Thm. 6.44], there exist constants M, δ depending only on $\|f\|_{L^p(D)}$, D, α, r_0 such that

$$\forall\, x, y \in \overline{\Omega}, \quad |h(x) - h(y)| \le M|x - y|^{\delta}.$$

Using $u = w - h$, we deduce the same property for u on $\overline{\Omega}$ and the inequality is obviously valid for all $x, y \in \overline{D}$ since $u = 0$ on $\overline{\Omega}^{\mathrm{c}}$. □

We will now deduce from Lemma 3.4.11 the following result:

Theorem 3.4.12 ([75]). *Let Ω_n be a sequence of open sets from the family $\mathcal{O}_{\alpha,r_0}$ which converges in the sense of Hausdorff to an open set Ω. Then for all $f \in H^{-1}(D)$, $u_n = u^f_{\Omega_n}$ converges to $u = u^f_{\Omega}$.*

Proof. As usual, we may assume $f \equiv 1$ and we know that there exists a subsequence (still denoted by u_n) that converges weakly to a function $u^\star \in H^1_0(D)$. The point is to show that $\tilde{u}^\star$ vanishes quasi-everywhere on Ω^{c}. We will in fact show that u (or $\tilde{u}$) is continuous and equal to zero everywhere outside Ω.

By Lemma 3.4.11, the functions u_n are Hölder continuous on $\overline{D}$ with the same constants M, δ. Since they are otherwise uniformly bounded (since they are nonnegative and bounded above by $w = u^1_D$), by Ascoli's theorem, this sequence is relatively compact for uniform convergence. We may therefore assume that u_n converges also uniformly on $\overline{D}$ to $u^\star$ (which is, in particular, continuous on $\overline{D}$).

Now let $x \in \Omega^{\mathrm{c}}$. Since Ω_n converges to Ω in the sense of Hausdorff, $x = \lim x_n$ where $x_n \in \Omega^{\mathrm{c}}_n$. In particular, $u_n(x_n) = 0$. By uniform convergence, we deduce $u^\star(x) = 0$. Thus $u^\star = 0$ everywhere outside Ω. □

Remark 3.4.13. In fact, the previous proof leads to the following more general result: Assume that the sequence Ω_n is such that $u_n = u^1_{\Omega_n}$ satisfies

(i) for all n, u_n is continuous on $\overline{D}$ and $u_n = 0$ everywhere outside Ω_n;

(ii) the sequence (u_n) is equicontinuous on $\overline{D}$.

Then u_n converges uniformly on $\overline{D}$ and in $H^1_0(D)$ to $u^\star = u^1_\Omega$.

It is classical in potential theory that $u^1_\omega \in C(\overline{\omega})$ and $u^1_\omega = 0$ everywhere outside ω if ω is *regular in the sense of Wiener*, that is, if (see [123], [4], [224], [177], etc.)

$$\forall\, x \in \partial\omega, \quad \lim_{r\to 0} w(x, r, 1, \omega) = +\infty, \tag{3.69}$$

where for $0 < r < R \le 1$,

$$w(x, r, R, \omega) := \int_r^R \frac{\operatorname{cap}\big(\omega^{\mathrm{c}} \cap B(x,t), B(x,2t)\big)}{\operatorname{cap}\big(B(x,t), B(x,2t)\big)} \frac{dt}{t}.$$

This is satisfied if $\omega \in \mathcal{O}_{\alpha,r_0}$ since then $w(x,r,r_0,\omega) \geq \log(r_0/r)$. In this case, condition (3.69) is even uniform with respect to $x \in \partial\omega$ and is satisfied uniformly in $\mathcal{O}_{\alpha,r_0}$.

It is natural to think that a uniform Wiener condition should guarantee the equicontinuity of the u_n without requiring the stronger capacity density introduced above, which in fact implies uniform Hölder continuity of the u_n. This is indeed the case and it is proved in [75], [65] that a Wiener condition, locally uniform in x and uniform in n, implies the expected equicontinuity and the convergence of $u^f_{\Omega_n}$ to u^f_Ω. The proof of this stronger and more natural result requires, however, much more work. We also refer to [140] where such conditions are considered.

To end this remark, let us emphasize that *condition (i) cannot be weakened to* the following one:

(i)′ u_n is continuous on $\overline{D}$ and $u_n = 0$ quasi-everywhere outside Ω_n^c.

Looking at the following example will convince the reader: $\Omega_n := D\backslash\{x_1, \dots ; x_n\}$, where $D \subset \mathbb{R}^2$ is the unit disk and (x_n) is a dense sequence in D. Then $u^f_{\Omega_n} \equiv u^f_D$ and Ω_n converges in the sense of Hausdorff to the empty set!

Application of Theorem 3.4.12 in dimension 2. It turns out that in dimension 2, the continuity for the Dirichlet problem with respect to Hausdorff convergence holds as soon as the "number of holes" in Ω_n is uniformly bounded. This is a very nice result due to Šverak [285]. Although it involves a topological statement that apparently has nothing to do with the notion of capacity, it turns out that it may be deduced from Theorem 3.4.12.

Let l be an integer ≥ 1. For any open subset $\Omega \subset D$, we will denote by $\sharp\Omega^c$ the number of connected components of the complement of Ω. We then define

$$\mathcal{O}_l = \{\Omega \subset D,\ \Omega \text{ open},\ \sharp\Omega^c \leq l\}.$$

Theorem 3.4.14 (Šverak). *Let Ω_n be a sequence of open sets in the class $\mathcal{O}_l$ which converges in the sense of Hausdorff to Ω. Then for all $f \in H^{-1}(D)$, $u^f_{\Omega_n}$ converges to u^f_Ω.*

Proof. As usual, there exists a subsequence such that $u^1_{\Omega_n}$ converges weakly in $H^1_0(D)$ and a.e. to $u^\star$ and the goal is to prove that $u^\star = 0$ q.e. on Ω^c. In general, one cannot find α, r_0 such that $\Omega_n \in \mathcal{O}_{\alpha,r_0}(D)$ since, for instance, it may happen that some of the connected components of Ω_n^c converge to a point, in which case Ω_n is not uniformly regular and does not meet the hypotheses of Theorem 3.4.12.

One may write $\overline{D}\backslash\Omega_n = F_n = F_n^1 \cup F_n^2 \cup \dots \cup F_n^l$, where the sets F_n^i are compact and connected, possibly empty. Up to a subsequence, one may assume that $F_n^j \xrightarrow{H} F_j$ for $j = 1, \dots, l$. Then there are three possibilities for the sets F_j:

First possibility: $F_j = \emptyset$, but then F_n^j is empty for j large enough. Let us denote by J_0 the subset of indices j for which this holds.

Second possibility: $F_j = \{x_j\}$ is reduced to only one point. Let us denote by J_1 the subset of indices for which this holds. Let us then consider $\Omega^\star = \Omega \cup \{x_i,\ i \in J_1\}$. We have $H_0^1(\Omega^\star) = H_0^1(\Omega)$ since a set of a finite number of points is of zero capacity (see Theorem 3.3.42).

Thus, showing that $u^\star \in H_0^1(\Omega)$ amounts to showing that $u^\star \in H_0^1(\Omega^\star)$. Let $I := \{1,\dots,l\}\backslash(J_0 \cup J_1)$. We will consider only $\Omega_n^\star = D\backslash \cup_{j\in I} F_j^n$, which converges in the sense of Hausdorff to $\Omega^\star$, which is exactly the third possibility.

Third possibility: for $j \in I$, F_j contains at least two points. Let a_j be their distance. They are limits of points of F_j^n whose distance is at least $a_j/2$ for n large. For all $x \in \partial\Omega_n^*$ and $j\ (= j(x)) \in I$ such that $x \in F_n^j$, one may write, by using Corollary 3.3.25, for all $r < a_j/4$,

$$\operatorname{cap}(\Omega_n^{*\,c} \cap B(x,r), B(x,2r)) \geq \operatorname{cap}(F_n^j \cap B(x,r), B(x,2r)) \geq k_2 > 0,$$

where k_2 is a "universal" constant. This shows that the open sets Ω_n^* belong to $\mathcal{O}_{\alpha,r_0}$ with $\alpha = k_2$, $r_0 = \min\{a_j, j \in I\}/4$.

By Theorem 3.4.12, $u^1_{\Omega_n^*}$ converges to $u^1_{\Omega^*} = u^1_\Omega$. But since $\Omega_n \subset \Omega_n^*$, by Proposition 3.1.22,

$$u^1_{\Omega_n^*} \geq u^1_{\Omega_n} \geq 0 \text{ a.e.},$$

and at the limit $u^1_\Omega \geq u^\star \geq 0$ a.e., which implies $u^\star \in H_0^1(\Omega)$ by Corollary 3.1.14. □

3.5 The γ-convergence

3.5.1 Definition

The previous analysis leads us to introduce a new topology on the family of open subsets of $\mathbb{R}^N$. The following definition is less geometric than those analyzed in Chapter 2: This so-called γ-convergence is nothing but the exact topology for which continuity holds for the solutions of the Dirichlet problem. Here D denotes again a given bounded open set of $\mathbb{R}^N$.

Definition 3.5.1. We say that a sequence of open subsets Ω_n of D γ-converges to the open subset $\Omega \subset D$ (and we write $\Omega_n \xrightarrow{\gamma} \Omega$) if for all $f \in H^{-1}(D)$, we have $u^f_{\Omega_n} \to u^f_\Omega$ in $H_0^1(D)$.

Remark 3.5.2. Introducing the resolvent operators $R_\Omega : f \mapsto u^f_\Omega$, γ-convergence exactly means the pointwise convergence of R_{Ω_n} to R_Ω. It turns out, using the compact embedding $H^1_0 \hookrightarrow L^2$, that this pointwise convergence implies strong convergence (in norm); see, e.g., [181, Thm. 2.3.2] and Lemma 4.7.3. Consequently, γ-convergence also implies convergence of the eigenvalues of the Laplace operator (with Dirichlet boundary conditions).

Remark 3.5.3. According to Theorem 3.2.5, in Definition 3.5.1, we may ask that the convergence be true only for $f \equiv 1$ and we may assume only that the convergence holds only in $L^2(D)$.

Note that the γ-limit is not necessarily unique since one can find distinct open sets that are equal q.e. and that therefore lead to the same H^1_0-space. But we will see in the next chapter that two open sets that are limits of the same sequence are equal q.e. (see also Proposition 3.3.44).

We actually saw in the previous sections several sufficient conditions guaranteeing γ-convergence of domains. For instance, Šverak's Theorem 3.4.14 or Chenais' Theorem 3.2.13 may be rephrased in terms of γ-convergence as follows.

In dimension 2, *if the sequence* $(\Omega_n)_n$ *Hausdorff-converges to* Ω *and if the number of connected components of* Ω_n^c *is uniformly bounded, then* $\Omega_n \xrightarrow{\gamma} \Omega$.

If $(\Omega_n)_n$ *is a sequence of uniformly Lipschitz open sets which Hausdorff-converges to* Ω*, the Lipschitz constant being uniform for the sequence, then* $\Omega_n \xrightarrow{\gamma} \Omega.i$

More generally, if the sequence $(\Omega_n)_n$ *Hausdorff-converges to* Ω *and satisfies the property of capacity density* (3.4.10), *then* $\Omega_n \xrightarrow{\gamma} \Omega$.

In fact, one may characterize γ-convergence rather precisely in terms of capacity: This is stated below in Proposition 3.5.6.

In the case when γ-convergence does not hold, one can nevertheless describe rather precisely the asymptotic behavior of the sequence $u^f_{\Omega_n}$. This will be one of the purposes of Chapter 7.

3.5.2 Link with Mosco convergence

It is useful to comment on the link between γ-convergence and Mosco convergence of the Sobolev spaces $H^1_0(\Omega_n)$ to the corresponding $H^1_0(\Omega)$. Let us first recall the definition of Mosco convergence.

Definition 3.5.4. Let A_n be a sequence of closed convex sets of a normed space X. We say that A_n Mosco-converges to A if the two following conditions hold:

(M1) For all $x \in A$, there exists a sequence x_n, $x_n \in A_n$ such that $x_n \to x$ (strong convergence).

(M2) For any sequence $(y_{n_k})_k$ of elements of A_{n_k} that converges weakly to some y, then $y \in A$.

The Sobolev spaces $H_0^1(\Omega_n)$ and $H_0^1(\Omega)$ being closed subspaces of $H_0^1(D)$, we may apply the above definition to $X := H_0^1(D)$, $A_n := H_0^1(\Omega_n)$, $A := H_0^1(\Omega)$. We then have the following:

Proposition 3.5.5. *Let $(\Omega_n)_n$ be a sequence of open subsets of D. Then Ω_n γ-converges to Ω if and only if $H_0^1(\Omega_n)$ Mosco-converges to $H_0^1(\Omega)$.*

Proof. Sufficient condition: Let $f \in H^{-1}(D)$ and $u_n = u^f_{\Omega_n}$. We know there exists a subsequence — which we still denote by u_n — that converges weakly in $H_0^1(D)$ to some $u^\star$. But by (M2), $u^\star \in H_0^1(\Omega)$.

Now let φ be a test function in $H_0^1(\Omega)$. By (M1), there exists a sequence $\varphi_n \in H_0^1(\Omega_n)$ that strongly converges to φ. The following holds:

$$\int_D \nabla u_n.\nabla\varphi_n\,dx = \int_{\Omega_n} \nabla u_n.\nabla\varphi_n\,dx = \int_{\Omega_n} f\varphi_n\,dx = \int_D f\varphi_n\,dx.$$

Passing to the limit in this equality (strongly in φ_n, weakly in u_n), we obtain

$$\begin{aligned} \forall\,\varphi \in H_0^1(\Omega), \quad &\int_D \nabla u^\star.\nabla\varphi\,dx = \int_D f\varphi\,dx,\\ \text{or} \quad &\int_\Omega \nabla u^\star.\nabla\varphi\,dx = \int_\Omega f\varphi\,dx. \end{aligned} \tag{3.70}$$

Thus $u^\star = u^f_\Omega$.

Necessary condition: Let us first prove (M1). Let $\varphi \in H_0^1(\Omega)$ and let $f = -\Delta\varphi$ computed in the sense of distributions in D so that $f \in H^{-1}(D)$. By construction, $\varphi = u^f_\Omega$. Let us introduce $\varphi_n := u^f_{\Omega_n}$. By γ-convergence of Ω_n to Ω, φ_n converges strongly to φ in $H_0^1(D)$. Whence (M1) since $\varphi_n \in H_0^1(\Omega_n)$.

Let us now prove (M2). Let φ_k be a sequence of functions in $H_0^1(\Omega_{n_k})$ that converges weakly in $H_0^1(D)$ to some function φ. The goal is to prove that $\varphi \in H_0^1(\Omega)$. Let us again set $f := -\Delta\varphi \in H^{-1}(D)$. Let us also introduce $u_k := u^f_{\Omega_{n_k}}$. By γ-convergence, u_k converges strongly to $u = u^f_\Omega$, which belongs to $H_0^1(\Omega)$. It remains to prove that u and φ are the same function. Passing to the limit in

$$\begin{aligned} \int_D \nabla(u_k - \varphi_k)\nabla u_k\,dx &= \int_{\Omega_{n_k}} \nabla(u_k - \varphi_k).\nabla u_k\,dx\\ &= \int_{\Omega_{n_k}} (u_k - \varphi_k)f\,dx = \int_D (u_k - \varphi_k)f\,dx, \end{aligned} \tag{3.71}$$

we obtain $\int_D \nabla(u-\varphi)\nabla u = \int_D (u-\varphi)f$. On the other hand, by definition of f, $\int_D \nabla(u-\varphi)\nabla\varphi = \int_D (u-\varphi)f$. This yields $\int_D |\nabla(u-\varphi)|^2 = 0$. □

There are various characterizations of Mosco convergence of $H_0^1(\Omega_n)$ to $H_0^1(\Omega)$ (see, e.g., [116], [117]). Let us recall here one nice characterization given in [62] whose proof may also be found in [65].

Proposition 3.5.6. *Given open subsets Ω_n, $n \geq 0$, Ω of a bounded open set D, then $H_0^1(\Omega_n)$ Mosco-converges to $H_0^1(\Omega)$ if and only if for all $x \in \mathbb{R}^N$,*

(C1) $\limsup_{n\to\infty} \operatorname{cap}\big(\Omega_n^c \cap \overline{B}(x,r), B(x,2r)\big) \leq \operatorname{cap}\big(\Omega^c \cap \overline{B}(x,r), B(x,2r)\big)$;

(C2) $\operatorname{cap}\big(\Omega^c \cap B(x,r), B(x,2r)\big) \leq \liminf_{n\to\infty} \operatorname{cap}\big(\Omega_n^c \cap B(x,r), B(x,2r)\big)$.

More precisely, (C1) is equivalent to (M1) in Definition 3.5.4 of Mosco convergence and (C2) is equivalent to (M2).

Let us also mention two more characterizations of Mosco convergence due to Attouch, Sonntag, and Tsukada (see [28]).

Theorem 3.5.7. *The following properties are equivalent:*

(i) $H_0^1(\Omega_n)$ Mosco-converges to $H_0^1(\Omega)$ (and therefore Ω_n γ-converges to Ω).

(ii) For all $\varphi \in H_0^1(D)$, $d(\varphi, H_0^1(\Omega)) = \lim_{n\to+\infty} d(\varphi, H_0^1(\Omega_n))$ (where $d(\varphi, X)$ denotes the $H_0^1(D)$-distance of φ to the convex set X).

(iii) For all $\varphi \in H_0^1(D)$, $\operatorname{proj}_{H_0^1(\Omega)}(\varphi) = \lim_{n\to+\infty} \operatorname{proj}_{H_0^1(\Omega_n)}(\varphi)$ (where $\operatorname{proj}_X(\varphi)$ denotes the $H_0^1(D)$-projection of φ onto the convex set X).

Remark 3.5.8. An important conclusion of this subsection is that the γ-convergence whose definition is given in terms of the Laplace operator Δ depends in fact only on the H^1-norm. This is clear in the above two characterizations, which are stated in terms of the convergence of the spaces $H_0^1(\Omega_n)$ or in terms (only) of the capacity associated with the H^1-norm. Thus we expect that this γ-convergence also implies good continuity properties for the solutions of problems that are well posed in H_0^1. We give such examples in the next subsection.

3.5.3 More operators associated with the $H_0^1\gamma$-convergence

We prove here that γ-convergence (as defined above for the Laplace operator with Dirichlet boundary conditions) implies a similar γ-convergence for the elasticity operator with homogeneous Dirichlet boundary conditions (see [71]).

Let us first recall the traditional notation for the linear elasticity problem (we follow here the notation of [104]).

Let Ω be an open subset of a ball D and let $f = (f_1, f_2, f_3) \in L^2(D;\mathbb{R}^3)$. We will denote by $H_0^1(\Omega;\mathbb{R}^3)$ the space of vector-valued functions $u = (u_1, u_2, u_3)$ such that $u_i \in H_0^1(\Omega)$ for $i = 1, 2, 3$. Given $u \in H_0^1(\Omega;\mathbb{R}^3)$, we denote by Du the Jacobian matrix[13] of u with entries $\frac{\partial u_i}{\partial x_j}$, $1 \le i, j \le 3$ and $\mathbf{e}(u) = \frac{1}{2}(Du +^t Du)$. When looking for (small) displacements u of an elastic body Ω under the action of an internal force f (and with a fixed boundary), we are led to solving the variational problem

$$\begin{cases} \text{find } u \in H_0^1(\Omega;\mathbb{R}^3) \text{ such that for all } v \in H_0^1(\Omega;\mathbb{R}^3), \\ \int_\Omega \{\lambda \operatorname{tr} \mathbf{e}(u) \operatorname{tr} \mathbf{e}(v)\, dx + 2\mu \mathbf{e}(u) : \mathbf{e}(v)\}\, dx = \int_\Omega f.v\, dx. \end{cases} \tag{3.72}$$

In (3.72), $\lambda, \mu \in (0, \infty)$ are called the Lamé constants and are specific to each material. Existence and uniqueness are provided by the Lax–Milgram theorem thanks to Korn's inequalities; see [104].

Proposition 3.5.9. *Let Ω_n, $n \ge 0$, Ω be open subsets of D and let us denote by u_n (resp. u) the corresponding solutions of* (3.72). *Then*

$$[\Omega_n \ \gamma\text{-converges to } \Omega] \Longrightarrow [u_n \longrightarrow u \text{ in } H_0^1(D;\mathbb{R}^3)].$$

Proof. Since Ω_n γ-converges to Ω, $H_0^1(\Omega_n)$ Mosco-converges to $H_0^1(\Omega)$. This immediately implies Mosco convergence of $H_0^1(\Omega_n;\mathbb{R}^3)$ to $H_0^1(\Omega;\mathbb{R}^3)$. The proof is now a copy of the proof of Proposition 3.5.5.

Replace v by u_n in the variational formulation (3.72). By ellipticity of the bilinear form and by the continuity of the linear form, it follows that the sequence u_n is bounded in $H_0^1(D;\mathbb{R}^3)$. Consider a subsequence (still denoted by u_n) converging weakly to some $u^\star$. By (M2) of Mosco convergence of $H_0^1(\Omega_n;\mathbb{R}^3)$ to $H_0^1(\Omega;\mathbb{R}^3)$, $u^\star \in H_0^1(\Omega;\mathbb{R}^3)$. It remains to prove that $u^\star$ satisfies (3.72) and that the convergence is strong. Let v be a test function in $H_0^1(\Omega;\mathbb{R}^3)$; by (M1), there exists $v_n \in H_0^1(\Omega_n;\mathbb{R}^3)$ converging strongly to v. Writing (3.72) on Ω_n leads to

$$\int_{\Omega_n} \{\lambda \operatorname{tr} \mathbf{e}(u_n) \operatorname{tr} \mathbf{e}(v_n)\, dx + 2\mu \mathbf{e}(u_n) : \mathbf{e}(v_n)\}\, dx = \int_{\Omega_n} f.v_n\, dx. \tag{3.73}$$

After extending all functions by 0 outside Ω_n, we may rewrite the above integrals with D in place of Ω_n. The quantities $\operatorname{tr} \mathbf{e}(u_n) \operatorname{tr} \mathbf{e}(v_n)$ on the one hand and $\mathbf{e}(u_n) : \mathbf{e}(v_n)$ on the other hand are linear combinations of products of terms $\frac{\partial u_{n,i}}{\partial x_j}$ with terms $\frac{\partial v_{n,k}}{\partial x_l}$. Thanks to the strong convergence of $\frac{\partial v_{n,i}}{\partial x_j}$ to $\frac{\partial v_i}{\partial x_j}$ and to the weak convergence of

[13] Carl **Gustav** Jacob JACOBI, 1804–1851, German, worked in very different domains: number theory, elliptic functions, functions of several variables, partial differential equations, quadratic forms.

$\frac{\partial u_{n,i}}{\partial x_j}$ to $\frac{\partial u_i^\star}{\partial x_j}$, it is possible to pass to the limit in (3.73). Thus we obtain that $u^\star$ is a solution of (3.72) (i.e., $u^\star = u$). Since u is the only possible limit point for the sequence u_n, the whole sequence converges to u. Finally, strong convergence may be obtained as usual. Choose $v = u_n$ as a test function in the variational formulation: If $a(u, v)$ denotes the bilinear form extended to D, from (3.72), we obtain that $a(u_n, u_n)$ converges to $a(u, u)$ and consequently that $a(u_n - u, u_n - u)$ tends to 0. Then strong convergence follows from Korn's inequality. □

Similarly, if we are interested in continuity with respect to the domain for a general elliptic operator, using Mosco convergence immediately leads to the following result.

Proposition 3.5.10. *Let D be a bounded open set and let*

$$Au := -\sum_{i,j=1}^{N} \frac{\partial}{\partial x_i}\left(a_{ij}(x)\frac{\partial u}{\partial x_j}\right)$$

be an elliptic operator with $a_{ij} \in L^\infty(D)$. If the open sets $\Omega_n \subset D$ γ-converge to an open set Ω, then for any $f \in H^{-1}(D)$, the solution u_n of

$$Au_n = f \quad \text{in } \Omega_n,\ u_n \in H_0^1(\Omega_n) \tag{3.74}$$

converges strongly in $H_0^1(D)$ to the solution u of

$$Au = f \quad \text{in } \Omega,\ u \in H_0^1(\Omega).$$

3.5.4 Remarks about nonlinear operators

When working with homogeneous Dirichlet boundary conditions, the same method as above may be similarly developed in many different situations. For instance, one may be willing to work with nonlinear operators like the so-called p-Laplacian $u \to \operatorname{div}\left(|\nabla u|^{p-2}\nabla u\right)$, $p > 1$. This requires working with the Sobolev space $W_0^{1,p}(D)$ and the capacity associated with its norm. Then the analysis may be made as above, replacing H_0^1 by $W_0^{1,p}$, and the results are very similar. We refer, for example, to [117], [119], [65], and to their own references.

3.6 Quantitative estimates

Up to now, we have given essentially *qualitative* convergence results. In some cases, we may need more precise *quantitative* estimates on the rate of convergence. They may involve geometric quantities like the Hausdorff distance.

In some simple situations, such estimates may be obtained by using only the maximum principle.

Proposition 3.6.1. *Let $(\Omega_n)_n$ be a sequence of open subsets of a given bounded open set Ω that we assume to be of class C^1. Let $f \in L^p(\Omega)$ where $p > N$ and denote $u_n = u^f_{\Omega_n}$, $u = u^f_{\Omega}$. Then the following L^∞-estimate holds:*

$$\|u_n - u\|_{L^\infty(\Omega)} \le C d_{\mathrm{H}}(\Omega_n, \Omega), \tag{3.75}$$

where C depends only on Ω, u and where $d_{\mathrm{H}}(\Omega_n, \Omega)$ is the Hausdorff distance between Ω_n and Ω.

Note that no regularity is assumed here on the Ω_n.

Proof. The assumption $f \in L^p(\Omega)$, $p > N$ and the C^1-regularity of Ω guarantee that $u \in C^1(\overline{\Omega})$ (see, e.g., [151]). Since all Ω_n are included in Ω, by subtracting the two equations in u_n, u, we have

$$\begin{cases} \Delta(u - u_n) = 0 & \text{in } \Omega_n, \\ u - \tilde{u}_n = u & \text{q.e. on } \Omega_n^c. \end{cases} \tag{3.76}$$

At least formally by using the maximum principle, denoting $\mathcal{C}_n = \overline{\Omega}\setminus\Omega_n$, we expect to have

$$\|u_n - u\|_{L^\infty(\Omega_n)} \le \sup_{\partial\Omega_n} |u| \le \sup_{\mathcal{C}_n} |u|.$$

This can be justified by using that with $m := \sup_{\mathcal{C}_n} |u|$, we have $[(u - u_n) - m]^+ = 0$ q.e. on Ω_n^c and therefore $[(u - u_n) - m]^+ \in H^1_0(\Omega_n)$. Thus

$$0 = \int_{\Omega_n} \nabla(u - u_n)\nabla[(u - u_n) - m]^+ = \int_{[(u-u_n)>m]} |\nabla(u - u_n)^+|^2,$$

and consequently $u - u_n \le m$ a.e. Similarly $u_n - u \le m$ a.e.

Since $u \in C^1(\overline{\Omega})$, if $M := \sup_{x\in\overline{\Omega}} |\nabla u(x)|$, we have by the mean value property and since $u = 0$ on $\partial\Omega$,

$$\forall\, x \in \mathcal{C}_n, \quad |u(x)| \le M d(x, \partial\Omega) \le M d_{\mathrm{H}}(\Omega_n, \Omega).$$

□

The previous results may be immediately generalized to the case when u^f_Ω is Lipschitz continuous, no matter the reason why this regularity holds.

Generalizations can also be made in different directions: We may assume less regularity on the domains or we may drop the (strong) assumption that the Ω_n are included in Ω. We give below two examples of such results. They may be found in [269] to which we refer for a proof.

In the first result below, we assume that all domains satisfy the ε-cone condition (see Chapter 2).

Theorem 3.6.2 (Savaré–Schimperna). *Let Ω_n, Ω be open subsets of a ball D satisfying the ε-cone condition (for the same ε). Assume $f \in L^2(D)$. Then there exists $C = C(\varepsilon, f)$ such that*

$$\|u^f_{\Omega_n} - u^f_{\Omega}\|_{H^1(D)} \le C\,(d_{\mathrm{H}}(\Omega_n, \Omega))^{1/2}. \tag{3.77}$$

Obviously, this estimate gives, as a corollary, a new proof of Theorem 3.2.13.

Remark 3.6.3. This kind of estimate is important in numerical analysis when we choose to replace the original domain Ω by a polygonal domain Ω_h, for example to perform some finite element method. Inequality (3.77) shows that we have to choose a polygonal domain whose Hausdorff distance to Ω is of order h^2 if we want to keep an H^1 error in $O(h)$ as predicted by the theory when Ω is, for example, a C^2-domain. More precisely, in this case, by choosing a finite number of points on the boundary of Ω whose respective distances are of order h, and if we construct the polyhedra Ω_h (or polygon in 2 dimensions) having these points as vertices, it can be proved that the Hausdorff distance satisfies

$$d_{\mathrm{H}}(\Omega_h, \Omega) \le Ch^2,$$

where C is a constant depending only on Ω (more precisely, on the maximum of the curvature of $\partial\Omega$).

One may want to weaken the hypotheses of Theorem 3.6.2 by assuming that only the limit open set Ω satisfies the ε-cone condition. But in this case, the Hausdorff distance between Ω_n and Ω is not enough to control the distance between the solutions. Recall that (see Chapter 2)

$$d_{\mathrm{H}}(\Omega_n, \Omega) = d^{\mathrm{H}}(\Omega_n^{\mathrm{c}}, \Omega^{\mathrm{c}}) = \max\{\rho(\Omega_n^{\mathrm{c}}, \Omega^{\mathrm{c}}), \rho(\Omega^{\mathrm{c}}, \Omega_n^{\mathrm{c}})\},$$

where $\rho(F_1, F_2) = \sup_{x \in F_1} d(x, F_2)$. In fact, one must replace $\rho(\Omega^{\mathrm{c}}, \Omega_n^{\mathrm{c}})$ by $\rho(\overline{\Omega}_n, \overline{\Omega})$. We then have the following nonsymmetric estimate.

Theorem 3.6.4 (Savaré–Schimperna). *Let Ω be an open set satisfying the ε-cone condition. Let Ω_n be any sequence of open subsets of D and let $f \in L^2(D)$. Then there exists $C = C(\varepsilon, f)$ such that*

$$\|u^f_{\Omega_n} - u^f_{\Omega}\|_{H^1(D)} \le C\left(\max\left\{\rho(\Omega_n^{\mathrm{c}}, \Omega^{\mathrm{c}}), \rho(\overline{\Omega}_n, \overline{\Omega})\right\}\right)^{1/2}. \tag{3.78}$$

The previous results are valid for elliptic operators more general than the Laplacian.

3.7 Continuity for the Neumann problem

3.7.1 Introduction

Throughout this section, we use a fixed open ball B. With each open subset Ω such that $\overline{\Omega} \subset B$, we associate the function u_Ω (or more simply u if it is not ambiguous), which is the solution of the problem

$$\begin{cases} u \in H^1(\Omega) \text{ and for all } v \in H^1(\Omega), \\ \int_\Omega \nabla u . \nabla v + \int_\Omega uv = \int_\Omega fv, \end{cases} \tag{3.79}$$

where $f \in L^2(B)$ is given. By the Lax–Milgram theorem, problem (3.79) has a unique solution. Moreover, if Ω is regular enough, this solution u satisfies

$$\begin{cases} -\Delta u + u = f & \text{in } \Omega, \\ \frac{\partial u}{\partial \nu} = 0 & \text{on } \partial\Omega, \end{cases} \tag{3.80}$$

where $\frac{\partial u}{\partial \nu}$ denotes the exterior normal derivative to $\partial\Omega$.

As in the previous sections, we consider a sequence of open subsets Ω_n of B and our goal is to study the convergence of u_{Ω_n} to u_Ω when Ω_n converges to Ω in some sense. As for the Dirichlet problem, there exists a uniform H^1-bound on the $u_n = u_{\Omega_n}$. More precisely, thanks to (3.79) applied with $v = u_n$, we have

$$\|u_n\|^2_{H^1(\Omega_n)} = \int_{\Omega_n} |\nabla u_n|^2 + \int_{\Omega_n} u_n^2 = \int_{\Omega_n} f u_n.$$

Thus by the Cauchy–Schwarz inequality,

$$\|u_n\|_{H^1(\Omega_n)} \leq \|f\|_{L^2(\Omega_n)} \leq \|f\|_{L^2(B)}. \tag{3.81}$$

We will denote by $\overline{u}_n$, $\overline{\nabla} u_n$ the extensions by 0 to $B \backslash \Omega_n$ of u_n, ∇u_n (this notation is somehow ambiguous since the extension operator involved here depends on n, but it will be enough for us here). Thus

$$\overline{u}_n, \overline{\nabla} u_n \text{ are bounded in } L^2(B). \tag{3.82}$$

In particular, a subsequence of $\overline{u}_n$ converges weakly in $L^2(B)$ to $u^\star \in L^2(B)$. If continuity holds, then $u^\star_{|\Omega} = u_\Omega$. If moreover $f \equiv 1$, we check by using (3.79) that $u_n \equiv 1$ on Ω_n and $u \equiv 1$ on Ω, that is, $\overline{u}_n \equiv \chi_{\Omega_n}$, $\overline{u} \equiv \chi_\Omega$. This implies that the sequence of characteristic functions χ_{Ω_n} converges weakly in $L^2(B)$ to χ_Ω and therefore strongly as well (see Chapter 2). Consequently we have the following necessary condition for the continuity.

Proposition 3.7.1. *Assume that, for all* $f \in L^2(B)$, $\overline{u}_{\Omega_n}$ *converges weakly in* $L^2(B)$ *to* $\overline{u}_\Omega$. *Then* χ_{Ω_n} *converges strongly in* $L^2(B)$ *to* χ_Ω.

Therefore it is natural to assume at least that the open sets Ω_n converge to Ω in the sense of characteristic functions. We first follow the analysis made in [95].

In the previous section devoted to homogeneous Dirichlet conditions, we worked with the Sobolev space $H_0^1(\Omega)$. No matter the regularity of Ω, $H_0^1(\Omega)$ can be considered a subspace of $H_0^1(B)$ by extending all functions by 0 outside Ω.

In the case of Neumann conditions, the situation is far from being so simple. Indeed, if the open set Ω is not regular, $H^1(\Omega)$ functions may not have any extension in $H^1(B)$. Let us for instance take $\Omega = (1, 0) \cup (0, 1)$ in dimension $N = 1$: A function in $H^1(\Omega)$ such that $u(0^-) \neq u(0^+)$ cannot be the restriction to Ω of a function of $H^1(-1, 1)$, which has to be continuous at 0. (See also Exercises 3.11 and 3.12.)

But to be able to speak of the convergence of u_{Ω_n} to u_Ω, it is highly useful to have such an extension. This is why we start by working with open sets having at least such regularity. We will moreover need that the extension operator be uniform in some sense for all open sets involved.

More precisely, we will consider the class $\mathcal{S}_k$ of open sets defined as follows for each $k \in (0, \infty)$:

$$\begin{cases} \text{for all } \Omega \in \mathcal{S}_k, \text{ there exists a linear continuous extension operator} \\ P_\Omega \text{ of } H^1(\Omega) \text{ into } H^1(B) \text{ with } \|P_\Omega\|_{H^1(\Omega)\to H^1(B)} \leq k. \end{cases} \tag{3.83}$$

A sufficient condition to have such an extension is, for instance, the ε-cone property (see Chapter 2).

Proposition 3.7.2 (Chenais). *Let* $\epsilon > 0$ *and*

$$\mathcal{O}_\epsilon = \{\Omega \subset B,\ \Omega \text{ open with the } \varepsilon\text{-cone property}\}.$$

Then there exists $k \in (0, \infty)$ *such that* $\mathcal{O}_\epsilon \subset \mathcal{S}_k$.

3.7.2 A main convergence result

Theorem 3.7.3. *Let* $(\Omega_n)_n$ *be a sequence of open sets of the class* $\mathcal{S}_k$ *converging in the sense of characteristic functions to an open subset* $\Omega \in \mathcal{S}_k$. *Let us denote by* u_n, u *the solutions of the homogeneous Neumann problem* (3.79) *respectively on* Ω_n, Ω *and let* $\widehat{u}_n = P_{\Omega_n}(u_n)$ *be the extension of* u_n *to* B *according to* (3.83).

Then $\widehat{u}_n$ *converges strongly in* $L^2(B)$ *and weakly in* $H^1(B)$ *to* $u^\star \in H^1(B)$ *such that* $u^\star_{|\Omega} = u$. *Moreover,* $\nabla\hat{u}_n\chi_{\Omega_n}$ *converges strongly in* $L^2(B)^N$ *to* $\nabla u^\star\chi_\Omega$.

Thanks to Proposition 3.7.2 (see also Theorem 2.4.10), the following corollary is an immediate consequence of the theorem.

Corollary 3.7.4. *Let $(\Omega_n)_n$ be a sequence of open sets in the class $\mathcal{O}_\epsilon$ (see Proposition 3.7.2) converging in the sense of Hausdorff (or in the sense of characteristic functions) to some open subset Ω of B. Let u_n, u, $\widehat{u}_n$ be defined as in Theorem 3.7.3. Then $\widehat{u}_n$ converges to u in the sense of Theorem 3.7.3.*

Proof of Theorem 3.7.3. By (3.81) and since $\Omega_n \in \mathcal{S}_k$, the sequence $\widehat{u}_n$, satisfies

$$\|\widehat{u}_n\|_{H^1(B)} \leq \|P_{\Omega_n}\| \, \|u_n\|_{H^1(\Omega_n)} \leq k\|f\|_{L^2(B)}.$$

This shows that the sequence $\widehat{u}_n$ is bounded in $H^1(B)$. Up to a subsequence, it may be assumed that $\widehat{u}_n$ converges weakly in $H^1(B)$ and strongly in $L^2(B)$ to a function $u^\star \in H^1(B)$. Let us show that $u^\star_{|\Omega}$ satisfies the variational formulation (3.79) on Ω.

By definition of u_n, for all $v \in H^1(B)$,

$$\int_{\Omega_n} \nabla u_n \nabla v + \int_{\Omega_n} u_n v = \int_{\Omega_n} f v. \tag{3.84}$$

Introducing the characteristic functions $\chi_n := \chi_{\Omega_n}$ and $\chi := \chi_\Omega$, this may also be written

$$\int_B \chi_n \nabla \widehat{u}_n \nabla v + \int_B \chi_n \widehat{u}_n v = \int_B \chi_n f v. \tag{3.85}$$

Since by assumption χ_n converges to χ in $L^1(B)$ and a.e. (see Chapter 2), we have

$$\chi_n f v \to \chi f v \text{ in } L^1(B), \quad \chi_n v \to \chi v, \quad \text{and} \quad \chi_n \nabla v \to \chi \nabla v \text{ in } L^2(B).$$

Using also the weak convergence in $H^1(B)$ of $\widehat{u}_n$, one may pass to the limit in (3.85) and obtain

$$\int_B \chi \nabla u^\star \nabla v + \int_B \chi u^\star v = \int_B \chi f v,$$

or also

$$\int_\Omega \nabla u^\star . \nabla v + \int_\Omega u^\star v = \int_\Omega f v \quad \forall \, v \text{ in } H^1(B). \tag{3.86}$$

This is also valid for all $v \in H^1(\Omega)$ thanks to the extension property of $\Omega \in \mathcal{S}_k$. This proves that $u^\star_{|\Omega} = u$.

Since this proof is valid for any subsequence of $(\widehat{u}_n)$, it follows that the whole sequence $\widehat{u}_n$ has the same property.

In fact $\chi_n \nabla \hat{u}_n$ converges *strongly* in $L^2(B)$ to $\chi \nabla u^\star$. Indeed, taking $v = \hat{u}_n$ in (3.84) and $v = u^\star$ in (3.86), we obtain

$$\lim \int_{\Omega_n} |\nabla u_n|^2 + u_n^2 = \lim \int_B \chi_n f \hat{u}_n = \int_B \chi f u^\star = \int_\Omega |\nabla u^\star|^2 + u^{\star 2}.$$

It follows that the convergence is strong in $L^2(B)$ for $\chi_n \nabla \hat{u}_n$. □

Remark 3.7.5. One may obtain strong convergence in $H^1(B)$ of $\hat{u}_n$ itself in some cases. Indeed we have

$$\|P_{\Omega_n}(u_n) - P_{\Omega_n}(\chi_n u^\star)\|_{H^1(B)} \le k\|u_n - \chi_n u^\star\|_{H^1(\Omega_n)}.$$

Since the right-hand side tends to 0, strong convergence holds for $\hat{u}_n$ if the extension operators P_{Ω_n} satisfy the compatibility condition that $P_{\Omega_n}(\chi_n u^\star)$ converges strongly to $u^\star$.

3.7.3 More convergence results

The convergence of Ω_n to Ω in the sense of characteristic functions does not imply the convergence of u_{Ω_n} to u_Ω as one can see in simple examples.

Ideas for such examples may be found in dimension 2 in [262]. Let us give one example, even in dimension 1, that relies on the same ideas. Let

$$\Omega = (0,2), \quad a_n = 1 - 1/n, \quad \Omega_n = (0,a_n) \cup (1,2), \quad f(x) = \chi_{(0,1)} + 2\chi_{(1,2)}.$$

Then Ω_n converges in the sense of characteristic functions to Ω, but

$$u_n = \chi_{(0,a_n)} + 2\chi_{(1,2)} \longrightarrow u^\star = \chi_{(0,1)} + 2\chi_{(1,2)},$$

which is not equal to u_Ω since it does even not belong to $H^1(\Omega)$ because of its discontinuity at the point 1.

Let us explain this example with respect to the result of Theorem 3.7.3. Note that there do exist continuous linear extension operators from $H^1(\Omega_n)$ to $H^1(\Omega)$ since Ω_n is quite regular. But their norms blow up with n (we can easily check that $P_{\Omega_n}(u_n)$ cannot stay bounded in $H^1(\Omega)$, no matter the choice of the linear extension P_{Ω_n}; see Exercise 3.11).

In this elementary example, we see that the relevant limit for the sequence Ω_n is $\hat{\Omega} = (0,1) \cup (1,2)$, which is in fact the limit in the sense of Hausdorff. We indeed have $u^\star = u_{\hat{\Omega}}$. We could think that the situation is better for convergence in the sense of Hausdorff, but it is absolutely not the case as shown by the following example.

Let R be the rectangle $(-1,1)\times(0,1)$ and let $K_n = \cup_{k=1}^{n-1}[-1,0]\times\{k/n\}$. Consider the sequence of open sets $\Omega_n := R\backslash K_n$. It is easy to check that the open sets Ω_n converge in the sense of Hausdorff to the unit square $\Omega = (0,1)\times(0,1)$. Let us then consider the function $u(x,y) = \sin(\frac{\pi x}{2})$. This function is a solution of the problem

$$-\Delta u(x,y) + u(x,y) = \left(1 + \frac{\pi^2}{4}\right)\sin\left(\frac{\pi x}{2}\right) := f(x,y)$$

on all open sets Ω_n with homogeneous Neumann conditions on $\partial\Omega_n$. The function $u_n = u$ does not depend on n and therefore converges to u that is a solution on $\partial\Omega$ of

$$\frac{\partial u}{\partial n}(0,y) = -\frac{\partial u}{\partial x}(0,y) = -\frac{\pi}{2},$$

which is not the limit that we expected.

The above open sets Ω_n are such that their perimeter tends to ∞ with n. This partly explains the strange behavior of the solution of the corresponding Neumann problem. One can however slightly modify this example in such a way that the perimeter of the Ω_n stays bounded. For this, it is sufficient to replace K_n by $\hat{K}_n$ where, in the definition of K_n, we replace each segment $[-1, 0] \times \{k/n\}$ by the union of small segments $\cup_i [i2^{-n} - \epsilon_n, i2^{-n} + \epsilon_n] \times \{k/n\}$, $i = 0, 2^{-n}$, where ϵ_n is small enough so that $n2^n \epsilon_n$ tends to 0. We then have $u_{\Omega_n} = u_{\hat{\Omega}_n}$ and $\hat{\Omega}_n = R \backslash \hat{K}_n$ still converges in the sense of Hausdorff to Ω. But $P(\partial \hat{\Omega}_n)$ is now bounded. However this change creates new connected components in $\partial \hat{\Omega}_n$ whose number tends to ∞.

In fact, by adding constraints *at the same time* on the perimeter and on the number of connected components of $\partial \Omega_n$, we can obtain a convergence result as in [92]. We state it without proof. Since it is easier to state it in terms of one-dimensional Hausdorff measure rather than in terms of perimeter, let us recall its definition.

Let Ω be a bounded open subset of $\mathbb{R}^2$. For all $\varepsilon > 0$, let us consider all coverings of $\partial \Omega$ by balls of radius $r_i \le \epsilon$. Let us denote

$$\sigma(\varepsilon) = \inf \left\{ \sum_i r_i \right\}, \tag{3.87}$$

where the infimum is taken over all possible coverings.

Definition 3.7.6. We call monodimensional Hausdorff measure of $\partial \Omega$, denoted by $\mathcal{H}^1(\partial \Omega)$, the limit as ε tends to 0 of $\sigma(\varepsilon)$ where $\sigma(\varepsilon)$ is defined in (3.87):

$$\mathcal{H}^1(\partial \Omega) = \lim_{\varepsilon \to 0} \sigma(\varepsilon).$$

Note that this limit (finite or infinite) always exists by monotonicity of $\epsilon \to \sigma(\varepsilon)$.

There exist very closed but different definitions of Hausdorff measures where the coverings use any open sets rather than only balls (see, e.g., [135]). These different definitions, and others as well, are widely compared in [137].

Let us now state the convergence result for the Neumann problem proved in [92].

Theorem 3.7.7 (Chambolle–Doveri). *Let $\Omega_n \subset B$ be a sequence of planar open sets that converge in the sense of Hausdorff to an open set Ω. Assume that*

$$\sup_n \mathcal{H}^1(\partial \Omega_n) < +\infty$$

and that the number of connected components of $\partial \Omega_n$ is uniformly bounded. Denote by u_n (resp. u) the solution of the Neumann problem (3.79) *on Ω_n (resp. Ω). Then $(\overline{u}_n, \overline{\nabla} u_n)$ converges strongly to $(\overline{u}, \overline{\nabla} u)$ in $L^2(B)^3$.*

Remark 3.7.8. We saw on our previous example that assuming bounded perimeter is not enough for continuity. The same example shows also that the only assumption, of boundedness of the number of connected components of $\partial\Omega_n$, is not sufficient either. Indeed, we can always assume that K_n is connected in this example by adding $\{-1\} \times [0, 1]$ to it.

To end this discussion, we now state, without proof again, a result proved in [73] that can be seen in some sense as a generalization of the previous one (see the discussion in [73]) and that is also in the spirit of Šverak's result [285] (see also [74] for an extension to the operator $u \to -\Delta u + a(x)u$).

Let us recall that $\mathcal{O}_l$ denotes the family of open subsets of B whose complement has at most l connected components. The following holds.

Theorem 3.7.9 (Bucur–Varchon). *Let $(\Omega_n)_n$ be a sequence of* planar *open subsets of the class $\mathcal{O}_l$. Assume that Ω_n converges to Ω in the sense of Hausdorff. Denote by u_n (resp. u) the solution of the Neumann problem* (3.79) *on Ω_n (resp. Ω). Then $(\overline{u}_n, \overline{\nabla u_n})$ converges strongly to $(\overline{u}, \overline{\nabla u})$ in $L^2(B)^3$ if and only if $|\Omega_n|$ tends to $|\Omega|$.*

It is not true that mixed convergence in the sense of Hausdorff *and* in the sense of characteristic functions implies the convergence of u_{Ω_n} to u_Ω. This may be seen by looking at the following counterexample.

Let $\mathcal{O} = (-1, 1) \times (0, 1)$. Let Ω_n be the open set obtained from $\mathcal{O}$ by taking off the points $\{0\} \times \{k/n\}$, $k = 0, \dots, n$. Then Ω_n converges in the sense of Hausdorff and of the characteristic functions to $\Omega = \big((-1, 0) \times (0, 1)\big) \cup \big((0, 1) \times (0, 1)\big)$. But $u_{\Omega_n} = u_{\mathcal{O}}$ is independent of n and u_Ω has a normal derivative that vanishes on the vertical segment $\{0\} \times (0, 1)$, while it is not the case in general for $u_{\mathcal{O}}$. Thus u_Ω is not the limit of the u_{Ω_n}.

One may however prove that, if this mixed convergence holds and if moreover the open limit set is regular enough so that any $H^1(\Omega)$ function has an $H^1(D)$-extension, then the convergence of u_{Ω_n} to u_Ω holds (see Exercise 3.13).

3.7.4 γ-convergence and the Neumann problem

As we did in the case of the homogeneous Dirichlet condition, we may look at the continuity question for the Neumann problem in terms of γ-convergence. But is there equivalence between γ-convergence and Mosco convergence of the Sobolev spaces $H^1(\Omega_n)$ to $H^1(\Omega)$?

In fact, the question is not as easy as for the spaces $H_0^1(\Omega_n)$ (see, e.g., [107], [241]). First, what is the definition of γ-convergence in the context of the Neumann problem? As a consequence of our analysis in the previous section, it seems there are (at least) two different possible definitions. The first one (see D. Chenais' approach [95]) consists in using extension operators. In the other one (see Chambolle–Doveri

[92] or Bucur–Varchon [73]), all functions and their first derivatives are extended by 0 outside Ω_n and Ω. This approach consists in identifying the space $H^1(\Omega)$, where $\Omega \subset B$, to the space

$$\mathcal{V}(\Omega) = \Big\{(v_0, v_1, v_2, \dots, v_N) \in L^2(B)^{N+1};\ \forall\, i = 0, \dots, N,\ v_i = 0 \text{ outside } \Omega \\ \text{and } v_{j\,|\Omega} = \tfrac{\partial v_0}{\partial x_j} \text{ in } \Omega\ \forall\, j = 1, \dots, N\Big\},$$

where this space is equipped with the natural norm associated with $L^2(B)^{N+1}$. It may be identified to $H^1(\Omega)$ by the mapping

$$u \in H^1(\Omega) \mapsto (\overline{u}, \overline{\nabla u}) \in \mathcal{V}(\Omega),$$

where $(\overline{u}, \overline{\nabla u})$ is the extension by 0 on $B\backslash\Omega$ of $(u, \nabla u)$.

One main interest of this definition is that it allows us to look at all spaces $\mathcal{V}(\Omega)$ as closed subspaces of $L^2(B)^{N+1}$ and therefore to deal with their Mosco convergence. This was not the case with the Sobolev spaces $H^1(\Omega)$, which are not canonically embedded in the same normed vector space independent of Ω.

We can then state the following: *Assume that $\mathcal{V}(\Omega_n)$ Mosco-converges to $\mathcal{V}(\Omega)$. Let u_n (resp. u) be the solution of the Neumann problem* (3.79) *on Ω_n (resp. Ω). Then $(\overline{u}_n, \overline{\nabla u}_n)$ converges to $(\overline{u}, \overline{\nabla u})$ in $L^2(B)^{N+1}$.*

The proof of this claim is quite similar to the proof of Proposition 3.5.5 (see below for the details). The converse is also essentially true except that the linear forms to be considered should not be only like the $L^2(B)$ functions of (3.79). A correct viewpoint is to consider the family of Neumann problems on $H^1(\Omega_n)$ associated with a linear continuous form L on $L^2(B)^{N+1}$ defined by

$$u_n \in H^1(\Omega_n),\ \forall\, v \in H^1(\Omega_n),\quad \int_{\Omega_n} u_n v + \nabla u_n \nabla v = \overline{L}(v), \tag{3.88}$$

where $\overline{L}(v) = L(\overline{v}, \overline{\nabla v})$. We immediately check that $\overline{L}$ is indeed a linear continuous form on $H^1(\Omega_n)$ so that u_n does exist and is unique by the Lax–Milgram theorem. The problems considered in (3.79) are of this type with $\overline{L}(v) = \int_{\Omega_n} fv$, $f \in L^2(B)$. We then have the following necessary and sufficient condition.

Proposition 3.7.10. *The sequence $\mathcal{V}(\Omega_n)$ Mosco-converges to $\mathcal{V}(\Omega)$ if and only if, for any linear continuous form L on $L^2(B)^{N+1}$, the extension $(\overline{u}_n, \overline{\nabla u}_n)$ of the solution u_n of* (3.88) *converges strongly in $L^2(B)^{N+1}$ to the extension $(\overline{u}, \overline{\nabla u})$ of the corresponding solution on Ω.*

Proof. Let us denote $H = L^2(B)^{N+1}$.

Necessary condition: We assume Mosco convergence. Let L be a continuous linear form on H and let u_n be the solution of (3.88). Its norm in $H^1(\Omega_n)$ and therefore in $\mathcal{V}(\Omega_n)$ is bounded by the norm of $\overline{L}$, itself being bounded by the norm of L (see (3.81)). Thus there exists a subsequence u_{n_k} that converges weakly in H to a function $u^\star \in H$. But by (M2), we have $u^\star \in \mathcal{V}(\Omega) = H^1(\Omega)$.

Now let φ be a test function in $\mathcal{V}(\Omega)$. By (M1), there exists a sequence φ_n of functions in $\mathcal{V}(\Omega_n)$ that converges to φ in H. We have

$$\int_{\Omega_n} u_n \varphi_n + \nabla u_n . \nabla \varphi_n \, dx = \overline{L}(\varphi_n) = L(\overline{\varphi}_n, \overline{\nabla} \varphi_n). \tag{3.89}$$

Passing to the limit in the above equality (with strong limit for φ_n and weak limit for u_n), we obtain that $u^\star$ is solution of the expected Neumann problem on Ω. As usual, the convergence of the whole sequence follows as well as the strong convergence (replace φ_n by u_n in (3.89) and use that $\overline{L}(u_n) \to \overline{L}(u^*)$).

Sufficient condition: Here we assume the convergence of $(\overline{u}_n, \overline{\nabla} u_n)$ in H. Let us first prove that (M1) holds. Let $\varphi \in \mathcal{V}(\Omega)$. Consider the continuous linear form defined on H by

$$\forall\, v \in H, \quad L(v) = \int_\Omega \varphi v_0 + \sum_{1 \le i \le N} \frac{\partial \varphi}{\partial x_i} v_i .$$

By construction of L, φ is a solution of the Neumann problem (3.88) with Ω_n replaced by Ω. Let us introduce the solution φ_n of problem (3.88). By assumption, φ_n converges in H to the solution of the same problem on Ω, which is exactly the latter function φ. This proves (M1) since $\varphi_n \in \mathcal{V}(\Omega_n)$.

Let us now prove (M2). Let φ^k be a sequence of functions in $\mathcal{V}(\Omega_{n_k})$ that converges weakly in H to some function $\varphi = (\varphi_i)_{0 \le i \le N} \in H$. The goal is to prove that $\varphi \in \mathcal{V}(\Omega)$. Consider the continuous linear form L defined on H by

$$\forall\, v \in H, \quad L(v) = \int_B \varphi_0 v_0 + \sum_{1 \le i \le N} \varphi_i v_i .$$

Introduce the corresponding solution u_k of (3.88) on Ω_{n_k}. By hypothesis, $(\overline{u}_k, \overline{\nabla} u_k)$ converges strongly in H to the extension $(\overline{u}, \overline{\nabla} u)$ of the solution u associated with L on Ω. Let us prove that $u = \varphi$. Passing to the limit in

$$\int_{\Omega_{n_k}} u_k(u_k - \varphi_k) + \nabla u_k \nabla (u_k - \varphi_k) = \overline{L}(u_k - \varphi_k),$$

where

$$\overline{L}(u_k - \varphi_k) = \int_{\Omega_{n_k}} \varphi_0 (u_k - \varphi_k) + \sum_{1 \le i \le N} \varphi_i \frac{\partial (u_k - \varphi_k)}{\partial x_i},$$

we obtain (where we denote by $\overline{\frac{\partial u}{\partial x_i}}$ the extension by 0 on $B\backslash\Omega$ of $\partial u/\partial x_i$)

$$\int_B \overline{u}(\overline{u}-\varphi_0)+\sum_{1\le i\le N}\overline{\frac{\partial u}{\partial x_i}}\left[\overline{\frac{\partial u}{\partial x_i}}-\varphi_i\right]=\int_B \varphi_0(\overline{u}-\varphi_0)+\sum_{1\le i\le N}\varphi_i\left[\overline{\frac{\partial u}{\partial x_i}}-\varphi_i\right].$$

By difference, we obtain $\overline{u}=\varphi_0$, $\overline{\nabla}u=(\varphi_1,\dots,\varphi_N)$. □

3.8 The bi-Laplacian operator

We saw in the previous Section 3.5.3 how to use Mosco convergence to analyze continuity with respect to domains of second-order elliptic operators of Laplacian type, where the underlying Sobolev space was H^1.

In this section, we will give some hints on studying the same questions for the bi-Laplacian operator, which is a model example of operators of order 4 and for which the underlying Sobolev space is H^2 instead of H^1. There are many similarities with the Laplacian case. We will mainly emphasize the differences.

We will look at the fourth-order elliptic problem

$$\begin{cases}\Delta^2 u=\Delta(\Delta u)=f & \text{in } \Omega,\\ u=0 & \text{on } \partial\Omega,\\ \frac{\partial u}{\partial n}=0 & \text{on } \partial\Omega,\end{cases} \tag{3.90}$$

where Ω is a bounded open subset of $\mathbb{R}^N$ and $f\in H^{-2}(\mathbb{R}^N)$. The variational formulation of (3.90) is given by

$$\begin{cases}u\in H_0^2(\Omega)\ \forall\, v\in H_0^2(\Omega),\\ \int_\Omega \Delta u\Delta v\,dx=\int_\Omega fv\,dx.\end{cases} \tag{3.91}$$

By the Lax–Milgram theorem, this problem has a solution that is unique. We will essentially follow the same pattern as for the Laplacian. We will emphasize mainly the differences, in particular in dimensions $N=2, 3$ where H_0^2 can be embedded into C^0, without, however, being embedded into C^1. This leads to a new situation compared with the Laplacian case. For different questions or for more details on those that will be discussed here, we refer to [214], [229], [171], [175].

3.8.1 H^2-capacity

As for second-order operators, we define the capacity associated with this new "fourth-order" context, which is here the H^2-capacity. We only state the main results; for more details see, for example, [170], [232], [4].

In all of this section, D denotes a *fixed* regular bounded open subset of $\mathbb{R}^N$.
As in the H^1-case, we first define the capacity of compact sets.

Definition 3.8.1. For any compact subset K of D, we define

$$\operatorname{cap}_2(K) = \inf\Big\{ \int_D |\Delta v|^2\,dx;\ v \in C_0^\infty(D),\ v \ge 1 \text{ on } K\Big\}. \tag{3.92}$$

Next we define the capacity of open sets, then of any subset.

Definition 3.8.2. Let $\omega \subset D$ be open. We define

$$\operatorname{cap}_2(\omega) := \sup\{\operatorname{cap}(K);\ K \text{ compact},\ K \subset \omega\}. \tag{3.93}$$

Let E be any subset of D. We define

$$\operatorname{cap}(E) := \inf\{\operatorname{cap}(\omega);\ \omega \text{ open},\ E \subset \omega\}. \tag{3.94}$$

Let us recall some Sobolev embeddings (see, e.g., [151]). *Assume $N \le 3$.*

- *Then $H_0^2(D)$ can be compactly embedded into the space $C^0(\overline{D})$ of continuous functions on $\overline{D}$ equipped with the uniform norm.*
- *If $N = 1$, then $H_0^2(D)$ can be compactly embedded into the space $C^1(\overline{D})$ of continuously differentiable functions on $\overline{D}$ equipped with its natural norm.*

Thus, in dimension $N \le 3$, the capacity of a point is strictly positive. It is equal to zero when $N \ge 4$. More generally, a regular manifold of dimension d is of zero capacity if and only if $d \le N - 4$.

Most properties of H^1-capacity carry over to this H^2-capacity. It is the case for points (1), (2), (3) of Proposition 3.3.9. One main difference is that *the space H^2 is not stable under truncations*. In other words, $[u \in H^2]$ does not imply $[\inf\{u, 1\} \in H^2]$. Thus point (4) of Proposition 3.3.9 is not valid, but subadditivity nevertheless holds (see, e.g., [4]). It is not true either that the capacitary potential is less than or equal to 1. However, it can be shown to be bounded, which is enough for many properties (again, see [4]).

The notion of quasi-continuous functions can be defined in the same way as in Definition 3.3.28 and Theorems 3.3.29 and 3.3.33 are valid.

There is also a characterization of H_0^2 similar to the one given in Theorem 3.3.42 for H_0^1.

Theorem 3.8.3. *Let ω be an open subset of D and let $u \in H_0^2(D)$. Then*

$$u \in H_0^2(\omega) \iff \begin{cases} \tilde{u} = 0 \ H^2\text{-q.e. on } \omega^c \ (= D\setminus\omega), \\ \widetilde{\nabla u} = 0 \ H^1\text{-q.e. on } \omega^c, \end{cases} \tag{3.95}$$

where $\tilde{u}$ is the H^2-quasi-continuous representative of u and $\widetilde{\nabla u}$ is the H^1-quasi-continuous representative of ∇u. If $N \leq 3$, $\tilde{u}$ is continuous and the following holds:

$$(\tilde{u} = 0 \ H^2\text{-q.e. on } \omega^c) \Leftrightarrow (\tilde{u} = 0 \text{ everywhere on } \omega^c).$$

3.8.2 Continuity with respect to domains

Here we assume that $N = 2$ or $N = 3$.

This is a new intermediate situation that did not appear in the context of H^1. Indeed, in dimension $N = 1$, everything is easy thanks to the compact embedding of H_0^2 into $\mathcal{C}^1$ (this is discussed in Exercise 3.14). In dimension $N \geq 4$, the situation is similar to the Laplacian case and we will not analyze it further here. In dimension $N = 2, 3$, since H_0^2 is compactly embedded into $\mathcal{C}^0$, an H^2-bound implies compactness for uniform convergence and therefore good stability of the supports of the solutions of (3.91) on variable open sets.

Let us consider a sequence of open subsets Ω_n of D that converges in the sense of Hausdorff to an open subset $\Omega \subset D$. Let us denote by u_n the solution of problem (3.90) or (3.91) set on Ω_n and by u the solution of the problem on Ω. Like in Proposition 3.2.1, we prove

Proposition 3.8.4. *There exists $u^\star$ in $H_0^2(D)$ and a subsequence u_{n_k} such that $u_{n_k} \rightharpoonup u^\star$ in $H_0^2(D)$ (weak convergence) and u_{n_k} converges uniformly to $u^\star$ in D. Moreover, $u^\star$ satisfies*

$$\forall\, v \in H_0^2(\Omega), \qquad \int_\Omega \Delta u^\star \Delta v = \langle f, v \rangle_{H^{-2}(\Omega)\times H_0^2(\Omega)}. \tag{3.96}$$

The uniform convergence of u_{n_k} to $u^\star$ obviously follows from the compact embedding of H_0^2 into $\mathcal{C}^0$ in dimension 2 or 3.

As in (3.32), the main question is the following: Under which condition can one claim that $u^\star \in H_0^2(\Omega)$, that is, $u^\star = u$? If it is the case, we will have, exactly as in Proposition 3.2.4, that the whole sequence u_n converges and that the convergence is strong in $H_0^2(D)$.

Thanks to the uniform convergence of u_{n_k}, it is easy to see that the limit $u^\star$ belongs to the space

$$V_0(\Omega) = \left\{ v \in H_0^2(D);\ \tilde{v} = 0 \text{ everywhere outside } \Omega \right\}.$$

In fact, we have the more precise following result (cf. [175]), where u_n, u are as above.

Proposition 3.8.5. *Let Ω be an open subset of D. Then u_n converges to u in $H_0^2(D)$ for all $f \in H^{-2}(D)$ and for any sequence of open sets Ω_n converging in the sense of Hausdorff to Ω if and only if*

$$V_0(\Omega) = H_0^2(\Omega). \tag{3.97}$$

Remark 3.8.6. It is interesting to notice that the fact that the expected convergence holds depends only on the limit Ω and not on the Ω_n. Obviously, condition (3.97) contains some kind of regularity for Ω. This is discussed in the next section.

Proof of Proposition 3.8.5. Assume (3.97) holds. It is sufficient to prove that the function $u^\star$ of Proposition 3.8.4 satisfies $u^\star \in V_0(\Omega)$. Let $x \in \Omega^c$. By the properties of convergence in the sense of Hausdorff, there exists a sequence $x_n \in \Omega_n^c$ converging to x. Since $\tilde{u}_{n_k}(x_{n_k}) = 0$ and since the convergence of $\tilde{u}_{n_k}$ to $\tilde{u}^\star$ holds uniformly, it follows that $\tilde{u}^\star(x) = 0$. Thus $u^\star \in V_0(\Omega)$.

Conversely, assume there exists $w \in V_0(\Omega)\setminus H_0^2(\Omega)$. Let $\Omega_n = D\setminus\Gamma_n$, where Γ_n denotes the set of the first n points of a dense sequence in $D\setminus\Omega$. Thus $\Omega \subset \Omega_n$ and

$$d^{\mathrm{H}}(\Omega, \Omega_n) = \max\left\{d(x, \overline{D}\setminus\Omega_n);\ x \in \overline{D}\setminus\Omega\right\} = \sup\left\{d(x, \Gamma_n);\ x \in D\setminus\Omega\right\}$$

converges to 0 as n tends to ∞.

If now $f = \Delta^2 w \in H^{-2}(D)$, we have $u_n = w$. Indeed, by restriction to Ω_n, we have $\Delta^2 w = f$ on Ω_n and also $w \in H_0^2(\Omega_n)$ since, by hypothesis, $\tilde{w} = 0$ everywhere outside Ω and therefore everywhere outside Ω_n. On the other hand, since Γ_n is of zero H^1-capacity, $\widetilde{\nabla w} = 0$ H^1-q.e. outside Ω_n.

In summary, the open sets Ω_n converge to Ω in the sense of Hausdorff and $u_n = w$ does not converge to u since w does not even belong to $H_0^2(\Omega)$ (by assumption). Whence a contradiction with the existence of such a w. □

One may easily check that open sets with cracks do not generally satisfy condition (3.97). We may think of a (two-dimensional) square where a segment has been removed or of a (three-dimensional) cube where a plane square has been removed. It is interesting to notice that property (3.97) is very much connected to H^1-stability that was discussed earlier. Indeed, we have the following result, which essentially says that H^1-stability guarantees H^2-stability.

Proposition 3.8.7. *Let Ω be an open set satisfying one of the equivalent properties of Theorem 3.4.6 (i.e., Ω is H^1-stable). Then Ω satisfies* (3.97).

Proof. Let $v \in V_0(\Omega)$. Let us show that ∇v vanishes H^1-quasi-everywhere on $D\setminus\Omega$. As v is identically zero on the open set $D\setminus\overline{\Omega}$, so is its gradient. But since Ω is H^1-stable,

$$\nabla v = 0 \text{ on } D\setminus\overline{\Omega} \implies \nabla v = 0\ H^1\text{-q.e. on } D\setminus\Omega.$$

□

We refer to [171], [175] for more necessary or sufficient stability conditions. One can find, for instance, a proof that the equality $H_0^1(\Omega) \cap H_0^2(D) = H_0^2(\Omega)$ is sufficient for (3.97) or also that it is sufficient that the set of "singular points" of $\partial\Omega$ be of H^1-capacity zero.

3.9 Exercises

Exercise 3.1. Let D be any open subset of $\mathbb{R}^N$ and let $u \in H_0^1(D)$, $\varphi \in \mathcal{C}^\infty(\mathbb{R}^N, \mathbb{R}^N)$. Prove that

$$\int_D \varphi \cdot \nabla(u^2) = -\int_D u^2 \operatorname{div} \varphi.$$

Deduce that any function of $H_0^1(D)$ that is constant on connected components of D is identically zero. (Hint: choose $\varphi(x) = x$.)

Exercise 3.2. Show that property (2) of Proposition 3.3.9 is not valid for any set.

Exercise 3.3. Let $S \subset B(0, 1/2) \subset \mathbb{R}^2$ be a segment of length a.

(1) If $D_1 = B(0, 1) \subset \mathbb{R}^2$, prove that there exists $k \in (0, \infty)$ such that $\operatorname{cap}_{D_1}(S) \leq -k/\log a$. (Hint: bound the capacity of S from above by the capacity of a square, then by the smallest disk around this square.)

(2) Let $K \subset S$ be a compact set of positive one-dimensional Hausdorff measure. Show that $\operatorname{cap}_{D_1}(K) > 0$. (Hint: prove and use that $\exists C > 0$ such that for all $u \in \mathcal{C}_0^\infty(D_1)$, $\int_S u^2 \leq C \int_{D_1} |\nabla u|^2$.)

Exercise 3.4. Let $u_n(x, y, z) = v_n(x, y)w(z)$, where $w \in \mathcal{C}_0^\infty(-2, 2)$, $w = 1$ on $(-1, 1)$, and where v_n is the capacitary potential of the disk $D_n = B(0, 1/n) \subset \mathbb{R}^2$. Prove that ∇u_n tends to 0 in $L^2(D_n \times (-2, 2))$. Deduce that the capacity of a segment is zero in dimension 3. Give a generalization of this property in dimension N.

Exercise 3.5. Let D be a bounded open subset of $\mathbb{R}^N$ and let $v \in \mathcal{C}_0^\infty(D)$.

(1) Show that $\int_D |\nabla v|^2 \leq \int_D |v||\Delta v| \leq \|v\|_\infty \int_D |\Delta v|$.

(2) Let $u \in \mathcal{C}^\infty(D)$ and $\varphi \in \mathcal{C}_0^\infty(D)$ with $0 \leq \varphi \leq 1$. Prove

$$\int_D \varphi |\nabla u|^2 \leq \|u\|_\infty \left[\int_{\text{Support}\,\varphi} |\Delta u| + \|u\|_\infty \int_D |\Delta \varphi|/2 \right].$$

(3) Deduce that a superharmonic function (i.e., $-\Delta u \geq 0$) that is also bounded on D belongs to $H^1_{\text{loc}}(D)$.

Exercise 3.6. Let $(x_n)_n$ be a dense sequence in the unit ball B_1 of $\mathbb{R}^3$ and let α_n be the generic term of a convergent series of nonnegative real numbers. Set

$$u(x) = \sum_{n=1}^{\infty} \alpha_n \frac{1}{|x - x_n|}$$

and $v = \inf(1, u)$. Check that $u \in L^1(B_1)$ and that u and v are superharmonic. Show that $v \in H^1(B_1)$ (one may use Exercise 3.5). Deduce that the set $A = \{x \in B_1,\ v(x) < 1\}$ is a quasi-open set that is not equal a.e. to an open set (assume that $\sum \alpha_n$ is small enough so that A is not empty).

Exercise 3.7. Let $D_2 = B(0,1) \subset \mathbb{R}^2$ and let D_1 be the set obtained by removing from D_2 the closed segment joining $(0,0)$ to $(1,0)$. Consider $\omega_n = B(0,1/n) \cap D_1$.

(1) Show that $\mathrm{cap}_{D_2}(\omega_n)$ tends to 0 as $n \to \infty$ while $\mathrm{cap}_{D_1}(\omega_n) = +\infty$.

(2) Check that the radial function $r \to [-\log(r)]^{1/4}$ belongs to $H_0^1(D_2)$, but is not "$H_0^1(D_1)$-quasi-continuous" in the sense of Remark 3.3.31.

Exercise 3.8. The goal of this exercise is to prove Proposition 3.2.11. Introduce

$$D = (0,1)^2, \quad f \in L^2(D), \quad x_{ij} = \big(i/n, j/n\big), \quad 0 < i, j < n,$$
$$\Omega_n = D \setminus \bigcup_{0<i,j<n} \overline{B}(x_{ij}, r_n), \quad u_n = u^f_{\Omega_n}.$$

Let $r_n = e^{-dn^2}$, $d > 0$.

(1) Show that there exists $u^\star \in H_0^1(D)$ and a subsequence of $(u_n)_n$ converging weakly to $u^\star$ in $H_0^1(D)$ and strongly in $L^2(D)$.

We call a (small) square centered at x_{ij} and of side $1/n$ a "cell". It contains a circular "hole" with the same center and of radius r_n. We denote by C_{ij}^n the circular ring built from a disk with same center and radius $1/2n$ where we removed the circular hole. Let us introduce the sequence of functions $z_n \in H^1(D)$ defined on each cell by

$$z_n = 0 \text{ on the hole}, \quad \Delta z_n = 0 \text{ on the ring } C_{ij}^n,$$

and $z_n = 1$ everywhere else.

(2) Write z_n explicitly in each cell. Check that, in each cell, $-\Delta z_n = \mu_n - \nu_n$ where μ_n, ν_n are nonnegative measures with support respectively in the (large) circle of radius $1/2n$ and in the (small) circle of radius r_n.

(3) Prove that z_n converges strongly to 1 in $L^2(D)$, that ∇z_n converges weakly to 0 in $L^2(D)^N$, and that μ_n converges weakly to the constant function $2\pi/d$ in $H^{-1}(D)$.

(4) Check that $u^\star$ is a solution of

$$u^\star \in H_0^1(D) \quad \text{and} \quad -\Delta u^\star + \frac{2\pi}{d} u^\star = f,$$

this equation being satisfied in a variational sense. (Hint: use that, if $\varphi \in C_0^\infty(D)$, then $\varphi z_n \in H_0^1(\Omega_n)$.)

(5) Adapt the above proof to deduce the first two points of Proposition 3.2.11.

Exercise 3.9. Construct a compact subset K of $(0,1)$ that is of empty interior and such that for all $x \in K$ and any $\epsilon > 0$, the measure of $K \cap (x-\epsilon, x+\epsilon)$ is strictly positive. (Hint: choose $K = [0,1]\backslash\omega$, where ω is the union of the open intervals centered at $k2^{-n}$, $0 \le k \le 2^n$ and of radius $r_{k,n} = 2^{-2n}$, $n \ge 2$.)

Set $\Omega := (0,1) \cap \omega$. Show that Ω is not stable but that

$$H_0^1(\Omega) = \left\{u \in H_0^1(0,1);\ u = 0 \text{ a.e. on } (0,1)\backslash\Omega\right\}.$$

Exercise 3.10. It is said that an open set Ω has the *corkscrew property* at a point x of its boundary if there are constants $r_0 > 0$, $\lambda \in (0,1)$ such that for all $r \in (0, r_0)$, there exists $y \in B(x,r)$ with $B(y, \lambda r) \subset B(x,r) \cap \Omega^c$.

Show that if this condition is satisfied for all $x \in \partial\Omega$ with constants r_0, λ independent of x, then Ω satisfies the property of capacity density.

Exercise 3.11. Let

$$\Omega = (0,2), \quad a_n = 1 - 1/n, \quad \Omega_n = (0, a_n) \cup (1,2), \quad u_n = \chi_{(0,a_n)} + 2\chi_{(1,2)}.$$

Denote by $\hat{u}_n$ the extension of u_n to Ω defined by

$$\forall x \in [a_n, 1], \quad \hat{u}_n(x) = u_n(a_n) + \alpha_n(x - a_n) \quad \text{with } \alpha_n = (u_n(1) - u_n(a_n))/(1 - a_n).$$

(1) Show that for any extension $\tilde{u}_n \in H^1(\Omega)$ of u_n, we have $\|\tilde{u}_n'\|_{L^2(\Omega)} \ge \|\hat{u}_n'\|_{L^2(\Omega)}$.

(2) Deduce that for any $k > 0$, we have $\Omega_n \notin \mathcal{S}_k$ (in the sense of definition (3.83)).

Exercise 3.12. Let Ω be the open subset of $\mathbb{R}^2$ defined by

$$\Omega = \{(x,y),\ 0 < x < 1,\ 0 < y < x^2\}.$$

(1) Is the open set Ω Lipschitz? Consider $v(x,y) = x^{1-\beta}$ with $\beta < 3/2$ but close to $3/2$.

(2) Show that $v \in H^1(\Omega)$.

(3) Let B be a ball containing Ω. Show that there is no function in $H^1(B)$ whose restriction to Ω is v.

Exercise 3.13. Let Ω_n be a sequence of open subsets of a ball $B \subset \mathbb{R}^N$ that converges in the sense of Hausdorff and of the characteristic functions to an open set Ω. Let u_n (resp. u) be the solution of the Neumann problem (3.79) on Ω_n (resp. Ω) and let $\overline{u}_n$, $\overline{\nabla u_n}$ (resp. $\overline{u}$, $\overline{\nabla u}$) be the usual extensions by 0 to the ball B.

(1) Let $(u^\star, W)$ be a weak limit in $L^2(B)^{N+1}$ of a subsequence of $(\overline{u}_n, \overline{\nabla u_n})$. Show that $(u^\star, W) = 0$ a.e. outside Ω, that $W = \nabla u^\star$ on Ω, and that $u^\star_{|\Omega} \in H^1(\Omega)$.

(2) Assume Ω is sufficiently regular that any function of $H^1(\Omega)$ is the restriction to Ω of an $H^1(B)$ function. Show that $u^\star_{|\Omega} = u$ and that all the sequence $(\overline{u}_n, \overline{\nabla u_n})$ converges strongly in $L^2(B)^{N+1}$ to $(\overline{u}, \overline{\nabla u})$.

Exercise 3.14. Consider the bi-Laplacian problem (3.90) in dimension $N = 1$. Let Ω_n be a sequence of open subsets of a fixed interval converging in the sense of Hausdorff to an open set Ω. Let u_n (resp. u) be the corresponding solutions of (3.90). Using the compact embedding of H_0^2 in C^1, prove that u_n converges to u in H_0^2.

Chapter 4

Existence of optimal shapes

4.1 Some geometric problems

4.1.1 Isoperimetric problems

Let us start with the classical isoperimetric problems, or minimal surface problems, that are mentioned in the introduction (Chapter 1). We use the notion of perimeter introduced in Chapter 2. Let D be an open subset of $\mathbb{R}^N$ and $V_0 \in (0, |D|)$. We consider the problems

$$P(\Omega^*) = \min\{P(\Omega);\ \Omega \subset D \text{ measurable},\ |\Omega| = V_0\}, \tag{4.1}$$

$$P_D(\Omega^*) = \min\{P_D(\Omega);\ \Omega \subset D \text{ measurable},\ |\Omega| = V_0\}, \tag{4.2}$$

where the perimeter $P(\Omega)$ and relative perimeter $P_D(\Omega)$ are defined in Section 2.3.1. We then have

Proposition 4.1.1. *Assume* $|D| < +\infty$ *and* $V_0 \in (0, |D|)$. *Then, problems* (4.1) *and* (4.2) *have a solution. If D can contain a ball of Lebesgue measure V_0, any solution of* (4.1) *is precisely a ball of measure V_0.*

Remark 4.1.2. Problems (4.1) and (4.2) are essentially equivalent to their "dual" version, which in fact look more like *isoperimetric problems*, namely,

$$|\Omega^*| = \max\{|\Omega|;\ \Omega \subset D \text{ measurable},\ P(\Omega) = P_0\}, \tag{4.3}$$

$$|\Omega^*| = \max\{|\Omega|;\ \Omega \subset D \text{ measurable},\ P_D(\Omega) = P_0\}. \tag{4.4}$$

We refer to Exercise 4.3 for a direct study of these problems and their connection with (4.1), (4.2).

Proof of Proposition 4.1.1. Let us first check that the family of "admissible" Ω is not empty by constructing an open subset of D of finite perimeter and of measure V_0. Since D is open, it is equal to a denumerable union of balls, that is, $D = \cup_p B_p$. For k large enough, the union of the first k balls $\omega_k := \cup_{p \le k} B_p$ has a measure that is greater than or equal to V_0 and ω_k is an open set of finite perimeter. But there exists $r \in (0, +\infty)$ such that the measure of $\omega_k \cap B(0, r)$ is exactly V_0, and this intersection is still of finite perimeter.

Now let Ω_n be a minimizing sequence for problem (4.1). Since $P(\Omega_n) = \|\nabla \chi_{\Omega_n}\|_1$ is bounded, there exist $\Omega^* \subset D$ measurable and a subsequence such that $\chi_{\Omega_{n_k}}$

converges in $L^1(D)$ to χ_{Ω^*} as $k \to \infty$ (see Theorem 2.3.11). We have $|\Omega^*| = V_0$ and, by lower semicontinuity, $P(\Omega^*) \le \lim P(\Omega_n)$. It follows that Ω^* meets the minimum in (4.1).

The proof is exactly the same for problem (4.2).

Let us recall the *isoperimetric inequality*, which says that, for all measurable sets $A \subset \mathbb{R}^N$, we have $P(A) \ge c_N |A|^{(N-1)/N}$ with $c_N = NV_N^{1/N}$, where V_N is the measure of the unit ball. Moreover, equality holds if and only if A is a ball. Thus, if D can contain a ball of measure m, this ball is the minimal shape and any other minimum is the same ball. □

Remark 4.1.3. Let us remark that, if D is not of finite measure, problem (4.1) may not have solutions. Let us, for instance, consider in $\mathbb{R}^2$ the domain

$$D = \{(x, y) \in \mathbb{R}^2;\ |y| < |2\pi^{-1} \arctan x|\},$$

and $V_0 = \pi$ = the area of the unit disk. By the isoperimetric inequality, the infimum inf in (4.1) is greater than or equal to 2π. But, if

$$\Omega_n := B\big((n+1, 0), 2\pi^{-1} \arctan n\big) \cup B\big((2n, 0), \epsilon_n\big),$$

where $\pi\epsilon_n^2 + \pi[2\pi^{-1} \arctan n]^2 = \pi$, we check that $\lim P(\Omega_n) = 2\pi$, which means that Ω_n is a minimizing sequence and inf $= 2\pi$. But D does not contain any disk of radius 1: therefore this infimum is not reached.

The same remark is valid for problem (4.2) (see Exercise 4.1 or also [65] where more examples may be found).

Obviously, if D is of measure $+\infty$ but contains a ball of measure V_0, then this ball is a solution of problem (4.1).

If D is a half-space, we check that the solution of problem (4.2) is a half ball that sticks to the boundary of D (see Exercise 4.2 or [99] for more general results and such isoperimetric inequalities outside a convex domain in $\mathbb{R}^N$). More generally, we may notice that the solutions of (4.2) want to "stick" to the boundary of D in order to minimize their relative perimeter.

Note that we may replace the volume constraint by a more general constraint of the kind $\int_\Omega f = V_0$ where $f \in L^1(D)$ is given: results and approaches are similar if the family of admissible sets is not empty. More generally, there is an important amount of literature for weighted isoperimetric inequalities; a good starting point to enter this domain could be [84], [85].

4.1.2 A generalization

The important point in the previous proof is the compactness of the characteristic functions $(\chi_{\Omega_n})_{n\ge 0}$ which comes from the bound on $P(\Omega_n)$. With the same approach, we may prove existence for a quite wider class of functionals. We give below a generic example.

Let us assume that we have a process (through partial differential equations or some other means) that with any measurable $\Omega \subset D$ associates a function $y(\Omega) \in L^2(D)$ in such a way that

$$\chi_{\Omega_n} \xrightarrow{L^1} \chi_\Omega \quad \Longrightarrow \quad y(\Omega_n) \longrightarrow y(\Omega) \quad \text{in } L^2(D). \tag{4.5}$$

Given $f \in L^2(D)$, $K \in L^1(D)$, and $\tau > 0$, we set

$$J(\Omega) = \int_D (y(\Omega) - f)^2 \, dx + \int_\Omega K + \tau P(\Omega). \tag{4.6}$$

Let us then consider the two problems

$$\min\{J(\Omega),\ \Omega \subset D\}, \tag{4.7}$$

$$\min\{J(\Omega),\ \Omega \subset D,\ |\Omega| = V_0\} \quad \text{with fixed } V_0. \tag{4.8}$$

Theorem 4.1.4. *Assume* $|D| < +\infty$ *and* $V_0 \in (0, |D|)$. *Then, problems* (4.7) *and* (4.8) *have a solution.*

Proof. For the two problems, the functional J is bounded from below by $-\|K\|_{L^1(D)}$ and the family of admissible Ω such that $J(\Omega) < +\infty$ is not empty (we may check it as in the proof of Proposition 4.1.1).

Let Ω_n be a minimizing sequence and M be an upper bound of the sequence $J(\Omega_n)$. Then, $|\Omega_n| \le |D|$ and $\tau P(\Omega_n) \le M + \|K\|_{L^1(D)}$. Theorem 2.3.11 implies the existence of $\Omega^* \subset D$ and of a subsequence $(\Omega_{n_k})_{k \ge 0}$ such that $\chi_{\Omega_{n_k}} \to \chi_{\Omega^*}$ in $L^1(D)$. This implies that $|\Omega^*| = V_0$ for problem (4.8) and, according to assumption (4.5), that $y(\Omega_{n_k}) \longrightarrow y(\Omega^*)$ in $L^2(D)$. Adding to this that $P(\cdot)$ is l.s.c., we deduce

$$J(\Omega^*) \le \liminf J(\Omega_n) = \inf J.$$

This proves that Ω^* is a minimizer of J. □

4.1.3 Capillary surfaces

Let us return to the capillary surfaces discussed in Chapter 1: One wants to describe the surface of a liquid at rest in a vessel by minimizing the total energy of the system. This energy is given (at least when data are regular enough) by

$$E(\Omega) = \text{surface}(\partial\Omega \cap D) + (\cos\gamma)\,\text{Area}(\partial\Omega \cap \partial D) - \int_\Omega g x_3 \, dx,$$

where D is a bounded open set corresponding to the vessel and $\Omega \subset D$ is the open subset describing the volume occupied by the liquid. These terms are respectively the surface tension, the capillarity, and the gravity energies.

Let us rewrite this functional in terms of perimeters of the measurable set Ω. Thus, we more precisely set

$$E(\Omega) = P_D(\Omega) + \cos\gamma\,(P(\Omega) - P_D(\Omega)) + \int_\Omega K(x)\,dx.$$

Here, $K(\cdot)$ denotes more generally a given function of $L^1(D)$.

Proposition 4.1.5. *Assume* $V_0 < |D| < +\infty$ *and* $P(D) < +\infty$. *Then, the problem*

$$\min\{E(\Omega),\ \Omega \textit{ measurable} \subset D,\ |\Omega| = V_0\}$$

has a solution.

Proof. The assumption on D guarantees that the family of admissible Ω with $E(\Omega) < +\infty$ is not empty.

Let us first assume $\cos\gamma \geq 0$. In this case, it clear that E is bounded from below by $-\|K\|_{L^1}$ (recall that $P_D(\Omega) \leq P(\Omega)$). Let Ω_n be a minimizing sequence. Then $P_D(\Omega_n)$ is bounded. Up to a subsequence, we may assume (see Theorem 2.3.11) that there exists a measurable $\Omega^* \subset D$ such that $\chi_{\Omega_n} \to \chi_\Omega^*$ in $L^1(D)$. Since $\cos\gamma \geq 0$ and $1 - \cos\gamma \geq 0$, by lower semicontinuity of $P(\cdot)$ and $P_D(\cdot)$, we deduce that Ω^* meets the expected minimum of $E(\cdot)$ (note that $|\Omega^*| = V_0$ by convergence in $L^1(D)$).

Let us now assume $\cos\gamma \leq 0$. We then reduce this situation to the previous one as follows. Set $\widetilde{\Omega} := D\backslash\Omega$ and remark that

$$E(\Omega) = \tilde{E}(\widetilde{\Omega}) = P_D(\widetilde{\Omega}) - \cos\gamma\,[P(\widetilde{\Omega}) - P_D(\widetilde{\Omega})] - \int_{\widetilde{\Omega}} K + M,$$

where $M = \int_D K + \cos\gamma\, P(D)$. This expression of $\tilde{E}$ is obtained by using that $P_D(\Omega) = P_D(\widetilde{\Omega})$ (since $\nabla\chi_\Omega = -\nabla\chi_{\widetilde{\Omega}}$ in D) and by using also

$$P(\Omega) - P_D(\Omega) = P(D) - [P(\widetilde{\Omega}) - P_D(\widetilde{\Omega})].$$

We in fact see that this identity says that the area of the wet part of the boundary is equal to its total area from which one subtracts the area of its dry part (this may be proved rigorously by going back to the definitions). □

Remark 4.1.6. We may drop the assumption $P(D) < +\infty$ when $\cos\gamma \geq 0$, but not when $\cos\gamma < 0$, since then $E(\cdot)$ is not necessarily bounded from below.

4.2 Examples of nonexistence

We are interested here in minimizing functionals of the type

$$\Omega \to \int_{\Omega} F(x, u_{\Omega}, \nabla u_{\Omega}),$$

where F is a "good" regular function and where $u_{\Omega} = u_{\Omega}^{f}$ is the solution of the Dirichlet problem (see Proposition 3.1.20)

$$u_{\Omega}^{f} \in H_0^1(\Omega),\ \forall\, v \in H_0^1(\Omega), \qquad \int_D \nabla u_{\Omega}^{f} \nabla v = \int_D u_{\Omega}^{f} v,$$

with f given in $L^2(D)$.

The goal of this section is to show that existence does not hold in general for this problem. We exhibit two examples in this direction. We also give a counterexample for a similar problem arising in elasticity theory. Let us also mention some more examples by Yu. Osipov and A. P. Suetov [249].

Example 4.2.1. The description of the first example needs Proposition 3.2.11 and Exercise 3.8. Let $D = (0,1) \times (0,1)$, $f \in L^2(D)$, $f > 0$ a.e. on D, $d \in (0, \infty)$, and w be the solution of

$$\begin{cases} w \in H_0^1(D) \cap H^2(D), \\ -\Delta w + \frac{2\pi}{d} w = f \quad \text{in } D. \end{cases}$$

We consider the functional J defined by

$$J(\Omega) = \int_{\Omega} (u_{\Omega}^{f} - w)^2\, dx.$$

It is a least square functional, rather classical and much used in applications. Let a be fixed in $(0,1)$ and let us consider

$$\mathcal{O}_a = \{\Omega \subset D,\ \Omega \text{ open set},\ |\Omega| \geq a\}.$$

Then the problem $\left[\min_{\Omega \in \mathcal{O}_a} J(\Omega)\right]$ *does not have any solutions.*

Indeed, we saw in Exercise 3.8 that setting

$$\Omega_n = D \backslash \bigcup_{i,j} \overline{B}\left(x_{ij}, e^{-dn^2}\right),$$

then $u_{\Omega_n}^{f}$ converges to w weakly in $H_0^1(D)$ and strongly in $L^2(D)$. Thus $J(\Omega_n) = \|u_{\Omega_n} - w\|_{L^2(\Omega_n)}^2$ converges to 0. Since it is clear that the measure of Ω_n tends to 1 $(1 > a)$, the open sets Ω_n are in $\mathcal{O}_a$ for large n so that $\inf J = 0$.

If this infimum were reached for some open set Ω, we would have $u_\Omega = w$ in $L^2(\Omega)$. Since $-\Delta u_\Omega = f$ in $\mathcal{D}'(\Omega)$, we would also have $-\Delta w = f$ in $\mathcal{D}'(\Omega)$. But, by definition of w, $-\Delta w + \frac{2\pi}{d} w = f$ a.e. in D. Therefore we should have $w = 0$ on Ω, which contradicts $f > 0$ a.e. on D (together with $|\Omega| > 0$).

The above argument can easily be adapted to show that there is no quasi-open set solution either. □

Example 4.2.2. Let us now give another counterexample suggested by G. Buttazzo (see, e.g., [65]).

Given a bounded open set D and $f \in L^2(D)$, we want to minimize the functional

$$J(\Omega) = \int_D (u_\Omega^f - u_0)^2 \, dx,$$

where u_0 is given in $L^2(D)$. A physical interpretation of this problem could be the following: D is a box or a room that is heated by some heat source f and $D \backslash \Omega$ represents a subdomain that we fill with ice. The goal is to determine this "frozen zone" in such a way that the temperature of the box be as close as possible to some given temperature u_0.

Let us prove that this problem does not have any solution in general, even for the following "simple" data: $f \equiv 1$, $u_0 \equiv c \equiv$ constant, and $D =$ the unit ball in $\mathbb{R}^2$.

By the maximum principle, for all $\Omega \subset D$ we have

$$0 \le u_\Omega^1 \le u_D^1 = \frac{1 - r^2}{4} \le \frac{1}{4}.$$

If $c \ge \frac{1}{4}$, we have

$$u_\Omega^1 - c \le u_D^1 - c \le 0,$$

so that

$$J(\Omega) = \int_D (u_\Omega^1 - c)^2 \, dx \ge \int_D (u_D^1 - c)^2 \, dx = J(D),$$

which shows that $\Omega = D$ reaches the minimum of J (this is quite intuitive: to get a high temperature, we should cool as little as possible!).

Let us now consider the case $0 < c < \frac{1}{8}$. It is easy to see that D is not the minimum of J anymore. Indeed, denoting by B_R the disk centered at O and of radius $R < 1$, we have $u_{B_R}^1 = \frac{R^2 - r^2}{4}$ for $r < R$ so that

$$\begin{aligned} J(B_R) &= 2\pi \int_0^R \left(\frac{R^2 - r^2}{4} - c \right)^2 r \, dr + 2\pi \int_R^1 (0 - c)^2 r \, dr \\ &= \frac{\pi}{48} \left(R^6 - 12cR^4 + 48c^2 \right). \end{aligned}$$

An elementary calculation shows that $J(B_R) < J(D)$ for $R = \sqrt{8c} < 1$.

Let us prove that J cannot have a (regular) minimum in this case. Assume by contradiction that Ω is such a minimum. We just proved that it is different from D, so that $|\Omega| < |D|$. Assume its closure is different from D (which is the case if it is regular enough so that $|\overline{\Omega}| = |\Omega| < |D|$). Let B_ε be a ball of radius ε included in $D\backslash\overline{\Omega}$. Let us set $\Omega_\varepsilon = \Omega \cup B_\varepsilon$ and let us show that, for ε small enough, Ω_ε is an open set "better" than Ω. Since Ω_ε is built out of two disjoint connected components Ω and B_ε, the solution u_{Ω_ε} may be computed separately in each of these components. On Ω, we have $u_{\Omega_\varepsilon} = u_\Omega$ while on B_ε, u_{Ω_ε} can be explicitly computed (it is radial with respect to the center of the ball). In particular, it is easy to see that, for ε small enough, we have $0 < u_{\Omega_\varepsilon} < c$ on B_ε. Let us now compare $J(\Omega_\varepsilon)$ to $J(\Omega)$. We have

$$\begin{aligned} J(\Omega_\varepsilon) &= \int_{\Omega_\varepsilon} (u_{\Omega_\varepsilon} - c)^2\,dx + \int_{D\backslash\Omega_\varepsilon} c^2\,dx \\ &= \int_\Omega (u_\Omega - c)^2\,dx + \int_{B_\varepsilon} (u_{B_\varepsilon} - c)^2\,dx + \int_{D\backslash\Omega} c^2\,dx - \int_{B_\varepsilon} c^2\,dx \\ &= J(\Omega) + \int_{B_\varepsilon} (u_{B_\varepsilon} - c)^2 - c^2\,dx. \end{aligned}$$

But, for ϵ small enough, $0 < u_{\Omega_\varepsilon} < c$, so that $(u_{\Omega_\varepsilon} - c)^2 < c^2$, which implies $J(\Omega_\varepsilon) < J(\Omega)$. This proves that J cannot have a regular minimum. One can get rid of this a priori regularity assumption on a possible minimum and show that there is indeed no measurable minimum at all, but the complete proof requires more tools (see the comments in [65] and also Chapter 7).

□

Example 4.2.3 (Minimization of the compliance). Let us now give another example of nonexistence, typical in solid mechanics optimization, where an homogenization phenomenon also leads to nonexistence of optimal shapes. Let us consider a membrane Ω included in the unit square $D = (0,1) \times (0,1)$. Assume that we pull on the left part $\Gamma_0 = \{0\} \times [0,1]$ and on the right part $\Gamma_1 = \{1\} \times [0,1]$ of the membrane with a constant and horizontal force with magnitude equal to 1 (see Figure 4.1). For simplicity, we choose here a scalar model. In the books [12], [13], we may find more counterexamples with the full elasticity models. The space of admissible displacements is $V_\Omega = H^1(\Omega)/\mathbb{R}$ (displacements are defined up to an additive constant). The energy of such a displacement $v \in V_\Omega$ is given by

$$\mathcal{E}_\Omega(v) := \frac{1}{2}\int_\Omega |\nabla v|^2 - \int_{\Gamma_0\cup\Gamma_1} fv, \tag{4.9}$$

where $f = -1$ on Γ_0 and $f = +1$ on Γ_1. For fixed Ω, the equilibrium position is obtained for the displacement v_Ω that minimizes $\mathcal{E}_\Omega$. Existence and uniqueness of

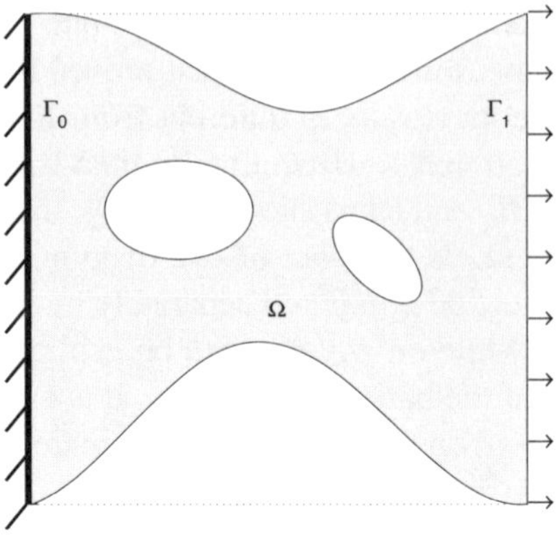

Figure 4.1. A membrane is pulled on its left and right sides by a constant force.

v_Ω may be obtained like in Proposition 3.1.20. Similarly, we easily show that v_Ω is a solution of

$$v_\Omega \in V_\Omega,\ \forall\,\psi \in V_\Omega, \qquad \int_\Omega \nabla v_\Omega \cdot \nabla\psi = \int_{\Gamma_0\cup\Gamma_1} \psi f, \tag{4.10}$$

which means, when Ω is a regular open set, that

$$\begin{cases} \Delta v_\Omega = 0 & \text{in } \Omega,\\ \frac{\partial v_\Omega}{\partial \nu} = -1 & \text{on } \Gamma_0,\\ \frac{\partial v_\Omega}{\partial \nu} = 1 & \text{on } \Gamma_1,\\ \frac{\partial v_\Omega}{\partial \nu} = 0 & \text{on } \partial\Omega\backslash(\Gamma_0\cup\Gamma_1). \end{cases} \tag{4.11}$$

With the compatibility condition $\int_{\partial\Omega} f\,ds = 0$ being satisfied, (4.11) has a unique solution in V_Ω.

A significant quantity in elasticity is the *compliance* or, in other words, the work of exterior forces. We denote it here by $\mathcal{C}(\Omega)$ and it is equal to

$$\mathcal{C}(\Omega) := \int_{\Gamma_0\cup\Gamma_1} f v_\Omega\,ds = \int_\Omega |\nabla v_\Omega|^2. \tag{4.12}$$

The second equality in (4.12) is obtained by the variational formulation (4.10). Note that this equality also provides

$$\mathcal{C}(\Omega) = -2\mathcal{E}_\Omega(v_\Omega). \tag{4.13}$$

One is naturally led to look for the most rigid membrane under the action of this applied force. This leads to the following shape optimization problem:

$$\text{find } \Omega \subset (0,1)^2 \text{ with } \Gamma_0\cup\Gamma_1 \subset \partial\Omega,\ |\Omega| = a, \text{ minimizing } \mathcal{C}(\Omega). \tag{4.14}$$

We are going to show that problem (4.14) does not have any solutions. In [93], A. Chambolle and C. Larsen show that when adding volume and perimeter, as extra penalization terms, to the compliance, then not only does one have existence of an optimal shape (see Section 4.6 for similar results), but this optimal shape is a C^∞-open set!

Let us choose $a = 1/2$ and let us denote by

$$\mathcal{O}_{\text{ad}} = \{\Omega \subset (0,1)\times(0,1) \text{ open}, |\Omega| = 1/2, \Gamma_0 \cup \Gamma_1 \subset \partial\Omega\}$$

the class of admissible open sets. For all $\Omega \in \mathcal{O}_{\text{ad}}$, the function $w_\alpha(x,y) = \alpha x$ belongs to the space V_Ω. Therefore, it is admissible for the variational formulation defining v_Ω. Thus

$$\mathcal{E}_\Omega(v_\Omega) \le \mathcal{E}_\Omega(w_\alpha) = \tfrac{1}{2}|\Omega|\alpha^2 - \alpha = \tfrac{1}{4}\alpha^2 - \alpha. \tag{4.15}$$

The last function in (4.15) reaches its minimum for $\alpha = 2$ for which its value is -1. We then deduce from (4.13) and (4.15),

$$\forall\, \Omega \in \mathcal{O}_{\text{ad}}, \quad \mathcal{C}(\Omega) \ge 2. \tag{4.16}$$

The second step consists in verifying that 2 is the infimum of $\mathcal{C}(\Omega)$ on $\mathcal{O}_{\text{ad}}$, that is, there exists a sequence of open sets Ω_n in $\mathcal{O}_{\text{ad}}$ such that $\mathcal{C}(\Omega_n) \to 2$. In the rectangle $[0,1]\times[-\frac{\varepsilon}{2},\frac{\varepsilon}{2}]$, with $\varepsilon < \frac{1}{6}$, let us define the following tubular domain (this is a construction suggested by A. Chambolle):

$$\mathcal{T}_\varepsilon = (x,y) \text{ such that } \begin{cases} -\frac{\varepsilon^2}{2(x+\varepsilon)} < y < \frac{\varepsilon^2}{2(x+\varepsilon)} & \text{if } 0 < x < \varepsilon(\frac{1}{k_\varepsilon}-1), \\ -\frac{k_\varepsilon \varepsilon}{2} < y < \frac{k_\varepsilon \varepsilon}{2} & \text{if } \varepsilon(\frac{1}{k_\varepsilon}-1) < x < 1-\varepsilon(\frac{1}{k_\varepsilon}-1), \\ -\frac{\varepsilon^2}{2(1-x+\varepsilon)} < y < \frac{\varepsilon^2}{2(1-x+\varepsilon)} & \text{if } 1-\varepsilon(\frac{1}{k_\varepsilon}-1) < x < 1; \end{cases}$$

see Figure 4.2, where k_ε is a constant (depending on ε) that we are going to fix now. We construct the minimizing sequence Ω_n by juxtaposing n tubes $\mathcal{T}_{1/n}$. To fit the

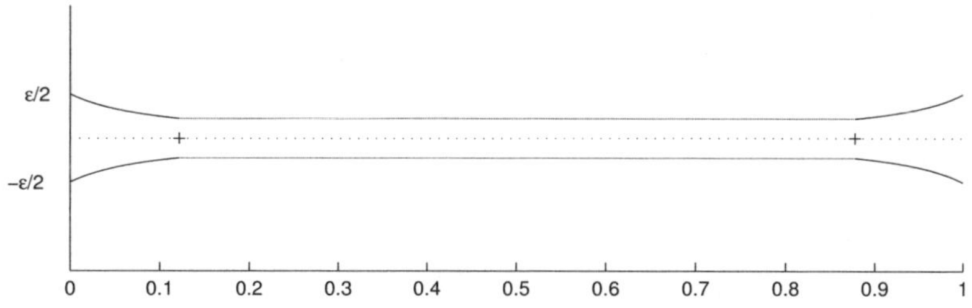

Figure 4.2. A tube $\mathcal{T}_\varepsilon$.

area constraint, we want that the area of $\mathcal{T}_\varepsilon$ is equal to $\varepsilon/2$. But the value of this area A is given by

$$A = 4\left(\int_0^{\varepsilon\left(\frac{1}{k_\varepsilon}-1\right)} \frac{\varepsilon^2\,dx}{2(x+\varepsilon)} + \frac{k_\varepsilon \varepsilon}{2}\left(\frac{1}{2} - \varepsilon\left(\frac{1}{k_\varepsilon} - 1\right)\right)\right)$$
$$= 2\varepsilon\left(\frac{k_\varepsilon}{2} + \varepsilon k_\varepsilon - \varepsilon - \varepsilon \log k_\varepsilon\right).$$

It is elementary to check that for $\varepsilon \in (0, \frac{1}{6})$, the equation $\frac{x}{2} + \varepsilon x - \varepsilon - \varepsilon \log x = \frac{1}{4}$ has a unique solution in $[\frac{1}{4}, \frac{1}{2}]$. This is what we define to be k_ε. We also set $\delta_\varepsilon = \varepsilon(\frac{1}{k_\varepsilon} - 1)$.

We define the open set Ω_n as being the union of n tubes piled up on each other like in Figure 4.3.

Figure 4.3. A sequence of open sets Ω_n minimizing the compliance.

To compute, or more precisely to bound from above the compliance of Ω_n, we will use the following principle. For all open sets $\Omega \subset D$, we have

$$C(\Omega) = \int_\Omega |\nabla v_\Omega(X)|^2 = \min_{\sigma \in H_\Omega} \int_\Omega \sigma.\sigma, \tag{4.17}$$

where the affine space H_Ω is defined by

$$H_\Omega = \left\{\sigma \in L^2(\Omega)^2,\ \forall \psi \in H^1(\Omega),\ \int_\Omega \nabla\psi \cdot \sigma = \int_{\Gamma_0 \cup \Gamma_1} \psi f\right\}. \tag{4.18}$$

If Ω is regular enough, this means

$$H_\Omega = \left\{ \sigma \in L^2(\Omega)^2 \text{ such that } \left| \begin{array}{ll} \operatorname{div}\sigma = 0 & \text{in } \Omega, \\ \sigma.\nu = -1 & \text{on } \Gamma_0, \\ \sigma.\nu = 1 & \text{on } \Gamma_1, \\ \sigma.\nu = 0 & \text{on } \partial\Omega\setminus(\Gamma_0 \cup \Gamma_1). \end{array} \right. \right\}$$

It is easy to check (thanks to (4.10), (4.18)) that the minimum in (4.17) is indeed reached for the vector field $\sigma = \nabla v_\Omega$. We will choose a specific field to bound the compliance from above. Since Ω_n is defined as a disjoint union of identical open sets, it is sufficient to define σ on the tube with basis $\mathcal{T}_\varepsilon$ and to repeat it by translation. Let us set

$$\sigma(x,y) = \begin{cases} \begin{pmatrix} 1 + x/\varepsilon \\ -y/\varepsilon \end{pmatrix} & \text{if } x \in (0, \delta_\varepsilon], \\ \begin{pmatrix} 1 + \delta_\varepsilon/\varepsilon \\ 0 \end{pmatrix} & \text{if } x \in (\delta_\varepsilon, 1-\delta_\varepsilon), \\ \begin{pmatrix} 1 + (1-x)/\varepsilon \\ y/\varepsilon \end{pmatrix} & \text{if } x \in [1-\delta_\varepsilon, 1]. \end{cases} \tag{4.19}$$

We check that $\operatorname{div}\sigma = 0$ in $\mathcal{T}_\varepsilon$, that $\sigma.\nu = -1$ on Γ_0 (where $x = 0$), that $\sigma.\nu = 1$ on Γ_1 (where $x = 1$), and finally that $\sigma.\nu = 0$ on the pieces of hyperbolas defining the boundary $\mathcal{T}_\varepsilon$, as well as on the horizontal part of this same boundary. Repeating this σ field n times provides an admissible test function for the variational formulation defining the compliance of Ω_n and we have

$$\mathcal{C}(\Omega_n) \le n \int_{\mathcal{T}_\varepsilon} \sigma.\sigma, \tag{4.20}$$

in which we set $\varepsilon = 1/n$. Computing this integral leads to

$$\int_{\mathcal{T}_\varepsilon} \sigma.\sigma\, dX = \varepsilon^2 \left(\frac{1}{k_\varepsilon^2} - \frac{11 + k_\varepsilon^2}{12} \right) + \frac{\varepsilon}{k_\varepsilon} - \frac{2\varepsilon\delta_\varepsilon}{k_\varepsilon} = \frac{\varepsilon}{k_\varepsilon} + O(\varepsilon^2). \tag{4.21}$$

Multiplying by n, we obtain

$$\mathcal{C}(\Omega_n) \le \frac{1}{k_{1/n}} + O(1/n). \tag{4.22}$$

It is clear in the definition of k_ε that it tends to $1/2$ when ε tends to 0. Thus, the upper bound in (4.22) tends to 2 as $n \to +\infty$, which is what we needed to check.

Now that we have proved that $\inf_{\Omega \in \mathcal{O}_{\text{ad}}} \mathcal{C}(\Omega) = 2$, let us check that this infimum cannot be reached. If it was for an admissible open set Ω, we would have

$$\mathcal{E}_\Omega(v_\Omega) = -\tfrac{1}{2}\mathcal{C}(\Omega) = -1 = \mathcal{E}_\Omega(w_2)$$

(see (4.15)). By uniqueness of v_Ω, we then could deduce that $v_\Omega(x,y) = w_2(x,y) = 2x$. This is impossible since v_Ω must verify $\frac{\partial v_\Omega}{\partial n} = 1$ on Γ_1. Whence the nonexistence of a minimum for problem (4.14).

□

Conclusion of this Section 4.2. After looking at the above counterexamples, we understand that existence of optimal shapes requires extra assumptions, either on the set of admissible domains or on the functional to be optimized. In the following sections, we will give existence results under various successive assumptions:

- uniform regularity of the boundary: property of the ε-cone in Section 4.3;
- capacitary constraints in Section 4.4;
- the case of Dirichlet energy in Section 4.5;
- bounded perimeter in Section 4.6;
- monotonicity of the functional in Section 4.7.

4.3 Uniform regularity of admissible shapes

Let us start by reducing the class of admissible sets to uniformly Lipschitz open sets (or equivalently to sets satisfying the property of the ε-cone; see Chapter 2).

Let $D \subset \mathbb{R}^d$ be open and bounded and let $f \in L^2(D)$. For all $\Omega \subset D$, let us consider the solution u_Ω of the Dirichlet problem or of the Neumann problem on Ω, that is (see (3.23), (3.79) for the precise definition),

$$u_\Omega \in H_0^1(\Omega), \quad -\Delta u_\Omega = f \quad \text{in } \Omega, \tag{4.23}$$

or

$$\begin{cases} u_\Omega \in H^1(\Omega), & -\Delta u_\Omega + u_\Omega = f \quad \text{in } \Omega, \\ \text{“}\frac{\partial u_\Omega}{\partial \nu} = 0 & \text{on } \partial\Omega\text{”}. \end{cases} \tag{4.24}$$

Let $j : D \times \mathbb{R} \times \mathbb{R}^N \to \mathbb{R}$ be measurable in the three variables, continuous in (r, p) a.e. x, and such that there exists a constant C with

$$|j(x, r, p)| \le C(1 + r^2 + |p|^2) \quad \forall\, x \in D,\ r \in \mathbb{R},\ p \in \mathbb{R}^N. \tag{4.25}$$

For all open sets $\Omega \subset D$, we set

$$J(\Omega) = \int_\Omega j(x, u_\Omega(x), \nabla u_\Omega(x))\, dx. \tag{4.26}$$

This functional J is well defined since, thanks to the assumptions on j, the function under the integral is measurable and integrable and we have

$$|J(\Omega)| \le C \int_\Omega 1 + |u_\Omega(x)|^2 + |\nabla u_\Omega(x)|^2\, dx = C(\|u_\Omega(x)\|_{H^1} + |\Omega|) < +\infty.$$

Let us give some "least square" functionals that are often used in applications and that satisfy (4.25):

- $j(x, r, p) = (r - g(x))^2$ or $j(x, r, p) = rg(x)$, where g is given in $L^2(D)$;
- $j(x, r, p) = |p - p_0(x)|^2$, where $p_0 \in L^2(D)^N$;
- $j(x, r, p) = |p|^2 - 2rf(x)$, where $f \in L^2(D)$.

Note that the definition of J includes the case when integration holds only on a strict subset Ω' of Ω: indeed, in this case we use the characteristic function $\chi_{\Omega'}$ as a factor in j.

The minimizing problem is generally not set on the whole family $\mathcal{O}_\varepsilon$ of open subsets of D with the ε-cone property. Most of the time, extra constraints are added. Thus, in the following we will assume that the family $\mathcal{O}_{\mathrm{ad}}$ of admissible sets satisfies

$$\begin{cases} \mathcal{O}_{\mathrm{ad}} \subset \mathcal{O}_\varepsilon \text{ and } \mathcal{O}_{\mathrm{ad}} \text{ is closed for one of the three types} \\ \text{of convergence: in the sense of Hausdorff, of the characteristic} \\ \text{functions, or of compact sets (see Chapter 2).} \end{cases} \tag{4.27}$$

In particular, this is satisfied for the following examples:

$$\mathcal{O}_{\mathrm{ad}} = \{\Omega \in \mathcal{O}_\epsilon;\ |\Omega| = m\}; \qquad \mathcal{O}_{\mathrm{ad}} = \{\Omega \in \mathcal{O}_\epsilon;\ |\Omega| \le m\};$$
$$\mathcal{O}_{\mathrm{ad}} = \{\Omega \in \mathcal{O}_\epsilon;\ K \subset \Omega\}, \quad K \text{ a given compact set;}$$

and their intersections. We may then state

Theorem 4.3.1. *Let $\mathcal{O}_{\mathrm{ad}}$ be a nonempty family of open sets satisfying* (4.27)*, let j be a function satisfying* (4.25)*, and let J be defined by* (4.26)*. Then there exists $\Omega \in \mathcal{O}_{\mathrm{ad}}$ that minimizes J.*

Proof. The functional J is obviously bounded from below. Let Ω_n be a minimizing sequence. By Theorem 2.4.10, there is a subsequence Ω_{n_k} that converges as $k \to \infty$ to $\Omega^* \in \mathcal{O}_\epsilon$ for the three topologies of (4.27). The closure assumption of (4.27) implies that $\Omega^* \in \mathcal{O}_{\mathrm{ad}}$.

For the Dirichlet problem, we obtain (see Theorem 3.2.13) that $u_{\Omega_{n_k}}$ converges strongly in $H_0^1(D)$ to u_{Ω^*}. The assumptions on J guarantee that $J(u_{\Omega^*}) = \lim J(u_{\Omega_{n_k}})$. Thus the limit Ω^* is the expected minimum.

For the Neumann problem, we apply Theorem 3.7.3: indeed, we can prove that the uniform Lipschitz property allows us to define linear extension operators from $H^1(\Omega_n)$ into $H^1(D)$ with uniformly bounded norm (see [95], [96]). Thus we obtain strong convergence of

$$\chi_{\Omega_{n_k}} u_{\Omega_{n_k}},\ \chi_{\Omega_{n_k}} \nabla u_{\Omega_{n_k}} \quad \text{respectively to} \quad \chi_\Omega u_{\Omega^*},\ \chi_\Omega \nabla u_{\Omega^*}.$$

We again deduce that $J(u_{\Omega^*}) = \lim J(u_{\Omega_{n_k}})$ and that Ω^* is the expected minimum. □

4.4 Constraints of capacitary type

We can significantly weaken the uniform regularity assumptions a priori required on the admissible open sets to obtain minimal shapes, by using the capacitary tools developed in Chapter 3. For instance, we have the following result (see Section 3.4.3 for the definitions):

Proposition 4.4.1. *Let J be defined by* (4.26) *where u_Ω is the solution of the Dirichlet problem* (4.23). *Then there exists a solution of*

$$\Omega^* \in \mathcal{O}_{\alpha,r_0}, \quad J(\Omega^*) = \min\{J(\Omega),\ \Omega \in \mathcal{O}_{\alpha,r_0}\}.$$

Theorem 4.4.2 (Šverak). *If $N = 2$ and J is defined by* (4.26), *where u_Ω is the solution of the Dirichlet problem* (4.23), *then there exists a solution of*

$$\Omega^* \in \mathcal{O}_l, \quad J(\Omega^*) = \min\{J(\Omega),\ \Omega \in \mathcal{O}_l\}$$

(we recall that $\mathcal{O}_l$ denotes the family of open subsets of a fixed ball whose complement — in this ball — has a number of connected components less than or equal to l).

Existence may still hold when adding some more constraints, like for instance $|\Omega| = m$.

The proof is obtained thanks to the continuity of the Dirichlet problem under capacitary constraints (Theorem 3.4.12) and thanks to the compactness of any bounded sequence of open sets for Hausdorff convergence (see Theorem 2.2.25 and Corollary 2.2.26). It remains to check that the family $\mathcal{O}_{\alpha,r_0}$ is closed for Hausdorff convergence and this is proved in the following lemma.

Lemma 4.4.3. *The class $\mathcal{O}_{\alpha,r_0}$ is closed and therefore compact for Hausdorff convergence.*

Proof. Let $(\Omega_n)_{n\geq 0}$ be a sequence of open sets in the class $\mathcal{O}_{\alpha,r_0}$ that converges to an open set Ω in the sense of Hausdorff. Let us denote $F_n = \overline{D}\backslash\Omega_n$ and $F = \overline{D}\backslash\Omega$. Let us also set $F_\varepsilon = \bigcup_{x\in F} \overline{B}(x,\varepsilon)$. By Hausdorff convergence, we have (by coming back to the definition)

$$\forall\, \varepsilon > 0,\ \exists\, N_\varepsilon, \quad n \geq N_\varepsilon \Longrightarrow F_n \subset F_\varepsilon.$$

Let $r < r_0$ and let x be a fixed point of $\partial\Omega$. Let x_n be a sequence of points of $\partial\Omega_n$ converging to x (see Chapter 2). For $\varepsilon > 0$ fixed, let us choose $x_n \in \partial\Omega_n$ such that $|x_n - x| < \varepsilon/2$ and $F_n \subset F_{\varepsilon/2}$. Let us denote by τ the translation of vector $x_n - x$. Since $\tau(B(x,r)) = B(x_n, r)$ and since the capacity is invariant by translation, we have

$$\mathrm{cap}_{B(x,2r)}(F_\varepsilon \cap \overline{B}(x,r)) = \mathrm{cap}_{B(x_n,2r)}(\tau(F_\varepsilon) \cap \overline{B}(x_n,r)).$$

Now, since $F_n \subset F_{\varepsilon/2}$ and $|x_n - x| < \varepsilon/2$, we have $F_n \subset \tau(F_\varepsilon)$. Therefore, by monotonicity of the capacity,

$$\mathrm{cap}_{B(x,2r)}(F_\varepsilon \cap \overline{B}(x,r)) \geq \mathrm{cap}_{B(x_n,2r)}(F_n \cap \overline{B}(x_n,r)) \geq \alpha,$$

this last inequality coming from the definition of the class $\mathcal{O}_{\alpha,r_0}$. Since the capacity is continuous for decreasing sequences of compact sets and since $\bigcap_{\varepsilon>0} F_\varepsilon = F$, by passing to the limit as $\varepsilon \to 0$, we have

$$\mathrm{cap}_{B(x,2r)}(F \cap \overline{B}(x,r)) \geq \alpha. \qquad \square$$

4.5 Minimization of the Dirichlet energy

It turns out that we get existence of an optimal shape in the particular case — important for application (see Chapter 1) — where the functional to be minimized is the Dirichlet functional itself, under volume constraint, namely,

$$J(\Omega) = \int_\Omega \frac{1}{2}|\nabla u_\Omega|^2 - f u_\Omega, \tag{4.28}$$

where $u_\Omega = u_\Omega^f$ is the solution of the Dirichlet problem on Ω associated with f. In this case, we obviously have

$$J(\Omega) = \inf\left\{\int_\Omega \tfrac{1}{2}|\nabla v|^2 - f v,\ v \in H_0^1(\Omega)\right\}. \tag{4.29}$$

In fact the optimal shape will exist only in *the class $\mathcal{A}(D)$ of quasi-open subsets of D*. This requires extending the notion of the Dirichlet problem to the case of quasi-open sets. Let us recall that for $\Omega \in \mathcal{A}(D)$, we define (see Definition 3.3.43 and Proposition 3.3.44)

$$H_0^1(\Omega) = \left\{v \in H_0^1(D),\ \tilde{v} = 0 \text{ q.e. on } D\backslash\Omega\right\},$$

where $\tilde{v}$ is the quasi-continuous representative of v. For $f \in H^{-1}(D)$, we set

$$G(v) = \int_D \frac{1}{2}|\nabla v|^2 - \langle f, v\rangle_{H^{-1}\times H_0^1}.$$

Proposition 4.5.1. *Let $D \subset \mathbb{R}^N$ be open and let $\Omega \in \mathcal{A}(D)$ with finite measure. For $f \in H^{-1}(D)$, there exists u_Ω^f, a unique solution of*

$$u_\Omega^f \in H_0^1(\Omega),\ \forall\, v \in H_0^1(\Omega), \quad \int_D \nabla u_\Omega^f \nabla v = \langle f, v\rangle_{H^{-1}\times H_0^1}.$$

Moreover, u^f_Ω *is the unique element of* $H^1_0(D)$ *satisfying*

$$G(u^f_\Omega) = \min\left\{G(v);\ v \in H^1_0(\Omega)\right\}.$$

Existence of u^f_Ω follows from the Lax–Milgram theorem, from the Poincaré inequality (see Lemma 4.5.3), and from the fact that $H^1_0(\Omega)$ is a closed nonempty subspace of $H^1_0(D)$. Its characterization in terms of minimization comes from the strict convexity of $v \to G(v)$.

□

We then have the following optimal shape existence result.

Theorem 4.5.2. *Let D be an open subset of $\mathbb{R}^N$, let $f \in H^{-1}(D)$, $0 < m < |D|$, and $J(\Omega) = G(u^f_\Omega)$. Then, there exists $\Omega^* \in \mathcal{A}(D)$ a solution of*

$$|\Omega^*| = m, \quad J(\Omega^*) = \min\{J(\Omega);\ \Omega \in \mathcal{A}(D),\ |\Omega| \leq m\}. \tag{4.30}$$

We are going to give three different proofs of the existence of an optimal shape for this model problem. They all have a general interest that goes quite beyond this specific example. The first one follows the approach in [115]. The second one is a consequence of Proposition 4.5.5. The third one (see Remark 4.7.14) will be an application of the general existence result for monotone functionals with respect to inclusion, which is due to Buttazzo–Dal Maso (Theorem 4.7.6).

Proof of Theorem 4.5.2. We give here a first proof that follows the approach of [115], which consists in considering the auxiliary problem

$$G(u) = \min\left\{G(v);\ v \in H^1_0(D),\ |\Omega_v| \leq m\right\}. \tag{4.31}$$

(Let us recall that Ω_v denotes the quasi-open set $\Omega_v := \{x \in D;\ \tilde{v}(x) \neq 0\}$.) Let us first assume the existence of u, a solution of (4.31). We have $u \in H^1_0(\Omega_u)$ and $|\Omega_u| \leq m$ so that $u = u^f_{\Omega_u}$ since, by (4.31), $G(u) \leq G(v)$ for all $v \in H^1_0(\Omega_u)$. On the other hand, we also have

$$G(u) \leq \inf\left\{G(u^f_\Omega);\ \Omega \in \mathcal{A}(D),\ |\Omega| \leq m\right\} = \inf\{J(\Omega);\ \Omega \in \mathcal{A}(D),\ |\Omega| \leq m\}.$$

Thus

$$J(\Omega_u) = \min\{J(\Omega);\ \Omega \in \mathcal{A}(D),\ |\Omega| \leq m\}.$$

If $|\Omega_u| = m$ (generic case), then $\Omega^* = \Omega_u$ is a solution of problem (4.30). If $|\Omega_u| < m$, we consider a quasi-open set Ω^* satisfying the two conditions $\Omega^* \supset \Omega_u$, $|\Omega^*| = m$ (there exists such a set: for instance, we may choose $\Omega_u \cup B(0, r) \cap D$, where $r > 0$ is well chosen). Then, since $u \in H^1_0(\Omega^*)$, we still have $u = u^f_{\Omega^*}$ (by (4.31) again,

$G(u) \le G(v)$ for all $v \in H_0^1(\Omega^*)$). We check as before, by using (4.31), that Ω^* is a solution of (4.30).

The main point in proving the existence of a solution u to (4.31) is the Poincaré inequality stated in the following lemma, which is valid for all $v \in H_0^1(D)$ satisfying the constraint $|\Omega_v| \le m$, even if D is not bounded. Thus since

$$2J(v) \ge \int_D |\nabla u|^2 - C\|f\|_{H^{-1}(D)}\|v\|_{H_0^1(D)},$$

we deduce that $J(v)$ is bounded from below on the family of such v. It follows that a minimizing sequence is bounded in $H_0^1(D)$. Thus, up to a subsequence, it converges to some $u \in H_0^1(D)$ weakly in H^1 and a.e. At the limit $|\Omega_u| \le m$ and by lower semicontinuity of $v \to \int_D |\nabla v|^2$, we obtain (4.31) and this ends the proof of the theorem. □

Lemma 4.5.3. *There exists $C = C(N)$ such that for all $v \in H_0^1(D)$ satisfying $|\Omega_v| \le m$, we have*

$$\int_D v^2 \le C m^{2/N} \int_D |\nabla v|^2.$$

Proof. We use a rearrangement argument: Let $v^* : [0, +\infty) \to [0, +\infty)$ be the nonincreasing radial function such that (see Chapter 6)

$$\forall\, a \in [0, +\infty), \quad |v^* \ge a| = |v \ge a|.$$

We denote by D^* the ball centred at 0 with Lebesgue measure $|D|$. We then know that (see Theorem 6.1.4 and references therein)

$$\int_{D^*} v^{*2} = \int v^2, \qquad \int_{D^*} |\nabla v^*|^2 \le \int_D |\nabla v|^2.$$

Thus it is sufficient to check the inequality of the lemma when $D = D^*$ and when v is nonincreasing and radial. This may be done as follows (we denote by $c = NV_N$ the perimeter of the unit ball in $\mathbb{R}^N$ and by R the radius of the ball of measure m, that is, $m = V_N R^N$):

$$\begin{aligned}
\int_D v^2 &= c\int_0^R r^{N-1} v^2(r)\, dr = c\int_0^R dr\, v(r) \int_R^r v'(s)\, ds \\
&= c\int_0^R ds\, v'(s) \int_0^s r^{N-1} v(r)\, dr \\
&\le -c\int_0^R ds\, v'(s) \Big[\int_0^s r^{N-1} v^2(r)\, dr\Big]^{1/2} \Big[\int_0^s r^{N-1}\, dr\Big]^{1/2}
\end{aligned}$$

$$\leq c\left[\int_0^R r^{N-1}v^2(r)\,dr\right]^{1/2}\int_0^R N^{-1/2}|v'(s)|s^{N/2}\,ds$$
$$\leq c\left[\int_0^R r^{N-1}v^2(r)\,dr\right]^{1/2}\left[\int_0^R s^{N-1}|v'(s)|^2\,ds\right]^{1/2}[R^2/2N]^{1/2}.$$

The inequality of Lemma 4.5.3 follows. □

Remark 4.5.4. The situation "$|\Omega_u| < m$" of the proof of Theorem 4.5.2 can happen: it is the case if and only if there exists w a solution of

$$w \in H_0^1(D),\ |\Omega_w| < m,\ -\Delta w = f \text{ in the whole box } D. \tag{4.32}$$

Indeed, on the one hand, such a w is a solution of (4.31) since it minimizes $G(v)$ among *all* $v \in H_0^1(D)$. On the other hand, if $|\Omega_u| < m$, for all real numbers t and for all $\varphi \in C_0^\infty(D)$ whose support is of measure at most $m - |\Omega_u|$, we have $G(u) \leq G(u + t\varphi)$. We deduce from this, after differentiating at $t = 0$, that u satisfies (4.32).

Now, we are going to give a second proof of Theorem 4.5.2 by using a quite different and elementary approach (it is also mentioned in [131]).

For a measurable set $\Omega \subset D$, we introduce

$$\hat{H}_0^1(\Omega) = \left\{u \in H_0^1(D),\ u = 0 \text{ a.e. on } D\backslash\Omega\right\}.$$

Obviously $H_0^1(\Omega) \subset \hat{H}_0^1(\Omega)$ and the inclusion is often strict. *Equality means that Ω has some kind of regularity.* For instance, equality holds if Ω is stable in the sense of Theorem 3.4.6. The space $\hat{H}_0^1(\Omega)$ is closed in $H_0^1(D)$. This implies existence (and uniqueness) of $\hat{u}_\Omega$ a solution of

$$\hat{u}_\Omega \in \hat{H}_0^1(\Omega), \quad G(\hat{u}_\Omega) = \min\left\{G(v);\ v \in \hat{H}_0^1(\Omega)\right\}.$$

And it is also the unique solution of

$$\hat{u}_\Omega \in \hat{H}_0^1(\Omega),\ \forall\, v \in \hat{H}_0^1(\Omega), \quad \int_D \nabla\hat{u}_\Omega \nabla v = \langle f, v\rangle_{H^{-1}\times H_0^1}.$$

Note that if u_Ω denotes the solution of the usual Dirichlet problem associated with $H_0^1(\Omega)$, we have $G(\hat{u}_\Omega) \leq G(u_\Omega)$ since $H_0^1(\Omega) \subset \hat{H}_0^1(\Omega)$.

We obtain the following optimality result.

Proposition 4.5.5. *Let D be an open subset of $\mathbb{R}^N$, let $m \in (0, |D|)$ and $f \in H^{-1}(D)$. Then, there exists a quasi-open set $\hat{\Omega} \subset D$ such that*

$$|\hat{\Omega}| = m, \quad G(\hat{u}_{\hat{\Omega}}) = \min\{G(\hat{u}_\Omega);\ \Omega \subset D,\ |\Omega| \leq m\}. \tag{4.33}$$

And $\hat{\Omega}$ is also solution of problem (4.30).

Proof. Thanks to the Poincaré inequality applied in D and thanks to the identity $\int_D |\nabla \hat{u}_\Omega|^2 = \langle f, \hat{u}_\Omega \rangle_{H^{-1}\times H^1_0}$, we obtain the existence of a constant $C = C(m, D)$ such that $\|\hat{u}_\Omega\|_{H^1_0(D)} \le C\|f\|_{H^{-1}}$.

Let Ω_n be a minimizing sequence for (4.33). By the previous remark, $u_n = \hat{u}_{\Omega_n}$ is bounded in $H^1_0(D)$. Up to a subsequence, we may assume that u_n converges weakly in $H^1_0(D)$ and a.e. to some function u. At the limit

$$G(u) \le \inf\{G(\hat{u}_\Omega);\ \Omega \subset D,\ |\Omega| \le m\}, \qquad |\Omega_u| \le \liminf |\Omega_{u_n}| \le m.$$

We have, in particular, $G(u) \le G(\hat{u}_{\Omega_u}) \le G(u_{\Omega_u})$ and since $u \in H^1_0(\Omega_u)$, $G(u_{\Omega_u})$ is even smaller than or equal to $G(u)$, whence the equality. By the strict convexity of $G(\cdot)$, we have $u = u_{\Omega_u} = \hat{u}_{\Omega_u}$.

If $|\Omega_u| = m$, then $\hat{\Omega} := \Omega_u$ satisfies (4.33). If $|\Omega_u| < m$, we consider a quasi-open subset of D satisfying the two conditions $\hat{\Omega} \supset \Omega_u$, $|\hat{\Omega}| = m$. Then $G(u) \le G(\hat{u}_{\hat{\Omega}}) \le G(u_{\hat{\Omega}})$ and again since $u \in H^1_0(\hat{\Omega})$ and by uniqueness, $u = \hat{u}_{\hat{\Omega}} = u_{\hat{\Omega}}$ and $\hat{\Omega}$ is therefore a solution of (4.33).

Since $u = u_{\hat{\Omega}}$ in all cases, it follows that the quasi-open set $\hat{\Omega}$ is also a solution of problem (4.30). □

Remark 4.5.6. It is a little surprising that we can obtain a solution of the initial problem (4.30) that is associated with the spaces $H^1_0(\Omega)$ by using an "incorrect" definition of these spaces, namely $\hat{H}^1_0(\Omega)$. In fact, we even obtain the following better result.

Proposition 4.5.7. *Under the assumptions of Proposition 4.5.5, one may choose $\hat{\Omega}$ in such a way that $\hat{H}^1_0(\hat{\Omega}) = H^1_0(\hat{\Omega})$.*

Remark 4.5.8. This proposition says that problem (4.30) has a solution that is a little bit regular. In fact, as we will see in the proof, any solution of (4.30) is equal a.e. to a quasi-open set having at least that much regularity.

However, one may check that in the general case, even if the constraint is saturated (i.e., $|\Omega_u| = m$), the optimal quasi-open set has poor regularity and may even not be open or a.e. equal to an open set (see [171], [172], and Exercise 4.5). On the other hand, if f is more regular, for instance if it is a nonnegative bounded function, it can be proved that the optimal shape is a regular open set, but this requires much work (see [56], [57]).

Proof of Proposition 4.5.7. Let us first show that, given a quasi-open set $\omega \subset D$, there exists another unique quasi-open set $\Omega^* \subset D$ such that

$$\Omega^* = \omega \text{ a.e.}, \quad \hat{H}^1_0(\omega) = H^1_0(\Omega^*)\ (= \hat{H}^1_0(\Omega^*)). \tag{4.34}$$

Uniqueness follows from Proposition 3.3.44. For the existence part, let us set $\Omega^* := \cup_n[v_n \neq 0] \cup \omega$, where (v_n) is a dense sequence in $\hat{H}_0^1(\omega)$. Then Ω^* is quasi-open as a denumerable union of quasi-open sets and we have $\Omega^* = \omega$ a.e. In particular,

$$H_0^1(\Omega^*) \subset \hat{H}_0^1(\Omega^*) = \hat{H}_0^1(\omega). \tag{4.35}$$

Let $v \in \hat{H}_0^1(\omega)$. There exists a subsequence $(v_{n_k})_k$ converging to v in $H_0^1(D)$ and quasi-everywhere (for their quasi-continuous representatives). Since for all k, $v_{n_k} = 0$ q.e. outside $[v_{n_k} \neq 0]$ and therefore outside Ω^*, we also have $v = 0$ q.e. outside Ω^* and consequently $v \in H_0^1(\Omega^*)$. Thus equality holds in (4.35).

Let us apply (4.34) to $\omega := \hat{\Omega}$ obtained in Proposition 4.5.5. We then have

$$|\Omega^*| = |\hat{\Omega}| = m, \quad u_{\Omega^*} = \hat{u}_{\hat{\Omega}} = u_{\hat{\Omega}},$$

and Ω^* is also a solution of (4.33). □

Remark 4.5.9. By (4.34), we have

$$\inf\{G(u_\Omega);\ \Omega \subset D \text{ quasi-open set}, |\Omega| \le m\}$$
$$= \inf\{G(\hat{u}_\Omega);\ \Omega \subset D \text{ quasi-open set}, |\Omega| \le m\}.$$

This explains why we may equally work with $H_0^1(\Omega)$ or with $\hat{H}_0^1(\Omega)$ for this minimization problem.

We are going to use the same idea to solve the problem with surface tension (or with a perimeter constraint), namely,

$$J(\Omega^*) = \min\{J(\Omega);\ \Omega \subset D, \text{ measurable}, |\Omega| = m\}, \tag{4.36}$$

where, for given $\sigma > 0$,

$$J(\Omega) = G(u_\Omega) + \sigma P(\Omega).$$

As we saw in Chapter 1, this functional plays an important role in applications, as well as problem (4.36). To simplify, we assume that D is bounded.

Proposition 4.5.10. *Let D be an open set, let $f \in H^{-1}(D)$, $m \in (0, |D|)$, and $\sigma > 0$. Then there exists Ω^* measurable a solution of* (4.36).

Proof. Let us set (with the notation of Proposition 4.5.5)

$$\hat{J}(\Omega) = G(\hat{u}_\Omega) + \sigma P(\Omega).$$

By the proof of Proposition 4.1.1, we know there exists a measurable set $\Omega \subset D$ such that $|\Omega| = m$, $P(\Omega) < +\infty$ and therefore such that $\hat{J}(\Omega) < +\infty$. As for $\sigma = 0$, we show that $\hat{J}(\cdot)$ is bounded from below.

Let Ω_n be a minimizing sequence for $\hat{J}(\cdot)$ with $|\Omega_n| = m$. Since $P(\Omega_n)$ is bounded and since $u_n = \hat{u}_{\Omega_n}$ is bounded in $H_0^1(D)$, up to a subsequence, we may assume that u_n converges weakly in $H_0^1(D)$ and a.e. to some function u and that there exists $\hat{\Omega} \subset D$ measurable such that χ_{Ω_n} converges in $L^1(D)$ to $\chi_{\hat{\Omega}}$. We then have

$$|\hat{\Omega}| = m, \quad P(\hat{\Omega}) \le \liminf P(\Omega_n), \quad G(u) \le \liminf G(u_n),$$
$$\sigma P(\hat{\Omega}) + G(u) \le \inf\{\sigma P(\Omega) + G(\hat{u}_\Omega);\ \Omega \subset D \text{ measurable}, |\Omega| = m\}.$$

But since $(1 - \chi_{\Omega_n})u_n = 0$ a.e. in D, we obtain that $(1 - \chi_{\hat{\Omega}})u = 0$ a.e. in D, which means that $u \in \hat{H}_0^1(\hat{\Omega})$. Thus we have, in particular,

$$\sigma P(\hat{\Omega}) + G(u) \le \sigma P(\hat{\Omega}) + G(\hat{u}_{\hat{\Omega}}),$$

which implies $G(u) \le G(\hat{u}_{\hat{\Omega}})$, that is, $u = \hat{u}_{\hat{\Omega}}$. We deduce

$$\hat{J}(\hat{\Omega}) = \min\left\{\hat{J}(\Omega);\ \Omega \subset D, \text{ measurable}, |\Omega| = m\right\}.$$

If $|\Omega_u| = m$, we have $\Omega_u = \hat{\Omega}$ a.e. By (4.34), there exists Ω^* quasi-open such that $\hat{H}_0^1(\hat{\Omega}) = H_0^1(\Omega^*)$, $\Omega^* = \hat{\Omega}$ a.e. Since $u = u_{\Omega^*}$, we may check that Ω^* is a solution of problem (4.36) *and it is even quasi-open.*

Let us now assume that $|\Omega_u| < m$. Since $\Omega_u \subset \hat{\Omega}$ a.e., one may choose a representative Ω^* of $\hat{\Omega}$ such that $\Omega_u \subset \Omega^*$ q.e. (we may take $\Omega^* := \hat{\Omega} \cup \Omega_u$) so that $u \in H_0^1(\Omega_u) \subset H_0^1(\Omega^*) \subset \hat{H}_0^1(\Omega^*) = \hat{H}_0^1(\hat{\Omega})$. Thus again $u = u_{\Omega^*} = \hat{u}_{\hat{\Omega}}$ and Ω^* is a solution of problem (4.36). □

Remark 4.5.11. Note that "most of the time", the solution is at least quasi-open: it is the case when $|\Omega_u| = m$. As for the problem without a perimeter, it is however possible that $|\Omega_u| < m$. It is the case if $f \equiv 0$: we are then back to the isoperimetric problem (4.1). We see that claiming existence of an optimal quasi-open set for this problem is already a regularity result! (In fact, we may prove that there exists an open optimal shape in this specific case.)

To end this section, let us mention a result on the "exterior shaping problem" described in Section 1.3.1. Then the variable domain Ω is an *exterior domain*, which means that it is the complement of a compact set, and the measure constraint is made on this compact complement (which represents the domain occupied by the liquid). Intuitively, we see that it is more difficult to obtain existence of an optimal shape since the involved forces want to push the liquid away to infinity. Indeed, the absolute minimum of the energy functional is reached when the liquid is sent to infinity. An important consequence is that the equilibrium shapes (which do exist; see, e.g., [174]) have to be found among the *local minima of the energy*. The following result may be proved (see [187]).

Proposition 4.5.12. *Let $f \in L^2(\mathbb{R}^2)$ have compact support, let $\mathcal{O}$ be the family of open subsets of $\mathbb{R}^2$ whose complement is compact, and let $\mathcal{W}(\Omega)$ be the closure of $C_0^\infty(\Omega)$ for the norm $w \to \left[\int_\Omega |\nabla w|^2\right]^{1/2}$. We set*

$$E(\Omega, w) = \int_\Omega \frac{1}{2}|\nabla w|^2 - fw \quad \forall\, \Omega \in \mathcal{O} \text{ and } w \in \mathcal{W}(\Omega),$$
$$c(m) = \inf\left\{E(\Omega, w);\ \Omega \in \mathcal{O},\ |\Omega^c| = m,\ w \in \mathcal{W}(\Omega)\right\}.$$

Then

$$\left(c(m) < +\infty\right) \Leftrightarrow \left(\int_{\mathbb{R}^2} f = 0\right).$$

Moreover, for all $m > 0$,

$$c(m) = \inf\left\{E(\mathbb{R}^2, w);\ w \in \mathcal{W}(\mathbb{R}^2)\right\},$$
$$c(m) < E(\Omega, w) \quad \forall\, \Omega \in \mathcal{O} \text{ and } w \in \mathcal{W}(\Omega).$$

4.6 Effects of perimeter constraints

We have seen in several examples in the previous sections that the presence of perimeter terms $P(\Omega)$ or $P_D(\Omega)$ in functionals adds compactness properties to minimizing sequences and therefore contributes to the existence of optimal shapes.

Nevertheless it is not true that adding $P(\Omega)$ to functionals of the type $\Omega \to \int_\Omega F(x, u_\Omega, \nabla u_\Omega)$ (see 4.2) implies existence of a minimal shape: the homogenization phenomenon may still appear. We saw that in the first example of Section 4.2, $P(\Omega_n) \le Cn^2 e^{-2dn^2} \to 0$ when $n \to \infty$. Thus, Ω_n is still a minimizing sequence for the functional $J(\cdot) + P(\cdot)$.

However, there are examples where perimeter terms do bring enough compactness to lead to existence of an optimal shape: this is the case for the example studied now.

The goal is to find the distribution of two conducting materials in a bounded open set $D \subset \mathbb{R}^N$ that has the least energy. Let us denote by α and β the conductivity (assumed to be constant) of each of these materials ($0 < \alpha < \beta$). The problem is to find a domain $\Omega \subset D$ (it is in fact the domain occupied by the material of smaller conductivity α) that minimizes the functional

$$J(\Omega) = -\int_D f(x) u_\Omega(x)\, dx,$$

where $f \in L^2(D)$ represents the electrical source density and u_Ω is the electrostatic potential solution of the problem

$$\begin{cases} -\operatorname{div}\left((\alpha\chi_\Omega + \beta\chi_{D\setminus\Omega})\nabla u_\Omega\right) = f & \text{in } D, \\ u_\Omega = 0 & \text{on } \partial D. \end{cases} \tag{4.37}$$

We know that, for a given measurable set $\Omega \subset D$, there exists a unique solution u_Ω of problem (4.37) in $H_0^1(D)$ that minimizes the functional

$$v \in H_0^1(D) \to J(v,\Omega) := \int_D (\alpha\chi_\Omega + \beta\chi_{D\setminus\Omega})|\nabla v|^2\,dx - 2\int_D fv\,dx \tag{4.38}$$

(and we have $J(\Omega) = J(u_\Omega, \Omega)$). Thus one may reformulate the shape optimization problem by directly minimizing the double functional $(v,\Omega) \to J(v,\Omega)$ on $H_0^1(D) \times \mathcal{O}_{\mathrm{ad}}$, where $\mathcal{O}_{\mathrm{ad}}$ is the family of admissible domains.

This example is a particular case of a class of functionals of the type

$$(v,\Omega) \to \int_D j(x, \chi_\Omega, v, \nabla v),$$

and we know that, even for reasonable j and $\mathcal{O}_{\mathrm{ad}}$, there does not exist in general optimal shapes associated with these functionals. There is most often a homogenization phenomenon (see [215], [244], [12], [13]). On the other hand, adding a perimeter will generally imply existence. This is what we are going to show for the significant example considered in [18], which contains the case of minimal energy just mentioned.

For all subsets Ω of D, we denote by $\alpha_\Omega(x)$ the function defined on D by

$$\alpha_\Omega(x) = \alpha\chi_\Omega(x) + \beta\chi_{D\setminus\Omega}(x).$$

We are given two functions g_1 and g_2 defined on $D \times \mathbb{R}$ and with values in $\mathbb{R}$ ($g_i = g_i(x,z)$). We assume they are lower semicontinuous with respect to the second variable (a.e. x) and integrable with $\sup_{|z|\le r}\{|g_1(\cdot,z)| + |g_2(\cdot,z)|\}$ integrable for all r. We set

$$g_\Omega(x,s) = \chi_\Omega(x)g_1(x,s) + \chi_{D\setminus\Omega}(x)g_2(x,s).$$

We first consider the following minimization problem for a given measurable $\Omega \subset D$:

$$\min\left\{\textstyle\int_D [\alpha_\Omega(x)|\nabla u(x)|^2 + g_\Omega(x,u)]\,dx,\ u \in H_0^1(D)\right\}. \tag{4.39}$$

We easily show the existence of a minimum under reasonable hypotheses, like, for instance,

Lemma 4.6.1. *Assume, moreover, that g_1 and g_2 satisfy*

$$g_i(x,s) \ge \gamma(x) - ks^2, \quad i = 1,2, \tag{4.40}$$

where $\gamma \in L^1(D)$ and $k < \alpha\lambda_1$ with λ_1 the first eigenvalue of the Dirichlet–Laplacian on D. Then, the minimization problem (4.39) *possesses (at least) one solution u_Ω. We denote by $E(\Omega)$ the minimum value of the functional defined in* (4.39).

We are now interested in the following shape optimization problem where $\sigma > 0$ is given:

$$\min \{E(\Omega) + \sigma P_D(\Omega);\ \Omega \subset D \text{ measurable}\}\,. \tag{4.41}$$

Theorem 4.6.2. *Under the assumptions of Lemma 4.6.1, the minimization problem* (4.41) *has a solution.*

Proof. Let Ω_n be a minimizing sequence. The relative perimeter of Ω_n is bounded. By Theorem 2.3.11, we can extract a subsequence, still denoted by Ω_n, such that χ_{Ω_n} converges in $L^1(D)$ to χ_Ω, where Ω is a measurable subset of D. Let us show that Ω is a solution of problem (4.41). Let us denote by u_n a solution of (4.39) associated with Ω_n. Since

$$E(\Omega_n) = \int_D [\alpha_{\Omega_n}(x)|\nabla u_n(x)|^2 + g_{\Omega_n}(x, u_n)]\,dx \le M,$$

using (4.40) and the definition of α_Ω, we have

$$\int_D \alpha|\nabla u_n(x)|^2 + \gamma(x) - k|u_n(x)|^2\,dx \le M.$$

We deduce, using the bound on k,

$$\alpha\int_D |\nabla u_n(x)|^2 - \lambda_1|u_n(x)|^2\,dx \le M - \int_D \gamma(x)\,dx = M_1.$$

The Poincaré inequality (whose best constant is precisely $1/\lambda_1$) shows that the sequence u_n is bounded in $H_0^1(D)$. Therefore, it converges (up to a subsequence) weakly in H_0^1 and strongly in L^2 to some function $u \in H_0^1(D)$. By l.s.c. of the norm and of the function g_Ω and by the pointwise convergence of χ_{Ω_n}, we have

$$\int_D \alpha_\Omega(x)|\nabla u(x)|^2\,dx \le \liminf_n \int_D \alpha_{\Omega_n}(x)|\nabla u_n(x)|^2\,dx$$

and

$$\int_D g_\Omega(x, u)\,dx \le \liminf \int_D g_{\Omega_n}(x, u_n)\,dx.$$

Since we have by definition

$$E(\Omega) \le \int_D [\alpha_\Omega(x)|\nabla u(x)|^2 + g_\Omega(x, u)]\,dx,$$

we deduce

$$E(\Omega) \le \liminf E(\Omega_n),$$

and the theorem follows by the l.s.c. of the perimeter with respect to L^1 convergence; see Proposition 2.3.7. □

Remark 4.6.3. We chose here the case where the functional is the same as the one that is minimized by the state function u_Ω. We could decouple the two (as is the case in many applications) and consider a functional of the form

$$\Omega \to \int_D j(x, \chi_\Omega, u_\Omega, \nabla u_\Omega),$$

where u_Ω is the solution of (4.37). Under reasonable hypotheses, in particular of l.s.c. type for j, a similar existence result of an optimal shape may be obtained. The extra point is to check that the solution u_{Ω_n} converges strongly in $H^1_0(D)$ to u_Ω when α_{Ω_n} converges pointwise to α_Ω. This is explained in detail in [65].

In all these minimization problems, one also gets existence when adding the volume constraint $|\Omega| = m$ since it is preserved by strong convergence of the χ_{Ω_n}.

4.7 Monotonicity of the functional

In this section, we are going to give a general existence result for an optimal shape for functionals that are nonincreasing with respect to inclusion:

$$J(\Omega_1) \ge J(\Omega_2) \quad \text{when } \Omega_1 \subset \Omega_2 \subset D, \tag{4.42}$$

where D is a fixed bounded open set. If $J(\cdot)$ is also assumed to be l.s.c. for γ-convergence, then there exists a *quasi-open set* minimizing $J(\cdot)$ with prescribed measure. This result is due to G. Buttazzo and G. Dal Maso; see [79].

We are going to work with the following class that we introduced earlier in this book:

$$\mathcal{A}(D) = \{\Omega \subset D;\ \Omega \text{ quasi-open}\}.$$

We recall (see Proposition 4.5.1) that for all $f \in H^{-1}(D)$, the problem

$$\begin{cases} u^f_\Omega \in H^1_0(\Omega)\ \forall\, v \in H^1_0(\Omega), \\ \int_\Omega \nabla u^f_\Omega . \nabla v\, dx = \langle f, v\rangle_{H^{-1}\times H^1_0}, \end{cases} \tag{4.43}$$

has a unique solution. We introduce the resolvent operator

$$R_\Omega : H^{-1}(D) \to H^1_0(\Omega) \subset H^1_0(D), \tag{4.44}$$

which to each $f \in H^{-1}(D)$ associates the unique solution u_Ω^f of (4.43). Choosing $v = u_\Omega^f$ in (4.43), we see that R_Ω is a linear continuous operator from $H^{-1}(D)$ into $H_0^1(\Omega)$. Moreover R_Ω is symmetric:

$$\forall f, g \in H^{-1}(D), \quad \langle f, R_\Omega(g)\rangle_{H^{-1}\times H_0^1} = \langle R_\Omega(f), g\rangle_{H^{-1}\times H_0^1}.$$

With this definition, the notion of γ-convergence takes the following form (see Definition 3.5.1 and the remarks following it):

Definition 4.7.1. Let Ω_n be a sequence of quasi-open sets and let Ω be another quasi-open set, each included in D. We say that Ω_n γ-converges to Ω if $R_{\Omega_n}(f)$ converges to $R_\Omega(f)$ strongly in $H_0^1(D)$ for all $f \in H^{-1}(D)$.

This notion of γ-convergence for the quasi-open sets has essentially all the same properties as those we saw for the open sets. Let us summarize them in the following proposition.

Proposition 4.7.2. *With the above notation,*

(i) there exists a constant $C = C(D)$ such that

$$\forall \Omega \in \mathcal{A}(D), \quad \|R_\Omega(f)\|_{H_0^1(D)} \le C\|f\|_{H^{-1}(D)};$$

(ii) if $R_{\Omega_n}(f)$ converges to $R_\Omega(f)$ in $L^2(D)$, the convergence holds strongly in $H_0^1(D)$;

(iii) if the γ-limit exists, it is unique (up to a set of zero capacity);

(iv) (Šverak) Ω_n γ-converges to Ω if and only if $R_{\Omega_n}(1)$ converges to $R_\Omega(1)$ in $L^2(D)$.

Proof. Points (i) and (ii) may be proved exactly as for open sets (see Proposition 3.2.1 and Corollary 3.2.2).

For (iii), we remember that $R_{\Omega_n}(f)$ is the projection onto $H_0^1(\Omega_n)$ of $R_D(f)$ since

$$\forall v \in H_0^1(\Omega_n), \quad \int_D (R_D(f) - R_{\Omega_n}(f))v = 0.$$

Thus if Ω_n γ-converges to Ω and to $\hat{\Omega}$, the projections of $R_D(f)$ on $H_0^1(\Omega)$ and on $H_0^1(\hat{\Omega})$ coincide for all f. It follows that these two spaces are equal and therefore that $\Omega = \hat{\Omega}$ by Proposition 3.3.44.

For (iv), let $(u^\star)^f$ be the weak limit in $H_0^1(D)$ of $R_{\Omega_n}(f)$ (or of a subsequence). We first remark that $(u^\star)^f \in H_0^1(\Omega)$. Indeed, if $f \in L^\infty$, we have $|R_{\Omega_n}(f)| \le \|f\|_\infty R_{\Omega_n}(1)$

so that, at the limit, $|(u^\star)^f| \le \|f\|_\infty R_\Omega(1)$, which implies $(u^\star)^f \in H_0^1(\Omega)$. We finish by using the density in $H_0^1(\Omega)$ of the space $\{R_\Omega(f);\ f \in L^\infty(D)\}$ (see Proposition 3.3.44).

It remains to prove that $(u^\star)^f$ satisfies the required variational equality to be identified with $R_\Omega(f)$. Let $\varphi \in H_0^1(\Omega)^+$. Since $\varphi_n = \inf\{\varphi, qR_{\Omega_n}(1)\} \in H_0^1(\Omega_n)$ $(q > 0)$, we have

$$\int_D \nabla u_{\Omega_n}^f \nabla\varphi_n = \langle f, \varphi_n\rangle_{H^{-1}\times H_0^1}.$$

But, by strong convergence of φ_n in H_0^1, if we set $\Phi = \inf\{\varphi, qR_\Omega(1)\}$, we also have

$$\int_D \nabla(u^\star)^f \nabla\Phi = \langle f, \Phi\rangle_{H^{-1}\times H_0^1}. \tag{4.45}$$

If $\varphi = R_\Omega(g)$ with $g \in L^\infty(D)$, then $\varphi \le \|g\|_\infty R_\Omega(1)$, so that $\Phi = \varphi$ as soon as $q \ge \|g\|_\infty$, whence (4.45) with Φ replaced by φ. We finish by density like above. □

We will also use the following characterization of γ-convergence.

Lemma 4.7.3. *The sequence Ω_n γ-converges to Ω if and only if R_{Ω_n} converges to R_Ω in the uniform operator topology $\mathcal{L}(L^2(D))$, that is,*

$$\lim_{n\to\infty} \sup_{\|f\|_{L^2(D)}\le 1} \|R_{\Omega_n}(f) - R_\Omega(f)\|_{L^2(D)} = 0.$$

Proof. Convergence in $\mathcal{L}(L^2(D))$ implies γ-convergence, for instance by (iv) of the previous proposition. Conversely, assume that Ω_n γ-converges to Ω. The operators R_{Ω_n}, R_Ω are continuous from H^{-1} to H_0^1 and consequently from L^2 into L^2. There exists f^n in the unit ball of $L^2(D)$ such that

$$\sup_{\|f\|_{L^2(D)}\le 1} \|R_{\Omega_n}(f) - R_\Omega(f)\|_{L^2(D)} = \|R_{\Omega_n}(f^n) - R_\Omega(f^n)\|_{L^2(D)}.$$

Indeed, if f_k is a maximizing sequence, there exists a subsequence weakly converging to some f^n that also belongs to the unit ball of $L^2(D)$. Since the embedding of $L^2(D)$ into $H^{-1}(D)$ is compact (as the adjoint of the compact embedding of $H_0^1(D)$ into $L^2(D)$), we may deduce the above equality when $k \to \infty$.

Let us now repeat the same process with the sequence f^n: there exists f in the unit ball of $L^2(D)$ such that f^n (or a subsequence) converges weakly in $L^2(D)$ and strongly in $H^{-1}(D)$ to f. Then let n_1 be such that for $n \ge n_1$, we have at the same time,

$$\|f^n - f\|_{H^{-1}(D)} \le \frac{\varepsilon}{4} \quad\text{and}\quad \|R_{\Omega_n}(f) - R_\Omega(f)\|_{L^2(D)} \le \frac{\varepsilon}{2},$$

the second inequality coming from the very definition of γ-convergence of Ω_n to Ω. We then have

$$\begin{aligned}\sup_{\|f\|_{L^2(D)}\le 1} \|R_{\Omega_n}(f) - R_\Omega(f)\|_{L^2(D)} &= \|R_{\Omega_n}(f^n) - R_\Omega(f^n)\|_{L^2(D)} \\ &\le \|R_{\Omega_n}(f) - R_\Omega(f)\|_{L^2(D)} \\ &\quad + \|R_{\Omega_n}(f^n - f) - R_\Omega(f^n - f)\|_{L^2(D)} \\ &\le \frac{\varepsilon}{2} + \|R_{\Omega_n} - R_\Omega\|_{\mathcal{L}(H^1,H_0^1)}\frac{\varepsilon}{4} \\ &\le \frac{\varepsilon}{2} + 2\frac{\varepsilon}{4} = \varepsilon,\end{aligned}$$

which proves the result. □

As a corollary, we see that γ-convergence implies convergence of eigenvalues; see also [200].

Corollary 4.7.4. *Let k be an integer with $k \ge 1$. Then the mapping $\Omega \longrightarrow \lambda_k(\Omega)$, which associates to a quasi-open set Ω the kth eigenvalue of the Dirichlet–Laplacian operator on Ω, is continuous in the topology of γ-convergence and is nonincreasing for the inclusion (the eigenvalues are indexed according to their multiplicity).*

Proof. Let us first remark that spectral theory on a quasi-open set is exactly the same as on an open set: the resolvent operator R_Ω is positive, compact, self-adjoint on $L^2(D)$. Its restriction to $L^2(\Omega)$ is one to one: indeed, if $R_\Omega(f) = 0$ for $f \in L^2(\Omega)$ (extended by 0 in D), then $\int_\Omega fv = 0$ for all $v \in H_0^1(\Omega)$. This implies $f \equiv 0$ by Proposition 3.3.44. If the capacity of the quasi-open set Ω is not zero, the space $L^2(\Omega)$ is not reduced to $\{0\}$ and is of infinite dimension. The eigenvalues of the Dirichlet–Laplacian operator are the inverses of the eigenvalues of the restriction of R_Ω to $L^2(\Omega)$, which has a sequence of positive eigenvalues that decreases to 0.

The continuity in the topology of γ-convergence may then be deduced from a classical continuity result of the eigenvalues of a compact operator for the uniform convergence of operators (see, e.g., Dunford–Schwartz [133, Vol. 2]).

Note that if Ω is of zero capacity, then $H_0^1(\Omega) = \{0\}$. By convention, we will set $\lambda_k(\Omega) = +\infty$ for all k. This definition allows us to make the mapping $\Omega \longrightarrow \lambda_k(\Omega)$ continuous in all cases, including the case when the sequence of quasi-open sets Ω_n converges to a quasi-open set Ω of zero capacity. Indeed, in this case, R_{Ω_n} converges uniformly to 0 and a result already recalled implies that the eigenvalues of R_{Ω_n} also converge to 0 so that their inverses converge to ∞.

Monotonicity is a consequence of the Courant–Fischer formula

$$\lambda_k(\Omega) = \min_{E_k \in \mathcal{H}_k} \max_{v \in E_k \setminus \{0\}} \frac{\int_\Omega |\nabla v|^2}{\int_\Omega v^2}, \tag{4.46}$$

where $\mathcal{H}_k$ denotes the family of linear subspaces of dimension k of $H_0^1(\Omega)$. This formula is well known for open sets (see, e.g., [123] or [263]). We may check that it is preserved for quasi-open sets, for instance by using that a quasi-open set is the decreasing limit of the open sets $\Omega \cup \omega_n$, where ω_n is a decreasing sequence of open sets whose capacity tends to 0. This implies γ-convergence of $\Omega \cup \omega_n$ to Ω and therefore the convergence of the $\lambda_k(\cdot)$. □

Let us now give one more example of a functional that is continuous in the topology of γ-convergence. Let $g : D \times \mathbb{R} \times \mathbb{R}^N \longrightarrow \overline{\mathbb{R}}$ be a measurable function such that, for a.e. $x \in D$, $g(x, \cdot, \cdot)$ is lower semicontinuous on $\mathbb{R} \times \mathbb{R}^N$ and, moreover,

$$g(x, s, \xi) \geq -\alpha(x) - \beta(s^2 + |\xi|^2), \tag{4.47}$$

with $\alpha \in L^1(D)$ and $\beta \in \mathbb{R}$, both being given. We are also given $f \in H^{-1}(D)$. With all $\Omega \in \mathcal{A}(D)$, we associate the function $u_\Omega = R_\Omega(f)$ as defined above. Then we have

Proposition 4.7.5. *The functional J defined on $\mathcal{A}(D)$ by*

$$J(\Omega) = \int_D g(x, u_\Omega(x), \nabla u_\Omega(x))\, dx$$

is lower semicontinuous in the topology of γ-convergence.

This comes, on the one hand, from the continuity of the mapping $\Omega \longrightarrow u_\Omega$ when $\mathcal{A}(D)$ is equipped with the topology of γ-convergence and when $H_0^1(D)$ is equipped with its strong topology, and, on the other hand, from the lower semicontinuity of the mapping $u \longrightarrow \int_D g(\cdot, u, \nabla u)$ in the strong topology of $H_0^1(D)$, which is a consequence of the assumptions on g and of Fatou's lemma.

Let us now state the main result of this section.

Theorem 4.7.6 (Buttazzo–Dal Maso). *Let $J : \mathcal{A}(D) \to (-\infty, +\infty]$ be a functional satisfying*

(i) J is lower semicontinuous in the topology of γ-convergence;

(ii) J is nonincreasing for the inclusion "if $\Omega_1 \subset \Omega_2$, then $J(\Omega_1) \geq J(\Omega_2)$".

Then for all $c \in (0, |D|)$, the problem

$$\min \{J(\Omega);\ \Omega \in \mathcal{A}(D),\ |\Omega| = c\}$$

has a solution.

With respect to applications, obviously assumption (ii) is the most restrictive. We will give below two interesting applications of this theorem that cover a rather wide range of problems.

The proof we will give here is not exactly the same as the one given in the original paper by Buttazzo and Dal Maso. Rather it borrows ideas from [67]. It became clear in Chapter 3 that $\mathcal{A}(D)$ is not compact in the topology of γ-convergence. This is why we introduce here the notion of *weak γ-convergence* for which $\mathcal{A}(D)$ will be sequentially compact.

For all $\Omega \in \mathcal{A}(D)$, let us denote by $w_\Omega = R_\Omega(1)$ the solution of the Dirichlet problem (4.43) with data $f \equiv 1$.

Definition 4.7.7. We will say that a sequence of quasi-open sets Ω_n of $\mathcal{A}(D)$ γ-converges *weakly* to a quasi-open set Ω if the sequence of functions w_{Ω_n} converges weakly in $H_0^1(D)$ to some function w such that $\Omega = \{x \in D;\ w(x) > 0\}$.

Remark 4.7.8.

- It is easy to check that γ-convergence implies weak γ-convergence (see [67]). One may check it as follows. Assume w_{Ω_n} converges strongly in H_0^1 to w_Ω. Let us check that
$$H_0^1(\Omega) = H_0^1([w_\Omega > 0]). \tag{4.48}$$
Then by the uniqueness property of Proposition 3.3.44, this will imply that $\Omega = [w_\Omega > 0]$ q.e., whence the weak γ-convergence.

  We obviously have $[w_\Omega > 0] \subset \Omega$ q.e. If now $v \in H_0^1(\Omega)$, by the same Proposition 3.3.44, v is the limit of $R_\Omega(g_p)$ for some $g_p \in L^\infty(D)$. In particular, $|R_\Omega(g_p)| \leq \|g\|_\infty w_\Omega$, which implies that $R_\Omega(g_p) \in H_0^1([w_\Omega > 0])$. This property is preserved at the limit for v. Whence the equality in (4.48).

- In general the w of Definition 4.7.7 does not coincide with w_Ω. This is in fact true only if Ω_n γ-converges to Ω.

- It can be proved (see [79]) that if Ω_n γ-converges weakly to Ω, then $H_0^1(\Omega)$ contains all weak limits of sequences of $H_0^1(\Omega_n)$. It means that property (M2) of Mosco convergence does hold.

- Weak γ-convergence is sequentially compact. Indeed if Ω_n is a sequence of quasi-open sets in $\mathcal{A}(D)$, we immediately have that w_{Ω_n} is bounded in $H_0^1(D)$. Thus there exists a subsequence converging to some function w. Setting $\Omega = \{x \in D;\ w(x) > 0\}$, we then obtain that the sequence Ω_n γ-converges weakly to Ω.

The key point in the proof of Theorem 4.7.6 is the following.

Proposition 4.7.9. *Let $J : \mathcal{A}(D) \to (-\infty, \infty]$ be a functional that is nonincreasing for the inclusion. Then J is lower semicontinuous in the topology of γ-convergence if and only if it is lower semicontinuous in the topology of weak γ-convergence.*

Proof of Theorem 4.7.6. It is a direct consequence of the above proposition. Note first that J is bounded below by $J(D) > -\infty$. Assume that $J \not\equiv +\infty$ (otherwise any quasi-open set is a solution). Let Ω_n be a minimizing sequence. It γ-converges weakly (up to a subsequence) to a quasi-open set Ω^* which, thanks to the l.s.c. properties of $J(\cdot)$ and $|\cdot|$, satisfies

$$|\Omega^*| \le c, \quad J(\Omega^*) \le \inf\{J(\Omega),\ \Omega \subset D \text{ quasi-open } |\Omega| = c\}.$$

If $|\Omega^*| = c$, then Ω^* is a solution. If $|\Omega^*| < c$, since J is nonincreasing, any quasi-open subset of D containing Ω^* and of measure c is a solution (and there exist such quasi-open sets). □

Proposition 4.7.9 is essentially a consequence of Lemma 4.7.11 stated below. This lemma is itself a consequence of Lemma 4.7.10 stated just before it. They both are stated in the original paper [79] as well as in [67]. Both are interesting for themselves. We will prove them later (with a new proof for the first one).

Lemma 4.7.10. *Assume w_{Ω_n} converges weakly in $H_0^1(D)$ to w. Let $v_n \in H_0^1(\Omega_n)$ converge weakly in $H_0^1(D)$ to v. Then $v \in H_0^1(\Omega_w)$.*

Lemma 4.7.11. *Let Ω_n be a sequence of quasi-open subsets of D such that w_{Ω_n} converges weakly in $H_0^1(D)$ to $w \in H_0^1(\Omega)$, where Ω is a quasi-open subset of D. Then there exist a sequence of integers $(n_k)_{k\ge 0}$ and a sequence of quasi-open sets $(C_k)_{k\ge 0}$ that γ-converges to Ω with $\Omega_{n_k} \subset C_k \subset D$.*

We postpone the proofs of Lemmas 4.7.10 and 4.7.11 and we now prove Proposition 4.7.9.

Proof of Proposition 4.7.9. According to the first point of Remark 4.7.8, it is sufficient to prove that lower semicontinuity in the topology of γ-convergence implies lower semicontinuity in the topology of weak γ-convergence.

Thus let Ω_n be a sequence of quasi-open subsets of D that γ-converges weakly to the quasi-open set Ω. Let $L := \liminf J(\Omega_n)$. Up to a subsequence, we may assume that $L = \lim J(\Omega_n)$. By definition, w_{Ω_n} converges weakly to $w \in H_0^1(D)$ with $[w > 0] = \Omega$, so that $w \in H_0^1(\Omega)$. Let $(n_k, C_k)_{k\ge 0}$ be associated to the sequence $(\Omega_n)_{n\ge 0}$ according to Lemma 4.7.11. By monotonicity of J and thanks to its lower semicontinuity in the topology of strong γ-convergence, we have

$$J(\Omega) \le \liminf J(C_k) \le \liminf J(\Omega_{n_k}).$$

□

Proof of Lemma 4.7.10. As it is sufficient to prove that $\tilde{v} = \inf\{|v|, k\} \in H_0^1(\Omega_w)$ for all $k > 0$ and as $\tilde{v}$ is the weak limit of $\inf\{|v_n|, k\} \in H_0^1(\Omega_n)$, we may assume that v_n, v are nonnegative and bounded.

Let us denote by k a uniform upper bound of $\|v_n\|_\infty$ and let us introduce for all $\lambda > 0$ the variational solution of

$$v_n^\lambda \in H_0^1(\Omega_n), \quad v_n^\lambda - \lambda\Delta v_n^\lambda = v_n \text{ in } \Omega_n,$$

that is, $\lambda v_n^\lambda = R_{\Omega_n}(v_n - v_n^\lambda)$, and also

$$\forall\, v \in H_0^1(\Omega_n), \quad \int_D v_n^\lambda v + \lambda \nabla v_n^\lambda \nabla v = \int_D v_n v. \tag{4.49}$$

Let us show that

$$\|v_n^\lambda\|_{H_0^1(D)} \le C, \quad \|v_n^\lambda\|_\infty \le k, \tag{4.50}$$

where C is independent of n, λ. We apply (4.49) with $v := v_n^\lambda - v_n \in H_0^1(\Omega_n)$ to obtain

$$\int_D (v_n^\lambda - v_n)^2 + \lambda |\nabla(v_n^\lambda - v_n)|^2 = -\lambda \int_D \nabla v_n \nabla(v_n^\lambda - v_n). \tag{4.51}$$

Using the Schwarz inequality in the last integral, we deduce that $\nabla(v_n^\lambda - v_n)$ is bounded in $L^2(D)$ by the norm of ∇v_n in $L^2(D)$. Whence the existence of C in (4.50) depending only on $\sup_n \|v_n\|_{H_0^1(D)}$. Then after subtracting k on both sides of (4.49), we choose $v := (v_n^\lambda - k)^+ \in H_0^1(\Omega_n)$ to obtain

$$\int_D [(v_n^\lambda - k)^+]^2 + \lambda |\nabla(v_n^\lambda - k)^+|^2 = \int_D (v_n - k)(v_n^\lambda - k)^+ \le 0.$$

Whence the L^∞-bound of v_n^λ in (4.50). We deduce from this bound, together with $\lambda v_n^\lambda = R_{\Omega_n}(v_n - v_n^\lambda)$, that $v_n^\lambda \le 2k\lambda^{-1} w_{\Omega^n}$. Let v^λ be a weak-$H_0^1(D)$ limit point of $n \to v_n^\lambda$. Then $0 \le v^\lambda \le 2k\lambda^{-1} w$. This implies that $v^\lambda \in H_0^1(\Omega_w)$ (see Definition 3.3.43). But we also deduce from (4.51) that

$$\|v_n^\lambda - v_n\|_{L^2(D)}^2 \le C\lambda, \quad \text{which implies } \|v^\lambda - v\|_{L^2(D)}^2 \le C\lambda.$$

This implies that v^λ tends to v in $L^2(D)$ as $\lambda \to 0$. Since it is bounded in $H_0^1(D)$, the convergence also holds weakly in $H_0^1(D)$ so that $v \in H_0^1(\Omega_w)$ as well. □

Proof of Lemma 4.7.11. We introduce the quasi-open sets

$$\Omega^\epsilon = [w_\Omega > \epsilon], \quad \Omega_n^\epsilon = \Omega_n \cup \Omega^\epsilon.$$

Up to a subsequence, we may assume that $w_{\Omega_n^\epsilon}$ converges weakly in $H_0^1(D)$ to some w^ϵ. The point is to prove that

$$(w_\Omega - \epsilon)^+ \le w^\epsilon \le w_\Omega. \tag{4.52}$$

Assume it is true and let us end the proof. Estimate (4.52) and the fact that w^ϵ is bounded in $H_0^1(D)$ imply that w^ϵ converges to w_Ω weakly in H^1 and strongly in L^2. Given a sequence ϵ_k decreasing to 0, we may find a subsequence n_k such that $w_{\Omega_{n_k}^{\epsilon_k}}$ converges to w_Ω in L^2 and weakly in H^1: thus $C_k := \Omega_{n_k}^{\epsilon_k}$ γ-converges to Ω.

To prove the first inequality of (4.52), let us remark that

$$w_{\Omega_n^\epsilon} \geq w_{\Omega^\epsilon} = (w_\Omega - \epsilon)^+,$$

and the inequality is preserved at the limit for w^ϵ.

For the second inequality, we introduce $v^\epsilon = \epsilon^{-1}(\epsilon - w_\Omega)^+$ and $v_n = \inf\{w_{\Omega_n^\epsilon}, v^\epsilon\}$. Then $v_n \in H_0^1(\Omega_n)$ and converges weakly to $v = \inf\{w^\epsilon, v^\epsilon\}$. By Lemma 4.7.10, $v = 0$ q.e. outside Ω. Since $v^\epsilon = 1$ outside Ω, we deduce $w^\epsilon = 0$ q.e. outside Ω, that is, $w^\epsilon \in H_0^1(\Omega)$. By the definitions of $w_{\Omega_n^\epsilon}$, w_Ω,

$$\int_D \nabla w_{\Omega_n^\epsilon} \nabla[(w_{\Omega_n^\epsilon} - w_\Omega)^+] = \int_D (w_{\Omega_n^\epsilon} - w_\Omega)^+,$$
$$\int_D \nabla w_\Omega \nabla[(w^\epsilon - w_\Omega)^+] = \int_D (w^\epsilon - w_\Omega)^+.$$

Passing to the limit in n and subtracting the two identities, we obtain $(w^\epsilon - w_\Omega)^+ = 0$, which is the second inequality of (4.52). □

Let us now consider some consequences of Theorem 4.7.6.

Corollary 4.7.12. *For all integer $k \geq 1$ and all $c \in (0, |D|)$, the problem*

$$\min\{\lambda_k(\Omega);\ \Omega \in \mathcal{A}(D),\ |\Omega| = c\} \tag{4.53}$$

(where λ_k is the kth eigenvalue of the Dirichlet–Laplacian operator) has a solution. More generally, if $\Phi : \mathbb{R}^m \to \mathbb{R}$ is nondecreasing and lower semicontinuous, the problem

$$\min\{\Phi(\lambda_{k_1}(\Omega), \lambda_{k_2}(\Omega), \dots, \lambda_{k_m}(\Omega));\ \Omega \in \mathcal{A}(D),\ |\Omega| = c\}$$

has a solution.

This result follows from Theorem 4.7.6 after using Corollary 4.7.4.

Note that we also obtain the existence of a solution for minimization problems with penalized constraints like

$$\min\{\lambda_k(\Omega) + \big||\Omega| - c\big|,\ \Omega \in \mathcal{A}(D)\}.$$

Here, the functional is not monotone anymore, but it is still l.s.c. in the topology of weak γ-convergence as the sum of two l.s.c. functionals. This is sufficient thanks to the compactness of $\mathcal{A}(D)$ for this convergence.

Optimization problem (4.53) is easy to state but leads to quite difficult questions. For instance, what are the qualitative properties of the minimum? What is its regularity? Is it possible to identify it in some cases? We refer to Section 1.2.3, as well as to [65], [180], [181], [182] for a discussion on these questions. See also Chapter 6 for some more information.

Theorem 4.7.6 also applies to the following functional involving the solution of the Dirichlet problem.

Corollary 4.7.13. *Let $f \in H^{-1}(D)$, $f \geq 0$ and let $g : D \times \mathbb{R} :\longrightarrow [-\infty, +\infty]$ be a measurable function such that a.e. $x \in D$, $g(x, \cdot)$ is lower semicontinuous on $\mathbb{R}^N \times \mathbb{R}$, nonincreasing, and satisfies*

$$g(x, s) \geq -\alpha(x) - \beta s^2, \tag{4.54}$$

for some $\alpha \in L^1(D)$, $\beta \in \mathbb{R}$. With each quasi-open set $\Omega \in \mathcal{A}(D)$, we associate the function $u_\Omega = R_\Omega(f)$ and we consider the functional J defined on $\mathcal{A}(D)$ by

$$J(\Omega) = \int_D g(x, u_\Omega(x))\, dx.$$

Then the minimization problem

$$\min \{J(\Omega);\ \Omega \in \mathcal{A}(D),\ |\Omega| = c\}$$

has a solution.

Indeed, hypothesis (i) of the theorem is satisfied thanks to Proposition 4.7.5 and hypothesis (ii) follows from the maximum principle (here $\Omega \to u_\Omega$ is nondecreasing since $f \geq 0$) and from the monotonicity of g.

Remark 4.7.14. This example contains as a particular case the Dirichlet energy considered in Section 4.5 where $g(x, s) = -f(x)s$. In fact, assuming $f \geq 0$ is not necessary here. Indeed, we have

$$J(\Omega) = -\int_D f u_\Omega = \inf \left\{ \textstyle\int_D |\nabla v|^2 - 2fv,\ v \in H_0^1(\Omega) \right\}.$$

Since $\Omega \longrightarrow H_0^1(\Omega)$ is nondecreasing, $\Omega \to J(\Omega)$ is nonincreasing, no matter the sign of f. This provides a new proof of Theorem 4.5.2, at least when D is bounded.

4.8 Unbounded class of domains

A key point in most of the existence results we have stated in this chapter is that we had to deal with a class of admissible domains that are contained in some fixed bounded domain (e.g., a big ball). Now, there are many situations where this assumption is quite restrictive, since the natural class would be any (bounded or with a given measure) domain $\Omega \subset \mathbb{R}^N$. The main difficulty in this case is the lack of compactness of a minimizing sequence. Several techniques have been developed to circumvent this difficulty and this section presents three of them:

- a concentration–compactness principle for domains, inspired by this general principle due to P. L. Lions to describe the possible behavior of any bounded sequence in the Sobolev space $H^1(\mathbb{R}^N)$;
- the introduction of the notion of shape subsolution;
- a surgery argument that consists in cutting long tails to construct another bounded minimizing sequence.

4.8.1 A concentration–compactness argument

The concentration–compactness principle was introduced by P. L. Lions in [228] to describe the possible behavior of any bounded sequence in $H^1(\mathbb{R}^N)$ and to replace the lack of compactness of the embedding $H^1(\mathbb{R}^N) \hookrightarrow L^2(\mathbb{R}^N)$. More precisely, let $(u_n)_{n\in\mathbb{N}}$ be a bounded sequence in $H^1(\mathbb{R}^N)$ and assume that $\int_{\mathbb{R}^N} u_n^2 \to \lambda > 0$. Then there exists a subsequence u_{n_k} such that one of the following three situations occurs:

(i) **Compactness:** there exists a sequence $y_k \in \mathbb{R}^N$ such that

$$\forall\, \varepsilon > 0,\ \exists\, R < \infty, \qquad \int_{y_k + B(0,R)} u_{n_k}^2 \, dx \geq \lambda - \varepsilon;$$

(ii) **Vanishing:**

$$\lim_{k\to\infty} \sup_{y\in\mathbb{R}^N} \int_{y+B(0,R)} u_{n_k}^2 \, dx = 0, \qquad \forall\, R < \infty;$$

(iii) **Dichotomy:** there exists $\alpha \in (0, \lambda)$, there exists $k_0 \geq 1$, and there exist two sequences u_k^1, u_k^2 bounded in $H^1(\mathbb{R}^N)$ satisfying for $k \geq k_0$,

$$\|u_{n_k} - (u_k^1 + u_k^2)\|_{L^2(\mathbb{R}^N)} \leq \delta(\varepsilon) \to 0 \quad \text{for } \varepsilon \to 0^+,$$

$$\left| \int_{\mathbb{R}^N} (u_k^1)^2 \, dx - \alpha \right| \leq \varepsilon \quad \text{and} \quad \left| \int_{\mathbb{R}^N} (u_k^1)^2 \, dx - (\lambda - \alpha) \right| \leq \varepsilon,$$

$$\operatorname{dist}(\operatorname{supp} u_k^1, \operatorname{supp} u_k^2) \to \infty \quad \text{for } k \to \infty,$$

$$\liminf_{n\to\infty} \int_{\mathbb{R}^N} |\nabla u_{n_k}|^2 - |\nabla u_k^1|^2 - |\nabla u_k^2|^2 \, dx \geq 0,$$

where $\mathrm{dist}(\Omega_1, \Omega_2) = \inf\{|x_1 - x_2|;\ x_1 \in \Omega_1,\ x_2 \in \Omega_2\}$. Now, let us consider an unbounded sequence of domains (open or quasi-open) Ω_n of bounded measure $|\Omega_n| \leq C$. We associate to this sequence of domains the sequence of resolvent operators R_{Ω_n} as defined in (4.44). We also consider the sequence of functions $w_n \in H_0^1(\Omega_n)$ defined as the solution of $-\Delta w_n = 1$ in Ω_n (in the sense of (3.23)), namely, $w_n = R_{\Omega_n}(1)$. The first result can be seen as a compactness result from $\bigcup_{n\in\mathbb{N}} H_0^1(\Omega_n) \hookrightarrow L^2(\mathbb{R}^N)$; see [63, Thm. 2.1].

Theorem 4.8.1. *Let (Ω_n) be a sequence of open or quasi-open sets of uniformly bounded measure and let w_n be the solution of $-\Delta w_n = 1$, $w_n \in H_0^1(\Omega_n)$. If w_n strongly converges in $L^2(\mathbb{R}^N)$ to some function w, then for any sequence (u_n) such that $u_n \in H_0^1(\Omega_n)$ and such that u_n converges weakly to some u in $H^1(\mathbb{R}^N)$, we have $u_n \to u$ in $L^2(\mathbb{R}^N)$.*

Now the idea is to apply the previous concentration–compactness principle to the sequence (w_n). First of all, thanks to the Poincaré inequality proved in Lemma 4.5.3 (where the constant depends only on the measure of the set), it is easy to check that the sequence (w_n) is bounded in $H^1(\mathbb{R}^N)$. This allows us to prove the following concentration–compactness principle for sets; see [63, Thm. 2.23], [182].

Theorem 4.8.2. *Let (Ω_n) be a sequence of open or quasi-open sets of uniformly bounded measure. There exists a subsequence (denoted with the same index) such that one of the following situations occurs:*

Compactness. *There exist a sequence of vectors $(y_n) \subset \mathbb{R}^N$ and a nonnegative Borel[1] measure μ vanishing on sets of zero capacity, such that $y_n + \Omega_n$ γ-converges to the measure μ and the sequence of resolvent operators $R_{y_n+\Omega_n}$ converges in the uniform operator topology of $L^2(\mathbb{R}^N)$ to R_μ.*

Dichotomy. *There exists a sequence of subsets $\widetilde{\Omega}_n \subseteq \Omega_n$ such that*

$$\widetilde{\Omega}_n = \Omega_n^1 \cup \Omega_n^2 \quad \textit{and} \quad \|R_{\Omega_n} - R_{\widetilde{\Omega}_n}\|_2 \to 0$$

with $\mathrm{dist}(\Omega_n^1, \Omega_n^2) \to +\infty$ *and* $\liminf_{n\to\infty} |\Omega_n^i| > 0$ *for* $i = 1, 2$.

Remark 4.8.3. Let us explain the compactness situation in more detail. The γ-convergence for measures in a bounded domain D is described in detail later in Section 7.2.2 (see Definition 7.2.4). The definitions when D is replaced by $\mathbb{R}^N$ are essentially the same. Let us briefly recall them here. Given a nonnegative Borel

[1] Felix Edouard Justin **Emile** BOREL, 1871–1956, French, student then professor at École Normale Supérieure in Paris, and the author of a pioneering work in measure theory and modern function theory. He also made some interesting contributions to divergent series and game theory, and was the author of philosophical and popular science writings as well.

measure on $\mathbb{R}^N$, we introduce $L^2_\mu(\mathbb{R}^N)$ as the space of (classes of) measurable functions u such that $\int_{\mathbb{R}^N} u^2\,d\mu < \infty$. The resolvent operator R_μ is the operator that maps any $f \in L^2(\mathbb{R}^N)$ to the solution $u \in H^1(\mathbb{R}^N) \cap L^2_\mu(\mathbb{R}^N)$ of $\left[-\Delta u + \mu u = f \text{ in } \mathbb{R}^N\right]$. By definition, γ-convergence of $y_n + \Omega_n$ to μ means that, for any $f \in L^2(\mathbb{R}^N)$, the solution of $\left[u_n \in H^1_0(y_n + \Omega_n),\ -\Delta u_n = f\right]$ converges strongly in $L^2(\mathbb{R}^N)$ to the solution $u \in H^1(\mathbb{R}^N) \cap L^2_\mu(\mathbb{R}^N)$ of $[-\Delta u + \mu u = f]$ (i.e., $R_{y_n+\Omega_n}(f) \to R_\mu(f)$ in $L^2(\mathbb{R}^N)$). The strong convergence of the resolvent operators implies the convergence of the eigenvalues (see Corollary 4.7.4 or [133, Lem. XI.9.5]).

Remark 4.8.4. If one compares this alternative for sets to the one stated above for functions, we see that the vanishing situation does not appear. In fact, the vanishing situation enters into the compactness situation since in this case, we can prove that $R_{\Omega_n} \to 0$ in $L^2(\mathbb{R}^N)$, which corresponds to $\mu = 0$ (or the γ-limit which is the empty set).

Theorem 4.8.2 can been used in several situations where no boundedness assumption for the class of domains is made. Let us give a simple but very significant example in dimension 2 for eigenvalues.

Theorem 4.8.5. *Let $C > 0$ be given. Then the problem*

$$\min\left\{\lambda_1(\Omega) + \lambda_2(\Omega),\ \Omega \subset \mathbb{R}^2,\ |\Omega| \le C\right\} \tag{4.55}$$

has a solution.

Here $\lambda_1(\Omega)$ and $\lambda_2(\Omega)$ denote the two first eigenvalues of the Laplacian with Dirichlet boundary conditions and Ω is any quasi-open set.

Proof. Without loss of generality, we consider the case where $C = \pi$. First of all, let us remark that the infimum of $\lambda_1 + \lambda_2$ is less than or equal to $(j_0)^2 + (j_1)^2 = 20.465\ldots$, which is the value of this sum for the unit disk, where j_ν denotes the first zero of the Bessel function J_ν (see [114], [293], or [181]).

Let Ω_n be a minimizing sequence for this problem. According to Theorem 4.8.2, the sequence Ω_n can have two behaviors: compactness or dichotomy. Let us prove that dichotomy cannot occur. Since $\|R_{\Omega_n} - R_{\widetilde{\Omega}_n}\|_2 \to 0$, we have $|\lambda_k(\Omega_n) - \lambda_k(\widetilde{\Omega}_n)| \to 0$ for any k (and in particular for $k = 1, 2$) and therefore, $\widetilde{\Omega}_n$ is also a minimizing sequence. Now, $\widetilde{\Omega}_n$ being disconnected, we determine its eigenvalues from the eigenvalues of each of its components Ω_n^1 and Ω_n^2. Thus, for the two first eigenvalues, two (and only) two situations are possible:

(i) $\lambda_1(\widetilde{\Omega}_n) = \lambda_1(\Omega_n^1)$ and $\lambda_2(\widetilde{\Omega}_n) = \lambda_2(\Omega_n^1)$ (it would be the same with Ω_n^2);

(ii) $\lambda_1(\widetilde{\Omega}_n) = \lambda_1(\Omega_n^1)$ and $\lambda_2(\widetilde{\Omega}_n) = \lambda_1(\Omega_n^2)$ (or vice versa).

In case (i), we can replace $\widetilde{\Omega}_n$ by $t_n\Omega_n^1$ with $t_n = [\pi/|\Omega_n^1|]^{1/2}$, getting a contradiction with the minimization property of $\widetilde{\Omega}_n$, since $\limsup t_n > 1$ by the property of the measures of Ω_n^1 and Ω_n^2 (recall that $(\lambda_1 + \lambda_2)(t_n\Omega_n^1) = t_n^{-2}(\lambda_1 + \lambda_2)(\Omega_n^1)$). In case (ii), we use the Faber–Krahn inequality: we can replace Ω_n^1 and Ω_n^2 by balls of the same measure respectively, thus decreasing $\lambda_1 + \lambda_2$. Now a direct computation shows that, for any couple of disjoint balls $B_{R_1} \cup B_{R_2}$ with $R_1^2 + R_2^2 = 1$, the minimum of $\lambda_1(B_{R_1}) + \lambda_1(B_{R_2})$ is achieved for $R_1 = R_2 = 1/\sqrt{2}$ and is equal to $4(j_0)^2$. But it turns out that the following inequality holds:

$$(j_0)^2 + (j_1)^2 = 20.465\ldots < 4(j_0)^2 = 23.132\ldots.$$

Together with the first point of this proof, this shows that the two disjoint equal balls of radius $1/\sqrt{2}$ yield a larger value for $\lambda_1 + \lambda_2$ than the ball of radius R, showing that this situation cannot provide a minimizing sequence.

It remains to consider the *compactness* situation. Then (up to some translation that does not modify the eigenvalues), the minimizing sequence γ-converges to some measure μ as explained in Remark 4.8.3. In particular, $\lambda_1(\mu) + \lambda_2(\mu) = \inf \lambda_1 + \lambda_2$. Let us introduce the functions $w_n = R_{y_n+\Omega_n}(1)$ and $w_\mu = R_\mu(1)$. By γ-convergence, (w_n) converges strongly in $L^2(\mathbb{R}^2)$ (and therefore almost everywhere up to some subsequence) to w_μ. Let us also introduce the set $A_\mu = \{x \in \mathbb{R}^2,\ w_\mu(x) > 0\}$ as in (7.27). This set is quasi-open according to Proposition 3.3.41 and corresponds to the *regular set* of measure μ. In particular, μ is infinite on $\mathbb{R}^2 \backslash A_\mu$ (see Proposition 7.2.7 in the bounded case and [69]). Let us prove that $|A_\mu| \le \pi$. If $x \in A_\mu$, $w_\mu(x) > 0$ and then $w_n(x) > 0$ for n large enough by pointwise convergence, which implies that $x \in y_n + \Omega_n$. In other words, $\chi_{A_\mu} \cdot \chi_{y_n+\Omega_n}$ converges a.e. to χ_{A_μ}. By Fatou's lemma,

$$|A_\mu| = \int_{\mathbb{R}^2} \chi_{A_\mu} \le \liminf_{n\to\infty} \int_{\mathbb{R}^2} \chi_{A_\mu} \cdot \chi_{y_n+\Omega_n} \le \liminf_{n\to\infty} |\Omega_n| = \pi.$$

We now use the Courant–Fischer variational characterization of the eigenvalues of μ and A_μ, namely (see (4.46), [69], [182]),

$$\begin{aligned} \lambda_k(A_\mu) &= \min_{E_k \in \mathcal{H}_k} \left\{ \max_{v \in E_k \backslash \{0\}} \frac{\int_{A_\mu} |\nabla v|^2}{\int_{A_\mu} v^2} \right\} \\ &\le \min_{E_k \in \mathcal{H}_k} \left\{ \max_{v \in E_k \backslash \{0\}} \frac{\int_{A_\mu} |\nabla v|^2 + v^2 d\mu}{\int_{A_\mu} v^2} \right\} \\ &= \lambda_k(\mu), \end{aligned}$$

where $\mathcal{H}_k$ denotes the family of linear subspaces with dimension k of $H_0^1(A_\mu)$. It follows that A_μ is a solution of the minimization problem (4.55). □

For more examples of the application of the same kind of technique we refer to [69], which provides a proof of existence of a minimizer for the third eigenvalue, and to [44] where existence is proved for the classical energy with a weight. See also [182].

4.8.2 Notion of shape subsolution

For a quasi-open set Ω of bounded measure, we consider the classical energy (as considered, e.g., in Section 4.5)

$$E(\Omega) = \min\left\{\int_\Omega \tfrac{1}{2}|\nabla v|^2 - v,\ v \in H_0^1(\Omega)\right\}. \tag{4.56}$$

The torsion function w_Ω is the function that realizes the minimum in (4.56). Following D. Bucur (see [64], [182]), let us introduce the notion of shape subsolution.

Definition 4.8.6. We say that a set Ω of finite measure is a **shape subsolution** for the energy if there exists $\Lambda > 0$ such that for every subset $\widehat{\Omega} \subset \Omega$ we have

$$E(\Omega) + \Lambda|\Omega| \le E(\widehat{\Omega}) + \Lambda|\widehat{\Omega}|. \tag{4.57}$$

Then, in the same paper [64], the following is proved.

Theorem 4.8.7. *Assume Ω is a shape subsolution for the energy. Then Ω is bounded and has finite perimeter.*

The proof is not easy and involves an estimate of Alt–Caffarelli type for some free boundary problems; see [17].

This theorem does not help directly to prove existence in an unbounded setting. Nevertheless, it was a key point in [64] to prove existence for the problem

$$\min\left\{\lambda_k(\Omega),\ \Omega \subset \mathbb{R}^N,\ |\Omega| = C\right\}. \tag{4.58}$$

The existence is proved in three steps:

- An inequality relating the eigenvalues and the energy is established: if $\Omega_1 \subset \Omega_2$,

$$0 \le \frac{1}{\lambda_k(\Omega_2)} - \frac{1}{\lambda_k(\Omega_1)} \le C_k \lambda_k(\Omega_2)^{N/2}\left(E(\Omega_1) - E(\Omega_2)\right),$$

  where C_k is an (explicit) constant depending only on k.

- This inequality allows us to prove that any minimizer of (4.58) is in fact a shape subsolution for the energy and therefore is bounded.

- To conclude, the author uses an induction argument together with the concentration–compactness principle explained in Theorem 4.8.2. Assuming that there exists a solution Ω_j^* to problem (4.58), with $C = 1$ and for any integer $j \le k$

(this is true for $k = 1$ by the Faber–Krahn inequality), these solutions are bounded according to the previous point.

Now we want to prove existence for $k + 1$ and we consider a minimizing sequence Ω_n.

According to Theorem 4.8.2, we have two possible situations: If we are in the compactness situation, we can conclude existence as in the proof of Theorem 4.8.5. If dichotomy occurs, then the sequence given by $\widetilde{\Omega}_n = \Omega_n^1 \cup \Omega_n^2$ with $\mathrm{dist}(\Omega_n^1, \Omega_n^2) \to \infty$ and $\lim |\Omega_n^i| = c_i > 0$ for $i = 1, 2$ is also minimizing. Now, following Wolf–Keller's remark (see [293], [69], [181, Thm. 5.2.1], or Theorem 6.4.2 in the present book), one can prove that, up to a subsequence, each component Ω_n^i, $i = 1, 2$ is in fact a minimizing sequence for eigenvalues of strictly inferior rank j and $k+1-j$ over sets of prescribed measure. More precisely, there exists $1 \le j \le k$ such that (up to a subsequence)

$$\lim_{n\to\infty} \lambda_j(\Omega_n^1) = \lim_{n\to\infty} \lambda_{k+1-j}(\Omega_n^2) = \lim_{n\to\infty} \lambda_{k+1}(\widetilde{\Omega_n}) = \min \lambda_{k+1},$$
$$\lim_{n\to\infty} \lambda_j(\Omega_n^1) = \lambda_j(|c_1|^{1/N}\Omega_j^*), \qquad \lim_{n\to\infty} \lambda_{k+1-j}(\Omega_n^2) = \lambda_{k+1-j}(|c_2|^{1/N}\Omega_{k+1-j}^*).$$

Since $\omega_1^* := |c_1|^{1/N}\Omega_j^*$ and $\omega_2^* := |c_2|^{1/N}\Omega_{k+1-j}^*$ are bounded, we can assume by translation that they are disjoint and it follows that

$$\lambda_{k+1}(\omega_1^* \cup \omega_2^*) \le \max\{\lambda_j(\omega_1^*), \lambda_{k+1-j}(\omega_2^*)\} = \min \lambda_{k+1}.$$

Since $|\omega_1^* \cup \omega_2^*| = c_1 + c_2 = 1$, then $\omega_1^* \cup \omega_2^*$ is a minimum for λ_{k+1}. □

4.8.3 A surgery argument

When we deal with domains with fixed volume but that are not constrained to lie in some fixed bounded box, a natural question is what the behavior of minimizing sequences can be. In particular, if such a sequence is not uniformly bounded, one typical possible behavior is that the sequence develops long and thin parts (we can call them *tails* or *tentacles*). Therefore, a possible strategy is to show that we can cut these tails without losing (or losing too much) for the functional under consideration. This strategy has been used by D. Mazzoleni and A. Pratelli in [231] to solve problem (4.58). More precisely, they prove the following interesting theorem, which shows that we can replace any domain by another one uniformly bounded with a control on the first k eigenvalues.

Theorem 4.8.8. *For every $K > 0$, there exists a constant $R = R(k, K, N)$ such that the following holds. If $\Omega \subset \mathbb{R}^N$ is an open set of unit volume with $\lambda_k(\Omega) \le K$, there exists another open set $\widehat{\Omega}$, still of unit volume but contained in a cube of side R, and with $\lambda_i(\widehat{\Omega}) \le \lambda_i(\Omega)$ for every $1 \le i \le k$.*

The existence of a minimizer for $\lambda_k(\Omega)$ follows easily, by combining Theorem 4.8.8 (which allows us to deal with a minimizing sequence contained in a fixed bounded domain) with Buttazzo–Dal Maso's Theorem 4.7.6; see also Corollary 4.7.12.

The proof of Theorem 4.8.8 proceeds in two steps. First the authors prove the boundedness of the tails: one can replace any domain by another one with lower eigenvalues, and such that any tail of the new domain has a horizontal projection of uniformly bounded length. Then they do an analogous construction to prove the boundedness of the domain itself. The proof needs to carefully control the Rayleigh quotient of the various parts of the domains.

Another technique that has been recently used in [72] consists in looking at the local behavior of the torsion function (the solution of $-\Delta u = 1$, $u \in H^1_0(\Omega)$). Then if this function is small enough in some region, it is proved that one can cut out a piece of the domain there, simultaneously controlling the first eigenvalues, the volume, and even the perimeter.

Finally, let us mention another idea that can be successful and seems to have been introduced by E. de Giorgi in [128]. First prove existence of a minimizer in the class $\{\Omega \subset B_R\}$, where B_R is a ball of radius R, for any $R > 0$. Then, prove that for R large enough, the optimal domain does not touch the boundary of the ball. It can be done in some cases by a careful analysis of the optimality conditions; see, for example, [70].

4.9 Exercises

Exercise 4.1. Show that, if $|D| = +\infty$, problem (4.2) may have no solution. (Hint: look at the example of Remark 4.1.3.)

Exercise 4.2. Assume D is a half-space. Prove by a symmetry argument that a solution to problem (4.2) is given by a half-ball "stuck" to the boundary.

Exercise 4.3. Let us consider the two isoperimetric problems

$$|\Omega^*| = \max\{|\Omega|;\ \Omega \subset D \text{ measurable},\ P(\Omega) \le P_0\}, \tag{4.59}$$

$$|\Omega^*| = \max\{|\Omega|;\ \Omega \subset D \text{ measurable},\ P_D(\Omega) \le P_0\}. \tag{4.60}$$

Prove that they have a solution that, if different from D, satisfies $P(\Omega^*) = P_0$ (resp. $P_D(\Omega^*) = P_0$) and coincides with the solution of (4.1) (resp. of (4.2)) for suitable values of V_0.

Exercise 4.4. Set $y(\Omega) := \chi_\Omega$. What happens in problems (4.7) and (4.8) when choosing $\tau = 0$ in the functional J defined by (4.6)?

Exercise 4.5. Let us consider the minimization problem (4.30) in the unit ball B of $\mathbb{R}^3$.

(1) Assume $w \in H_0^1(B)$ is such that $-\Delta w = f$ in B and $|\Omega_w| = m$. Then prove that Ω_w is a quasi-open solution of (4.30). Assume also that there exists an *open set* Ω^*, a solution of (4.30). Prove that $w = u_{\Omega^*}$, $\Omega_w = \Omega^*$ a.e.

Let $(x_n)_{n\geq 1}$ be a dense sequence in B and define

$$v(x) = \sum_{n\geq 1} \alpha_n |x - x_n|^{-1}, \qquad z = \inf\{1, v\},$$

where $\alpha_n > 0$, $\sum_n \alpha_n = \alpha < (16\pi)^{-1}$.

(2) Prove that (see also Exercise 3.6)

$$z = 1 \text{ in } B(x_n, \alpha_n), \qquad |[z < 1]| > 0,$$
$$-\Delta v \geq 0 \text{ in } B, \qquad -\Delta z \geq 0 \text{ in } B, \qquad v \in H^1(B).$$

(3) Let $\psi \in C_0^\infty(B)$ with $0 \leq \psi \leq 1$, $\psi \equiv 1$ on $B_{1/2}$, and $\eta > 0$ small enough so that $|[v < 1 - \eta]| > 0$. We set $w = \psi(1 - \eta - z)^+$. Prove that Ω_w is a quasi-open set that is not a.e. equal to an open set.

(4) Prove that $f \in H^{-1}(B)$ and $m \in (0, 4\pi/3)$ may be chosen so that problem (4.30) has no open solution.

Exercise 4.6. Prove that any solution of problem (4.36) converges to a solution of problem (4.30) as $\sigma \to 0$.

Exercise 4.7. Prove that if Ω_n γ-converges weakly to a quasi-open set of zero capacity (i.e., the "empty" quasi-open set), then it γ-converges strongly. Consider the sequence of sets in the unit square of $\mathbb{R}^2$ obtained by removing all disks centered at $(i/n, j/n)$, $1 \leq i, j \leq n - 1$ and of radius r_n. Prove that this sequence has the above property. (See Exercise 3.8.)

Exercise 4.8. Prove Lemma 4.6.1.

Exercise 4.9. Is Corollary 4.7.12 true if the eigenvalues of the Dirichlet–Laplacian operator are replaced by those of the Neumann–Laplacian operator?

Exercise 4.10. Let D be the union of two disjoint open disks D_1, D_2 with radii $R_1 > R_2$. Let $m \in (\pi R_1^2, \pi(R_1^2 + R_2^2))$ and let $\omega \subset D_2$ be a quasi-open set of measure $m - \pi R_1^2$.

Show that $\Omega^* = \omega \cup D_1$ is a solution of the problem (see problem (4.53))

$$|\Omega^*| = m, \quad \lambda_1(\Omega^*) = \min\{\lambda_1(\Omega);\ \Omega \in \mathcal{A}(D),\ |\Omega| = m\}.$$

Exercise 4.11. Let D be a bounded open subset of $\mathbb{R}^N$ and let $m \in (0, |D|)$. Show with the help of Theorem 4.7.6, that the following problem has a solution:

$$\min \left\{ \operatorname{cap}_D(F);\ F \subset D \text{ quasi-closed}, |F| = m \right\}.$$

Exercise 4.12. Let K be a compact subset of $\mathbb{R}^N$ and let $m > |K|$. Show that

$$\min \left\{ \operatorname{cap}_\Omega(K);\ \Omega \text{ quasi-open}, |\Omega| = m \right\}$$

has a solution.

Chapter 5

Differentiating with respect to domains

5.1 Introduction

In this chapter, we study how one can write optimality conditions of first and second order for shape functionals. Like in any optimization problem, they are needed for several purposes. First, they provide interesting features of the minimum (which is a priori not known) and help to find it. For instance, we know that, in finite dimensions, one of the methods for finding the minimum of a functional $J : \mathbb{R}^N \to \mathbb{R}$ consists in solving the (generally nonlinear) equation $\nabla J(x) = 0$ in $\mathbb{R}^N$. Next, one makes a selection to find the solutions that provide a minimum. When the variable is a shape, this equation leads most of the time to a free boundary problem which is said to be *overdetermined* in the sense that, besides the state equation, which by itself is a well-posed problem, the optimality condition provides an extra equation on the boundary for the solution.

On the other hand, several free boundary problems may appear as the Euler equation of some optimization problem. It is then said to be a variational formulation of the initial free boundary problems and one may use this formulation either to prove existence of a solution or to compute it numerically, and this is one more goal of the optimality conditions and of the shape differentiation.

Indeed, when in finite dimensions, in order to find numerically the minimum of a functional, one may (among other approaches)

- either, as explained above, solve the system of equations $\nabla J(x) = 0$; this may be done using Newton's method which then requires J to be differentiated once more and involves the Hessian (= matrix of the second derivatives of J);

- or use a descent method of gradient type or "quasi-Newton" type and in each case, it is necessary to compute the derivative of J at each step of the iteration.

The same approach may be used to compute optimal shapes.

The second-order optimality conditions allow us to analyze whether a critical shape — that is, a shape at which the first derivative vanishes — is a minimal shape by looking at the positivity of the second derivative (like usual in variational calculus).

For all these reasons, and many others, it is important to be able to compute the derivatives of the functional we want to optimize. However, it is not an easy task for shape functionals. Indeed, the classical notion of differentiability requires a normed vector space framework, but there is no canonical one when dealing, for instance, with a family of open subsets or domains of $\mathbb{R}^N$. So, what can be done?

Let $E : \mathcal{O} \longrightarrow X$, where $\mathcal{O}$ is a family of subsets of $\mathbb{R}^N$ and X is a normed vector space. A rather convenient approach is to consider the application $\theta \longrightarrow \mathcal{E}(\theta) = E\big((I + \theta)(\Omega)\big)$, where θ varies around 0 in a normed vector space Θ of applications from $\mathbb{R}^N$ into itself. We then speak of *Fréchet differentiability in the classical sense*[1] for the application $\theta \in \Theta \longrightarrow \mathcal{E}(\theta) \in X$. This point of view is particularly interesting and efficient for proving the regularity properties of the usual shape functionals, for actually computing the derivatives, and for ably detecting the structure of these so-called "shape derivatives".

However, it is known that, in order to compute Fréchet derivatives (which can here quickly become rather technical and complex), it is often more pleasant to compute derivatives with respect to a real variable (it is known that for all $\xi \in \Theta$, $\mathcal{E}'(0)\xi = \frac{d}{dt}_{|t=0}\mathcal{E}(t\xi)$). Thus, we shall also widely analyze derivatives of functions like $t \in [0,T) \longrightarrow E(\Omega_t)$, where $\Omega_t = \Phi(t)(\Omega)$ and $\Phi(t) \in \Theta$. A classical example consists in taking $\Phi(t) = I + t\theta$, where θ is a regular vector field from $\mathbb{R}^N$ into $\mathbb{R}^N$.

No matter the point of view, it is necessary to get around the following difficulty: most of the involved functions are naturally defined on the variable domain Ω_t (or $\Omega_\theta = (I + \theta)(\Omega)$) and they are a priori not defined on the same domain for small t (or θ). For instance, how should a function $t \longrightarrow u_t \in H^1(\Omega_t)$, where Ω_t is a variable open subset, be differentiated?

In fact, the answer depends on the situation and, rather than giving a definition, let us indicate some general principles:

- If u_t may "naturally" be extended to $\mathbb{R}^N$, we shall use the derivative of its extension: this is the case, for instance, if $u_t \in H_0^1(\Omega_t)$ (which may be extended by 0).
- The function $v_t = u_t \circ \Phi(t)$ is always defined on the fixed domain Ω. In the above example, it belongs to the fixed space $H^1(\Omega)$ (we assume Φ is regular) and one can then look at the differentiability of $t \longrightarrow v_t \in H^1(\Omega)$. If the goal is to differentiate a function of u_t, this may be sufficient since we would transfer everything on Ω through $\Phi(t)$.
- If this is not sufficient or if one really wants to deal with u_t, one may compose v_t with $\Phi(t)^{-1}$ again ($u_t = v_t \circ \Phi(t)^{-1}$) and obtain in this way some useful information on the differentiability of $t \longrightarrow u_{t\,|K}$, the restriction to any compact subset K of the open set Ω (note that $K \subset \Omega_t$ for small t). As Ω is the union of the compact sets K, this provides a definition of the derivative for $t \longrightarrow u_t$ on the whole of Ω.

This approach becomes simpler when there exists a linear continuous extension operator P that extends functions from Ω to $\mathbb{R}^N$, for instance from $H^1(\Omega)$ to $H^1(\mathbb{R}^N)$.

[1]Maurice FRÉCHET, 1878–1973, French, a professor in Poitiers, Strasbourg, and Paris, has widely contributed to topology and functional analysis.

We then use the differentiability of $t \longrightarrow P(v_t)\circ\Phi(t)^{-1}$ and we find again the properties of $t \longrightarrow u_{t\,|K}$.

Let us finally remark that, in most applications, $[t \longrightarrow v_t]$ is more regular than $[t \longrightarrow u_t]$ and it is often more strategic to study $[t \longrightarrow v_t]$.

There are lots of contributions concerned with domain differentiation and it is not so easy to present all of them in a simple way. After the pioneer paper in 1907 by J. Hadamard[2] [162], one must mention [270], [148], then, with the new interest in the 1970s, the contributions [90], [87], [88] by J. Céa et al., and [89] more recently. Let us also mention [233], and next the important series of papers by F. Murat–J. Simon [242], [243], [277], [278], the more recent work [129], [130], [76], [158], [159], [161], [160], [90], [248], [120], [121], [221], [49], [50], as well as the books [280], [131]. This chapter borrows a lot from all of these articles: we chose to present what appeared to us to be at the same time essential and simple enough for good use of shape differentiation, and with a self-contained presentation. We did not consider the numerical aspects of shape differentiation although they are important and although they have seen spectacular development in recent years.

5.2 Integrals on variable domains

5.2.1 Introduction

Let us start with the main question arising in many applications: how to differentiate

$$t \longrightarrow I(t) = \int_{\Omega_t} f(t,x)\,dx,$$

where $\Omega_t = \Phi(t,\Omega)$ is the image of a fixed measurable subset $\Omega \subset \mathbb{R}^N$ by a variable diffeomorphism $\Phi(t,\cdot) : \mathbb{R}^N \longrightarrow \mathbb{R}^N$ defined for $t \in [0,T)$ with $\Phi(0,y) = y$ for all $y \in \mathbb{R}^N$. For simplicity, assume that f is real valued. The computation may be done through a change of variable by setting $x = \Phi(t,y)$, $y \in \Omega$ so that

$$I(t) = \int_{\Omega} f(t,\Phi(t,y))J(t,y)\,dy, \tag{5.1}$$

where

$$J(t,y) = \det\big(D_y\Phi(t,y)\big)$$

is the Jacobian of $\Phi(t,\cdot)$, that is, the determinant of the Fréchet derivative in y of $\Phi(t,\cdot)$ (here $\Phi(t)$ is close to the identity and $\det(D_y\Phi(t,y)) > 0$). The main progress is

[2] Jacques Salomon HADAMARD, 1865–1963, French, was professor in Bordeaux, then at the Sorbonne, at the École Polytechnique, and at the Collège de France. He has been credited with the prime numbers theorem, but also with innovative work on integral equations and the variational analysis of partial differential equations.

that the integration domain is now independent of t and, assuming enough regularity on the data, after setting $V(0,y) = \frac{\partial \Phi}{\partial t}(t,y)_{|t=0}$, we obtain

$$I'(0) = \int_\Omega \left[\frac{\partial f}{\partial t}(0,y) + \nabla_y f(0,y)V(0,y) + f(0,y)\frac{\partial}{\partial t}_{|t=0} J(t,y) \right] dy. \tag{5.2}$$

A simple computation (see below) shows that $\frac{\partial}{\partial t}_{|t=0} J_\Phi(t,y) = \operatorname{div}_y V(0,y)$, so that (5.2) may be rewritten

$$I'(0) = \int_\Omega \left[\frac{\partial f}{\partial t}(0,y) + \operatorname{div}_y(fV)(0,y) \right] dy, \tag{5.3}$$

or also, after integrating by parts when possible,

$$I'(0) = \int_\Omega \frac{\partial f}{\partial t}(0,y)\,dy + \int_{\partial\Omega} (fV)(0,y).n(y)\,d\mathcal{H}^{N-1}(y), \tag{5.4}$$

where $n(\cdot)$ denotes the exterior unit normal vector to $\partial\Omega$ and $\mathcal{H}^{N-1}$ is the surface measure on $\partial\Omega$ (or the $(N-1)$-Hausdorff measure in general).

We will now give a more precise statement of (5.3).

5.2.2 Notation

We denote by $W^{1,\infty}(\mathbb{R}^N, \mathbb{R}^N)$ (or more simply by $W^{1,\infty}$ when not ambiguous) the linear space of bounded and Lipschitz continuous applications from $\mathbb{R}^N$ into itself equipped with the norm

$$\|\theta\|_{1,\infty} = \sup_{y,\hat{y}\in\mathbb{R}^N, y\neq\hat{y}} \left\{ |\theta(y)| + |\theta(y) - \theta(\hat{y})|/|y-\hat{y}| \right\} \quad \forall\, \theta \in W^{1,\infty},$$

where $\mathbb{R}^N$ is equipped with the Euclidean norm $|\cdot|$. The identity in $\mathbb{R}^N$ is denoted by I.

We recall (see, e.g., [135]) that this space may be identified with the subspace of functions in $L^\infty(\mathbb{R}^N)$ whose partial derivatives (in the sense of distributions) are functions of $L^\infty(\mathbb{R}^N)$. Moreover, functions from $W^{1,\infty}$ are differentiable a.e. and

$$\|\theta\|_{1,\infty} = \|\theta\|_\infty + \operatorname{supess}_{y\in\mathbb{R}^N} \|D_y\theta(y)\| \quad \forall\, \theta \in W^{1,\infty},$$

where the norms of the Fréchet derivatives are understood as linear operators from $\mathbb{R}^N$ into itself. Readers who would like to work with classical derivatives could replace $W^{1,\infty}$ by $C^{1,\infty} := C^1 \cap W^{1,\infty}$ in the following without losing any essential ideas. Working with $W^{1,\infty}$ is, however, interesting when perturbing open sets with a Lipschitz boundary.

If $\|\theta\|_{1,\infty} < 1$, by a classical fixed-point argument, $I+\theta$ is invertible, $(I+\theta)^{-1}-I \in W^{1,\infty}$ and we have

$$\begin{cases} \|(I+\theta)^{-1} - I\|_{1,\infty} \le \|\theta\|_{1,\infty}(1 - \|\theta\|_{1,\infty})^{-1}, \\ \|(I+\theta)^{-1} - I + \theta\|_{\infty} \le \|\theta\|_{1,\infty}\|I - (I+\theta)^{-1}\|_{\infty}. \end{cases} \tag{5.5}$$

Thus, $\theta \in W^{1,\infty} \to (I+\theta)^{-1} - I \in W^{1,\infty}$ is continuous at 0 and $\theta \in W^{1,\infty} \to (I+\theta)^{-1} - I \in L^{\infty}$ is differentiable at 0, its derivative being $-I$.

Let $\Phi(t) : \mathbb{R}^N \mapsto \mathbb{R}^N$ be defined for all $t \in [0,T)$ and such that

$$\begin{cases} t \in [0,T) \to \Phi(t) - I \in W^{1,\infty}(\mathbb{R}^N) \text{ is differentiable at 0,} \\ \Phi(0) = I,\ \Phi'(0) := \frac{d}{dt}(\Phi(t) - I) = V. \end{cases} \tag{5.6}$$

Since $\Phi(t) - I$ is close to zero in $W^{1,\infty}$ when t is close to 0, $\Phi(t)$ is invertible and, up to choosing T smaller, according to (5.5),

$$\begin{cases} [t \in [0,T) \to \Phi(t)^{-1} - I \in W^{1,\infty}] \text{ is continuous at 0,} \\ [t \in [0,T) \to \Phi(t)^{-1} \in L^{\infty}] \text{ is differentiable at 0 with derivative } -V. \end{cases} \tag{5.7}$$

We will write either $\Phi(t)(y)$ or $\Phi(t,y)$ (and similarly for all functions). We denote by $J(t,y) = \det(D_y\Phi(t)(y))$ the Jacobian of $\Phi(t)$ at y (which is defined, as we recall, a.e.$y \in \mathbb{R}^N$).

Remark 5.2.1 (The choice of $\Phi(t)$). A frequent choice for the functions $\Phi(t)$ is

$$\Phi(t)(x) = x + t\theta(x) \quad \text{with } \theta \in W^{1,\infty}(\mathbb{R}^N).$$

This choice will turn out to be very important when we talk of differentiability with respect to $\theta \in W^{1,\infty}$ of functions defined on $\Omega_\theta = (I+\theta)(\Omega)$. For second derivatives, it will also be interesting to choose $\Phi(t) = I + t\theta + t^2\hat{\theta}$. We refer to Remark 5.2.9 for more choices of functions $\Phi(t)$.

Let $\Omega \subset \mathbb{R}^N$ measurable be given. We set $\Omega_t := \Phi(t)(\Omega)$ for all $t \in [0,T]$. We can easily check that Ω_t is measurable and that, if Ω is open, so is Ω.

Let $f(t,\cdot) \in L^1(\Omega_t)$ for all $t \in [0,T)$ and let us consider the function

$$t \in [0,T) \to I(t) = \int_{\Omega_t} f(t,x)\,dx = \int_{\Omega} f(t,\Phi(t,y))J(t,y)\,dy. \tag{5.8}$$

Note that a proof of this change of variable formula may be found in [135] in this Lipschitz situation. In the following, we will often omit to indicate the integration variable and write more simply

$$\int_{\Omega_t} f(t) = \int_{\Omega} f(t,\Phi(t))J(t).$$

5.2.3 The differentiation formula

Let us start with the easier case where $f(t,\cdot)$ is everywhere defined on $\mathbb{R}^N$. The derivatives at 0 are right-derivatives.

Theorem 5.2.2. *Let Φ satisfy* (5.6). *Assume the following two hypotheses:*

$$t \in [0,T) \to f(t) \in L^1(\mathbb{R}^N) \text{ is differentiable at } 0 \text{ (with derivative } f'(0)), \tag{5.9}$$

$$f(0) \in W^{1,1}(\mathbb{R}^N). \tag{5.10}$$

Then, $t \to I(t) = \int_{\Omega_t} f(t)$ is differentiable at 0 *and*

$$I'(0) = \int_\Omega f'(0) + \operatorname{div}[f(0)V]. \tag{5.11}$$

If, moreover, Ω is an open set with a Lipschitz boundary, then

$$I'(0) = \int_\Omega f'(0) + \int_{\partial\Omega} f(0)n.V. \tag{5.12}$$

Remark. Note that in this theorem, the differentiability of I and formula (5.11) are obtained under the assumption that Ω *is measurable* only. Formula (5.12) requires a little more regularity on Ω: in fact, it is enough that the trace of $f(0)V$ on $\partial\Omega$ be defined and H^{N-1}-integrable. As $f(0)V \in W^{1,1}$, it is the case if Ω is an open set with a Lipschitz boundary (see [135]).

Now, if $f(t) \equiv f \in L^1(\mathbb{R}^N)$ only, it is not true in general that $t \to \int_{\Omega_t} f$ is differentiable: as an elementary example, we may choose

$$\Omega = (0,1), \qquad \Omega_t = (t,1), \qquad f(x) = x^{-1/2}\chi_\Omega.$$

A very frequent situation is when $f(t,\cdot)$ is defined only on the variable domain Ω_t. Obviously, Theorem 5.2.2 still applies if f has an extension to $\mathbb{R}^N$ that satisfies the above assumptions. However, it is often more interesting to make the assumptions on the function $(t,y) \to f(t,\Phi(t,y))$ rather than on f itself: indeed, this function is defined on a fixed domain and is often more regular than f in applications.

We have the following corollaries where $f(t)$ is defined only on Ω_t:

Corollary 5.2.3. *Let Φ satisfy* (5.6) *and let $t \in [0,T) \to f(t) \in L^1(\Omega_t)$. Let us assume that*

$$t \in [0,T) \to F(t) = f(t,\Phi(t,\cdot)) \in L^1(\Omega) \quad \text{is differentiable at 0,} \tag{5.13}$$

and that there exists a linear continuous extension operator $P : L^1(\Omega) \to L^1(\mathbb{R}^N)$ such that $P(f(0)) \in W^{1,1}(\mathbb{R}^N)$.

Then, there exists an extension $t \in [0,T) \to \tilde{f}(t) \in L^1(\mathbb{R}^N)$ *of* $t \to f(t)$ *that is differentiable at* $t = 0$ *with*

$$\tilde{f}'(0) = F'(0) - \nabla P(f(0)).V.$$

Moreover, $t \to I(t) = \int_{\Omega_t} f(t)$ *is differentiable at* 0 *and we have* (5.11) *after setting* $f'(0)(x) := \tilde{f}'(0)(x)$ *a.e.* $x \in \Omega$.

Remark 5.2.4. If Ω is an open set with a regular boundary, such an extension operator does exist (see, e.g., [54] in the C^1 case). If Ω is open, for all $z \in \Omega$, there exist $t_z \in (0,T)$ and $r_z > 0$ such that

$$\forall t \in (0,t_z), \quad B(z,r_z) \subset \Omega_t, \text{ so that a.e. } x \in B(z,r_z),\ f(t,x) = \tilde{f}(t,x).$$

Thus, for any compact set $K \subset \Omega$, $t \to f(t)_{|K} \in L^1(K)$ is differentiable at 0 and the definition depends only on f. In fact, even if Ω is measurable only, we may prove that this definition does not depend on the choice of the extension $\tilde{f}$ of f: see Exercise 5.11.

If Ω is not open, we must understand that $\mathrm{div}[f(0)V]$ is obtained by first differentiating $P(f(0))V$ in $\mathbb{R}^N$ and then taking its restriction to Ω. Like for the time derivative, we check that it does not depend on the choice of the extension.

Corollary 5.2.5. *Let* Φ *satisfy* (5.6), *let* Ω *be open, and let* $t \in [0,T) \to f(t) \in L^1(\Omega_t)$. *Let us assume*

$$t \in [0,T) \to F(t) = f(t,\Phi(t,\cdot)) \in L^1(\Omega) \text{ is differentiable at } 0, \tag{5.14}$$

and $f(0) \in W^{1,1}(\Omega)$. *Then,* $t \to I(t) = \int_{\Omega_t} f(t)$ *is differentiable at* 0; *for all* $K \subset \Omega$ *compact,* $t \to f(t)_{|K} \in L^1(K)$ *is differentiable at* 0, $f'(0) = F'(0) - \nabla f(0).V \in L^1(\Omega)$, *and we have* (5.11).

Remark. One should understand that $f'(0)$ is defined here by extension to Ω of the derivatives at $t = 0$ of $t \to f(t)_{|K}$. We prove that it is equal to $F'(0) - \nabla f(0).V$: in particular, it belongs to $L^1(\Omega)$.

5.2.4 The proofs

Let us start with a lemma.

Lemma 5.2.6. *Let* $g \in W^{1,1}(\mathbb{R}^N)$ *and let* $\Psi : [0,T) \to W^{1,\infty}$ *be continuous at* 0 *with* $\Psi(0) = I$ *and* $t \to \Psi(t) \in L^\infty$ *differentiable at* 0. *Then,*

$$t \to G(t) := g \circ \Psi(t) \in L^1(\mathbb{R}^N)$$

is differentiable at 0 *and* $G'(0) = \nabla g.\Psi'(0)$.

Proof. Preliminary remark: We will use here and later on that, under the assumptions of Lemma 5.2.6,

$$\lim_{t\to 0} h \circ \Psi(t) = h \text{ in } L^1(\mathbb{R}^N) \quad \forall\, h \in L^1(\mathbb{R}^N). \tag{5.15}$$

Indeed, we may approximate h in $L^1(\mathbb{R}^N)$ by a sequence $h^p \in C_0^\infty(\mathbb{R}^N)$ and, introducing $h^p \circ \Psi(t)$, we see by change of variable $x = \Psi(t)y$ (whose Jacobian is uniformly bounded as $t \to 0$) that

$$\|h \circ \Psi(t) - h\|_{L^1} \le C\|h - h^p\|_{L^1} + \|h^p \circ \Psi(t) - h^p\|_{L^1}.$$

Since $h^p \in C_0^\infty$, this last term tends to 0 for p fixed as $t \to 0$ so that

$$\limsup_{t\to 0} \|h \circ \Psi(t) - h\|_{L^1} \le C\|h - h^p\|_{L^1} \quad \forall\, p.$$

Let us set $Z := \Psi'(0)$ and

$$\eta_t = t^{-1}\|g(\Psi(t)) - g - t\nabla g.Z\|_{L^1}.$$

The goal is to prove that

$$\lim_{t\to 0} \eta_t = 0. \tag{5.16}$$

Let us first assume that $g \in C_0^\infty$. Recall that, for all $k \in \mathbb{R}^N$,

$$g(y+k) - g(y) - \nabla g(y).k = \int_0^1 [\nabla g(y+sk) - \nabla g(y)].k\, ds.$$

Let us apply this with $k = \Psi(t,y) - y = tZ(y) + t\epsilon(t,y)$, where $\epsilon(t,\cdot)$ tends to 0 in L^∞ and let us integrate with respect to y on $\mathbb{R}^N$. We deduce

$$\eta_t \le \|\nabla g\|_{L^1}\|\epsilon(t)\|_\infty + K\|e(t,g)\|_{L^1}, \tag{5.17}$$

where K is a uniform upper bound for $\|Z + \epsilon(t)\|_\infty$ and where

$$e(t,g)(y) := \int_0^1 |\nabla g((1-s)y + s\Psi(t,y)) - \nabla g(y)| ds.$$

Let us bound the L^1-norm of $e(t,g)$ in two ways. First, from

$$\|e(t,g)\|_{L^1} \le \|\nabla g\|_{L^1} + \left\| y \to \int_0^1 |\nabla g((1-s)y + s\Psi(t,y))|\, ds \right\|_{L^1}$$

and by using the change of variable $x = (1-s)y + s\Psi(t,y)$, it easily follows that

$$\|e(t,g)\|_{L^1} \le C_\Psi \|\nabla g\|_{L^1}, \qquad C_\Psi = C(\|\Psi(t)\|_{W^{1,\infty}}). \tag{5.18}$$

On the other hand, writing

$$e(t,g)(y) = \int_0^1 \left| \int_0^1 D^2 g(y + \sigma s[\Psi(t,y) - y]) s(\Psi(t,y) - y)\, d\sigma \right| ds,$$

and using again an obvious change of variable, it follows that

$$\|e(t,g)\|_{L^1} \le C_\Psi \|D^2 g\|_{L^1} \|\Psi(t) - I\|_\infty, \qquad C_\Psi = C(\|\Psi(t)\|_{W^{1,\infty}}). \tag{5.19}$$

Now let $g \in W^{1,1}$ and let $g^p \in C_0^\infty$ converge to g in $W^{1,1}$. Inequality (5.17) remains valid for $g \in W^{1,1}(\mathbb{R}^N)$. On the other hand, writing

$$e(t,g) \le e(t, g - g^p) + e(t, g^p),$$

it follows from the two inequalities (5.18) and (5.19) that, for all p,

$$\|e(t,g)\|_{L^1} \le C_\Psi \|g - g^p\|_{W^{1,1}} + C_\Psi \|D^2 g^p\|_{L^1} \|\Psi(t) - I\|_\infty.$$

This implies that, for all p,

$$\limsup_{t \to 0} \|e(t,g)\|_{L^1} \le C_\psi \|g - g^p\|_{W^{1,1}}.$$

Together with (5.17), it follows that $\lim_{t\to 0} \eta_t = 0$, which yields (5.16) and the conclusion of the lemma. □

Proof of Theorem 5.2.2. By (5.6),

$$\text{a.e. } y \in \mathbb{R}^N, \quad D_y\Phi(t,y) = D_y\Phi(0,y) + tD_yV(y) + t\epsilon(t,y),$$

where $\epsilon(t)$ tends to 0 in L^∞. Here $D_y\Phi(0,y) = I$ is the identity in $\mathbb{R}^N$ and we remember that the determinant mapping $M \longrightarrow \det(M)$, where M are $N \times N$ matrices, is differentiable (since it is multilinear and we are in a finite dimension). Moreover, its Fréchet derivative around the identity is $M \longrightarrow \operatorname{tr}(M)$. Thus,

$$\text{a.e. } y \in \mathbb{R}^N, \quad \det(D_y\Phi(t,y)) = 1 + t\operatorname{tr}(D_yV(y)) + t\epsilon_1(t,y),$$

with $\epsilon_1(t)$ tending to 0 in L^∞. We may also write

$$J(t,y) = 1 + t \operatorname{div}_y V(y) + t\epsilon_1(t,y). \tag{5.20}$$

Now, let us decompose $[I(t) - I(0)]/t$ as the sum of three terms (see (5.8)):

$$I_1 = \frac{1}{t}\int_\Omega [f(t,\Phi(t)) - f(0,\Phi(t))]J(t),$$

$$I_2 = \frac{1}{t}\int_\Omega [f(0,\Phi(t)) - f(0)]J(t), \qquad I_3 = \int_\Omega f(0)\frac{1}{t}[J(t) - J(0)].$$

By (5.20) and by dominated convergence, I_3 tends to $\int_\Omega f(0)\,\mathrm{div}_y V$ as $t \longrightarrow 0$. Coming back on Ω_t, we see that

$$I_1 = \frac{1}{t}\int_{\Omega_t} [f(t) - f(0)] = \frac{1}{t}\int_{\mathbb{R}^N} \chi_{\Omega_t}[f(t) - f(0)].$$

Therefore I_1 converges to $\int_\Omega f'(0)$ by (5.9) and by the fact that χ_{Ω_t} tends strongly to χ_Ω (this is easy to check; see Exercise 5.11). For I_2, we apply Lemma 5.2.6 with $g = f(0)$ and $\Psi(t) = \Phi(t)$ to find that the limit is equal to $\int_\Omega \nabla f(0).V$. This ends the proof of the differentiability of $I(\cdot)$ and of the formula (5.11). □

For the corollaries, we use the following lemma:

Lemma 5.2.7. *Let $t \in [0, T) \longrightarrow h(t) \in L^1(\mathbb{R}^N)$ be differentiable at* 0 *with $h(0) \in W^{1,1}(\mathbb{R}^N)$. Then, under assumption* (5.6), *$t \longrightarrow g(t) = h(t) \circ \Phi(t)^{-1} \in L^1(\mathbb{R}^N)$ is differentiable at* 0 *and $g'(0) = h'(0) - \nabla g(0).V$.*

Proof. Let us set $\psi_t = \Phi(t)^{-1}$. We write $[g(t) - g(0)]/t = A(t) + B(t) + C(t)$ with

$$A(t) = \left[\frac{h(t) - h(0)}{t} - h'(0)\right] \circ \psi_t, \quad B(t) = h'(0) \circ \psi_t, \quad C(t) = \left[\frac{h(0) \circ \psi_t - h(0)}{t}\right].$$

By a change of variable, $\|A(t)\|_{L^1} \le C\|\frac{h(t)-h(0)}{t} - h'(0)\|_{L^1}$ tends to 0. The second term $B(t)$ tends to $h'(0)$ by (5.15). For the third term $C(t)$, we apply Lemma 5.2.6, using (5.7) and $h(0) \in W^{1,1}$ to conclude that it converges to $-\nabla h(0).V$. □

Proof of Corollary 5.2.3. Let us set $\tilde{f}(t) := P(F(t)) \circ \Phi(t)^{-1} \in L^1(\mathbb{R}^N)$. We have $\tilde{f}(t)_{|\Omega_t} = f(t)$. Since by composition $t \longrightarrow P(F(t))$ is differentiable at 0 and since $P(f(0)) \in W^{1,1}$, we apply Lemma 5.2.7 to obtain the differentiability of $t \longrightarrow \tilde{f}(t)$.

We finally apply Theorem 5.2.2 to $\tilde{f}$ to obtain the corollary. □

Proof of Corollary 5.2.5. Let $\zeta \in C_0^\infty(\Omega)$ with $\zeta \equiv 1$ on a neighborhood of K. By Lemma 5.2.7 and (5.7), $t \longrightarrow (\zeta f(t)) \circ \Phi(t)^{-1} \in L^1(\mathbb{R}^N)$ is differentiable at 0. Its restriction to K, which is $f(t)_{|K}$ for t small, is therefore differentiable in $L^1(K)$ and we have on K that $f'(0) = [F'(0) - \nabla f(0).V]_{|K}$. This allows us to define $f'(0)$ on the whole Ω and we have the formula announced for $f'(0)$.

Since $I(t) = \int_\Omega F(t)J(t)$, it follows that $t \longrightarrow I(t)$ is differentiable and we have

$$I'(0) = \int_\Omega F'(0) + f(0)\,\mathrm{div}\, V = \int_\Omega f'(0) + \nabla f(0).V + f(0)\,\mathrm{div}\, V,$$

which yields formula (5.11). □

5.2.5 Differentiating on intervals and first applications

Let us now state one of the main tools for differentiating on intervals of R.

Corollary 5.2.8. *Let us assume*

$$\Phi \in C^1([0,T); W^{1,\infty}(\mathbb{R}^N)), \qquad f \in C^1([0,T); L^1(\mathbb{R}^N)) \cap C([0,T); W^{1,1}(\mathbb{R}^N)).$$

We set $V(t,x) = \frac{\partial \Phi}{\partial t}(t, \Phi(t)^{-1}(x))$ *for all* $x \in \Omega_t$.

Then $t \in [0,T) \to I(t)$ *is continuously differentiable on* $[0,T)$ *and we have*

$$I'(t) = \int_{\Omega_t} \frac{\partial f}{\partial t}(t) + \operatorname{div}[fV](t). \tag{5.21}$$

Proof. According to the definition of V, we have

$$\forall\, (t,x) \in [0,T) \times \mathbb{R}^N, \quad \frac{\partial \Phi}{\partial t}(t,x) = V(t, \Phi(t,x)), \quad \Phi(0,x) = x. \tag{5.22}$$

Let $t_0 \in [0,T)$. We set

$$\overline{\Phi}(t) = \Phi(t+t_0) \circ \Phi(t_0)^{-1}, \qquad \overline{f}(t,x) = f(t+t_0, x),$$

and we apply Theorem 5.2.2 to $\overline{\Phi}, \overline{f}$ to obtain that I is differentiable at t_0 together with the corresponding formula. Note that the definition of V implies that

$$\frac{\partial \overline{\Phi}}{\partial t}(0) = \frac{\partial \Phi}{\partial t}(t_0, \Phi(t_0)^{-1}(\cdot)) = V(t_0, \cdot).$$

It is easy to check the continuity of $t \to I'(t)$ on the formula. □

Remark 5.2.9. In this corollary, Φ is given, and V is defined in terms of Φ. Very often, V is given first and Φ is defined next as being the flow associated with the vector field V according to the equation (5.22). Both approaches are completely equivalent. Indeed, if we are given $V \in C([0,T); W^{1,\infty}(\mathbb{R}^N))$, then equation (5.22) has a unique solution on $[0,T)$. The solution $\Phi(t)$ does belong to the required space since $Z(t,y) = D_y\Phi(t,y)$ is a solution of $\frac{\partial Z}{\partial t} = D_y V(t, \Phi(t,x)).Z$: thus Z and $\frac{\partial Z}{\partial t}$ are continuous from $[0,T)$ into $L^\infty(\mathbb{R}^N)$.

Note that if $\Phi(t)(x) = x + t\theta(x)$, where $\theta \in W^{1,\infty}(\mathbb{R}^N)$, then

$$V(0) = \theta, \qquad V(t) = \theta \circ (I + t\theta)^{-1}. \tag{5.23}$$

We end this section with a few applications.

Let $f \in W^{1,1}(\mathbb{R}^N)$ and $\Phi \in C^1([0,T); W^{1,\infty})$. Then, $t \in [0,T) \to \int_{\Omega_t} f$ is differentiable and we have

$$\frac{d}{dt}\int_{\Omega_t} f = \int_{\Omega_t} \operatorname{div}[fV] \quad \Big(= \int_{\partial\Omega_t} fV.n \text{ if } \Omega_t \text{ is a regular open set}\Big). \tag{5.24}$$

This is a direct application of Corollary 5.2.8. We deduce, for instance, the volume preservation law for a mapping Φ.

Corollary 5.2.10. *Let $\Phi \in C^1([0,T); W^{1,\infty})$. Then $|\Omega_t| = |\Omega|$ for all measurable sets $\Omega \subset \mathbb{R}^N$ if and only if $\operatorname{div}_x V = 0$ on $\mathbb{R}^N$.*

Indeed, we apply the previous formula with $f \equiv 1$. We see that the volume is preserved if and only if $\int_\Omega \operatorname{div}_x V = 0$ for all measurable sets Ω. This is equivalent to $\operatorname{div}_x V = 0$ a.e.

In the same way, we obtain the highly important "continuity equation" or "conservation of mass equation" for continuous media.

Corollary 5.2.11. *Under the assumptions of Corollary 5.2.8, we have*

$$\int_{\Omega_t} f = \int_\Omega f \quad \forall \text{ measurable } \Omega$$

if and only if

$$\frac{\partial f}{\partial t} + \operatorname{div}_x[fV] = 0 \text{ on } (0,T) \times \mathbb{R}^N.$$

Next, we will need a differentiation formula that we have not seen yet: indeed, we will need to differentiate integrals of the form $\int_{\Omega_t} gu(t)$, where $u(t) \in H_0^1(\Omega_t)$, but where g belongs only to $L^2(\mathbb{R}^N)$. When $g \in H^1(\mathbb{R}^N)$ and Ω is a regular open set, we obtain

$$\frac{d}{dt}_{|t=0} \int_{\Omega_t} gu(t) = \int_\Omega gu'(0) + \int_{\partial\Omega} gu(0)V(0).n = \int_\Omega gu'(0).$$

It turns out that this remains valid when $g \in L^2$ only and for all measurable Ω (we refer to the previous chapter for the definition of $H_0^1(\Omega)$ when Ω is only measurable or quasi-open).

Lemma 5.2.12. *Let Φ satisfy (5.6), let $g \in L^2(\mathbb{R}^N)$, and $t \in [0,T) \to u(t) \in H_0^1(\Omega)$ be differentiable at 0 for the norm of $L^2(\mathbb{R}^N)$. Then, $t \to \int_{\Omega_t} gu(t)$ is differentiable at 0 and we have*

$$\frac{d}{dt}_{|t=0} \int_{\Omega_t} gu(t) = \int_\Omega gu'(0).$$

Proof. It is sufficient to remark that $u(t) \in H_0^1(\Omega)$ allows us to write

$$I(t) = \int_{\Omega_t} gu(t) = \int_{\mathbb{R}^N} gu(t).$$

Thus

$$\frac{I(t) - I(0)}{t} = \int_{\mathbb{R}^N} g\frac{u(t) - u(0)}{t}$$

converges to $\int_{\mathbb{R}^N} gu'(0)$ as t tends to 0. □

5.3 A model PDE problem

5.3.1 Presentation of the problem

We introduce again $t \to \Phi(t)$ satisfying (5.6) together with $\Omega \subset \mathbb{R}^N$ measurable *bounded* and $\Omega_t = \Phi(t,\Omega)$. We are given $f \in L^2_{\mathrm{loc}}(\mathbb{R}^N)$, $\lambda \geq 0$. We consider the solution of the Dirichlet problem

$$\begin{cases} -\Delta u + \lambda u = f & \text{in } \Omega, \\ u = 0 & \text{on } \partial\Omega. \end{cases}$$

Remark. We refer to Proposition 4.5.1 for the definition of the solution u when Ω is only measurable: we can indeed extend the definition of $H^1_0(\Omega)$ to this more general situation. As proved in Proposition 3.3.44, if necessary we can always assume Ω is a quasi-open set without modifying $H^1_0(\Omega)$ (and in a unique way).

Let us consider the solution u_t of the following Dirichlet problem for t small:

$$\begin{cases} -\Delta u_t + \lambda u_t = f & \text{in } \Omega_t, \\ u_t = 0 & \text{on } \partial\Omega_t, \end{cases} \tag{5.25}$$

which is more precisely defined through its variational formulation

$$u_t \in H^1_0(\Omega_t),\ \forall\, \varphi_t \in \Omega_t, \qquad \int_{\Omega_t} \nabla u_t \nabla \varphi_t + \lambda u_t \varphi_t = \int_{\Omega_t} f \varphi_t. \tag{5.26}$$

As a typical functional, we will consider in this section the following one:

$$J(\Omega_t) := a \int_{\Omega_t} |\nabla u_t - \nabla v_0|^2 + b \int_{\Omega_t} |u_t - v_1|^2, \tag{5.27}$$

where a and b are fixed real numbers and $v_0 \in H^2_{\mathrm{loc}}(\mathbb{R}^N)$ and $v_1 \in H^1_{\mathrm{loc}}(\mathbb{R}^N)$ are given.

To compute the derivative of $t \to J(\Omega_t)$, it may be useful to differentiate $t \to u_t$ in the right way. As we saw above, the situation is quite different depending on whether the function has a good extension to $\mathbb{R}^N$. Since here $u_t \in H^1_0(\Omega_t)$, it may trivially be extended by 0 to an $H^1(\mathbb{R}^N)$ function. We shall adopt this point of view and we will not distinguish between $u_t \in H^1_0(\Omega_t)$ and $u_t \in H^1(\mathbb{R}^N)$.

5.3.2 A formal computation

Let us suppose for a moment that Ω is an open set with regular boundary and that $t \to u_t$ has good differentiability properties (we denote by u' its derivative at 0). Then we may differentiate (5.25), first inside Ω, then at its boundary, by differentiating the following identity with respect to t and for fixed x:

$$\forall\, x \in \partial\Omega, \quad u_t(\Phi(t,x)) = 0.$$

We then obtain

$$\text{differentiation in the interior:} \qquad -\Delta u' + \lambda u' = 0 \quad \text{in } \Omega, \tag{5.28}$$

$$\text{differentiation at the boundary:} \qquad u' + \nabla u.V = 0 \quad \text{on } \partial\Omega. \tag{5.29}$$

Thus, u' is characterized as being the solution of a boundary value problem of nonhomogeneous Dirichlet type on Ω. One may continue the computation in a formal way and apply formula (5.11) to differentiate $t \longrightarrow j(t) = J(\Omega_t)$, which gives

$$\begin{aligned} j'(0) = &\int_\Omega 2a\nabla u'.(\nabla u - \nabla v_0) + 2bu'(u - v_1) \\ &+ \int_{\partial\Omega} [a|\nabla u - \nabla v_0|^2 + b|u - v_1|^2]V(0).n. \end{aligned} \tag{5.30}$$

This formula presents several difficulties. First, we will see that u' is only in $L^2(\mathbb{R}^N)$ in general: this leads to a problem in the first integral of (5.30) for the product $\nabla u'.\nabla u$. On the other hand, $\nabla u \in L^2(\Omega)$ does not generally have a trace on $\partial\Omega$, which leads to a problem when trying to write the second integral in (5.30). Coming back to the nonintegrated form $\int_\Omega \operatorname{div}[\|\nabla u - \nabla v_0\|^2 V]$ does not really help since ∇u does not belong to H^1: thus, writing (5.30), and a fortiori justifying it, lead to real difficulties, at least without regularity on Ω. In fact, we can progress in the question by using that

$$\int_{\Omega_t} |\nabla u_t|^2 + \lambda u_t^2 = \int_{\Omega_t} f u_t. \tag{5.31}$$

This may be obtained through the variational formulation (5.26) by taking $\varphi_t = u_t$. Thus, we may rewrite

$$j(t) = \int_{\Omega_t} a\big(f u_t - \lambda u_t^2 - 2\nabla u_t.\nabla v_0 + |\nabla v_0|^2\big) + b(u_t^2 - 2u_t v_1 + v_1^2).$$

But, since $u_t \in H_0^1(\Omega_t)$, this may also be written

$$j(t) = \int_{\mathbb{R}^N} a f u_t + (b - \lambda a)u_t^2 + 2a u_t \Delta v_0 - 2b u_t v_1 + \int_{\Omega_t} a|\nabla v_0|^2 + b v_1^2. \tag{5.32}$$

We immediately see that, as soon as one knows that $t \longrightarrow u_t \in L^2(\mathbb{R}^N)$ is differentiable at 0, then $j'(0)$ exists and the computation is straightforward since the variable domain appears only in the last integral.

5.3.3 The two main results

Theorem 5.3.1. *Let Ω be measurable, bounded and $f \in H^1(\mathbb{R}^N)$. Assume that Φ satisfies* (5.6). *Then, $t \to u_t \in L^2(\mathbb{R}^N)$ is differentiable at* 0.

If Ω is open, its derivative u' is the unique solution of the problem

$$u' + \nabla u.V \in H_0^1(\Omega), \quad -\Delta u' + \lambda u' = 0 \text{ in } \Omega. \tag{5.33}$$

In all cases, the function j is differentiable at 0 *and*

$$j'(0) = \int_\Omega u'[af + 2(b - a\lambda)u + 2a\Delta v_0 - 2bv_1] + \operatorname{div}[a|\nabla v_0|^2 + bv_1^2]. \tag{5.34}$$

Remark. It is also possible to write a weak variational formulation of $-\Delta u' + u' = 0$ when Ω is only measurable: see Exercise 5.3.

The point that is probably the most important in proving the differentiability of $t \to u_t$ is the use of the function $v_t = u_t \circ \Phi(t)$, which is defined on the fixed domain Ω. First, it turns out that this function is differentiable with values in $H_0^1(\Omega)$. The differentiability of u_t may then be deduced by composition since $u_t = v_t \circ \Phi(t)^{-1}$ (and we then have by direct differentiation that $u' = v' - \nabla u.V$). Next, the main general tool for proving the differentiability of v_t is the *implicit function theorem applied to the equation transferred onto the fixed domain* Ω.

At this point, it becomes more interesting to deal with Fréchet differentiability rather than t-differentiability with respect to the real variable $t \in [0, T]$. Thus, we are going to consider perturbations $I + \theta$ of the identity, where $\theta \in W^{1,\infty}(\mathbb{R}^N; \mathbb{R}^N)$ and is close to 0 in the norm of this space, so that $I + \theta$ is a bi-Lipschitz homeomorphism (see (5.5)).

We then introduce $\Omega_\theta = (I + \theta)(\Omega)$ and u_θ the solution of the problem

$$u_\theta \in H_0^1(\Omega_\theta),\ \forall\, \varphi_\theta \in H_0^1(\Omega_\theta), \quad \int_{\Omega_\theta} \nabla u_\theta \nabla \varphi_\theta + \lambda u_\theta \varphi_\theta = \int_{\Omega_\theta} f \varphi_\theta. \tag{5.35}$$

Next we set $v_\theta = u_\theta \circ (I + \theta)$. We then have the following key result.

Theorem 5.3.2. *Assume $f \in H^1(\mathbb{R}^N)$. Then, $\theta \in W^{1,\infty} \to v_\theta \in H_0^1(\Omega)$ is of class C^1 in a neighborhood of* 0. *Moreover, $\theta \in W^{1,\infty} \to u_\theta \in L^2(\mathbb{R}^N)$ is differentiable at* 0.

Remark. Note that Ω *is only measurable* and bounded. We will see later that, when f is C^∞, then the mapping $\theta \to v_\theta$ is even of class $\mathcal{C}^\infty$, this *for any measurable* Ω! On the other hand, if f is only in L^2, then $\theta \to v_\theta \in H_0^1$ is not necessarily differentiable at 0 (see Remarks 5.3.4 and 5.3.6).

5.3.4 The proofs

The proofs of the various results above use the following technical lemma that will be also useful later on. We denote by J_θ the Jacobian of $I + \theta$ ($J_\theta = \det(I + D\theta)$). For a function $\theta \to G(\theta)$, we will denote by $G'(\theta)$ its Fréchet derivative at θ.

Lemma 5.3.3. *Let $g \in W^{1,p}(\mathbb{R}^N)$, $1 \le p < +\infty$. Then, the mapping*

$$G : \theta \in W^{1,\infty} \to g \circ (I + \theta) \in L^p(\mathbb{R}^N)$$

is of class C^1 on a neighborhood of 0 and

$$\forall \xi \in W^{1,\infty}, \quad G'(\theta).\xi = [\nabla g \circ (I + \theta)].\xi.$$

More generally, if $\theta \in W^{1,\infty} \to \Psi(\theta) \in W^{1,\infty}$ is continuous at 0 with $\Psi(0) = I$ and $\theta \in W^{1,\infty} \to (g(\theta), \Psi(\theta)) \in L^p \times L^\infty$ differentiable at 0 with $g(0) \in W^{1,p}$ and $g'(0) : W^{1,\infty} \to W^{1,p}$ continuous, then the mapping

$$\mathcal{G} : \theta \in W^{1,\infty} \to g(\theta) \circ \Psi(\theta) \in L^p(\mathbb{R}^N)$$

is differentiable at 0 and

$$\forall \xi \in W^{1,\infty}, \quad \mathcal{G}'(0)\xi = g'(0)\xi + \nabla g(0).\Psi'(0)\xi. \tag{5.36}$$

Remark 5.3.4. If $g \in L^2(\mathbb{R}^N)$ only, it is not true that $\theta \in W^{1,\infty} \to g \circ (I+\theta) \in H^{-1}$ is differentiable at 0 (see [280] for a counterexample). On the other hand, $\theta \in W^{1,\infty} \to g \circ (I + \theta) J_\theta \in H^{-1}$ (which naturally appears in the change of variables) is weakly differentiable in a neighborhood of 0. Indeed, for $v \in H^1(\mathbb{R}^N)$,

$$\int v[g \circ (I + \theta)J_\theta - g] = \int g[v \circ (I + \theta)^{-1} - v],$$

and by the previous lemma, $\theta \in W^{1,\infty} \to v \circ (I+\theta)^{-1} \in L^2$ is differentiable. The term J_θ is important since this property is not true for $\theta \in W^{1,\infty} \to g \circ (I + \theta) \in H^{-1}$ (see Exercise 5.5). If one replaces $W^{1,\infty}$ by $C^{k,\infty}$, $k \ge 2$, the two are weakly differentiable, but not strongly in general.

Remark 5.3.5. The second part of Lemma 5.3.3 applies in particular to $\Psi(\theta) = (I + \theta)^{-1}$ by (5.5). We use it to prove differentiability *on compact subsets K of the open set Ω*: if $\theta \to g(\theta) \in W^{1,p}(\Omega)$ satisfies the same assumptions as above on Ω (rather than on $\mathbb{R}^N$) and if $\zeta \in C_0^\infty(\Omega)$, $\zeta \equiv 1$ on a neighborhood of $K \subset \Omega$ compact, by applying Lemma 5.3.3 to $\theta \to (\zeta g(\theta), (I + \theta)^{-1})$, we deduce, as in the proof of Corollary 5.2.5, that

$$\theta \in W^{1,\infty} \to g(\theta) \circ (I + \theta)^{-1}_{|K} \in L^p(K)$$

is differentiable at 0. If $g(\theta) = f(\theta) \circ (I + \theta)$, where $f(\theta) \in W^{1,p}(\Omega_\theta)$, we obtain the differentiability of $\theta \to f(\theta)$ inside Ω.

Let us prove Theorem 5.3.1, assuming for a moment that Theorem 5.3.2 and Lemma 5.3.3 are proved.

Proof of Theorem 5.3.1. By Theorem 5.3.2 and by composition, $t \longrightarrow u_t = u_{\Phi(t)-I} \in L^2(\mathbb{R}^N)$ is differentiable at 0 and

$$u_t = v_t \circ \Phi(t)^{-1} \Rightarrow u' = v' - \nabla u.V.$$

This proves that $u' + \nabla u.V \in H_0^1(\Omega)$.

If Ω is open and $\varphi \in C_0^\infty(\Omega)$, for small t, we also have $\varphi \in C_0^\infty(\Omega_t)$ and therefore

$$\int_{\mathbb{R}^N} f\varphi = \int_{\Omega_t} f\varphi = \int_{\Omega_t} \nabla u_t.\nabla\varphi + \lambda u_t\varphi = \int_{\mathbb{R}^N} -u_t\Delta\varphi + \lambda u_t\varphi.$$

By differentiation, we obtain

$$0 = \int_\Omega -u'\Delta\varphi + \lambda u'\varphi \quad \forall\, \varphi \in C_0^\infty(\Omega).$$

And for formula (5.34), it has already been proved as a consequence of (5.32). □

Proof of Lemma 5.3.3. The result on G is a consequence of the result on $\mathcal{G}$: we fix θ_0 small and we choose $g(\theta) \equiv g \circ (I + \theta_0)$ and $\Psi(\theta) = I + (I + \theta_0)^{-1}\theta$. We then see the continuity of $\theta_0 \longrightarrow G'(\theta_0)$ on the formula (we use the approach of (5.15) for the continuity of $\theta_0 \longrightarrow \nabla g \circ (I + \theta_0)$).

For the differentiability of $\mathcal{G}$ at 0, let us show

$$\|g(\theta) \circ \Psi(\theta) - g(0) - \nabla g(0).\Psi'(0)\theta - g'(0)\theta\|_p = o(\|\theta\|_{1,\infty}).$$

We write this expression as the sum of four terms:

$$\begin{aligned}
A(\theta) &= [g(\theta) - g(0) - g'(0)\theta] \circ \Psi(\theta),\\
B(\theta) &= \nabla g(0).[\Psi(\theta) - \Psi(0) - \Psi'(0)\theta],\\
C(\theta) &= g(0) \circ \Psi(\theta) - g(0) - \nabla g(0).[\Psi(\theta) - \Psi(0)],\\
D(\theta) &= (g'(0).\theta) \circ \Psi(\theta) - g'(0)\theta.
\end{aligned}$$

By change of variable for A,

$$\begin{aligned}
\|A(\theta)\|_p &\le \|g(\theta) - g(0) - g'(0)\theta\|_p \|\Psi(\theta)\|_{1,\infty} = o(\|\theta\|_{1,\infty}),\\
\|B(\theta)\|_p &\le \|g\|_{1,p} \|\Psi(\theta) - \Psi(0) - \Psi'(0)\theta\|_\infty = o(\|\theta\|_{1,\infty}).
\end{aligned}$$

For C, let us set $g := g(0)$, $\psi := \Psi(\theta) - \Psi(0)$. We have, as in the proof of Lemma 5.2.6,

$$\|g(I+\psi)-g-\nabla g.\psi\|_p = \left\| \int_0^1 ds\,[\nabla g(I+s\psi)-\nabla g].\psi \right\|_p \leq \|\psi\|_\infty \|e(g)\|_p, \quad (5.37)$$

where $e(g) = \int_0^1 |\nabla g(I+s\psi)-\nabla g|\,ds$. We approximate g in $W^{1,p}$ by $g^k \in C_0^\infty$, and like in the proof of Lemma 5.2.6, we have

$$\|e(g)\|_p \leq \|e(g-g^k)\|_p + \|e(g^k)\|_p \leq 2\|g-g^k\|_{1,p}[1+\|\psi\|_{1,\infty}] + C_k\|\psi\|_\infty,$$

where C_k depends only on g_k. Since ψ tends to 0 in L^∞ and is bounded in $W^{1,\infty}$ when $\|\theta\|_{1,\infty}$ tends to 0, we deduce that $\|e(g)\|_p$ also tends to 0. Next by (5.37), $C(\theta) = o(\|\theta\|_{1,\infty})$ since $\|\psi\|_\infty \leq C\|\theta\|_{1,\infty}$. Finally, we have for D,

$$\|D(\theta)\|_p \leq \|g'(0)\theta\|_{1,p}\|\Psi(\theta)-I\|_\infty \leq C\|\theta\|_{1,\infty}^2.$$

□

Proof of Theorem 5.3.2. The main point is that v_θ is a solution of

$$v_\theta \in H_0^1(\Omega), \quad -\operatorname{div}\left(A(\theta)\nabla v_\theta\right) + \lambda J_\theta v_\theta = [f\circ(I+\theta)]J_\theta, \quad (5.38)$$

where

$$A(\theta) = J_\theta(I+D\theta)^{-1}(I+^t D\theta)^{-1}. \quad (5.39)$$

This can be seen by a change of variable. Indeed, we have $\varphi \in H_0^1(\Omega)$, so that $\varphi_\theta := \varphi\circ(I+\theta)^{-1} \in H_0^1(\Omega_\theta)$ and we have

$$\nabla\varphi_\theta = [(I+^t D\theta)^{-1}\nabla\varphi]\circ(I+\theta)^{-1}.$$

By definition of u_θ,

$$\int_{\Omega_\theta} \nabla u_\theta \nabla\varphi_\theta + \lambda u_\theta\varphi_\theta = \int_{\Omega_\theta} f\varphi_\theta \quad \forall\,\varphi \in H_0^1(\Omega). \quad (5.40)$$

By the change of variable $x = (I+\theta)(y)$, recalling that $\nabla v_\theta = (I+^t D\theta)[\nabla u_\theta\circ(I+\theta)]$, this can also be written as

$$\forall\,\varphi \in H_0^1(\Omega),$$
$$\int_\Omega \{[(I+^t D\theta)^{-1}\nabla v_\theta][(I+^t D\theta)^{-1}\nabla\varphi] + \lambda v_\theta\varphi\}J_\theta = \int_\Omega [f\circ(I+\theta)]\varphi J_\theta.$$

We deduce (5.38).

We now consider the operator[3]

$$F:(\theta,v)\in W^{1,\infty}\times H_0^1(\Omega)\to -\operatorname{div}\big(A(\theta)\nabla v\big)+\lambda J_\theta v-[f\circ(I+\theta)]J_\theta\in H^{-1}(\Omega).$$

Note that, with this notation, (5.38) may be rewritten

$$F(\theta,v_\theta)=0 \quad (\text{and } v_0=u_0). \tag{5.41}$$

The mapping F is of class C^1 for small θ: indeed, $\theta\in W^{1,\infty}\to J_\theta=\det(I+D\theta)\in L^\infty$ is of class C^∞ since $\theta\in W^{1,\infty}\to I+D\theta\in L^\infty(\mathbb{R}^N,\mathcal{M}_N)$ is linear and continuous and therefore C^∞ (where we denote by $\mathcal{M}_N$ the space of $N\times N$ square matrices). As already noticed, the mapping $M\to\det(M)$ is multilinear and therefore differentiable since the dimension of $\mathcal{M}_N$ is finite. Similarly, $\theta\in W^{1,\infty}\to (I+D\theta)^{-1}=\sum_{q\ge0}(-1)^q D\theta^q\in L^\infty(\mathbb{R}^N,\mathcal{M}_N)$ is also C^∞ (its Fréchet derivative is $\xi\to-(I+D\theta)^{-1}D\xi(I+D\theta)^{-1}$). Thus

$$\theta\in W^{1,\infty}\to A(\theta)\in L^\infty(\mathbb{R}^N,\mathcal{M}_N) \quad \text{is of class } C^\infty, \tag{5.42}$$

and the mapping

$$(A,v)\in L^\infty(\mathbb{R}^N,\mathcal{M}_N)\times H_0^1(\Omega)\to -\operatorname{div}(A\nabla v)\in H^{-1}(\Omega)$$

is of class C^∞ since it is bilinear and continuous. Finally, by Lemma 5.3.3,

$$\theta\to k(\theta)=[f\circ(I+\theta)]J_\theta\in L^2(\mathbb{R}^N)\subset H^{-1}(\Omega)$$

is of class C^1.

The operator $D_vF(0,v_0)$ is an isomorphism from $H_0^1(\Omega)$ onto $H^{-1}(\Omega)$ since

$$\forall\varphi\in H_0^1(\Omega),\quad D_vF(0,v_0)\varphi=-\Delta\varphi+\lambda\varphi.$$

Thus, by the implicit function theorem, there exists $\theta\in W^{1,\infty}\to v(\theta)\in H_0^1(\Omega)$ of class C^1 on a neighborhood of 0 such that $F(\theta,v(\theta))\equiv0$ and $v(0)=u_0$. By uniqueness for problems (5.40) and (5.38), we deduce $v(\theta)=v_\theta$, whence the announced regularity for v_θ.

For the regularity of u_θ, we write $u_\theta=v_\theta\circ(I+\theta)^{-1}$. We then apply the second part of Lemma 5.3.3 with $p=2$, $g(\theta)=v_\theta$ — which satisfies the required assumption by the previous analysis — and with $\Psi(\theta)=(I+\theta)^{-1}$, which meets the hypotheses by (5.5). □

[3]Comment: If Ω is open, this definition is clear. If Ω is only measurable, $H^{-1}(\Omega)$ is the dual of $H_0^1(\Omega)$ and we define the differentiation operator by duality as

$$\forall\varphi\in H_0^1(\Omega),\quad \langle-\operatorname{div}(A(\theta)\nabla v),\varphi\rangle=\int_\Omega A(\theta)\nabla v.\nabla\varphi.$$

Remark 5.3.6. We see that the regularity of the mapping F is in fact only limited by the regularity of the data f. By exploiting the above approach, we can easily obtain differentiability properties of higher order. This will be done in the next section.

On the other hand, if f is only in L^2, we see that $\theta \to v_\theta \in H_0^1(\Omega)$ is not differentiable in general. Indeed, if it was differentiable, then $\theta \to \operatorname{div}(A(\theta)\nabla v_\theta) \in H^{-1}(\Omega)$ would be differentiable as well, since so is $\theta \to A(\theta) \in L^\infty$, and the other operators are linear and continuous. But, for $\lambda = 0$, this is equal to $f \circ (I+\theta)J_\theta$ on Ω and there exist functions $f \in L^2$ for which this is not differentiable (see Remark 5.3.4).

One may also wonder what the regularity of $\theta \to u_\theta$ is in a neighborhood of 0 and not only exactly at 0. We see that this depends on the regularity of $(I+\theta)^{-1}$ and one must be very careful (see Exercise 5.2). But we should also remember that recomposing by $(I+\theta)^{-1}$ the function $v_\theta = u_\theta \circ (I+\theta)$ generates artificial difficulties (Exercises 5.1 and 5.2).

5.3.5 Differentiability of higher order

Proposition 5.3.7. *We assume $f \in H^k(\mathbb{R}^N)$ with $k \geq 1$ an integer. Then the mapping*

$$\theta \in W^{1,\infty} \to v_\theta \in H_0^1(\Omega)$$

is of class C^k on a neighborhood of 0. *If $f \in C^\infty(\mathbb{R}^N)$, it is of class C^∞* (this is true for all bounded measurable sets Ω).

Proof. It is sufficient to prove that the function F that we introduced in the previous proof is of class C^k and to apply the implicit function theorem as above. Since the first piece of F is C^∞, it is sufficient to prove that $\theta \to [f \circ (I+\theta)]J_\theta \in L^2(\Omega)$ is of class C^k. This may easily be proved by induction starting from Lemma 5.3.3 (see also Lemma 5.3.9 below). □

Remark. Obviously, without any extra assumption, one should not expect as much regularity for $\theta \to u_\theta = v_\theta \circ (I+\theta)^{-1}$, since the first Fréchet derivative already involves $\nabla v_\theta \circ (I+\theta)^{-1}$ (which is not even differentiable with values in H^{-1} since ∇v_θ is only in L^2 in general). On the other hand, we may, for instance, deduce that

$$\theta \in W^{1,\infty} \to \int_{\Omega_\theta} u_\theta = \int_\Omega v_\theta J_\theta \quad \text{is } C^\infty,$$

since so are $\theta \to v_\theta$, J_θ. It is the same for the Dirichlet energy functional that we have widely analyzed in the previous chapter.

Corollary 5.3.8. *Let us assume $f \in H^k(\mathbb{R}^N)$, $k \geq 1$. Then, the mapping*

$$\theta \in W^{1,\infty} \to j(\theta) = \frac{1}{2}\int_{\Omega_\theta} |\nabla u_\theta|^2 + \lambda u_\theta^2 - \int_\Omega f u_\theta$$

is of class C^k in a neighborhood of 0 *and we have*

$$\forall\, \xi \in W^{1,\infty}, \quad j'(0).\xi = -\frac{1}{2}\int_\Omega fu' = -\frac{1}{2}\int_\Gamma |\nabla u|^2 \xi.n \quad (\textit{if } \Omega \textit{ is regular enough}).$$

Proof. We remember that $\int_{\Omega_\theta} |\nabla u_\theta|^2 + \lambda u_\theta^2 = \int_{\Omega_\theta} f u_\theta$, so that

$$j(\theta) = -\frac{1}{2}\int_{\Omega_\theta} f u_\theta = -\frac{1}{2}\int_\Omega f \circ (I+\theta) v_\theta J_\theta.$$

This last formula allows us to see that j is of class C^k by Proposition 5.3.7. We may differentiate $j(\theta) = -\frac{1}{2}\int_{\mathbb{R}^N} f u_\theta$ to obtain

$$\forall\, \xi \in W^{1,\infty}, \quad j'(0).\xi = -\frac{1}{2}\int_{\mathbb{R}^N} fu' = -\frac{1}{2}\int_\Omega fu',$$

the last equality being true since $u' = 0$ a.e. on Ω^c (recall Lemma 3.1.8 and that $u' + \nabla u \cdot V \in H_0^1(\Omega)$).

If Ω is open, we have $f = -\Delta u + \lambda u$ on Ω and if Ω has a regular boundary,

$$2j'(0).\xi = -\int_\Omega u'(-\Delta u + \lambda u) = \int_\Gamma u' \nabla u.n,$$

by using $-\Delta u' + \lambda u' = 0$. Since $u' = -\nabla u.\xi$ at the boundary and since ∇u is collinear to n, we deduce the formula of the corollary. □

5.3.6 Differentiability in regular functional spaces

Let us end this section with some regularity results in more regular spaces. Let us first gather, in a lemma, several higher-order differentiability properties. For all integer $k \geq 1$, we denote by $W^{k,\infty}(\mathbb{R}^N, \mathbb{R}^N)$ (or more simply $W^{k,\infty}$) the space of functions of class C^{k-1} whose derivatives up to order $(k-1)$ are bounded and whose $(k-1)$th derivative is in $W^{1,\infty}$, and we equip this space with a naturally associated norm.

Lemma 5.3.9. *Assume $g \in H^{m+k}(\mathbb{R}^N)$ with integers $m \geq 0$, $k \geq 1$. Then,*

$$\theta \in W^{\max\{m,1\},\infty} \to g \circ (I+\theta) \in H^m(\mathbb{R}^N)$$

is of class C^k.

If $t \in [0,T) \to G(t) \in H^{m+k}(\mathbb{R}^N)$ is of class C^k and if $\xi \in C_0^\infty(\mathbb{R}^N)$, then

$$t \in [0,T) \to G(t) \circ (I + t\xi)^{-1} \in H^m(\mathbb{R}^N)$$

is of class C^k in a neighborhood of 0.

Proof. Let us define $\mathcal{G}(\theta) = g \circ (I + \theta)$. First assume $m = 0$: we have to prove that, if $g \in H^k$, $k \geq 1$, then $[\theta \in W^{1,\infty} \to \mathcal{G}(\theta) \in L^2]$ is of class C^k. We already know that it is of class C^1 with $\mathcal{G}'(\theta)\xi = \nabla g \circ (I + \theta).\xi$ where $\nabla g \in H^{k-1}$. Thus, we obtain the result by induction for $k \geq 2$.

Now assume $m = 1$: we have to prove that, if $g \in H^{k+1}$, then $[\theta \in W^{1,\infty} \to \mathcal{G}(\theta) \in H^1]$ is of class C^k or that

$$\theta \in W^{1,\infty} \to \nabla\mathcal{G}(\theta) =^t (I + D\theta)\nabla g \circ (I + \theta) \in L^2$$

is of class C^k. But $\theta \in W^{1,\infty} \to D\theta \in L^\infty$ is of class C^∞ and $\nabla g \in H^k$. Therefore, the result follows from the previous case $m = 0$.

Next let $g \in H^{m+k}$ with $m \geq 2$. We have to prove that

$$\theta \in W^{m,\infty} \to \nabla\big(\mathcal{G}(\theta)\big) =^t (I + D\theta)\nabla g \circ (I + \theta) \in H^{m-1}$$

is of class C^k. Since $\theta \in W^{m,\infty} \to D\theta \in W^{m-1,\infty}$ is of class C^∞ and since multiplying by a function in $W^{m-1,\infty}$ is a continuous mapping from H^{m-1} into H^{m-1}, it is sufficient to prove that

$$\theta \in W^{m-1,\infty} \to \nabla g \circ (I + \theta) \in H^{m-1}$$

is of class C^k, knowing that $\nabla g \in H^{m-1+k}$. Thus, this may be obtained by induction on m starting at $m - 1 = 1$ for which we just gave the proof.

We deduce from the previous result that, for $m, k \geq 1$ and $l \geq \max\{m, 1\}$, the mapping

$$(g, \Theta) \in H^{m+k}(\mathbb{R}^N) \times W^{l,\infty} \to g \circ (I + \Theta) \in H^m(\mathbb{R}^N) \tag{5.43}$$

is of class C^k (thanks to its linearity with respect to g). Since the mapping

$$t \in [0, T) \to (G(t), (I + t\xi)^{-1} - I) \in H^{m+k} \times W^{l,\infty}$$

is of class C^k (we use $\xi \in C_0^\infty$ and we assume T small enough), it follows that $t \in [0, T) \to G(t) \circ (I + t\xi)^{-1} \in H^m(\mathbb{R}^N)$ is of class C^k. □

Proposition 5.3.10. *Let $m \geq 0$, $k \geq 1$. If $f \in H^{(m-1)^+ + k}(\mathbb{R}^N)$ and Ω is of class C^{m+1}, then*

$$\theta \in W^{m+1,\infty} \to v_\theta \in H_0^1(\Omega) \cap H^{m+1}(\Omega),$$

is of class C^k in a neighborhood of 0.

Let us moreover assume $f \in H^{(m+k-2)^+ + k}(\mathbb{R}^N)$ with $k \geq 1$, $m \geq 0$, Ω of class C^{m+k} and $\xi \in C_0^\infty$. Then, there exists an extension $\tilde{u}_t \in H^{m+k}(\mathbb{R}^N)$ of $u_t := u_{t\xi}$ such that

$$t \in [0, T) \to \tilde{u}_t \in H^m(\mathbb{R}^N)$$

is of class C^k.

Proof. The case $m = 0$ has already been treated in Proposition 5.3.7. Let us assume $m \geq 1$. Let us consider the restriction of the operator F defined in the proof of Theorem 5.3.2 to $X = W^{m+1,\infty} \times \big(H_0^1(\Omega) \cap H^{m+1}(\Omega)\big)$. Considered as being with values in $H^{m-1}(\Omega)$, this restriction is of class C^k: indeed,

$$\begin{aligned} &\theta \in W^{m+1,\infty} \to J_\theta \in W^{m,\infty} \\ \text{and} \quad &(\theta, v) \in X \to -\operatorname{div}(A(\theta)\nabla v) + \lambda J_\theta v \in H^{m-1} \end{aligned}$$

are of class C^∞ and $\theta \in W^{m+1,\infty} \to f \circ (I + \theta)J_\theta \in H^{m-1}$ is of class C^k by Lemma 5.3.9. On the other hand, $D_v F(0, v_0)$ is an isomorphism from $H_0^1(\Omega) \cap H^{m+1}(\Omega)$ into $H^{m-1}(\Omega)$ (see, e.g., [54] for the corresponding regularity results). Then the announced regularity of v_θ follows from the implicit function theorem.

The regularity of u_t also follows: since $f \in H^{(m'-1)^+ + k}$ with $m' = m + k - 1 \geq 0$ and since Ω is of class C^{m+k}, then $\theta \in W^{m+k,\infty} \to v_\theta \in H^{m+k}(\Omega)$ is of class C^k. By composition, $t \to v_{t\xi} \in H^{m+k}(\Omega)$ is also of class C^k. We then apply Lemma 5.3.9 with $G(t) = P(v_{t\xi})$, where P is a continuous linear extension of $H^{m+k}(\Omega)$ to $H^{m+k}(\mathbb{R}^N)$ (see [5]). It follows that $t \to \tilde{u}_t = P(v_{t\xi}) \circ (I + t\xi)^{-1} \in H^m(\mathbb{R}^N)$ is of class C^k and $\tilde{u}_t$ is indeed an extension of u_t. □

Remark. We can also state differentiability results in Hölder spaces $C^{k,\alpha}$: for this, we consider the restriction of the operator F introduced above to these subspaces and we apply Schauder[4] regularity theory to $D_v F(0, v_0)$. For instance, one may consider

$$F : C^{2,\alpha} \times \big(H_0^1 \cap C^{2,\alpha}\big) \to C^\alpha,$$

which is of class C^∞ if $f \in C^\infty$. By Schauder theory, if Ω is regular enough, $D_v F(0, v_0)$ is an isomorphism from $H_0^1 \cap C^{2,\alpha}$ into C^α. We deduce that $\theta \in C^{2,\alpha} \to v_\theta \in C^{2,\alpha}$ is of class C^∞.

5.4 Integrals on moving boundaries

Lots of problems lead to differentiating integrals defined on the boundary of moving sets (i.e., of the kind $\Omega \to \int_{\partial\Omega} f(\Omega, x)\, d\sigma(x)$, where $d\sigma$ is the area measure on $\partial\Omega$). This is the case, for instance, when considering Neumann-type boundary conditions in partial differential equations, or also when computing second derivatives, since first derivatives already contain boundary integrals. We give in this section several useful tools to do this kind of differentiation.

[4]Jules SCHAUDER, 1899–1943, Polish from the famous Lwów School, one of the Shoah victims, is among the contributors to the spectacular development of functional spaces and topological methods in analysis.

5.4.1 Boundary integrals: Definitions and properties

Throughout this section, we are given a bounded open set Ω which is at least of class C^1, which means that Ω has a Lipschitz boundary according to Definition 2.4.5 and the functions φ involved in this definition are of class C^1 (and not only Lipschitz continuous).[5] We denote $\Gamma = \partial\Omega$.

Thanks to compactness, we assume that Γ is represented by a finite set of graphs of class C^1 around points $x_i \in \Gamma$, $i = 1, \dots, p$. Thus, we assume that, for each i (see Definition 2.4.5), there exists a local orthonormal system of coordinates $x = (x', x_N) \in \mathbb{R}^{N-1} \times \mathbb{R}$ around $x_i = (0,0)$ and

$$\varphi_i : B(0, r_i) \subset \mathbb{R}^{N-1} \to (-a_i, a_i) \quad \text{of class } C^1, \ \varphi_i(0) = 0,$$

such that $\Gamma = \cup_i \Gamma_i$, $\Gamma_i = \{(x', \varphi_i(x'));\ x' \in B(0, r_i)\}$. If $\mathcal{O}_i := B(0, r_i) \times (-a_i, a_i)$, the functions ψ_i defined in local coordinates by

$$\forall\, x = (x', x_N) \in \mathcal{O}_i, \quad \psi_i(x) = \psi_i(x', x_N) := (x', \varphi_i(x') - x_N)$$

define C^1-diffeomorphisms from $\mathcal{O}_i$ into the open neighborhood $\psi_i(\mathcal{O}_i)$ of x_i — note that $\psi_i^{-1}(y', y_N) = (y', \varphi_i(y') - y_N)$. Moreover,

$$\psi_i\big(B(0, r_i) \times \{0\}\big) = \Gamma_i = \Gamma \cap \psi_i(\mathcal{O}_i), \qquad \psi_i\big(\mathcal{O}_i \cap \mathbb{R}^{N-1} \times (-\infty, 0)\big) = \Omega \cap \psi_i(\mathcal{O}_i).$$

With the open covering of Γ by the $\psi_i(\mathcal{O}_i)$, we associate a partition of unity $\xi_i \in C_0^\infty(\psi_i(\mathcal{O}_i))$, $\xi_i \geq 0$, $i = 1, \dots, p$ with $\sum_i \xi_i \equiv 1$ on a neighborhood of Γ. Any function $\gamma : \Gamma \to \mathbb{R}^q$, $q \geq 1$ can then be extended to the whole space $\mathbb{R}^N$ according to the formula

$$\forall\, x \in \mathbb{R}^N, \quad \gamma(x) = \sum_i \xi_i(x) \gamma(\psi_i \circ \pi_i \circ \psi_i^{-1}(x)), \tag{5.44}$$

where π_i is the orthogonal projection defined by $\pi_i(x', x_N) = (x', 0)$. This allows us, in particular, to define a continuous extension to the whole space $\mathbb{R}^N$ of the exterior unit normal vector n to Γ. This normal vector is defined on Γ_i, for instance, by

$$\forall\, x \in \Gamma_i, \quad n(x) = n(x', \varphi_i(x'))(\nabla_{x'}\varphi_i(x'), -1)/[1 + |\nabla\varphi_i(x')|^2]^{1/2}. \tag{5.45}$$

Given $f : \Gamma \to \mathbb{R}$, we define $f_i = f\xi_i$. As usual, we say that $f \in L^1(\Gamma)$ if, for all i, $[x' \to f_i(x', \varphi_i(x'))] \in L^1(B(0, r_i))$ and we set $\int_\Gamma f = \sum_i \int_{\Gamma_i} f_i$ with

$$\int_{\Gamma_i} f_i = \int_{B_i} f_i(x', \varphi_i(x'))[1 + |\nabla\varphi_i(x')|^2]^{1/2}\, dx'. \tag{5.46}$$

[5]Note that many of the properties that we are going to recall may be extended to a Lipschitz framework, but we choose to work in a C^1 setting to keep a simple presentation.

It is classical that this definition depends only on f and Γ as we may check, thanks to the following lemma (where we denote by e_N the Nth vector of the local basis around x_i).

Lemma 5.4.1. *Let ω be an open neighborhood of 0 in $\mathbb{R}^N$ and $\psi : \omega \to \psi_i(\mathcal{O}_i)$ a second C^1-diffeomorphism such that $\psi(\omega') = \Gamma_i$, where $\omega' = \omega \cap [\mathbb{R}^{N-1} \times \{0\}]$. Then*

$$\int_{\Gamma_i} f_i = \int_{\omega'} f_i \circ \psi |^t(D\psi)^{-1} e_N| \, |\det(D\psi)|.$$

Proof. Let us set in $\psi_i(\mathcal{O}_i)$,

$$(x', \varphi_i(x') - x_N) = \psi(\zeta) = \psi(\zeta', \zeta_N) = (\psi'(\zeta', \zeta_N), \psi_N(\zeta', \zeta_N)),$$

and, in particular, $(x', \varphi_i(x')) = (\psi'(\zeta', 0), \psi_N(\zeta', 0))$. We make the change of variable $x' = \psi'(\zeta', 0)$ in (5.46), which gives

$$\int_{\Gamma_i} f_i = \int_{\omega'} f_i \circ \psi [1 + |\nabla \varphi_i|^2]^{1/2} |\det[D_{\zeta'} \psi'(\zeta', 0)]| \, d\zeta'.$$

Since $x' = \psi'(\zeta', 0) = \psi'(\zeta', \zeta_N)$, we have $D_{\zeta_N} \psi' \equiv 0$ and therefore

$$\det D\psi = \det[D_{\zeta'} \psi'(\zeta', 0)] \frac{\partial \psi_N}{\partial \zeta_N}.$$

Now, let $\nu :=^t (D\psi)^{-1} e_N = (\nu', \nu_N)$, which means

$$^t D_{\zeta'} \psi' \nu' + \nu_N \nabla_{\zeta'} \psi_N = 0, \qquad \frac{\partial \psi_N}{\partial \zeta_N} \nu_N = 1.$$

By differentiating $\varphi_i(\psi'(\zeta', 0)) = \psi_N(\zeta', 0)$, we obtain

$$\nabla \varphi_i =^t D_{\zeta'} \psi'^{-1} . \nabla_{\zeta'} \psi_N,$$

and consequently $\nu' = -\nu_N \nabla \varphi_i$. Thanks to the above relations, we also have

$$|\det[D_{\zeta'} \psi'(\zeta, 0)]| \, [1 + |\nabla \varphi_i|^2]^{1/2} = |\nu_N \det(D\psi)| \big(1 + |\nu'|^2 / \nu_N^2\big)^{1/2} = |\det D\psi| |\nu|,$$

whence the lemma. □

Now let T be a C^1-diffeomorphism from $\mathbb{R}^N$ into itself and let Ω_T be the image of Ω by T (with boundary $\Gamma_T = T(\Gamma)$). We will denote by $[T']$ the Jacobian matrix of T with entries $[T']_{ij} = \frac{\partial T_i}{\partial x_j}$ and by $\mathrm{Jac}(T) = |\det[T']|$ the Jacobian of T.

Definition 5.4.2. We call the tangential Jacobian of T on Γ the quantity denoted by $\mathrm{Jac}_\Gamma(T)$ and defined by

$$\mathrm{Jac}_\Gamma(T) = |^t[T']^{-1} n| \, \mathrm{Jac}(T). \tag{5.47}$$

The tangential Jacobian is therefore a continuous function defined on Γ. Let us now state the change of variable formula.

Proposition 5.4.3. *Let $f : \Gamma_T \to \mathbb{R}$. Then $f \in L^1(\Gamma_T)$ if and only if $f \circ T \in L^1(\Gamma)$ and*

$$\int_{\Gamma_T} f = \int_\Gamma f \circ T \, \mathrm{Jac}_\Gamma(T). \tag{5.48}$$

Proof. We consider the various objects φ_i, ψ_i, $\mathcal{O}_i$, $B(0, r_i)$, ξ_i associated to Γ as above and we apply Lemma 5.4.1 to Γ_i, f_i, ω, ψ replaced by $\widetilde{\Gamma}_i$, $\widetilde{f}_i$, $\mathcal{O}_i$, $\widetilde{\psi}$, where

$$\widetilde{\Gamma}_i = T(\Gamma_i), \qquad \widetilde{f}_i = f.(\xi_i \circ T^{-1}), \qquad \widetilde{\psi} = T \circ \psi_i.$$

We obtain

$$\int_{T(\Gamma_i)} \widetilde{f}_i = \int_{B(0,r_i)} (f \circ T.\xi_i) \circ \psi_i |^t[T' \circ \psi_i]^{-1^t}(D\psi_i)^{-1} e_N| \, | \mathrm{Jac}(T) \circ \psi_i \det D\psi_i |.$$

We check that $\det D\psi_i = -1$ and $\psi_i^{-1}(y', y_N) = (y', \varphi_i(y') - y_N)$ so that

$$^t(D\psi_i)^{-1} e_N = (\nabla \varphi_i, -1) = [1 + |\nabla \varphi_i|^2]^{1/2} n.$$

Thus

$$\int_{T(\Gamma_i)} \widetilde{f}_i = \int_{B(0,r_i)} (f \circ T.\xi_i . \mathrm{Jac}_\Gamma(T)) \circ \psi_i [1 + |\nabla \varphi_i|^2]^{1/2},$$

which, according to Definition 5.46 also gives

$$\int_{T(\Gamma_i)} \widetilde{f}_i = \int_{\Gamma_i} f \circ T.\xi_i . \mathrm{Jac}_\Gamma(T).$$

Then, we sum over i to obtain the expected formula. □

5.4.2 A first statement

Let us consider a family $t \in [0, T) \to \Phi(t)$ of C^1-*diffeomorphisms* from $\mathbb{R}^N$ into itself satisfying (5.6). Let us denote by Γ_t the boundary of $\Omega_t = \Phi(t)(\Omega)$, where Ω is a fixed open set of class at least C^1. We are interested in differentiating expressions like $t \to G(t) = \int_{\Gamma_t} g(t)$, where $g(t) : \Gamma_t \to \mathbb{R}$ is given. By Proposition 5.4.3 and Lemma 5.4.1, we have to differentiate the quantity

$$t \to G(t) = \int_\Gamma g \circ \Phi(t) \, \mathrm{Jac}_\Gamma(\Phi(t)) = \int_\Gamma g \circ \Phi(t) \|^t D\Phi(t)^{-1} n \| \det D\Phi(t).$$

We will concentrate on differentiating each of these terms later on. However, it turns out to be possible to obtain rather general formulas by directly applying

Theorem 5.2.2 although it gives the derivative of an integral on Ω_t itself. Let us start with the "good" situation, where $g(t) = W(t).n_t$. Then, we apply formula (5.11) to $G(t) = \int_{\Gamma_t} W(t).n_t = \int_{\Omega_t} \operatorname{div} W(t)$, that is,

$$G'(0) = \int_\Omega \operatorname{div} W'(0) + \operatorname{div}[V \operatorname{div} W(0)] = \int_\Gamma W'(0).n + (V.n) \operatorname{div} W(0).$$

This formula is valid if $t \to W(t) \in W^{1,1}(\mathbb{R}^N, \mathbb{R}^N)$ is differentiable at 0 and $\operatorname{div} W(0) \in W^{1,1}$.

This approach may in fact be applied to any boundary integral, at least when assuming enough regularity on $\partial\Omega$, on g, and on the family $\Phi(t)$. Indeed, one may then consider that g and the unit normal vector n_t to Γ_t both have extensions to the whole space $\mathbb{R}^N$ (see (5.44)): we still denote them by g, n_t. Then we may write

$$G(t) = \int_{\Gamma_t} g(t) = \int_{\Gamma_t} g(t) n_t.n_t = \int_{\Omega_t} \operatorname{div}(g(t)n_t).$$

At least formally, by application of formula (5.11), we obtain

$$G'(0) = \int_\Omega \frac{\partial}{\partial t}_{|t=0} \big(\operatorname{div}(g(t)n_t)\big) + \operatorname{div}\big(V \operatorname{div}(g(0)n)\big).$$

After changing the order of differentiation and integrating by parts, we also obtain

$$G'(0) = \int_\Gamma n. \left[\frac{\partial}{\partial t}_{|t=0} (g(t)n_t) + V \operatorname{div}(g(0)n)\right].$$

Assume n_t is unitary: then at $t = 0$,

$$n.\frac{\partial}{\partial t} n_t = \frac{\partial}{\partial t} \frac{1}{2} n_t.n_t = 0.$$

Thus, if we denote $g'(0) = \frac{\partial}{\partial t}_{|t=0} g(t)$, we have

$$G'(0) = \int_\Gamma g'(0) + (V.n)\big(n.\nabla g(0) + g(0) \operatorname{div} n\big). \tag{5.49}$$

Thus we can state

Proposition 5.4.4. *Assume Ω is of class C^3, that $t \in [0, T) \to \Phi(t) \in C^2$ is differentiable at 0 with $\Phi(0) = I$, $\Phi'(0) = V$, and that $t \to g(t) \in W^{1,1}(\mathbb{R}^N)$ is differentiable at 0 with $g(0) \in W^{2,1}(\mathbb{R}^N)$.*

Then, $t \to G(t)$ is differentiable at 0 and we have formula (5.49).

Proof. We apply Theorem 5.2.2 to $G(t) = \int_{\Omega_t} \operatorname{div}(g(t)n_t)$. The hypotheses imply that $\operatorname{div} n \in C^1$ and $\nabla g(0) \in W^{1,1}$, so that $\operatorname{div}(g(0)n) \in W^{1,1}(\mathbb{R}^N)$. Thus, it is sufficient to verify that

$$t \longrightarrow \operatorname{div}(g(t)n_t) = g(t)\operatorname{div} n_t + \nabla g(t).n_t \ \in L^1(\mathbb{R}^N)$$

is differentiable at 0. Since the regularity of $t \longrightarrow \operatorname{div} n_t \in C^0$ is controlled by the regularity of $t \longrightarrow \Phi(t) \in C^2$ (see, in particular, definition (5.44) of the extension of n_t), this follows from the differentiability assumptions made in the lemma. □

An application. We can apply this to differentiate the perimeter $t \longrightarrow p(t) = P(\Omega_t) = \int_{\Gamma_t} 1$. Thus, under the above assumptions on Ω and $\Phi(t)$, we obtain

$$p'(0) = \int_\Gamma \operatorname{div} n(V.n).$$

The function $\operatorname{div} n$ does not depend on the choice of the unitary extension of n and depends only on the geometry of Ω: it is actually the mean curvature of Γ. This is proved in the next section where we recall some tools of differential geometry that are often useful when differentiating boundary integrals.

5.4.3 Some differential geometry

We say that a function $g : \Gamma \longrightarrow \mathbb{R}$ is of class C^1 on Γ if its extension defined by (5.44) is of class C^1. We denote by $C^1(\Gamma)$ the corresponding space.

Definition 5.4.5 (Tangential gradient). Let g be a function of class C^1 on Γ. We define its tangential gradient by

$$\operatorname{grad}_\Gamma g = \nabla_\Gamma g = \operatorname{grad} \tilde{g} - (\nabla \tilde{g}.n)n \quad \text{on } \Gamma,$$

where $\tilde{g} \in C^1(\mathbb{R}^N)$ is an extension of g.

We immediately see that this definition is independent of the extension $\tilde{g}$, since if $\tilde{g} = 0$ on Γ, by differentiating in local coordinates $\tilde{g}(x', \varphi_i(x')) = 0$, we obtain $\nabla_{x'}\tilde{g} + \partial_{x_N}\tilde{g}\nabla_{x'}\varphi_i = 0$ and therefore (see Section 5.4.1)

$$\nabla \tilde{g} = (\nabla_{x'}\tilde{g}, \partial_{x_N}\tilde{g}) = -n(1 + |\nabla_{x'}\varphi_i|^2)^{1/2}\partial_{x_N}\tilde{g} \quad (\text{so that } \nabla_\Gamma \tilde{g} = 0).$$

We denote by $W^{1,1}(\Gamma)$ the closure of $C^1(\Gamma)$ for the norm

$$\|g\|_{1,1} = \int_\Gamma |g| + |\nabla_\Gamma g|.$$

Thus, $\nabla_\Gamma g$ is defined by density for all $g \in W^{1,1}(\Gamma)$ and it belongs to $L^1(\Gamma, \mathbb{R}^N)$.

Definition 5.4.6 (Tangential divergence). Let $W \in C^1(\Gamma, \mathbb{R}^N)$. We define its tangential divergence by

$$\operatorname{div}_\Gamma W = \operatorname{div} \widetilde{W} - [\widetilde{W}']n.n, \tag{5.50}$$

where $\widetilde{W} \in C^1(\mathbb{R}^N, \mathbb{R}^N)$ is an extension of W.

This definition does not depend on the choice of the extension of W. Indeed, we check that

$$\operatorname{div} \widetilde{W} - [\widetilde{W}']n.n = \operatorname{tr}(D_\Gamma W), \tag{5.51}$$

where $D_\Gamma W$ is the matrix whose ith line is $\nabla \widetilde{W}_i - (\nabla \widetilde{W}_i.n)n = \nabla_\Gamma W_i$.

This definition also extends by density to the whole space $W^{1,1}(\mathbb{R}^N, \mathbb{R}^N)$.

Remark. For $x \in \Gamma$, let us choose an orthonormal coordinate system centered at x such that the hyperplane $x_N = 0$ is tangent to Γ (and therefore $n = \pm e_N$). We then have

$$\operatorname{div}_\Gamma W(x) = \sum_{j=1}^{N-1} \frac{\partial \widetilde{W}_j}{\partial x_j}(x). \tag{5.52}$$

Let us note the following formulas:

$$\forall f, g \in C^1(\Gamma), \quad \nabla_\Gamma(fg) = g\nabla_\Gamma f + f\nabla_\Gamma g. \tag{5.53}$$

$$\forall f \in C^1(\Gamma),\ \forall W \in C^1(\mathbb{R}^N, \mathbb{R}^N), \quad \operatorname{div}_\Gamma(fW) = f \operatorname{div}_\Gamma W + W.\nabla_\Gamma f. \tag{5.54}$$

The first formula directly follows from the definition of the tangential gradient. For the second one, we apply the first one to each line of the matrix $D_\Gamma(fW)$ and we take the trace (see (5.51)).

Definition 5.4.7. Assume Ω is of class C^2. We then define the mean curvature of Γ by $H = \operatorname{div}_\Gamma n$.

This mean curvature may be written in terms of the graphs φ_i that define Γ if we moreover assume $D\varphi_i(0) = 0$. Indeed, by (5.45) and (5.51) or (5.52), we then have at the center $x_i \in \Gamma$ of the coordinate system,

$$H(x_i) = \sum_{j=1}^{N-1} \partial_{x_j}\{\partial_{x_j}(\varphi_i)/[1 + |\nabla\varphi_i|^2]^{1/2}\} = \sum_{j=1}^{N-1} \frac{\partial^2 \varphi_i}{\partial x_j^2}(0).$$

We recover here the usual geometric meaning of the mean curvature as the sum of the curvatures of the sections of Γ by $N-1$ planes that are orthogonal to each other and orthogonal to the tangent hyperplane. Indeed, the intersection of Γ with the plane P_i defined by the axes x_i, x_N is the curve parametrized by

$$x_N = \varphi_i(0, \ldots, 0, t, 0, \ldots, 0)$$

(where t is in the ith slot). According to the assumptions on φ_i, the curvature of this curve at $t = 0$ is equal to $\frac{\partial^2 \varphi_i}{\partial x_i^2}(0)$, which provides the announced result since the family of planes P_i satisfies the required geometric condition.

Let us also note

Proposition 5.4.8. *Let Ω be of class C^2. Then, for all extensions N of n that are unitary and of class C^1, we have*

$$\operatorname{div} N = H \quad \text{on } \partial\Omega.$$

Proof. By differentiating $N^2 = 1$ with respect to each x_i, we obtain ${}^t[N'].n = 0$ on $\partial\Omega$. Since $H = \operatorname{div}_\Gamma n = \operatorname{div} N - [N']n.n$ and $[N']n.n = n.{}^t[N']n = 0$, we obtain the expected equality. □

Application: Decomposition of the tangential divergence. For all vector fields W defined on Γ, we define the tangential component of W, which we denote by W_Γ, as the orthogonal projection of W on the tangent hyperplane:

$$W_\Gamma := W - (W.n)n.$$

Proposition 5.4.9. *Let Ω be of class C^2. Let $f \in W^{1,1}(\Gamma)$, $W \in W^{1,1}(\Gamma, \mathbb{R}^N)$. Then*

$$\operatorname{div}_\Gamma(fn) = Hf, \qquad \operatorname{div}_\Gamma W = \operatorname{div}_\Gamma W_\Gamma + Hn.W, \tag{5.55}$$

$$\int_\Gamma \operatorname{div}_\Gamma W = \int_\Gamma Hn.W, \qquad \int_\Gamma W.\nabla_\Gamma f = \int_\Gamma -f \operatorname{div}_\Gamma W + HfW.n. \tag{5.56}$$

Proof. We may suppose f, W in C^1 and end by density. The first identity may be directly deduced from (5.54) (since $n.\nabla_\Gamma f = 0$). It implies the second one thanks to $W = W_\Gamma + (W.n)n$. The third one also follows with the help of the following lemma. The fourth one easily follows as well, by using (5.54) also. □

Lemma 5.4.10. *Let Ω be of class C^2. Let $V \in W^{1,1}(\Gamma, \mathbb{R}^N)$ such that $V.n = 0$ on Γ. Then*

$$\int_\Gamma \operatorname{div}_\Gamma V = 0.$$

Proof. We may assume that V is of class C^1. Using the partition of unity ξ_i of Section 5.4.1, it is sufficient to prove $\int_{\Gamma_i} \operatorname{div}_\Gamma(\xi_i V) = 0$, which means we may assume that V is compactly supported in Γ_i. We denote by $\widetilde{V}$ the extension of V defined in local coordinates by $\widetilde{V}(x', x_N) = V(x', \varphi_i(x'))$. We set $\alpha = [1 + |\nabla_{x'}\varphi_i|^2]^{1/2}$. We will verify next that

$$\operatorname{div}_\Gamma V = \alpha^{-1} \sum_{j=1}^{N-1} \partial_{x_j}(\alpha \widetilde{V}_j). \tag{5.57}$$

The expected result will then follow. Indeed,

$$\int_\Gamma \operatorname{div}_\Gamma V = \int_{B(0,r_i)} \sum_{j=1}^{N-1} \partial_{x_j}(\alpha \widetilde{V}_j) = 0,$$

since $[x' \to \widetilde{V}(x', \varphi_i(x'))]$ is compactly supported in $B(0, r_i)$.

For simplicity, let us set $\varphi = \varphi_i$. The definition of tangential divergence leads to

$$\operatorname{div}_\Gamma V = \sum_{j=1}^{N-1} \frac{\partial \widetilde{V}_j}{\partial x_j} - \sum_{j=1}^{N} \sum_{l=1}^{N-1} \frac{\partial \widetilde{V}_j}{\partial x_l} n_j n_l \quad \text{on } \Gamma_i.$$

Using the formula that gives the n_j (in (5.45)) and also using the hypothesis $V.n = 0$, which is written here as $\widetilde{V}_N = \sum_{j=1}^{N-1} \widetilde{V}_j \frac{\partial \varphi}{\partial x_j}$, we obtain

$$\begin{aligned}
\operatorname{div}_\Gamma V &= \sum_{j=1}^{N-1} \frac{\partial \widetilde{V}_j}{\partial x_j} - \frac{1}{\alpha^2} \sum_{j=1}^{N-1} \sum_{l=1}^{N-1} \frac{\partial \widetilde{V}_j}{\partial x_l} \frac{\partial \varphi}{\partial x_j} \frac{\partial \varphi}{\partial x_l} \\
&\quad + \frac{1}{\alpha^2} \sum_{l=1}^{N-1} \frac{\partial}{\partial x_l} \left(\sum_{j=1}^{N-1} \widetilde{V}_j \frac{\partial \varphi}{\partial x_j} \right) \frac{\partial \varphi}{\partial x_l} \\
&= \sum_{j=1}^{N-1} \frac{\partial \widetilde{V}_j}{\partial x_j} + \frac{1}{\alpha^2} \sum_{j=1}^{N-1} \sum_{l=1}^{N-1} \widetilde{V}_j \frac{\partial^2 \varphi}{\partial x_j \partial x_l} \frac{\partial \varphi}{\partial x_l} = \alpha^{-1} \sum_{j=1}^{N-1} \partial_{x_j}(\alpha \widetilde{V}_j),
\end{aligned}$$

whence (5.57). □

Remark. A more geometric proof would consist in introducing the solution of

$$\zeta(0, x) = x, \qquad \partial_t \zeta(t, x) = \widetilde{V}\big(\zeta(t, x)\big) \quad \forall t > 0.$$

Since $V.n = 0$ on Γ, then $[x \in \Gamma \Rightarrow \zeta(t, x) \in \Gamma]$ so that $\zeta(t, \Omega) = \Omega$. In particular, if J_t denotes the tangential Jacobian of $\zeta(t, \cdot)$, then $t \to \int_\Gamma J_t$ is constant. But, its derivative is exactly $\int_\Gamma \operatorname{div}_\Gamma V$ by Lemma 5.4.15, which is proved below.

The Laplacian of a function u is defined on an open set by the formula $\Delta u = \operatorname{div}(\operatorname{grad} u)$. If the function u is defined on a manifold (here on the boundary Γ of the open set Ω), we have a similar definition when it belongs to

$$W^{2,1}(\Gamma) := \left\{ u \in W^{1,1}(\Gamma);\ \nabla_\Gamma u \in W^{1,1}(\Gamma, \mathbb{R}^N) \right\}.$$

Definition 5.4.11 (Laplace–Beltrami operator). Let Ω be of class C^2. The Laplace–Beltrami[6] operator on Γ, denoted by Δ_Γ, is defined by

$$\forall\, u \in W^{2,1}(\Gamma), \quad \Delta_\Gamma u = \mathrm{div}_\Gamma[\nabla_\Gamma u]. \tag{5.58}$$

From this definition and Proposition 5.4.9, we deduce the following formula.

Proposition 5.4.12. *Let Ω be an open set of class C^2 and $u : \overline{\Omega} \to \mathbb{R}$ of class C^2. Then*

$$\Delta u = \Delta_\Gamma u + H\frac{\partial u}{\partial n} + \frac{\partial^2 u}{\partial n^2} \quad \text{on } \Gamma. \tag{5.59}$$

Remark. We recall that the normal derivative to the boundary is given by $\frac{\partial u}{\partial n} = \frac{d}{dt}_{|t=0} u(x + tn(x)) = \nabla u.n$ and the second normal derivative by

$$\begin{aligned}\frac{\partial^2 u}{\partial n^2} &= \frac{d^2}{dt^2}_{|t=0} u(x + tn(x)) \\ &= \frac{d}{dt}_{|t=0} \nabla u(x + tn(x)).n = (D^2 u.n).n = \sum_{i,j=1}^{N} \frac{\partial^2 u}{\partial x_i x_j} n_i n_j.\end{aligned}$$

Formula (5.59) can be extended by density to functions in $H^3(\Omega)$.

Proof of Proposition 5.4.12. Starting from (5.55), we have

$$\mathrm{div}_\Gamma\, \nabla u = \mathrm{div}_\Gamma\, \nabla_\Gamma u + H\nabla u.n = \Delta_\Gamma u + H\nabla u.n. \tag{5.60}$$

Thanks to formula (5.50), we also have

$$\mathrm{div}_\Gamma\, \nabla u = \mathrm{div}(\nabla u) - [(\nabla u)']n.n, \tag{5.61}$$

with $[(\nabla u)'] = D^2 u$, which is the Hessian matrix of u with entries $\frac{\partial^2 u}{\partial x_i \partial x_j}$. We then deduce formula (5.59). □

Let us give some more integration by parts formulas for integrals on Γ.

Theorem 5.4.13. *Let Ω be an open set of class C^2 with boundary Γ. For f, $g \in H^2(\Omega)$,*

$$\int_\Gamma \frac{\partial f}{\partial x_i} g = -\int_\Gamma f \frac{\partial g}{\partial x_i} + \int_\Gamma \left(\frac{\partial}{\partial n}(fg) + Hfg\right) n_i. \tag{5.62}$$

[6]Eugenio BELTRAMI, 1835–1900, Italian, professor in Bologna, Pisa, Roma, and Pavia, worked in differential geometry with, among other contributions, the famous proof of the independence of the axiom of the parallels in Euclidean geometry.

If f is in $H^2(\Omega)$ and g in $H^3(\Omega)$, then

$$\int_\Gamma \nabla f.\nabla g = -\int_\Gamma f\Delta g + \int_\Gamma \left(\frac{\partial f}{\partial n}\frac{\partial g}{\partial n} + f\frac{\partial^2 g}{\partial n^2} + Hf\frac{\partial g}{\partial n}\right) \tag{5.63}$$

and

$$\int_\Gamma \nabla_\Gamma f.\nabla_\Gamma g = -\int_\Gamma f\Delta_\Gamma g. \tag{5.64}$$

Let us remark that formula (5.64) is very similar to the usual ones on open sets, except that no "boundary term" appears since, precisely, the manifold Γ is "without boundary".

Proof. Starting from (5.56), we have

$$\int_\Gamma \operatorname{div}_\Gamma(fW) = \int_\Gamma HfW.n. \tag{5.65}$$

But $\operatorname{div}_\Gamma fW = W.\operatorname{grad}_\Gamma f + f\operatorname{div}_\Gamma W$ (see 5.54), so that by expanding the tangential gradient into $\operatorname{grad}_\Gamma f = \nabla f - \frac{\partial f}{\partial n}n$ and after replacing it in (5.65), we obtain

$$\int_\Gamma f\operatorname{div}_\Gamma W + \nabla f.W = \int_\Gamma \left(Hf + \frac{\partial f}{\partial n}\right)W.n.$$

Let us take $W = ge_i$ in this last formula, where e_i is the ith vector of the orthonormal basis. We obtain

$$\int_\Gamma f\operatorname{div}_\Gamma ge_i + \frac{\partial f}{\partial x_i}g = \int_\Gamma \left(Hf + \frac{\partial f}{\partial n}\right)gn_i. \tag{5.66}$$

We now use the definition of the tangential divergence to compute $\operatorname{div}_\Gamma ge_i$. Since $\operatorname{div} ge_i = \frac{\partial g}{\partial x_i}$ and $[(ge_i)']n.n = \sum_{k,l=1}^N \frac{\partial g}{\partial x_i}\delta_{i,k}n_k n_l = \frac{\partial g}{\partial n}n_i$, formula (5.62) may immediately be deduced from (5.66).

To obtain (5.63), it is sufficient to replace g by $\frac{\partial g}{\partial x_i}$ in (5.62), then to sum for $i = 1,\dots,N$.

And (5.64) may easily be obtained from (5.63) by using formula (5.59), as well as the identity

$$\nabla f.\nabla g = \left(\nabla_\Gamma f + \frac{\partial f}{\partial n}n\right).\left(\nabla_\Gamma g + \frac{\partial g}{\partial n}n\right) = \nabla_\Gamma f.\nabla_\Gamma g + \frac{\partial f}{\partial n}\frac{\partial g}{\partial n}.$$

□

5.4.4 Extension of the unit normal vector to a variable domain

We state here a result that is rather technical but very useful when differentiating boundary integrals. We set $C^{1,\infty} = C^1 \cap W^{1,\infty}(\mathbb{R}^N, \mathbb{R}^N)$ equipped with the norm of $W^{1,\infty}$. In the following, we are given

$$t \in [0,T) \to \Phi(t) \in C^{1,\infty} \quad \text{differentiable at 0, with } \Phi(0) = I,\ \Phi'(0) = V. \tag{5.67}$$

Thus, we have all properties (5.7) up to replacing $W^{1,\infty}$ by $C^{1,\infty}$ and assuming T small enough. And we define $\Omega_t = \Phi(t)(\Omega)$, $\Gamma_t = \partial\Omega_t$.

Proposition 5.4.14. *Let Ω be of class C^2. Let $n \in C^1(\mathbb{R}^N, \mathbb{R}^N)$ be an extension of the unit normal vector to Γ. Then*

$$n_t = w(t)/\|w(t)\|, \quad \textit{where } w(t) = ({}^tD\Phi(t)^{-1}n) \circ \Phi(t)^{-1}, \tag{5.68}$$

is normal to Γ_t and $t \to n_t$ is in $C^0(\mathbb{R}^N, \mathbb{R}^N)$ and is differentiable at 0.

For any extension $t \in [0,T) \to \tilde{n}_t \in C^0(\mathbb{R}^N, \mathbb{R}^N)$ differentiable at 0 with $\tilde{n}_0 \in C^1(\mathbb{R}^N, \mathbb{R}^N)$ and $\tilde{n}_0 = n$ on Γ, we have

$$\frac{\partial \tilde{n}_t}{\partial t}_{|t=0} = -\nabla_\Gamma(V.n) - (D\tilde{n}_0.n)V.n \quad \textit{on } \Gamma. \tag{5.69}$$

Proof. Let $n \in C^1(\mathbb{R}^N, \mathbb{R}^N)$ be an extension of the unitary normal vector to Γ. The function $t \to n_t = w(t)/\|w(t)\| \in C^0(\mathbb{R}^N, \mathbb{R}^N)$ is differentiable by composition (see (5.7)). Its restriction to Γ_t is indeed its unitary normal vector: let us check it by using the local representation of Γ as the graph of $\varphi = \varphi_i$ (see Section 5.4.1). If one defines in local coordinates $G(x) = \varphi(x') - x_N$, then Γ_t is locally defined as being the set of points $x \in \mathbb{R}^N$ such that $G(\Phi(t)^{-1}x) = 0$. Thus, we may obtain n_t locally around Γ_i for $t \in [0,T)$ as the normalized gradient of $[x \to G(\Phi(t)^{-1}x)]$, which gives formula (5.68).

Differentiating ${}^tD\Phi(t) \circ \Phi(t)^{-1}w(t) = n \circ \Phi(t)^{-1}$, we obtain

$$\begin{gathered} {}^tDV.n + w'(0) = -Dn.V \quad \text{or} \quad w'(0) = -\nabla(V.n) + ({}^tDn - Dn)V, \\ \frac{\partial n_t}{\partial t}_{|t=0} = w'(0) - n(w'(0).n)w'(0)_\Gamma = -\nabla_\Gamma(V.n) - Z_\Gamma, \end{gathered} \tag{5.70}$$

where $Z = ({}^tDn - Dn)V$. At this stage, in order to continue the computations it is necessary to make an extension of n explicit. And it is better to choose one so that Dn is symmetric.

This is the case for the gradient of the signed distance function to Γ. Let us recall its definition and its properties when using the local representation of Γ in Section 5.4.1. There exists τ_i such that

$$\delta : (x', \tau) \in B(0, r_i) \times (-\tau_i, \tau_i) \to (x', \varphi_i(x')) + \tau n(x', \varphi_i(x')), \quad \delta(0,0) = (0,0)$$

is a C^1-diffeomorphism around (0, 0) (we check that $D\delta(0,0)$ is invertible). Thus, we can set $d(\delta(x',\tau)) := \tau$. This function is a priori of class C^1; it is actually of class C^2. We may see it by differentiating the relation $x = \delta(x', d(x))$ with respect to x: if we set $\beta = [1 + |\nabla_{x'}\varphi_i|^2]^{-1/2}$, $N = \beta\nabla_{x'}\varphi_i$ so that $n = (N, -\beta)$, we obtain

$$[I] = \begin{bmatrix} D_x x' + N D_x d + d D_x N \\ D_{x'}\varphi_i D_x x' - \beta D_x d - d D_x \beta \end{bmatrix}.$$

We multiply on the left by $[D_{x'}\varphi_i \; -1]$. Using $D_{x'}\varphi_i D_x N + D_x\beta = 0$ and $x' = \pi_1(\delta^{-1}(x))$, we deduce

$$D_x d(x) = [\beta(D_{x'}\varphi_i, -1)] \circ \pi_1 \circ \delta^{-1}(x),$$

where π_1 is the projection $(x',\tau) \to x'$. We see that $D_x d$ is of class C^1. Moreover, $\nabla d =^t D_x d$ is a unitary vector that coincides with n on Γ. This extension $n := \nabla d$ may be made on the whole space $\mathbb{R}^N$ with the partition of unity ξ_i. In particular, $Dn = D^2 d$ is symmetric around Γ.

Coming back to (5.70), we deduce that $\frac{\partial n_t}{\partial t}_{|t=0} = -\nabla_\Gamma(V.n)$ for this extension, which is exactly formula (5.69) since, for all extensions n,

$$[n \text{ unitary on the neighborhood of } \Gamma] \Rightarrow [{}^t Dn.n \equiv 0]. \tag{5.71}$$

Since Dn is symmetric, we also have $Dn.n \equiv 0$.

Now let $\tilde{n}_t$ be another differentiable extension of the normal vector to Γ_t. We have, for all $x \in \Gamma$, $(\tilde{n}_t - n_t)(\Phi(t)x) = 0$ so that

$$\frac{\partial(\tilde{n}_t - n_t)}{\partial t}_{|t=0} + D(\tilde{n}_0 - n).V = 0.$$

But

$$D_\Gamma(\tilde{n}_0 - n) = 0 \Rightarrow D(\tilde{n}_0 - n).V = D(\tilde{n}_0 - n).n(V.n) = D\tilde{n}_0.n(V.n).$$

We deduce

$$\frac{\partial \tilde{n}_t}{\partial t}_{|t=0} = \frac{\partial n_t}{\partial t}_{|t=0} - D(\tilde{n}_0 - n).V = -\nabla_\Gamma(V.n) - (D\tilde{n}_0.n)(V.n).$$

□

5.4.5 A general formula for boundary differentiation

Let us now consider the case of a boundary integral

$$G(\theta) = \int_{\Gamma_\theta} g(\theta) = \int_\Gamma g(\theta) \circ (I + \theta) J^\theta, \tag{5.72}$$

where

$$J^\theta = \mathrm{Jac}_{\Gamma_\theta}(I+\theta) = \det(I+D\theta)\|{}^t(I+D\theta)^{-1}n\|. \tag{5.73}$$

Here, we can no longer work with $\theta \in W^{1,\infty}$ only since $D\theta$, J_θ would then be defined only a.e. and we would not be able to define their traces on $\Gamma = \partial\Omega$. The natural space is $C^{1,\infty} := C^1 \cap W^{1,\infty}$ which we equip with the same norm as $W^{1,\infty}$. Similarly, we introduce $C^{k,\infty} := C^k \cap W^{k,\infty}$ for integers $k \geq 1$.

We have a first lemma.

Lemma 5.4.15. *Let Ω be open, bounded, and of class C^1. The mapping $\theta \in C^{1,\infty} \to J^\theta \in C^0(\Gamma)$ is of class C^∞ on a neighborhood of 0 and we have*

$$\forall \xi \in C^{1,\infty}, \quad D_\theta J^\theta_{|\theta=0}\xi = \mathrm{div}_\Gamma\, \xi.$$

If $t \to \Phi(t) \in C^{1,\infty}$ is differentiable at 0 with derivative equal to V, then $t \to J^{\Phi(t)-I} \in C^0(\Gamma)$ is differentiable at 0 and we have

$$\frac{d}{dt}_{|t=0} J^{\Phi(t)-I} = \mathrm{div}_\Gamma\, V.$$

And we have the following corollary:

Corollary 5.4.16. *Let Ω be open, bounded, and of class C^1. Then, $\theta \in C^{1,\infty} \to P(\Omega_\theta)$ is of class C^∞. In particular, for all $\xi \in C^{1,\infty}$, $t \to p(t) = P\big((I+t\xi)(\Omega)\big)$ is infinitely differentiable on a neighborhood of 0 and we have $p'(0) = \int_\Gamma \mathrm{div}_\Gamma\, \xi$.*

The second derivative will be computed later. Note that, if Ω is of class C^2, by Proposition 5.4.9, we may also write

$$p'(0) = \int_\Gamma Hn.\xi.$$

It is also interesting to see what happens when Ω is less regular: Exercise 5.6 treats a model case where Ω is only piecewise C^1. An adequate general framework would actually be when Ω is only of finite perimeter: since $P(\Omega_\theta) = \int_\Gamma J^\theta$ and since J^θ is continuous on $\mathbb{R}^N$, one needs only to be able to define the integral on Γ of a continuous function. This is possible when Ω is of finite perimeter with the tools of geometric measure theory (see, e.g., [135]). Then, we still have C^∞ properties for $\theta \to P(\Omega_\theta)$ in this wider framework (see, e.g., [154]).

We now give the two main results concerning differentiation of boundary integrals. We consider the more difficult case when the function $g(\theta)$ under the integral is defined only on the moving domain and the assumptions are given in terms of the transported function on the fixed domain. A fortiori, the result is valid if $g(\theta)$ has a regular enough extension to $\mathbb{R}^N$.

Theorem 5.4.17. *Let Ω be open, bounded, and of class C^1. Let $\theta \in C^{1,\infty} \to g(\theta) \in W^{1,1}(\Omega_\theta)$ be such that $\theta \in C^{1,\infty} \to h(\theta) = g(\theta) \circ (I+\theta) \in W^{1,1}(\Omega)$ is differentiable at* 0. *Then $\theta \to \mathcal{G}(\theta) = \int_{\Gamma_\theta} g(\theta)$ is differentiable at* 0 *and we have*

$$\forall \xi \in C^{1,\infty}, \quad \mathcal{G}'(0)\xi = \int_\Gamma h'(0)\xi + g(0)\operatorname{div}_\Gamma \xi.$$

For all $K \subset \Omega$ compact, $\theta \in C^{1,\infty} \to g(\theta)_K \in L^1(K)$ is differentiable at 0 *and we have*

$$\forall \xi \in C^{1,\infty}, \quad g'(0)\xi = h'(0)\xi - \nabla g(0).\xi \in L^1(\Omega).$$

If, moreover, Ω is of class C^2 and $g(0) \in W^{2,1}(\Omega)$, then

$$\begin{aligned}\mathcal{G}'(0)\xi &= \int_\Gamma g'(0)\xi + \nabla g(0).\xi + g(0)\operatorname{div}_\Gamma \xi \\ &= \int_\Gamma g'(0)\xi + (\xi.n)\left[\frac{\partial g(0)}{\partial n} + Hg(0)\right].\end{aligned}$$

Proposition 5.4.18. *Let Ω be open, bounded, and of class C^2 and let Φ satisfy* (5.67). *Assume $t \to g(t) \circ \Phi(t) \in W^{1,1}(\Omega)$ is differentiable at* 0 *with $g(0) \in W^{2,1}(\Omega)$.*

Then $t \to G(t) = \int_{\Gamma_t} g(t)$ is differentiable at 0, *$t \to g(t)_{|\omega} \in W^{1,1}(\omega)$ is differentiable at* 0 *for all open $\omega \subset \overline{\omega} \subset \Omega$; thus, the derivative $g'(0)$ is then defined in $W^{1,1}(\Omega)$ and we have*

$$G'(0) = \int_\Gamma g'(0) + [V.n]\left[\frac{\partial g(0)}{\partial n} + Hg(0)\right].$$

Let us now give proofs of all these results.

Proof of Lemma 5.4.15. By composition of C^∞-mappings, the mapping $\theta \in C^{1,\infty} \to J^\theta = \det(I + D\theta)\,\|{}^t(I+D\theta)^{-1}n\| \in C^0(\Gamma)$ is also of class C^∞. We know that $D_\theta[\det(I+D\theta)]_{|\theta=0} \cdot \xi = \operatorname{div}\xi$. It remains to compute the derivative at $t = 0$ of

$$t \to w(t) = \|v(t)\| \quad \text{where } v(t) = {}^t(I + tD\xi)^{-1}n.$$

We have $w'(0) = n.v'(0)$ and we obtain $v'(0)$ by differentiating ${}^t(I + t\,D\xi)v(t) = n$, which gives ${}^tD\xi.n + v'(0) = 0$. We deduce $w'(0) = -n.({}^tD\xi.n) = -(D\xi.n).n$. Thus

$$D_\theta J^\theta_{|\theta=0} \cdot \xi = \operatorname{div}\xi - (D\xi.n).n = \operatorname{div}_\Gamma \xi.$$

The other part of the lemma follows by composition. □

Proof of Corollary 5.4.16. We have $P(\Omega_\theta) = \int_\Gamma J^\theta$. Since integrating on Γ is a continuous linear mapping from $C^0(\Gamma)$ into $\mathbb{R}$, regularity and formula may be obtained by composition starting from Lemma 5.4.15. □

Proof of Theorem 5.4.17. Since $\mathcal{G}(\theta) = \int_\Gamma h(\theta)J^\theta$, the differentiability of $\mathcal{G}$ as well as the expression of $\mathcal{G}'(0)$ can be obtained by composition starting from Lemma 5.4.15 and thanks to the continuity of the trace from $W^{1,1}(\Omega)$ into $L^1(\Omega)$. The second point is a direct consequence of Lemma 5.3.3 (see Remark 5.3.5). If $\nabla g(0)$ has a trace on Γ, we can obtain a new expression for $\mathcal{G}'(0)$ by using $h'(0)$: it is the case under the extra assumption that $g(0) \in W^{2,1}(\Omega)$ and Ω is of class C^2. The last formula for the derivative is obtained thanks to the identity

$$\nabla g(0).\xi + g(0)\,\mathrm{div}_\Gamma\,\xi = \frac{\partial g(0)}{\partial n}(\xi \cdot n) + \mathrm{div}_\Gamma(g(0).\xi),$$

and to (5.56). □

Proof of Proposition 5.4.18. Note that $h(t) = g(t)\circ\Phi(t)$. Since $G(t) = \int_\Gamma h(t)J^{\Phi(t)-I}$, the differentiability of G follows from Lemma 5.4.15 and we have $G'(0) = \int_\Gamma h'(0) + h(0)\,\mathrm{div}_\Gamma\,V$. The differentiability of $t \to g(t) \in W^{1,1}(\omega)$ is a consequence of Corollary 5.2.5 applied to $g(t)$ and to $\nabla g(t)$ and we have $g'(0) = h'(0) - \nabla g(0).V \in W^{1,1}(\Omega)$. We deduce

$$G'(0) = \int_\Gamma g'(0) + \nabla g(0).V + g(0)\,\mathrm{div}_\Gamma\,V,$$

and this leads to the announced formula by the same computation as in the previous proof. □

Remark. We recall that, for an open set of class C^1, there exists a linear continuous extension $P : W^{1,1}(\Omega) \to W^{1,1}(\mathbb{R}^N)$ such that $P(v)_{|\Omega} = v$ for all $v \in W^{1,1}(\Omega)$ (see [54]). Thus, under the assumptions of Proposition 5.4.18, $t \to P\big(g(t)\circ\Phi(t)\big) \in W^{1,1}(\mathbb{R}^N)$ is differentiable at 0. Therefore we can also apply Theorem 5.2.2 or its Corollary 5.2.3 to obtain the differentiability of

$$t \to P\big(g(t)\circ\Phi(t)\big)\circ\Phi(t)^{-1} \in L^1(\mathbb{R}^N),$$

and consequently the differentiability of the restrictions of $t \to g(t)$ to open subsets whose closure is included in Ω.

5.5 Differentiating the Neumann problem

The case of a Neumann boundary condition is not as easy as for a Dirichlet condition since the normal derivative itself depends on the variable domain. We give here the essential results.

In all of this section, Ω is open, bounded, and of class at least C^1 and we are given Φ satisfying (5.67). We are also given $f \in L^2(\Omega)$ and $g \in H^1(\mathbb{R}^N)$ so that

$g_{\Gamma_t} \in L^2(\Gamma_t)$ is well defined. We consider the solution of the Neumann problem that may be formally written as

$$\begin{cases} -\Delta u_t + u_t = f & \text{in } \Omega_t, \\ \frac{\partial u_t}{\partial n} = g & \text{on } \Gamma_t, \end{cases} \tag{5.74}$$

and whose exact variational formulation is

$$\begin{cases} u_t \in H^1(\Omega_t) \text{ and for all } v_t \in H^1(\Omega_t), \\ \int_{\Omega_t} \nabla u_t.\nabla v_t + u_t v_t = \int_{\Omega_t} f v_t + \int_{\Gamma_t} g v_t. \end{cases} \tag{5.75}$$

As for the Dirichlet problem, one may first *formally* write the equation satisfied by the derivative of $t \longrightarrow u_t$ at $t = 0$ that we denote by u' (and we drop the lower index "t" to avoid confusion). Inside Ω we have

$$-\Delta u' + u' = 0 \quad \text{in } \Omega.$$

At the boundary, we differentiate with respect to t the relation

$$\forall x \in \Gamma, \quad \nabla u_t(\Phi(t,x)).n_t(\Phi(t,x)) = g(\Phi(t,x)), \tag{5.76}$$

which gives

$$\nabla u'.n + \big(D^2u.V\big).n + \nabla u.\left[\frac{\partial n_t}{\partial t} + Dn.V\right] = \nabla g.V \quad \text{on } \Gamma. \tag{5.77}$$

Thus, u' is fully identified as being the solution of a new Neumann problem. One may make a little more explicit the value at the boundary. Using $V = (V.n)n + V_\Gamma$, we obtain

$$(D^2u.V).n + \nabla u.[Dn.V] - \nabla g.V = (V.n)\left[(D^2u.n).n + \nabla u(Dn.n) - \frac{\partial g}{\partial n}\right] + a_\Gamma,$$

with $a_\Gamma = (D^2u.V_\Gamma).n + \nabla u.(Dn.V_\Gamma) - \nabla g.V_\Gamma$. By differentiating $\nabla u.n = g$ tangentially in the direction V_Γ, we have $a_\Gamma = 0$. We then use Proposition 5.4.14 and (5.69) to transform (5.77) into

$$-\nabla u'.n = (V.n)\left[\frac{\partial^2 u}{\partial n^2} - \frac{\partial g}{\partial n}\right] - \nabla u.\nabla_\Gamma(V.n). \tag{5.78}$$

Note that this formula requires the ability to define traces on Γ of $\nabla u'$, D^2u, ∇g.

Let us first state the differentiability results for $\theta \longrightarrow u_\theta$, where u_θ is the solution of the Neumann problem associated with f, g on $\Omega_\theta = (I+\theta)(\Omega)$ with $\Gamma_\theta = \partial\Omega_\theta$. We set $U_\theta = u_\theta \circ (I+\theta)$.

Theorem 5.5.1. *Assume Ω is open, bounded, and of class C^1 and let $f \in H^1(\mathbb{R}^N)$, $g \in H^2(\mathbb{R}^N)$. Then, the mapping*

$$\theta \in C^{1,\infty}(\mathbb{R}^N, \mathbb{R}^N) \to U_\theta \in H^1(\Omega)$$

is of class C^1 on a neighborhood of 0.

If $f \in H^{\max\{m,1\}}(\mathbb{R}^N)$, $g \in H^{m+2}(\mathbb{R}^N)$ with $m \geq 0$, it is of class C^{m+1}. If, moreover, Ω is of class C^{m+1}, the mapping

$$\theta \in C^{m+1,\infty} \to U_\theta \in H^{m+1}(\Omega)$$

is then of class C^1 and there exists an extension $\tilde{u}_\theta \in H^{m+1}(\mathbb{R}^N)$ of u_θ such that

$$\theta \in C^{m+1}(\mathbb{R}^N, \mathbb{R}^N) \to \tilde{u}_\theta \in H^m(\mathbb{R}^N)$$

is of class C^1.

Proof. Like in the proof of Theorem 5.3.2, we start by writing the variational equation on the fixed domain Ω:

$$\int_{\Omega_\theta} \nabla u_\theta \nabla v_\theta + u_\theta v_\theta = \int_{\Omega_\theta} f v_\theta + \int_{\Gamma_\theta} g v_\theta \quad \forall\, v_\theta \in H^1(\Omega_\theta).$$

Choosing $v_\theta = v \circ (I + \theta)$ with $v \in H^1(\Omega)$, we obtain

$$\int_\Omega A(\theta) \nabla U_\theta \nabla v + U_\theta v J_\theta = \int_\Omega [f \circ (I + \theta)] v J_\theta + \int_\Gamma g \circ (I + \theta) v J^\theta,$$

where $A(\theta)$ is defined in (5.39) and J^θ in (5.73). We consider the mapping F that to $(\theta, U) \in C^{1,\infty} \times H^1(\Omega)$ associates

$$\mathcal{A}_\theta U + U J_\theta - [f \circ (I + \theta)] J_\theta - [g \circ (I + \theta)] J^\theta \in \big(H^1(\Omega)\big)',$$

where $\mathcal{A}_\theta$ is the operator from $H^1(\Omega)$ into its dual $\big(H^1(\Omega)\big)'$ defined by

$$\langle \mathcal{A}_\theta(U), v \rangle = \int_\Omega A(\theta) \nabla U \nabla v \quad \forall\, U, v \in H^1(\Omega).$$

Similarly, $[g \circ (I + \theta)] J^\theta$ is the element of $\big(H^1(\Omega)\big)'$ that to $v \in H^1(\Omega)$ associates $\int_\Gamma [g \circ (I + \theta)] J^\theta v$: it is continuous by the continuity of the trace from $H^1(\Omega)$ into $L^2(\Gamma)$.

We check, like in the proof of Theorem 5.3.2, that F is of class C^1 for small θ. The only new point to check is the differentiability of

$$\theta \in C^{1,\infty} \to [g \circ (I + \theta)] J^\theta \in \big(H^1(\Omega)\big)'.$$

But, since $g \in H^2(\mathbb{R}^N)$, by Lemma 5.3.9, $\theta \in C^1 \to g \circ (I+\theta) \in H^1(\mathbb{R}^N)$ is of class C^1 and by continuity of the trace, $\theta \in C^1 \to g \circ (I+\theta)_{|\Gamma} \in L^2(\Gamma)$ is also of class C^1. We end by using that $\theta \in C^1 \to J^\theta \in C^0(\Gamma)$ is C^∞. On the other hand, $\theta \in C^{1,\infty} \to f \circ (I+\theta)_{|\Omega} \in L^2(\Omega) \subset \big(H^1(\Omega)\big)'$ is differentiable since $f \in H^1(\mathbb{R}^N)$.

The operator $D_U F(0,u_0)$ is an isomorphism from $H^1(\Omega)$ onto $\big(H^1(\Omega)\big)'$ since

$$(D_U F(0,u_0)v).\hat{v} = \int_\Omega \nabla v \nabla \hat{v} + v\hat{v} \quad \forall\, v, \hat{v} \in H^1(\Omega).$$

By the implicit function theorem, there exists $\theta \in C^1 \to U(\theta) \in H^1(\Omega)$ of class C^1 on a neighborhood of 0 such that $F(\theta, U(\theta)) \equiv 0$ and $U(0) = u_0$. By uniqueness, we have $U(\theta) = U_\theta$. This proves the C^1-regularity of $\theta \to U_\theta$.

Let us now assume $f \in H^{m+1}(\mathbb{R}^N)$, $g \in H^{m+2}(\mathbb{R}^N)$. We also know that $(\theta, U) \in C^{1,\infty} \times H^1(\Omega) \to \mathcal{A}_\theta U + UJ^\theta \in \big(H^1(\Omega)\big)'$ is C^∞. The regularity of F is therefore limited only by the regularity

$$\theta \in C^{1,\infty} \to (f \circ (I+\theta)J_\theta, g \circ (I+\theta)J^\theta) \in (L^2(\Omega), H^1(\Omega)),$$

which is then of class C^{m+1} by Lemma 5.3.9. It follows that $\theta \in C^1 \to U_\theta \in H^1$ is also of class C^{m+1}.

Moreover, let us assume that Ω is of class C^{m+1} and $m \geq 1$. We then consider the restriction of F to $C^{m+1,\infty} \times H^{m+1}$. Again by Lemma 5.3.9, the above application is of class C^1 with values in $H^{m-1}(\Omega) \times H^m_{|\Gamma}$ and F is therefore of class C^1 with values in this subspace of $\big(H^1(\Omega)\big)'$. Moreover, $D_U F(0,u_0)$ is an isomorphism from H^{m+1} onto this subspace (see, e.g., [54] and the references in it for these regularity results). We deduce, again by the implicit function theorem, that $\theta \in C^{m+1,\infty} \to U_\theta \in H^{m+1}$ is of class C^1.

Finally, if P is a linear continuous extension from $H^{m+1}(\Omega)$ to $H^{m+1}(\mathbb{R}^N)$, we set $\tilde{u}_\theta := P(v_\theta) \circ (I+\theta)^{-1}$ and we deduce the last point of the theorem (see Lemma 5.3.3 and Exercise 5.2). □

We deduce the following result for problem (5.74).

Theorem 5.5.2. *Let Ω be open, bounded, and of class C^2, let $f \in H^1(\mathbb{R}^N)$, $g \in H^2(\mathbb{R}^N)$, and let Φ satisfy* (5.67). *We consider the solution u_t of problem* (5.74) *and $U_t = u_t \circ \Phi(t)$. Then, $t \to U_t \in H^1(\Omega)$ is differentiable at* 0 *and there exists an extension $\tilde{u}_t \in H^1(\mathbb{R}^N)$ of u_t with $t \to \tilde{u}_t \in L^2(\mathbb{R}^N)$ differentiable at* 0. *Its derivative $u' \in H^1(\Omega)$ is the unique solution of the following variational problem: for all $v \in H^2(\Omega)$,*

$$\int_\Omega \nabla u'.\nabla v + u'v + \int_\Gamma (\nabla_\Gamma u.\nabla_\Gamma v + uv)(V.n) = \int_\Gamma \left[f + Hg + \frac{\partial g}{\partial n}\right] v(V.n),$$

where $u \in H^2(\Omega)$. If Ω is of class C^3 and $V \in C^2$, then $u \in H^3(\Omega)$, $u' \in H^2(\Omega)$ and u' is the unique solution of

$$\begin{cases} -\Delta u' + u' = 0 & \text{in } \Omega, \\ \dfrac{\partial u'}{\partial n} = \left(\dfrac{\partial g}{\partial n} - \dfrac{\partial^2 u}{\partial n^2}\right) V.n + \nabla u.\nabla_\Gamma (V.n) & \text{on } \Gamma. \end{cases} \tag{5.79}$$

Proof. For the regularity of U_t, u_t, we apply Theorem 5.5.1, using that $U_t = U_{\Phi(t)-I}$, $u_t = u_{\Phi(t)-I}$ and we obtain, moreover, that $u' = U'(0) - \nabla u.V \in H^1(\Omega)$ since $u \in H^2(\Omega)$, according to the regularity of the data. By Lemma 5.2.7, one can also say that $t \to \nabla u_t \in L^2(\omega)$ is differentiable at 0 for all ω such that $\overline{\omega} \subset \Omega$. We are then able to differentiate the variational equation (5.75): given $v \in H^2(\Omega)$, since Ω is of class C^2, it may be extended to $H^2(\mathbb{R}^N)$ (we still denote the extension by v). Thus, we may choose $v_t \equiv v$ in (5.75). We differentiate by using Corollary 5.2.5 and Proposition 5.4.18 and we obtain

$$\int_\Omega \nabla u'.\nabla v + u'v + \int_\Gamma (\nabla u.\nabla v + uv)V.n = \int_\Gamma (V.n)\left[fv + Hgv + \frac{\partial (gv)}{\partial n}\right]. \tag{5.80}$$

Using the already-seen formula

$$\nabla u.\nabla v - \frac{\partial u}{\partial n}\frac{\partial v}{\partial n} = \nabla_\Gamma u.\nabla_\Gamma v,$$

and $g = \frac{\partial u}{\partial n}$, we obtain the first expression of the theorem.

Now choosing v in $C_0^\infty(\Omega)$, we immediately obtain that u' satisfies $-\Delta u' + u' = 0$ in the sense of distributions. If Ω is of class C^3 and $V \in C^2$, then $u \in H^3(\Omega)$ and $u' \in H^2(\Omega)$ and we may write

$$\int_\Omega \nabla u'.\nabla v + u'v = \int_\Gamma v\frac{\partial u'}{\partial n} + \int_\Omega (-\Delta u' + u') = \int_\Gamma v\frac{\partial u'}{\partial n}.$$

Using $u - f = \Delta u = \Delta_\Gamma u + H\frac{\partial u}{\partial n} + \frac{\partial^2 u}{\partial n^2}$, as well as the integration by parts formula (5.64),

$$\int_\Gamma (V.n)v\Delta_\Gamma u(V.n) = -\int_\Gamma \nabla_\Gamma u.\nabla_\Gamma (vV.n) = -\int_\Gamma (V.n)\nabla_\Gamma u.\nabla_\Gamma v + v\nabla_\Gamma u.\nabla_\Gamma (V.n),$$

and using that the traces on Γ of $H^2(\Omega)$ functions are dense in $L^2(\Gamma)$, we deduce the expression (5.79). □

5.6 How to differentiate boundary value problems

To exploit our training of the previous sections, let us now enumerate some general principles for the differentiation of a boundary value problem set on a variable domain

$\Omega_t = \Phi(t)(\Omega)$ (according to the previous notation). We emphasized the two model problems associated with the Laplace operator $-\Delta + \lambda I$, with either Dirichlet or Neumann boundary conditions. Let us consider now a general (linear or nonlinear) boundary value problem that may formally be written

$$A(t, u_t) = f \quad \text{in } \Omega_t, \qquad B(t, u_t) = g \quad \text{on } \Gamma_t, \tag{5.81}$$

where $A(t, \cdot)$, $B(t, \cdot)$ act on functions spaces that are defined on Ω_t, Γ_t. In the two previous examples, we had

$$A(t, u) = -\Delta u + \lambda u, \qquad B(t, u) = u, \qquad \text{or} \qquad B(t, u) = \nabla u.n_t,$$

where n_t is an extension to the whole $\mathbb{R}^N$ of the unit normal vector to Γ_t.

The first step is to prove the differentiability of $t \to u_t$. This may be done by applying the implicit function theorem to the transported operator on Ω, Γ. Thus, first we obtain the regularity of the function $t \to U_t = u_t \circ \Phi(t)$. We then deduce the differentiability of $t \to u_t$ at least inside Ω: this allows us to define $u'(0)$ on the whole of Ω and its regularity up to the boundary is generally deduced from the expression $u'(0) = U'(0) - \nabla u.\Phi'(0)$.

The second step is the actual computation of this derivative $u' = u'(0)$. For this, we "differentiate" the system (5.81). Under adequate assumptions, we obtain that the function u' is characterized as the solution of the new boundary value problem

$$\partial_t A(0, u) + \partial_u A(0, u).u' = 0 \quad \text{in } \Omega, \tag{5.82}$$

$$\partial_t B(0, u) + \partial_u B(0, u).u' = \frac{\partial}{\partial n}\big(g - B(0, u)\big)(V.n) \quad \text{on } \Gamma, \tag{5.83}$$

where the differentiation in the interior of Ω is generally easy to obtain in the sense of distributions. The differentiation at the boundary is obtained as follows: setting $Z_t := B(t, u_t) - g$, the goal is to differentiate at $t = 0$ the relation $Z_t \circ \Phi(t) = 0$ on Γ. We obtain on Γ,

$$Z'(0) + \nabla Z(0).V = 0 \quad \text{and} \quad \nabla_\Gamma Z(0) = 0,$$
$$\Rightarrow \nabla Z(0) = n\frac{\partial Z(0)}{\partial n} \quad \text{and} \quad Z'(0) = -(V.n)\frac{\partial Z(0)}{\partial n}.$$

Here $Z'(0) = \partial_t B(0, u) + \partial_u B(0, u).u'$, whence the expression (5.83).

Thus, if, $B(t, u) = \nabla u.n_t$, we have

$$\partial_t B(0, u) = \nabla u.n'(0), \qquad \partial_u B(0, u).u' = \nabla u'.n,$$
$$\frac{\partial}{\partial n} B(0, u) = (D^2u.n).n + \nabla u.(Dn.n),$$

and we do find again the computation made in Section 5.5 thanks to the expression of $n'(0)$ given in (5.69), which leads to

$$[\nabla u_t.n_t = g \text{ on } \Gamma_t] \Rightarrow \left[\frac{\partial u'}{\partial n} = \left(\frac{\partial g}{\partial n} - \frac{\partial^2 u}{\partial n^2}\right) V.n + \nabla u.\nabla_\Gamma(V.n) \text{ on } \Gamma\right]. \tag{5.84}$$

In (5.81), we may as well assume that f, g depend on t: it is then sufficient to sum their derivatives with respect to t in (5.82), (5.83).

We can find several more explicit statements of results in this spirit, with adequate assumptions in [242], [243], [277], or [280].

Now, we are going to apply all of these ideas to eigenvalue problems.

5.7 Differentiation of a simple eigenvalue

We are interested here in differentiating $t \longrightarrow \lambda_k(\Omega_t)$ = the kth eigenvalue of the Laplacian operator $\Omega_t = \Phi(t)(\Omega)$ with Dirichlet or Neumann boundary conditions, together with the corresponding eigenfunctions. In Chapters 3 and 4, we have already studied the continuity of $\Omega \mapsto \lambda_k(\Omega)$, which implies the continuity of $t \longrightarrow \lambda_k(\Omega_t)$ for a family of functions $\Phi(t)$ satisfying (5.6).

The question of differentiability is more delicate. Essentially, it will hold (in the usual sense) *only for simple eigenvalues.* We can understand why by looking at an elementary finite-dimensional example: let A_t be the 2×2 matrix given by

$$A_t = \begin{pmatrix} 1+t & 0 \\ 0 & 1-t \end{pmatrix}$$

The matrix $A_0 = I$ has a double eigenvalue. The first eigenvalue λ_1 of A_t (which means its smallest one) is $1-t$ if $t \geq 0$ and $1+t$ if $t \leq 0$, that is,

$$\lambda_1(A_t) = 1 - |t|, \qquad \lambda_2(A_t) = 1 + |t|.$$

Therefore, $t \longrightarrow \lambda_1(A_t)$ is not differentiable at $t = 0$. This bad situation does not happen around simple eigenvalues. Actually, when looking more carefully at this multiple situation for $t \longrightarrow \lambda_1(A_t)$, we realize that there are two "branches", which intersect at $t = 1$, and on each of them $t \longrightarrow \lambda(A_t)$ is differentiable (and even analytic). In other words, we may renumber the eigenvalues in such a way that they are differentiable. This is a general property and the reader interested in multiple values may have a look at [94], [109] in which the authors use a subdifferential technique, or [265] or [240], where a directional differentiability is proved. Finally, [202] remains a key reference for these questions.

Here, for simplicity, we will limit the study to simple eigenvalues for the Laplace operator with Dirichlet and Neumann boundary conditions. We will start by "guessing" the formula for the derivative through a formal computation. Next, we will rigorously prove the differentiability in the Dirichlet case, and justify the previous computations, by using the implicit function theorem.

We assume we are given an open set Ω and a family of applications $\Phi(t)$ satisfying (5.6). We denote by $\lambda_k(t)$ the kth eigenvalue of the operator $-\Delta$ on $\Omega_t = \Phi(t)(\Omega)$ with Dirichlet or Neumann boundary conditions. We assume it is simple and, k being fixed in the following, we denote by u_t an associated eigenfunction in $H_0^1(\Omega_t)$ (Dirichlet) or in $H^1(\Omega_t)$ (Neumann) that we normalize as

$$\int_{\Omega_t} u_t^2(x)\,dx = 1. \tag{5.85}$$

Let us start with the Dirichlet case. We have

$$\begin{cases} -\Delta u_t = \lambda_k(t) u_t & \text{in } \Omega_t, \\ u_t = 0 & \text{on } \Gamma_t := \partial\Omega_t. \end{cases} \tag{5.86}$$

Let us denote by u' the derivative at 0 of $t \to u_t$. Applying the procedure summarized in the previous section, the differentiation of (5.86), (5.85) leads to

$$\begin{cases} -\Delta u' = \lambda_k u' + \lambda_k' u & \text{in } \Omega, \\ u' = -\frac{\partial u}{\partial n} V.n & \text{on } \Gamma, \\ \int_\Omega u u' = 0. \end{cases} \tag{5.87}$$

Let us multiply (5.87) by u and integrate by parts on Ω. Using the normalization (5.85), we obtain

$$-\int_\Omega u\Delta u' = \lambda_k' \int_\Omega u^2 + \lambda_k \int_\Omega u u' = \lambda_k'. \tag{5.88}$$

By integration by parts and using (5.86), (5.87) again,

$$-\int_\Omega u\Delta u' = \int_\Gamma u'\frac{\partial u}{\partial n} - \int_\Omega u'\Delta u = \int_\Gamma u'\frac{\partial u}{\partial n} = -\int_\Gamma \left[\frac{\partial u}{\partial n}\right]^2 V.n.$$

Thus, we state with the previous notation,

Theorem 5.7.1 (Dirichlet case). *Let Ω be open, bounded, and of class C^2. We assume $\lambda_k(\Omega)$ is a simple eigenvalue. Then, the functions $t \to \lambda_k(t)$, $t \to u_t \in L^2(\mathbb{R}^N)$ are differentiable at $t = 0$ and $u' \in H^1(\Omega)$ is the unique solution of* (5.87) *with*

$$\lambda_k'(0) := -\int_\Gamma \left(\frac{\partial u}{\partial n}\right)^2 V.n. \tag{5.89}$$

A proof of this theorem will be given later. Note that formula (5.89) allows some qualitative analysis. For instance, if we consider a vector field V that shrinks the domain, that is, $V.n \le 0$ everywhere, then $\lambda'_k \ge 0$, which is consistent with the well-known fact that the eigenvalue should then increase.

Note also that the computation of $\lambda'_k(0)$ could have also been made by differentiating $\lambda_k(t) = \int_{\Omega_t} |\nabla u_t|^2$.

Let us now do the job for a Neumann condition. Similarly, by differentiating

$$\begin{cases} -\Delta u_t = \lambda_k(t) u_t & \text{in } \Omega_t, \\ \frac{\partial u_t}{\partial n} = 0 & \text{on } \Gamma_t, \end{cases}$$

as well as the normalization relation (5.85), and also using formula (5.84), we obtain

$$\begin{cases} -\Delta u' = \lambda_k u' + \lambda'_k u & \text{in } \Omega, \\ 0 = \int_\Gamma u^2(V.n) + 2\int_\Omega u u', \\ \frac{\partial u'}{\partial n} = -\frac{\partial^2 u}{\partial n^2} V.n + \nabla u.\nabla_\Gamma(V.n) & \text{on } \Gamma. \end{cases} \tag{5.90}$$

Differentiating $\lambda_k(t) = \int_{\Omega_t} |\nabla u_t|^2$, we obtain

$$\lambda'_k = \int_\Gamma |\nabla u|^2(V.n) + 2\int_\Omega \nabla u.\nabla u',$$

$$2\int_\Omega \nabla u.\nabla u' = -2\int_\Omega u'\Delta u = 2\lambda_k \int_\Omega u' u = -\lambda_k \int_\Gamma u^2(V.n).$$

Whence $\lambda'_k = \int_\Gamma \left(|\nabla u|^2 - \lambda_k u^2\right) V.n$, and we can finally state

Theorem 5.7.2 (Neumann case). *Let Ω be open and of class C^3. Assume $\lambda_k(\Omega)$ is a simple eigenvalue. Then, the functions $t \longrightarrow \lambda_k(t)$, $t \longrightarrow u_t \in L^2(\omega)$, where ω is an open subset such that $\overline{\omega} \subset \Omega$, are differentiable at $t = 0$ and $u' \in H^2(\Omega)$ is the unique solution of* (5.90) *with*

$$\lambda'_k(0) := \int_\Gamma \left(|\nabla u|^2 - \lambda_k u^2\right) V.n. \tag{5.91}$$

We are now going to prove Theorem 5.7.1. We refer for instance to [234] for a similar approach, but with different functional spaces. The proof given here is general and may be extended to many other similar situations. The starting point is the variational formulation of the spectral problem on the measurable set $\Omega_\theta = (I+\theta)(\Omega)$, where $\theta \in W^{1,\infty}(\mathbb{R}^N, \mathbb{R}^N)$:

$$\begin{cases} u_\theta \in H^1_0(\Omega_\theta),\ \int_{\Omega_\theta} u_\theta^2 = 1, \text{ and } \forall\, \varphi_\theta \in H^1_0(\Omega_\theta), \\ \int_{\Omega_\theta} \nabla u_\theta.\nabla\varphi_\theta = \lambda_k(\theta)\int_{\Omega_\theta} u_\theta\varphi_\theta \text{ (we set } \lambda_k(\theta) := \lambda_k(\Omega_\theta)). \end{cases} \tag{5.92}$$

(We refer to Chapter 4 for the definition of $H_0^1(\Omega)$ when Ω is only measurable and for the associated eigenvalue problem.) We assume that the kth eigenvalue $\lambda_k = \lambda_k(\Omega)$ is simple and we again denote by u an associated eigenfunction with $\int_\Omega u^2 = 1$. For θ close to 0 in $W^{1,\infty}$, $\lambda_k(\theta)$ is simple and we may choose the sign of u_θ so that

$$\theta \in W^{1,\infty} \to (u_\theta, \lambda(\theta)) \in H_0^1(\Omega_\theta) \times \mathbb{R} \subset H^1(\mathbb{R}^N) \times \mathbb{R}$$

is continuous (this follows from the study of continuity in Chapter 4).

We use the same notation and arguments as in the proof of Theorem 5.3.2. We transport this variational equation onto Ω. This gives, using the notation $v_\theta = u_\theta \circ (I + \theta)$,

$$v_\theta \in H_0^1(\Omega), \quad -\operatorname{div}\big(A(\theta)\nabla v_\theta\big) = \lambda(\theta) v_\theta J_\theta \quad \text{on } \Omega,$$

and the normalization

$$\int_\Omega v_\theta^2 J_\theta = 1.$$

We then consider the operator $\mathcal{F} : W^{1,\infty} \times H_0^1(\Omega) \times \mathbb{R} \to H^{-1}(\Omega) \times \mathbb{R}$ defined by[7]

$$\mathcal{F}(\theta, v, \lambda) = \left(-\operatorname{div}(A(\theta)\nabla v) - \lambda v J_\theta, \int_\Omega v^2 J_\theta - 1\right).$$

This operator is of class C^1 and even C^∞ as we have already seen. Moreover,

$$\forall\, (\hat{v}, \hat{\lambda}) \in H_0^1(\Omega) \times \mathbb{R}, \quad D_{v,\lambda}\mathcal{F}(0, u, \lambda_k)(\hat{v}, \hat{\lambda}) = \left(-\Delta\hat{v} - \hat{\lambda} u - \lambda_k \hat{v}, 2\int_\Omega u\hat{v}\right).$$

Let us show that $D_{v,\lambda}\mathcal{F}(0, u, \lambda_k)$ is an isomorphism from $H_0^1 \times \mathbb{R}$ onto $H^{-1} \times \mathbb{R}$: since it is continuous, by the Banach theorem, it is sufficient to prove that it is one-to-one, that is,

Lemma 5.7.3. *Assume that λ_k is a simple eigenvalue. Given $(Z, \Lambda) \in H^{-1}(\Omega) \times \mathbb{R}$, there exists a unique solution $(\hat{v}, \hat{\lambda}) \in H_0^1(\Omega) \times \mathbb{R}$ to the system*

$$\begin{cases} -\Delta\hat{v} - \hat{\lambda} u - \lambda_k \hat{v} = Z & \text{in } \Omega, \\ 2\int_\Omega u\hat{v} = \Lambda. \end{cases} \tag{5.93}$$

Proof. Again we should understand (5.93) in the variational sense in order to include the case where Ω is measurable only. By compactness of the operator

$$(-\Delta)^{-1} : X = \big(H_0^1(\Omega)\big)' \to H_0^1(\Omega) \subset \big(H_0^1(\Omega)\big)',$$

[7] If Ω is only measurable, this operator is defined through the corresponding variational formulation, like in Theorem 5.3.2.

we may apply the Fredholm[8] alternative to the operator $-\Delta - \lambda_k I$. Since, by assumption, the kernel of this operator is of dimension 1, $\varphi \in X$ will belong to its range if and only if it satisfies the orthogonality property $\langle \varphi, u \rangle_{X \times H_0^1} = 0$. When applied to $\varphi = Z + \hat{\lambda} u$, this gives

$$0 = \left\langle Z + \hat{\lambda} u, u \right\rangle = \hat{\lambda} + \langle Z, u \rangle .$$

This relation defines $\hat{\lambda}$ in a unique way. Moreover, any element in the preimage of $Z + \hat{\lambda} u$ by $-\Delta - \lambda_k I$ may be written $v_0 + su$, $s \in \mathbb{R}$, where v_0 is one of the elements of the preimage. But the relation $2 \int_\Omega u \hat{v} = \Lambda$ imposes

$$\Lambda = 2 \int_\Omega u(v_0 + su) = 2s + 2 \int_\Omega u v_0.$$

Thus, s is also defined in a unique way, which proves existence of a unique solution to (5.93). □

By the implicit function theorem, there exist $\theta \longrightarrow (v(\theta), \lambda(\theta)) \in H_0^1(\Omega) \times \mathbb{R}$ of class C^∞ on a neighborhood $\mathcal{V}$ of 0 in $W^{1,\infty}$ and a neighborhood $\mathcal{O}$ of $(0, u, \lambda_k)$ in $W^{1,\infty} \times H_0^1(\Omega) \times \mathbb{R}$ such that

$$v(0) = u, \qquad \lambda(0) = \lambda_k, \qquad \mathcal{F}^{-1}(\{0\}) \cap \mathcal{O} = \{(\theta, v(\theta), \lambda(\theta));\ \theta \in \mathcal{V}\}.$$

Thus, $\theta \longrightarrow (v(\theta), \lambda(\theta))$ necessarily coincides with the continuous function $\theta \longrightarrow (v_\theta, \lambda_k(\theta))$. We may then deduce

Theorem 5.7.4. *Let Ω be measurable and bounded and assume that $\lambda_k(\Omega)$ is simple. Then, the application $\theta \in W^{1,\infty} \longrightarrow (v_\theta, \lambda_k(\theta)) \in H_0^1(\Omega) \times \mathbb{R}$ is of class C^∞ on a neighborhood of* 0. *The mapping $\theta \in W^{1,\infty} \longrightarrow u_\theta \in L^2(\mathbb{R}^N)$ is differentiable at* 0 *and we have*

$$\forall \xi \in W^{1,\infty}, \quad D_\theta u_{\theta\,|\theta=0} \xi = D_\theta v_{\theta\,|\theta=0} \xi - \nabla u . \xi \in L^2(\mathbb{R}^N).$$

In particular, $t \longrightarrow \lambda_k(t) = \lambda_k(\Phi(t) - I)$ and $t \longrightarrow u_t = u_{\Phi(t)-I} \in L^2(\mathbb{R}^N)$ are differentiable at 0.

If Ω is of class C^2, then $u \in H^2(\Omega)$, $\lambda_k'(0)$ is given by (5.89), *and $u' \in H^1(\Omega)$ is the unique solution of* (5.87).

Proof. The C^∞-regularity was proved above. We compose with $t \longrightarrow \theta = \Phi(t) - I$ to get the differentiability of $t \longrightarrow \lambda_k(t)$, u_t and we have $u' = v' - \nabla u.V$, where $v' \in H_0^1(\Omega)$ is the derivative of $t \longrightarrow v_t = v_{\Phi(t)-I}$.

[8] Erik **Ivar** FREDHOLM, 1866–1927, Swedish, taught in Stockholm, is famous for his contributions to the theory of integral equations.

If Ω is of class C^2, the solution u is in $H^2(\Omega)$ (see [54]). Thus, by the above expression, $u' \in H^1(\Omega)$ and also $\nabla u.n \in L^2(\Gamma)$. Thus, the computations used for the proof of Theorem 5.7.1 are justified. □

5.8 Use of the adjoint state

In this section, we are going to show how the introduction of an auxiliary problem, or *adjoint problem*, sometimes allows us to simplify the expression of the derivatives of shape functionals. Its principle is to get rid of the derivative u' in the expression of the derivative of functionals. This can be used for many different purposes, including numerical ones.

Let us remark that the introduction of this auxiliary adjoint problem is not specific to shape optimization. It is widely used in the framework of optimal control.

Let us explain its principle with an example. Let us go back to the example of Section 5.3 and briefly recall the results that we obtained. The state was defined as the solution of the boundary value problem

$$\begin{cases} -\Delta u_t + u_t = f & \text{in } \Omega_t, \\ u_t = 0 & \text{on } \Gamma_t. \end{cases}$$

The functional to be minimized was given by

$$j(t) = J(\Omega_t) := a \int_{\Omega_t} |\nabla u_t - \nabla v_0|^2 + b \int_{\Omega_t} |u_t - v_1|^2, \tag{5.94}$$

and for its derivative, we had the formula

$$\begin{aligned} j'(0) = 2a \int_{\Omega} (\nabla u - \nabla v_0).\nabla u' + 2b \int_{\Omega} (u - v_1) u' \\ + \int_{\Gamma} \left[a|\nabla u - \nabla v_0|^2 + b|u - v_1|^2 \right] (V.n). \end{aligned} \tag{5.95}$$

(Here, to simplify, we assume we are dealing with regular data.)

Let us introduce the adjoint state p as being the solution of the problem

$$\begin{cases} -\Delta p + p = 2a(\Delta v_0 - \Delta u) + 2b(u - v_1) & \text{in } \Omega, \\ p = 0 & \text{on } \Gamma. \end{cases} \tag{5.96}$$

Multiplying by u' the equation that defines p and integrating over Ω lead to

$$-\int_\Omega u'\Delta p + \int_\Omega u'p = 2a\int_\Omega u'(\Delta v_0 - \Delta u) + 2b\int_\Omega u'(u - v_1),$$

that is, using Green's formula,

$$-\int_\Omega \Delta u'p - \int_\Gamma u'\frac{\partial p}{\partial n} + \int_\Omega u'p = -2a\int_\Omega \nabla u'.(\nabla v_0 - \nabla u) + 2a\int_\Gamma u'\frac{\partial(v_0 - u)}{\partial n} + 2b\int_\Omega u'(u - v_1).$$

Using the equation satisfied by u', namely $-\Delta u' + u' = 0$, we obtain

$$2a\int_\Omega \nabla u'.(\nabla u - \nabla v_0) + 2b\int_\Omega u'(u - v_1) = -\int_\Gamma u'\frac{\partial p}{\partial n} - 2a\int_\Gamma u'\frac{\partial(v_0 - u)}{\partial n}.$$

Introducing this in (5.95) (which gives $j'(0)$) and using the fact that u' is given on Γ by $u' = -\frac{\partial u}{\partial n}V.n$, we obtain another formula for the derivative of the functional J:

$$j'(0) = \int_\Gamma \left[\frac{\partial u}{\partial n}\frac{\partial p}{\partial n} + 2a\frac{\partial u}{\partial n}\frac{\partial(v_0 - u)}{\partial n} + a|\nabla u - \nabla v_0|^2 + b(u - v_1)^2\right] V.n.$$

One can still transform this a little bit by using that the gradient of u is carried by the normal vector, and, in particular, $\nabla u.\nabla v_0 = \frac{\partial u}{\partial n}\frac{\partial v_0}{\partial n}$. Thus we can state

Proposition 5.8.1. *The derivative of the functional J defined in* (5.94) *can be written*

$$j'(0) = \int_\Gamma \left[\frac{\partial u}{\partial n}\frac{\partial p}{\partial n} + a\left[|\nabla v_0|^2 - \left(\frac{\partial u}{\partial n}\right)^2\right] + b(u - v_1)^2\right] V.n, \tag{5.97}$$

where p is the solution of problem (5.96).

Some comments

- As we see, the introduction of the adjoint state provides an expression of the derivative in terms of u and p, and therefore without the derivative u'. More precisely, formula (5.97) involves, under the integral, a quantity that depends on u and p, namely,

$$\mathcal{F}(u, p) = \frac{\partial u}{\partial n}\frac{\partial p}{\partial n} + a\left[|\nabla v_0|^2 - \left(\frac{\partial u}{\partial n}\right)^2\right] + b(u - v_1)^2,$$

  which is multiplied by $V.n$. What is its interest? The functions u and p are completely independent of the displacement vector field so that the function $\mathcal{F}(u, p)$

may be computed a priori before choosing in which direction we are going to move the boundary $\partial\Omega$. Thus, when minimizing the functional J by using a descent method, we will for instance choose the displacement vector field V so that the derivative $j'(0)$ is as negative as possible. Thus, the choice of V such that

$$V.n = -\mathcal{F}(u, p)$$

is adequate for such a goal. This kind of choice would not be possible when using the expression (5.95) of J, since the definition of u' depends on $V.n$.

- The fact that we dealt with Dirichlet conditions in the above example is not specific. Most important is that the function u' (or $\frac{\partial u'}{\partial n}$ etc.) is known on the boundary of Ω (see below).

- There is something a bit mysterious in the approach we just described: How does one find the right equation to be satisfied by the adjoint state p? *The principle is to get rid of the integrals containing u'*. By the previous point, u' is known at the boundary. It remains to take care of the integrals on Ω. For this, *a technical* trick *consists in replacing u' by the test function v* in the integrals appearing in the expression of $j'(0)$ (see (5.95)) in order to build a variational formulation of the equation to be satisfied by p. Thus, in the present case, it is written

$$\begin{cases} p \in H_0^1(\Omega) \text{ and for all } v \in H_0^1(\Omega), \\ \int_\Omega \nabla p.\nabla v + pv = 2a \int_\Omega (\nabla u - \nabla v_0)\nabla v + 2b \int_\Omega (u - v_1)v. \end{cases} \tag{5.98}$$

More generally, for a functional $J(\Omega) = \int_\Omega F(x, u(x), \nabla u(x))\, dx$, where $F = F(x, u, q)$ is a regular function, as we have

$$j'(0) = \int_\Omega u'\partial_u F + \nabla_q F.\nabla u' + \int_\Gamma F V.n,$$

we will introduce the adjoint problem

$$p \in H_0^1(\Omega), \quad -\Delta p + p = \partial_u F - \operatorname{div}_x[\nabla_q F].$$

Then, multiplying by u' and integrating by parts as before,

$$-\int_\Gamma u' \frac{\partial p}{\partial n} = \int_\Omega u'\partial_u F + \nabla u'.\nabla_q F - \int_\Gamma u'\nabla_q F.n.$$

We deduce

$$j'(0) = \int_\Gamma (V.n) \left[\frac{\partial u}{\partial n}\frac{\partial p}{\partial n} + F - \frac{\partial u}{\partial n}\nabla_q F.n \right].$$

If one deals with the Neumann problem for u where (see (5.79)) u' is a solution of

$$-\Delta u' + u' = 0, \qquad \frac{\partial u'}{\partial n} = \nabla u.\nabla_\Gamma (V.n) + (V.n)\left[\frac{\partial g}{\partial n} - \frac{\partial^2 u}{\partial n^2}\right], \tag{5.99}$$

we will consider the adjoint problem

$$\frac{\partial p}{\partial n} = n.\nabla_q F \text{ on } \Gamma, \quad -\Delta p + p = \partial_u F - \operatorname{div}_x[\nabla_q F] \text{ in } \Omega.$$

By the same computation, we obtain

$$j'(0) = \int_\Gamma p\frac{\partial u'}{\partial n} + FV.n,$$

and here $\frac{\partial u'}{\partial n}$ is given by (5.99).

If now J contains boundary terms like $J(\Omega) = \int_\Gamma G(\sigma, u(\sigma), \nabla u(\sigma))\, d\sigma$, for instance, with $G = G(x, u, q)$ regular, we recall that (see Proposition 5.4.18)

$$j'(0) = \int_\Gamma u'\partial_u G + \nabla_q G.\nabla u' + (V.n)\left[\frac{\partial G}{\partial n} + HG\right].$$

Since by writing $\nabla u' = \nabla_\Gamma u' + \frac{\partial u'}{\partial n} n$ and using (5.56) we have

$$\int_\Gamma \nabla_q G.\nabla u' = \int_\Gamma (\nabla_q G.n)\frac{\partial u'}{\partial n} - u' \operatorname{div}_\Gamma G + Hu'\nabla_q G.n,$$

we then introduce the adjoint problem

$$\frac{\partial p}{\partial n} = \partial_u G - \operatorname{div}_\Gamma \nabla_q G + H\nabla_q G.n \quad \text{on } \Gamma, \qquad -\Delta p + p = 0 \quad \text{in } \Omega.$$

And as before, after multiplying by u' and after integration by parts, we obtain

$$j'(0) = \int_\Gamma (V.n)\left[\frac{\partial G}{\partial n} + HG\right] + [p + \nabla_q G.n]\frac{\partial u'}{\partial n}.$$

The principle is quite general with any linear differential operator A with linear boundary conditions. One then has to solve an adjoint problem associated with the adjoint differential operator A^* with nonhomogeneous terms chosen to "absorb" the part of u' that is not known at the boundary. We refer to [280] for more examples and other statements.

Conclusion. In all examples, it appears that the derivative of the functional J is a *linear form with respect to* $V.n$. This property is actually quite general, and it will be the purpose of the next section to analyze it.

5.9 Structure of shape derivatives

5.9.1 Introduction and notation

We call a "shape functional" any mapping $E : \mathcal{O} \to \mathbb{R}$, where $\mathcal{O}$ is a family of subsets of R^N. We have seen many examples where such functionals lead to differentiable applications $t \in [0,T) \to E(\Omega_t)$ along a subfamily $\Omega_t \in \mathcal{O}$ or to differentiable mappings $\theta \in W^{1,\infty}$ (or $\in C^{k,\infty}$) $\to E((I+\theta)(\Omega))$. As we noticed, these derivatives have particular structures due to the fact that they are defined through shape functionals. This section is dedicated to the analysis of these structures. Let us first describe the notation that will be used.

Notation. We continue to use the notation $W^{k,\infty}$, $C^{k,\infty}$ (see Section 5.4.5). For θ in these spaces and $\Omega \subset \mathbb{R}^N$, we also set $\Omega_\theta = (I+\theta)(\Omega)$, $\Gamma_\theta = \partial\Omega_\theta$. For $k \geq 1$ integer, we denote by $\mathcal{O}_k$ the family of open subsets of class C^k of $\mathbb{R}^N$.

The first and second Fréchet derivatives in θ of a mapping $\mathcal{E}$ from the spaces $W^{k,\infty}$, $C^{k,\infty}$ with values in $\mathbb{R}$ are denoted by $\mathcal{E}'(\theta)$, $\mathcal{E}''(\theta)$ and we write $\mathcal{E}'(\theta)\xi$, $\mathcal{E}''(\theta)(\xi,\zeta)$ for the results of their application to elements ξ, ζ of the linear functional space where the differentiation takes place.

Given $\Omega \subset \mathbb{R}^N$ and $\Gamma = \partial\Omega$, we continue to denote by $V_\Gamma = V - (V.n)n$ the tangential component of a vector field V on Γ. More generally, if A is a function from Γ with values in the space of matrices $p \times N$, we define

$$A_\Gamma := A - (An)^t n \quad \text{or} \quad \left(A_\Gamma\right)_{i,j} = A_{i,j} - (An)_i n_j.$$

With this definition, we have

$$\forall \xi \in C^0(\Gamma, \mathbb{R}^N), \quad A_\Gamma \xi = A\xi_\Gamma = A_\Gamma \xi_\Gamma. \tag{5.100}$$

We will also use the tangential operators defined in Section 5.4.3, that is,

$$\nabla_\Gamma,\ \mathrm{div}_\Gamma,\ \Delta_\Gamma \quad \text{and} \quad \text{for } V \in C^1(\Gamma, \mathbb{R}^p),\ D_\Gamma V := [DV]_\Gamma.$$

If $V = (V_1, \ldots, V_p)$, the ith line of the matrix $D_\Gamma V$ is ${}^t\nabla_\Gamma V_i$. Its application to $\xi \in C^0(\Gamma, \mathbb{R}^N)$ will be denoted simply by $D_\Gamma V \xi$. Let us recall that, for Ω of class C^2, if d is the distance function to Γ (see the proof of Proposition 5.4.14), then $n = \nabla d$ defines a unitary extension of class C^1 of n around Γ and therefore

$$D_\Gamma n = D^2 d \text{ is a symmetric matrix on } \Gamma. \tag{5.101}$$

Indeed, ${}^t Dn\, n = 0$ since n is unitary around Γ. Thus, $D_\Gamma n = Dn - (Dn\, n).n = D^2 d - n.({}^t Dn\, n) = D^2 d$.

5.9.2 A first structure result

Let us start with a first result on the structure of the first shape derivatives. For $k \geq 1$ integer, and $\Omega \subset \mathbb{R}^N$, we set

$$\mathcal{O}(\Omega, k) = \left\{(I+\theta)(\Omega);\ \theta \in C^{k,\infty},\ \|\theta\|_{k,\infty} < 1\right\}.$$

Proposition 5.9.1. *Let k be an integer, $k \geq 1$, $\Omega \subset \mathbb{R}^N$, $E : \mathcal{O}(\Omega, k\} \to \mathbb{R}$. We assume that the function*

$$\theta \in C^{k,\infty} \to \mathcal{E}(\theta) := E(\Omega_\theta)$$

is differentiable at 0. *Then,*

$$\xi \in C_0^\infty(\mathbb{R}^N, \mathbb{R}^N) \to \mathcal{E}'(0)\xi \in \mathbb{R}$$

defines a distribution of order less than or equal to k on $\mathbb{R}^N$ whose support is in $\Gamma = \partial\Omega$.

If, moreover, Ω is an open set of class C^1, then, for all $\xi \in C^{1,\infty}$ such that $\xi.n = 0$ on Γ, we have $\mathcal{E}'(0)\xi = 0$.

In other words, the derivative around Ω of a shape functional in a direction ξ depends only on the trace of ξ on the boundary of Ω and even, in the regular case, only on the normal component of this trace. This had already been noticed by Hadamard [162].

Proof. By definition, $\mathcal{E}'(0)$ is a continuous linear form on $C^{k,\infty}(\mathbb{R}^N, \mathbb{R}^N)$. Its restriction to C_0^∞ is therefore a distribution that is at most of order k since it does have a continuous extension to $C^{k,\infty}$.

Let $\xi \in C^{1,\infty}$ with support in $\mathbb{R}^N \backslash \Gamma$. Let us consider the flow associated with ξ, that is, the solution of

$$\forall\,(t,x) \in [0,+\infty) \times \mathbb{R}^N, \quad \frac{\partial \zeta}{\partial t}(t,x) = \xi\big(\zeta(t,x)\big), \quad \zeta(0,x) = x, \tag{5.102}$$

which exists globally in time since ξ is globally Lipschitz continuous. Let ω be an open neighborhood of Γ on which ξ vanishes. Then $\xi(\zeta(t,x)) \equiv 0$ for all $x \in \omega$, $t \geq 0$, so that $\zeta(t,x) \equiv x$. It follows that $\zeta(t,\Omega) = \Omega$ and $t \to \mathcal{E}(\zeta(t))$ is therefore constant. Its derivative at 0 is equal to zero and to $\mathcal{E}'(0)\xi$. Thus, the restriction to the open set $\mathbb{R}^N \backslash \Gamma$ of the distribution $\mathcal{E}'(0)$ is identically zero (so that its support is in Γ).

If Ω is of class C^1 and if $\xi.n = 0$ on Γ, we still have $\zeta(t,\Omega) = \Omega$ since Γ is invariant for the flow of ξ (see below). Whence the same conclusion. □

Indication. The invariance of Γ under this assumption is classical. We can re-prove it as follows: Let us use the local representation of Γ as the graph of φ_i (see Section 5.4.1) and let us set, *on a neighborhood of* Γ_i,

$$n = (\nabla_{x'}\varphi_i, -1)/\beta, \qquad \beta = (1 + \|\nabla_{x'}\|^2)^{1/2},$$
$$\tilde{\xi} := \xi - (\xi.n)n, \qquad G(x', x_N) = \varphi_i(x') - x_N.$$

Then, Γ is (locally) invariant for any flow $t \to \tilde{\zeta}(t)$ of the continuous vector field $\tilde{\xi}$ since $G(\tilde{\zeta})' = \beta n.\tilde{\xi} \equiv 0$. Thus $G(\tilde{\zeta}(t)) = G(\tilde{\zeta}(0)) = 0$ if $\tilde{\zeta}(0) \in \Gamma$. Since ξ and $\tilde{\xi}$ coincide on Γ, by uniqueness of the flow of ξ, $\zeta = \tilde{\zeta}$ for initial data on Γ.

Remark. The last point of the proposition applies more generally when $\partial\Omega$ is a manifold of class C^1, without being the boundary of an open set. We may think of the length of a closed curve in $\mathbb{R}^3$ or, even more generally, to the d-Hausdorff measure of a manifold without boundary. On the other hand, this does not include the case where Ω is a segment in $\mathbb{R}^2$ and $E(\cdot)$ is the length of the perturbed curves: in this case, the derivative must take into account the variations of the ends of the segment in the tangential direction (see [142]).

5.9.3 A selected list of first shape derivatives

We have seen several examples of first shape derivatives in the previous sections. Let us recall some of them. We denote by ξ a generic element of $C^{1,\infty}$ and we recall that $\mathcal{E}'(0)\xi = \frac{d}{dt}_{|t=0}\mathcal{E}(t\xi)$.

- $E(\Omega) = |\Omega| = \int_\Omega dx$: $\mathcal{E}'(0)\xi = \int_\Omega \operatorname{div}\xi = -\langle\nabla\chi_\Omega, \xi\rangle_{\mathcal{D}'\times C_0^\infty}$ (see (5.24)). Thus $\mathcal{E}'(0) = -\nabla\chi_\Omega$, which is indeed a distribution of order at most 1. We know that it is of order 0 if and only if Ω is locally of finite perimeter.
- One may more generally take $E(\Omega) = \int_\Omega f$, where $f \in W^{1,1}(\mathbb{R}^N)$. Then according to (5.24),

$$\mathcal{E}'(0)\xi = \int_\Omega \operatorname{div}(f\xi) = \int_\Omega \nabla f.\xi + f\operatorname{div}\xi = \langle\chi_\Omega\nabla f - \nabla(f\chi_\Omega), \xi\rangle_{\mathcal{D}'\times C_0^\infty},$$

  so that $\mathcal{E}'(0) = \chi_\Omega\nabla f - \nabla(f\chi_\Omega)$. When Ω is of class C^1, it may be more interesting to write $\mathcal{E}'(0)\xi = \int_\Gamma f\xi.n$. We see that this depends only on $\xi.n_{|\Gamma}$ and is a continuous linear form with respect to $\xi.n_{|\Gamma}$. It is proved later that this property is very general.
- $E(\Omega) = P(\Omega)$, which is differentiable at least when Ω is of class C^1 and we have (see Corollary 5.4.16)

$$\mathcal{E}'(0)\xi = \int_\Gamma \operatorname{div}_\Gamma \xi = \int_\Gamma H\xi.n,$$

the second version being valid when Ω is of class C^2. In this case, $\mathcal{E}'(0) = H(\mathcal{H}^{N-1})_{|\Gamma} n$, where $(\mathcal{H}^{N-1})_{|\Gamma}$ is the restriction to Γ of the $(N-1)$-Hausdorff measure.

- $E(\Omega) = \int_\Gamma g$, where $g \in W^{2,1}(\mathbb{R}^N)$ and Ω of class C^2. Then (see Proposition 5.4.18),
$$\mathcal{E}'(0)\xi = \int_\Gamma (\xi.n)\left(\frac{\partial g}{\partial n} + Hg\right).$$

- $E(\Omega) = \int_\Omega |\nabla u_\Omega|^2$, where u_Ω is the solution of the Dirichlet problem. Differentiability holds for all measurable Ω and we have (see Corollary 5.3.8)
$$\mathcal{E}'(0)\xi = \int_\Omega \operatorname{div}(|\nabla u|^2 \xi) = \int_\Gamma |\nabla u|^2 \xi.n, \tag{5.103}$$
the second formula being valid if Ω is of class C^1. In any case,
$$\mathcal{E}'(0) = \chi_\Omega \nabla |\nabla u|^2 - \nabla\big(|\nabla u|^2 \chi_\Omega\big). \tag{5.104}$$

- $E(\Omega) = \lambda_k(\Omega) =$ the kth eigenvalue of the Laplacian with Dirichlet boundary conditions. We have (if λ_k is simple, see (5.89))
$$\mathcal{E}'(0)\xi = -\int_\Omega \operatorname{div}(|\nabla u^2| \xi) = -\int_\Gamma |\nabla u|^2 \xi.n.$$
We then obtain the same formula as in (5.104) up to the sign.

- $E(\Omega) = \int_\Omega |\nabla u_\Omega|^2 + u_\Omega^2$, where u_Ω is the solution of the Neumann problem (5.74) on Ω of class C^2. Then,
$$\mathcal{E}'(0)\xi = \int_\Gamma (\xi.n)\left[|\nabla u|^2 + u^2 + 2u\left(\frac{\partial g}{\partial n} - \frac{\partial^2 u}{\partial n^2}\right)\right] + 2\nabla u \nabla_\Gamma(\xi.n).$$
Indeed, a direct differentiation gives
$$\mathcal{E}'(0)\xi = \int_\Omega 2\nabla u.\nabla u' + uu' + \int_\Gamma (\xi.n)\left[|\nabla u|^2 + u^2\right],$$
and, after integration by parts, the first integral may be written $2\int_\Gamma u \frac{\partial u'}{\partial n}$. We then use (5.79) to obtain the formula.

- $E(\Omega) = \lambda_k(\Omega) =$ the kth eigenvalue of the operator $-\Delta + I$ with Neumann boundary conditions. We have (if λ_k is simple, see (5.91))
$$\mathcal{E}'(0)\xi = \int_\Gamma \left[\lambda_k u^2 - |\nabla u|^2\right](\xi.n) = \int_\Omega \operatorname{div}\left[\left(\lambda_k u^2 - |\nabla u|^2\right).\xi\right].$$

- $E(\Omega) = -\frac{1}{2}\int_\Omega |\nabla u_\Omega|^2 + \tau P(\Omega) - \Lambda|\Omega|$: this example corresponds to the minimization problem with volume constraint that we considered in Chapter 4. If the minimum is reached at Ω, then the *Euler–Lagrange*[9] equation says that there exists $\Lambda \in \mathbb{R}$ such that the derivative of this $E(\cdot)$ is equal to zero at Ω. By linear combination of the previous examples, we obtain

$$0 = \int_\Gamma (\xi.n)\left[-\frac{1}{2}|\nabla u|^2 + \tau H - \Lambda\right],$$

or also, since ξ is arbitrary in $C^{1,\infty}$,

$$-\frac{1}{2}|\nabla u|^2 + \tau H = \Lambda \quad \text{on } \Gamma. \tag{5.105}$$

We saw in all these examples that $\mathcal{E}'(0)\xi = l(\xi.n)$, where l is a continuous linear form on $C^1(\Gamma)$. This a general property that we have already more or less pointed out in Proposition 5.9.1. We are going to prove it by a different approach that will also allow us to make explicit the structure of *second derivatives* of shape functionals and, if we want, the structure of even higher-order shape derivatives (see (5.117) below).

5.9.4 The structure theorem and its corollaries

The following result is proved in [248].

Theorem 5.9.2. *Let k be an integer, $k \geq 1$, $E : \mathcal{O}_k \to \mathbb{R}$, $\Omega \in \mathcal{O}_k$, and*

$$\forall\, \theta \in C^{k,\infty}, \quad \mathcal{E}(\theta) := E((I+\theta)(\Omega)).$$

(i) Let us assume $\Omega \in \mathcal{O}_{k+1}$ and $\mathcal{E} : C^{k,\infty} \to \mathbb{R}$ is differentiable at 0. Then, there exists a continuous linear form on $C^k(\Gamma)$ such that for all $\xi \in C^{k,\infty}$,

$$\mathcal{E}'(0)\xi = l_1(\xi_{|\Gamma}.n).$$

(ii) Let us assume $\Omega \in \mathcal{O}_{k+2}$ and $\mathcal{E} : C^{k,\infty} \to \mathbb{R}$ is twice differentiable at 0. Then, there exists a continuous symmetric bilinear form l_2 on $C^k(\Gamma) \times C^k(\Gamma)$ such that for all $\xi, \zeta \in C^{k+1,\infty}$,

$$\begin{cases} \mathcal{E}''(0)(\xi,\zeta) = l_2(\xi.n, \zeta.n) + l_1(Z), \\ \text{with } Z = (D_\Gamma n \zeta_\Gamma).\xi_\Gamma - \nabla_\Gamma(\xi.n).\zeta_\Gamma - \nabla_\Gamma(\zeta.n).\xi_\Gamma. \end{cases} \tag{5.106}$$

[9] Joseph Louis LAGRANGE, 1736–1813, born in Italy from a French native father, worked successively in Turin, Berlin, and Paris. Along with Euler, he is considered to be a founder of the calculus of variations, and we owe to him a lot of the theory of functions, but also physical models of waves and celestial mechanics.

Remark. We immediately see that if Ω is a *critical shape* for the functional E, that is, if $\mathcal{E}'(0) \equiv 0$ and thus $l_1 \equiv 0$, then the second derivative $\mathcal{E}''(0)(\xi, \zeta) = l_2(\xi.n, \zeta.n)$ depends only on the normal components of the traces of ξ, ζ on Γ. We also have the same simpler formula when ξ, ζ are normal to Γ since all terms in l_1 vanish when their tangential components vanish. On the other hand, the second derivative is not equal to zero for tangential displacements ξ, ζ.

Strategy for the computation of the second derivatives. Computing second shape derivatives is not an easy task. It can become much easier when taking into account the structure formula (5.106). Indeed, it is sufficient to identify the forms l_1 and l_2: for this, we may choose to differentiate along favorite paths, well adapted to the functional to be differentiated. One may also choose the displacements ξ, ζ. For instance, in order to know l_2, it is sufficient to compute the quadratic form

$$\xi_n \in C^k(\Gamma) \to \mathcal{E}''(0)(\xi_n n, \xi_n n),$$

which is nothing but the *second derivative of* $t \to \mathcal{E}(t\xi)$ *with* $\xi = \xi_n n$. We may, in particular, use the following corollaries which, on the one hand, show that knowledge of l_1, l_2 provides the value of second derivatives of $t \in [0, T) \to E(\Omega_t)$ *for any path* $t \to \Omega_t$, and which, on the other hand, conversely allow us to deduce the expressions of l_1, l_2 from elementary differentiation with respect to the real variable t and along freely chosen and well-adapted paths.

In the following two corollaries, assumption (ii) of Theorem 5.9.2 is supposed to hold.

Corollary 5.9.3. *Let* $j(t) = \mathcal{E}(\Phi(t) - I)$*, where* $\Phi'(t) = V(t, \Phi(t))$*,* $\Phi(0) = I$ *with* $V(\cdot, \cdot)$ *of class* C^k*. We set* $V = V(0)$*,* $V' = \partial_t V(t, \cdot)_{|t=0}$*. Then,*

$$j'(0) = l_1(V.n), \quad j''(0) = l_2(V.n, V.n) + l_1(Z),$$

with $Z = [V' + DV\,V].n + (D_\Gamma n\, V_\Gamma).V_\Gamma - 2\nabla_\Gamma(V.n).V_\Gamma$.

Corollary 5.9.4. *Let* $\xi \in \mathcal{O}_{k+1}$ *and* $e(t) = \mathcal{E}(t\xi)$*. Then,*

$$e'(0) = l_1(\xi.n), \quad e''(0) = l_2(\xi.n, \xi.n) + l_1(Z),$$

with $Z = (D_\Gamma n \xi_\Gamma).\xi_\Gamma - 2\nabla_\Gamma(\xi.n).\xi_\Gamma$.

Remark. We can also compute, like in [131], [76],

$$d^2E(\Omega, V, W) := \frac{\partial^2}{\partial t \partial s}_{|s,t=0} \left\{E\left(\Phi(t+s) \circ \Phi(t)^{-1} \circ \Psi(t)(\Omega)\right)\right\},$$

where $\Psi'(t) = W(t, \Psi(t))$, $\Psi(0) = I$. Notice that for fixed t, $X(s) := \Phi(t+s) \circ \Phi(t)^{-1}$ satisfies

$$X'(s) = V(t+s, X(s)), \quad X(0) = I.$$

We immediately obtain $d^2E(\Omega, V, W)$ by composition after noticing that it is the cross second derivative of $(s,t) \to \mathcal{E}(\Phi(t+s) \circ \Phi(t)^{-1} \circ \Psi(t) - I)$:

$$\begin{aligned} d^2E(\Omega, V, W) &= \partial_t\{\mathcal{E}'(\Psi(t) - I)V(t, \Psi(t))\} \\ &= \mathcal{E}''(0)(V(0), W(0)) + \mathcal{E}'(0)\big(V'(0) + DV(0)\, W(0)\big), \end{aligned}$$

and it is easy to write it in terms of l_1, l_2. In the *autonomous* case when $V(t)$, $W(t)$ are independent of t, we see that $d^2E(\Omega, V, W) = d^2E(\Omega, W, V)$ if and only if $\mathcal{E}'(0)(DW\, V) = \mathcal{E}'(0)(DV\, W)$ which, according to the structure of $\mathcal{E}'(0)$, holds in particular if $[DW\, V - DV\, W].n = 0$ on Γ. This property was noticed in [76], [131].

5.9.5 The proofs

Let us first prove the corollaries, assuming for the moment Theorem 5.9.2.

Proof of Corollary 5.9.3. By composition, we have $j'(t) = \mathcal{E}'(\Phi(t) - I)V(t, \Phi(t))$ and $j''(0) = \mathcal{E}''(0)(V, V) + \mathcal{E}'(0)\{V' + DV\, V\}$. We then apply (5.106). □

Proof of Corollary 5.9.4. We remember (see (5.23)) that $t \to \Phi(t) = t\xi$ is of the previous form with $V(t,x) = \xi \circ (I + t\xi)^{-1}$ for which $V' = -D\xi\, \xi$ so that $V' + DV\, V \equiv 0$. We deduce the formula of the corollary. □

The proof of Theorem 5.9.2 uses two ideas:

- The first one is that a small regular perturbation $(I + \theta)$ of a regular open set Ω can be represented in a unique way as a normal perturbation of its boundary (i.e., by $\Psi(\theta) \in C^1(\Gamma)$), together with a shift on its boundary. This representation is made here in a functional way so that we may differentiate it as many times as we want.

- The second idea is that, since a shift on the boundary does not perturb Ω, any function $\mathcal{E}(\theta) = E((I + \theta)(\Omega))$ actually depends only on the normal perturbation $\Psi(\theta)$. It may then be written as $\mathcal{E}(\theta) = \mathcal{F}(\Psi(\theta))$. Thus, $\mathcal{E}'(0)\xi = \mathcal{F}'(0)(\Psi'(0)\xi)$. And the only step left now is to compute the geometric term $\Psi'(0)\xi$, which is actually equal to $\xi.n$. Whence $\mathcal{E}'(0)\xi = l_1(\xi.n)$, where l_1 is a continuous linear form on $C^1(\Gamma)$. The structure of $\mathcal{E}''(0)$ is obtained by twice differentiating $\mathcal{E}(\theta) = \mathcal{F}(\Psi(\theta))$.

Let us start with the geometric representation lemma. Let us denote by $G^k(\Gamma, \Gamma)$ the mapping from $C^k(\Gamma, \mathbb{R}^N)$ with values in Γ (i.e., the C^k-shifts on Γ).

Lemma 5.9.5. *Let* $\Omega \in \mathcal{O}_k$.

(i) Then, for all $1 \leq l \leq k$, *there exist an open neighborhood* $\mathcal{U}_k$ *of* 0 *in* $C^{k,\infty}$ *and*

$$\theta \in \mathcal{U}_k \to (\Psi(\theta), G(\theta)) \in C^{k-l}(\Gamma) \times G^{k-l}(\Gamma, \Gamma)$$

of class C^l, *unique, such that* $G(0) = I$, $\Psi(0) = 0$, *and for all* $\theta \in \mathcal{U}_k$,

$$(I + \theta) \circ G(\theta) = I + \Psi(\theta) n \quad \text{on } \Gamma. \tag{5.107}$$

(ii) Moreover, for all ξ, $\zeta \in C^{k,\infty}$, *we have*

$$\begin{cases} \Psi'(0)\xi = \xi.n & \text{for } l \geq 1, \\ \Psi''(0)(\xi, \eta) = (D_\Gamma n \zeta_\Gamma).\xi_\Gamma - \nabla_\Gamma(\xi.n).\zeta_\Gamma - \nabla_\Gamma(\zeta.n).\xi_\Gamma & \text{for } l \geq 2. \end{cases} \tag{5.108}$$

This lemma says that, up to a (unique) shift $G(\theta)$, any regular perturbation of Ω into $(I + \theta)(\Omega)$ is completely represented, and in a unique way, by the value $\Psi(\theta)$ of the normal perturbation of the boundary Γ.

Proof of Lemma 5.9.5. Since Ω is of class C^k, there exist $\delta \in C^k(\mathbb{R}^N, \mathbb{R})$ and ω an open neighborhood of Γ such that

$$\Gamma = \{x \in \omega;\ \delta(x) = 0\}, \quad \Omega \cap \omega = \{x \in \omega,\ \delta(x) < 0\},$$
$$\text{and on } \Gamma : \nabla\delta \neq 0, \quad n = \nabla\delta / |\nabla\delta|.$$

Indication. Starting from the representation in Section 5.4.1 as the graph of the φ_i and from the partition of unity (ξ_i), we set $\delta := \sum_i \xi_i G_i$, where G_i is defined in local coordinates by $G_i(x', x_N) = \varphi_i(x') - x_N$.

We introduce $Z^l := C^{k-l}(\Gamma, \mathbb{R}^N) \times C^{k-l}(\Gamma, \mathbb{R})$ and

$$F : \begin{cases} C^{k,\infty} \times Z^l \to Z^l, \\ (\theta, (G, \Psi)) \to \big((I + \theta) \circ G - I - \Psi n, \delta \circ G\big). \end{cases}$$

We can easily check that F is of class C^l with $F(0, (I, 0)) = (0, 0)$ and that, for $(g, \psi) \in Z^l$,

$$D_{(G,\Psi)}(0, (I, 0))(g, \psi)) = (g - \psi n, |\nabla\delta| n.g).$$

Thus, $D_{(G,\Psi)}(0, (I, 0))$ is an isomorphism from Z^l onto itself since the solution of $D_{(G,\Psi)}(0, (I, 0))(g, \psi) = (\widehat{g}, \widehat{\psi})$ is unique and given by

$$\psi = \widehat{\psi} |\nabla\delta|^{-1} - \hat{g}.n, \qquad g = \hat{g} + \psi n.$$

By the implicit function theorem, there exist an open neighborhood $\mathcal{U}_k$ of 0 in $C^{k,\infty}$ and a function $\theta \in \mathcal{U}_k \to (\Psi(\theta), G(\theta)) \in Z^l$ of class C^l such that $F(\theta, \Psi(\theta), G(\theta)) \equiv 0$, which means

$$(I+\theta)\circ G(\theta) = I + \Psi(\theta)n, \qquad \delta(G(\theta)) = 0 \quad \text{on } \Gamma. \tag{5.109}$$

This implies that (5.107) is satisfied and that $G(\theta)$ takes its values in Γ. It remains to differentiate (5.109). For $\xi \in C^{k,\infty}$, we have on Γ,

$$\begin{cases} \xi(G(\theta)) + (I + D\theta)(G(\theta))G'(\theta)\xi = n\Psi'(\theta)\xi, \\ \delta'(G(\theta))G'(\theta)\xi = 0 = n(G(\theta)).G'(\theta)\xi. \end{cases} \tag{5.110}$$

We deduce that in $\theta = 0$, $\xi + G'(0)\xi = \Psi'(0)\xi n$, $n.G'(0)\xi = 0$, so that, multiplying the first relation by n,

$$\Psi'(0)\xi = \xi.n, \qquad G'(0)\xi = -\xi + (\xi.n)n = -\xi_\Gamma.$$

Again differentiating (5.110) if $l \geq 2$, we obtain for $\zeta \in C^{k,\infty}$ by using the previous line,

$$-D\xi\zeta_\Gamma - D\zeta\xi_\Gamma + G''(0)(\xi,\zeta) = n\Psi''(0)(\xi,\zeta), \qquad 0 = (D_\Gamma n\zeta_\Gamma).\xi_\Gamma + n.G''(0)(\xi,\zeta).$$

Multiplying the first equality by n, we obtain

$$-(D\xi\zeta_\Gamma).n - (D\zeta\xi_\Gamma).n - (D_\Gamma n\zeta_\Gamma).\xi_\Gamma = \Psi''(0)(\xi,\zeta). \tag{5.111}$$

Using that ${}^tD_\Gamma n\, n = 0$ (since $|n| = 1$ on Γ) and (5.100), as well as

$${}^tD_\Gamma \xi n = \nabla_\Gamma(\xi.n) -{}^t D_\Gamma n\xi = \nabla_\Gamma(\xi.n) -{}^t D_\Gamma n\xi_\Gamma,$$

we see that

$$(D\xi\zeta_\Gamma).n = (D_\Gamma\xi\zeta_\Gamma).n = \zeta_\Gamma.({}^tD_\Gamma\xi n) = \zeta_\Gamma.\nabla_\Gamma(\xi.n) - (D_\Gamma n\zeta_\Gamma)\cdot\xi_\Gamma,$$

and similarly by exchanging ξ and ζ. We deduce (5.108). Note that it may also be interesting to keep (5.111). □

Proof of Theorem 5.9.2. Let us show that $\mathcal{E}(\theta) = \mathcal{F}(\Psi(\theta))$ for a well-chosen function $\mathcal{F}$. As a preliminary, let us remark that for $\theta_1, \theta_2 \in C^{1,\infty}$ close to 0,

$$[\Gamma_{\theta_1} = \Gamma_{\theta_2}] \Rightarrow [\Omega_{\theta_1} = \Omega_{\theta_2}]. \tag{5.112}$$

Indeed, if $\Phi := (I+\theta_2)^{-1}\circ(I+\theta_1)$, we then have $\Phi(\Gamma) = \Gamma$ and consequently $\Phi(\Omega) = \Omega$ by an elementary connectivity argument. In the following, we assume that

Ω is of class C^{k+1}. Let us denote by P a continuous linear extension from $C^k(\Gamma, \mathbb{R}^N)$ to $C^k(\mathbb{R}^N, \mathbb{R}^N)$ (e.g., we may use formula (5.44)). Let us set

$$\forall \psi \in C^k(\Gamma, \mathbb{R}^N), \quad \mathcal{F}(\psi) := \mathcal{E}(P(\psi n)) = E\big((I + P(\psi n))(\Omega)\big).$$

By (5.112), this definition does not depend on the choice of P and, by composition, if $\mathcal{E}$ is once or twice differentiable at 0 on $C^{k,\infty}$, so is $\mathcal{F}$.

For part (i) of the theorem, we apply Lemma 5.9.5 with k replaced by $k+1$ and $l = 1$: we then obtain

$$\theta \in \mathcal{U}_{k+1} \to (\Psi(\theta), G(\theta)) \in C^k(\Gamma) \times G^k(\Gamma, \Gamma) \tag{5.113}$$

is of class C^1. The main point of the proof is that

$$\forall \theta \in \mathcal{U}_{k+1}, \quad \mathcal{E}(\theta) = \mathcal{F}(\Psi(\theta)),$$

which is true since, by (5.112) and relation (5.107), we have

$$\Gamma_\theta = (I + \theta)(\Gamma) = (I + \theta) \circ G(\theta)(\Gamma) = (I + \Psi(\theta)n)(\Gamma).$$

We deduce by composition and with the help of (5.108),

$$\forall \xi \in C^{k+1,\infty}, \quad \mathcal{E}'(0)\xi = \mathcal{F}'(0)(\Psi(0)\xi) = \mathcal{F}'(0)(\xi.n),$$

which proves (i) by setting $l_1 := \mathcal{F}'(0)$ and by density of $C^{k+1,\infty}$ in $C^{k,\infty}$.

For part (ii) of the theorem, we apply Lemma 5.9.5 with k replaced by $k+2$ and $l = 2$. Up to reducing the involved open sets, we may assume that $\mathcal{U}_{k+2} = \mathcal{U}_{k+1}$ and the mapping (5.113) is then of class C^2. After differentiating, we then have

$$\forall \xi, \zeta \in C^{k+2,\infty}, \quad \mathcal{E}''(0)(\xi, \zeta) = \mathcal{F}''(0)(\Psi'(0)\xi, \Psi'(0)\zeta) + \mathcal{F}'(0)(\Psi''(0)(\xi, \zeta)),$$

which, after setting $l_2 := \mathcal{F}''(0)$, gives (ii) by application of (5.108) and by density of $C^{k+2,\infty}$ in $C^{k+1,\infty}$. □

Remark (On the regularity assumptions in Theorem 5.9.2). The above proof is elementary, the only difficulty coming from following the regularity of the various mappings involved. It is not really possible to improve the regularity assumptions: we see that, for $k = 1$, the assumption "Ω is of class C^1" would not be sufficient for statement (i) of Theorem 5.9.2 since l_1 is only continuous on C^1 and n would then be only continuous.

We may better understand the difficulty by looking at the example $E(\Omega) = P(\Omega)$, where $\mathcal{E}$ is of class C^∞ on $C^{1,\infty}$ (see Corollary 5.4.16): in this case, $\mathcal{E}'(0)\xi = \int_\Gamma \operatorname{div}_\Gamma \xi$. When Ω is of class C^2, we may also write $\mathcal{E}'(0)\xi = \int_\Gamma H\xi.n = l_1(\xi.n)$. Saying that l_1

may be extended to $C^0(\Gamma)$ is equivalent to saying that the mean curvature is a bounded measure on Γ, which is not the case if Ω is only C^1. Similarly, one generally needs the assumption that Ω is of class C^3 in (ii) when $k = 1$. We may be convinced of this by looking at the example $E(\Omega) = \int_\Omega |\nabla u_\Omega|^2$, which will be computed later for the Dirichlet problem. Indeed, l_2 involves the trace on Γ of $D^2 u$ and it is not defined if Ω is only of class C^2. One may however show that, if l_1 can be extended as a continuous function on C^{k-1} (as is often the case), then $\Omega \in \mathcal{O}_{k+1}$ is sufficient and the identity (5.106) extends to all $\xi, \zeta \in C^{k,\infty}$. This is proved in [248].

5.9.6 A selected list of second shape derivatives

Let us describe l_2 in each of the following examples where we assume that Ω is of class C^2 or C^3:

- $E(\Omega) = |\Omega|$: $l_2(\varphi, \hat\varphi) = \int_\Gamma H\varphi\hat\varphi$.
- $E(\Omega) = \int_\Omega f$ with $f \in C^1(\mathbb{R}^N)$. Then

$$l_2(\varphi, \hat\varphi) = \int_\Gamma \varphi\hat\varphi \left[Hf + \frac{\partial f}{\partial n}\right].$$

- $E(\Omega) = P(\Omega)$:

$$l_2(\varphi, \hat\varphi) = \int_\Gamma \nabla_\Gamma\varphi . \nabla_\Gamma\hat\varphi + \varphi\hat\varphi \left[H^2 - \operatorname{tr}({}^t D_\Gamma n D_\Gamma n)\right].$$

  In the case $N = 2$, one simply has

$$l_2(\varphi, \hat\varphi) = \int_\Gamma \nabla_\Gamma\varphi . \nabla_\Gamma\hat\varphi.$$

- $E(\Omega) = \int_\Omega |\nabla u_\Omega|^2$ where $u_\Omega = u$ is the solution of the Dirichlet problem with Ω of class C^3:

$$l_2(\varphi, \hat\varphi) = \int_\Gamma -2V(\varphi)\frac{\partial V(\hat\varphi)}{\partial n} - \varphi\hat\varphi\left[2f\frac{\partial u}{\partial n} + H\left(\frac{\partial u}{\partial n}\right)^2\right],$$

  where $V(\varphi)$ is the solution of

$$\Delta V(\varphi) = 0 \quad \text{in } \Omega, \qquad V(\varphi) = -\varphi\frac{\partial u}{\partial n} \quad \text{on } \Gamma. \tag{5.114}$$

  Note that $\int_\Gamma V(\varphi)\frac{\partial V(\hat\varphi)}{\partial n} = \int_\Omega \nabla V(\varphi)\nabla V(\hat\varphi)$ is indeed symmetric.

- $E(\Omega) = \lambda_k(\Omega)$ = the kth eigenvalue of the Laplacian, assumed to be simple, and with Ω of class C^3:

$$l_2(\varphi, \hat{\varphi}) = \int_\Gamma 2V(\varphi)\frac{\partial V(\hat{\varphi})}{\partial n} + \varphi\hat{\varphi}H\left(\frac{\partial u}{\partial n}\right)^2,$$

where $V(\varphi)$ is the solution of

$$-\Delta V(\varphi) = \lambda_k V(\varphi) - u\int_\Gamma \left(\frac{\partial u}{\partial n}\right)^2 \varphi \quad \text{in } \Omega,$$
$$V(\varphi) = -\varphi\frac{\partial u}{\partial n} \quad \text{on } \Gamma, \qquad \int_\Omega uV(\varphi) = 0.$$

- $E(\Omega) = -\frac{1}{2}\int_\Omega |\nabla u_\Omega|^2 + \tau P(\Omega) - \Lambda|\Omega|$. We looked in (5.105) at the critical shapes for this functional when minimized with volume constraint (see Chapter 4). It is interesting to write down the second derivative of E around a critical shape in order to, for instance, study its positivity on the hyperplane tangent to the constraint. We obtain with

$$\beta^2 = 2(\tau H - \Lambda) = \left(\frac{\partial u}{\partial n}\right)^2, \quad V = V(\xi_n) \text{ is the solution of (5.114)}, \quad \xi_n = \xi \cdot n,$$
$$\int_\Gamma V\frac{\partial V}{\partial n} + \xi_n^2\left[f\beta + 2H(\tau H - \Lambda) - \tau\,\mathrm{tr}\{{}^t D_\Gamma n D_\Gamma n\}\right] + \tau|\nabla_\Gamma \xi_n|^2.$$

Note that $\int_\Gamma V\frac{\partial V}{\partial n} = \int_\Omega |\nabla V|^2$ does contribute to the positivity of this quadratic form. It may also be written $\langle \xi_n, \mathcal{A}\xi_n\rangle$ with $\mathcal{A} = \beta\mathcal{D}(\beta\cdot) - \alpha I - \tau\Delta_\tau$, where $\alpha \in C^0(\Gamma, \mathbb{R})$ and $\mathcal{D}$ is the pseudo-differential operator, sometimes called Steklov–Poincaré (or Dirichlet-to-Neumann), which to ξ_n associates the trace of the normal derivative of the harmonic extension of ξ_n to Ω . This is typical of the structure of second shape derivatives. We refer to [121], [120], [190], [189], [147], [134] for the study of the positivity of such quadratic forms.

Details of the computations of the selected second derivatives. We apply the principle that we mentioned above: namely, computing the second derivative at 0 of $t \to e(t) = \mathcal{E}(t\xi)$, where $\xi \in C^{1,\infty}$ with $\xi = n\xi_n$, $\xi_n \in C^1(\Gamma, \mathbb{R})$. Thus, ξ is normal to Γ and $\xi_\Gamma = 0$. The quadratic form $\mathcal{E}(t\xi) = l_2(\xi_n, \xi_n)$ allows us to identify the bilinear form l_2.

We refer to the previous analysis of each example for the notation and various properties involved. Let us recall the formula deduced from (5.21):

$$\frac{d}{dt}\int_{\Omega_t} f(t) = \int_{\Omega_t} \frac{\partial f}{\partial t}(t) + \mathrm{div}[f(t)\xi \circ \Pi(t)] \quad \text{with } \Pi(t) = (I + t\xi)^{-1}. \tag{5.115}$$

By a new application of this formula, we get after integration by parts (let us recall that $\Pi'(0) = -\xi$ and $(\xi \circ \Pi(t))'(0) = -D\xi\,\xi$),

$$\frac{d}{dt}_{|t=0} \int_{\Omega_t} \operatorname{div}[f(t)\xi \circ \Pi(t)] = \int_\Gamma [f'(0) + \operatorname{div}(f(0)\xi)]\xi_n - f(0)(D\xi\,\xi).n.$$

Using that $\xi = n\xi_n$ and $\operatorname{div}\xi - (D\xi n).n = \operatorname{div}_\Gamma \xi = H\xi_n$, we have

$$\frac{d}{dt}_{|t=0} \int_{\Omega_t} \operatorname{div}[f(t)\xi \circ \Pi(t)] = \int_\Gamma \xi_n f'(0) + \xi_n^2 \left[Hf(0) + \frac{\partial f(0)}{\partial n} \right]. \tag{5.116}$$

For $E(\Omega) = \int_\Omega f$, we thus obtain that $e''(0) = \int_\Gamma \xi_n^2 (Hf + \frac{\partial f}{\partial n})$. The associated bilinear form l_2 may then be deduced.

For $E(\Omega) = P(\Omega)$, we choose $f(t) = \operatorname{div} n_t$, where n_t is unitary. Thus, $\int_{\Omega_t} \frac{\partial f}{\partial t} = \int_{\Gamma_t} n_t.\frac{\partial n_t}{\partial t} = 0$ and $e''(0)$ is given by (5.116). According to the computations in the proof of Proposition 5.4.14 (see (5.71)), with $n = \nabla d$, we have on Γ,

$$\frac{\partial n_t}{\partial t}_{|t=0} = -\nabla_\Gamma \xi_n, \qquad Dn = D_\Gamma n, \qquad \frac{\partial(\operatorname{div} n)}{\partial n} = -\operatorname{tr}({}^t D_\Gamma n D_\Gamma n).$$

For this last identity, we use, for instance,

$$0 = \Delta(|\nabla d|^2)/2 = \nabla(\Delta d).\nabla d + \operatorname{tr}\left[(D^2 d)^2\right] = \nabla(\operatorname{div} n).n + \operatorname{tr}({}^t D_\Gamma n D_\Gamma n).$$

Since $\int_\Gamma -\xi_n \operatorname{div}(\nabla_\Gamma \xi_n) = \int_\Gamma -\xi_n \operatorname{div}_\Gamma(\nabla_\Gamma \xi_n) = \int_\Gamma |\nabla_\Gamma \xi_n|^2$, we deduce

$$e''(0) = \int_\Gamma |\nabla_\Gamma \xi_n|^2 + \xi_n^2 \left[H^2 - \operatorname{tr}\left({}^t D_\Gamma n D_\Gamma n\right)\right].$$

Whence the bilinear form associated to l_2. In dimension 2, we notice that

$$Dn\,n = 0 \Rightarrow 0 = 2\det Dn = (\Delta d)^2 - \operatorname{tr}\left[(D^2 d)^2\right] = H^2 - \operatorname{tr}\left({}^t D_\Gamma n D_\Gamma n\right).$$

For $E(\Omega) = \int_\Omega |\nabla u_\Omega|^2$, we choose $f(t) = |\nabla u(t)|^2$, where we denote $u(t) := u_{\Omega_t}$. First we notice that $\int_{\Omega_t} \nabla u(t) \nabla u'(t) = -\int_{\Omega_t} u(t)\Delta u'(t) = 0$. Again $e''(0)$ is then given by (5.116), which implies, since $u'(0) = V(\xi_n)$,

$$e''(0) = \int_\Gamma 2\xi_n \nabla u \nabla V(\xi_n) + \xi_n^2 \left[H|\nabla u|^2 + \nabla|\nabla u|^2.n\right].$$

We end by using the definition of $V(\xi_n)$ and

$$\nabla u = n\left(\frac{\partial u}{\partial n}\right), \qquad \nabla|\nabla u|^2.n = 2\frac{\partial u}{\partial n}\frac{\partial^2 u}{\partial n^2} = -2\left(H\frac{\partial u}{\partial n} + f\right)\frac{\partial u}{\partial n},$$

since on Γ, $-f = \Delta u = \Delta_\Gamma u + H\frac{\partial u}{\partial n} + \frac{\partial^2 u}{\partial n^2}$ and $\Delta_\Gamma u = 0$.

Finally, for $E(\Omega) = \lambda_k(\Omega)$, we choose $f(t) = |\nabla u(t)|^2$, where $u(t)$ is the corresponding eigenfunction on Ω_t. We notice, by using (5.88), that $\int_{\Omega_t} \nabla u(t)\nabla u'(t) = -\int_{\Omega_t} u(t)\Delta u'(t) = \lambda_k'(t)$. Thus, applying formula (5.115) tells us that $\lambda''(t)$ is given by the opposite of expression (5.116), namely,

$$e''(0) = -\int_\Gamma \xi_n 2\nabla u\nabla u' + \xi_n^2\left[H|\nabla u|^2 + \nabla|\nabla u|^2.n\right].$$

We deduce the expression of l_2 as in the previous example (here $0 = H\frac{\partial u}{\partial n} + \frac{\partial^2 u}{\partial n^2}$ on Γ).

The last example may be obtained by a linear combination of the previous ones.

5.9.7 Three final remarks

Remark (On the case when Ω is not regular). The structure Theorem 5.9.2 requires some regularity on Ω since computations are made at the boundary. However, we have seen several examples of shape functionals that are C^∞ around a set Ω that is only measurable. It is the case for

$$\theta \in C^{1,\infty} \to |\Omega_\theta|, \quad \int_{\Omega_\theta} |\nabla u_{\Omega_\theta}|^2, \quad \lambda_k(\Omega_\theta).$$

We may ask how to write the higher-order derivatives. Let us do it for $\mathcal{E}(\theta) = |\Omega_\theta|$. By (5.24),

$$\mathcal{E}'(\theta)\xi = (\mathcal{E}(\theta + t\xi))'_{t=0} = \int_{\Omega_\theta} \operatorname{div}(\xi \circ (I+\theta)^{-1}).$$

But, since the function under the integral is only continuous, it is difficult to continue. Thus, it is better to come back to an integral on Ω, namely $\mathcal{E}(\theta) = \int_\Omega \det(I + D\theta)$. Then,

$$\mathcal{E}'(\theta)\xi = \int_\Omega \det(I + D\theta)\operatorname{tr}((I + D\theta)^{-1}D\xi).$$

Next, we use $\mathcal{E}''(\theta)(\xi,\zeta) = (\mathcal{E}'(\theta + s\zeta)\xi)'_{s=0}$ to obtain, after setting $U_\theta := (I + D\theta)^{-1}$,

$$\mathcal{E}''(\theta)(\xi,\zeta) = \int_\Omega \det(I + D\theta)\left\{\operatorname{tr}(U_\theta D\zeta)\operatorname{tr}(U_\theta D\xi) - \operatorname{tr}(U_\theta D\zeta U_\theta D\xi)\right\}.$$

It is not difficult to continue, but obviously a little long. Let us at least write the third derivative at 0:

$$\begin{aligned}\mathcal{E}^{(3)}(0)(\xi,\zeta,\eta) = \int_\Omega &(\operatorname{tr} D\xi)(\operatorname{tr} D\zeta)(\operatorname{tr} D\eta) - (\operatorname{tr} D\xi)\operatorname{tr}(D\zeta D\eta)\\ &- (\operatorname{tr} D\zeta)\operatorname{tr}(D\eta D\xi) - (\operatorname{tr} D\eta)\operatorname{tr}(D\xi D\zeta)\\ &+ \operatorname{tr}(D\eta D\zeta D\xi + D\zeta D\eta D\xi).\end{aligned}$$

We also refer to [222] and [221] for more results on shape derivatives around irregular domains.

Remark (On higher-order derivatives). When Ω is regular, it is possible to write the higher-order derivatives on the boundary. The structure Theorem 5.9.2 can actually be extended to these derivatives. Indeed, the basic identity is that $\mathcal{E}(\theta) = \mathcal{F}(\Psi(\theta))$ and we can differentiate it as many times as we want. For instance, for the third derivative, we obtain a new trilinear form $l_3 = \mathcal{F}^{(3)}(0)$. Then, it is sufficient to identify $\Psi^{(3)}(0)$, which we can do by differentiating (5.110) with respect to all θ and by differentiating one more time at 0. The formula is a little long, but easy to obtain. We show, for instance, that if $\xi_\Gamma = \zeta_\Gamma = \eta_\Gamma = 0$, then the second and third derivatives of Ψ in these directions vanish. Thus

$$\mathcal{E}^{(3)}(0)(\xi, \zeta, \eta) = -\mathcal{F}^{(3)}(0)(\xi.n, \zeta.n, \eta.n). \tag{5.117}$$

This may lead to the computation of l_3 as for l_2. The complete expression of the third derivative may be deduced. We refer also to [159], [161], [160] for more results on higher-order derivatives.

Remark (On the functionals E with values in a Banach space). Theorem 5.9.2 remains valid for functionals E with values in a Banach space X and not only in $\mathbb{R}$. Indeed, by application of the geometric Lemma 5.9.5, we may still write $E\big((I+\theta)(\Omega)\big) = \mathcal{F}(\Psi(\theta))$, where $\mathcal{F}$ is defined in the same way, but with values in X. The same result may then be deduced with linear applications l_1, l_2 with values in X.

We may verify it for the example of the derivative of $\theta \in C^{1,\infty} \to u_\theta \in L^2(\mathbb{R}^N)$ the solution of the Dirichlet problem on Ω_θ. If Ω is of class C^2, one knows that $u' = u'(0)\xi$ is a solution of

$$-\Delta u' = 0 \text{ in } \Omega, \qquad u' = -(\xi.n)\frac{\partial u}{\partial n} \text{ on } \Gamma,$$

and $l_1 : \xi.n \in C^{1,\infty} \to u' \in L^2(\mathbb{R}^N)$ is indeed a continuous linear mapping.

5.9.8 Conclusion

We did not consider here all questions concerning differentiation with respect to domains. Let us mention some more:

- The case of variational inequalities or more generally of boundary value problems with the constraint of living in some given convex set. This generally leads to nondifferentiable problems in the classical sense, but for which a directional or conic differentiability may be defined.
- The differentiation of evolution problems set on variable domains.

- Differentiating through level sets.
- The use of shape derivatives and of their discretization for numerical computations of optimal shapes.
- And all this work on the *topological derivative*: its goal is to analyze what happens when deleting a small hole of size ϵ in a problem set on given domain Ω. This requires describing the behavior of this perturbation with respect to ϵ. Obviously, this is not in the scope of the present framework since one can no longer represent the perturbed problem in the form $(I+\theta)(\Omega)$, where $(I+\theta)$ is a homeomorphism. However, it is important for shape optimization to decide whether this kind of excision may improve the functional. This requires analyzing infinitesimal variations of the characteristic function as in [90]. See the book [247] and the articles [281], [89], [149] for an introduction to the topological derivative.

5.10 Exercises

Exercise 5.1. Let $\Omega = (0,1)$ and $f \equiv 1$. Compute, for all $\theta \in W^{1,\infty}(\mathbb{R})$ with $\theta(0) = 0$, the solution u_θ on $\Omega_\theta = (I+\theta)(\Omega)$ of

$$-u'' = f \quad \text{on } \Omega_\theta \text{ with } u(0) = 0, \qquad u((I+\theta)(1)) = 0.$$

Write $v_\theta = u_\theta \circ (I+\theta)$ explicitly. Compute the successive derivatives of $t \to u_{\theta+t\xi}$, where $\xi \in W^{1,\infty}(\mathbb{R})$ with $\xi(0) = 0$.

Perform the same computation with $f = \chi_{(0,1/2)}$.

Exercise 5.2. Prove that $\theta \in W^{1,\infty} \to \Psi(\theta) = (I+\theta)^{-1} \in L^\infty$ is differentiable in a neighborhood of 0 and that

$$\forall \xi \in W^{1,\infty}, \quad \Psi'(\theta)\xi = -D\Psi(\theta)[\xi \circ \Psi(\theta)],$$

where $D\Psi(\theta) = (I + D\theta)^{-1} \circ (I+\theta)^{-1} \in L^\infty(\mathbb{R}^N, \mathcal{M}_N)$. Check that Ψ is of class C^1 from $W^{1,\infty} \cap C^1$ into L^∞, but not from $W^{1,\infty}$ into L^∞.

Let $m \geq 1$ and let $\theta \in C^{m,\infty} \to v_\theta \in H^m(\Omega)$ be of class C^1 on a neighborhood of 0 and $u_\theta = v_\theta \circ (I+\theta)^{-1}$. Check that, for all $\xi \in C^{m,\infty}$,

$$v'(\theta)\xi = (u'(\theta)\xi) \circ (I+\theta) + \nabla u_\theta \circ (I+\theta).\xi,$$
$$\nabla u_\theta = ({}^t(I + D\theta)^{-1} \nabla v_\theta) \circ (I+\theta)^{-1}.$$

Deduce the regularity of $\theta \to u_\theta$.

Exercise 5.3. Let us choose the same assumptions and notation as in Theorem 5.3.1. Assume that Ω is quasi-open. Write an equation satisfied by u'.

Exercise 5.4. Let Φ be a Lipschitz continuous mapping from $\mathbb{R}^N$ into itself and let $K \subset \mathbb{R}^N$. Show that $\operatorname{cap}(K) = 0$ implies $\operatorname{cap}(\Phi(K)) = 0$. Prove that, if Ω is quasi-open, then $\Phi(\Omega)$ is quasi-open.

We assume that Φ is a bi-Lipschitz homeomorphism and Ω is measurable. Show that $v \in H_0^1(\Phi(\Omega))$ is equivalent to $v \circ \Phi \in H_0^1(\Omega)$.

Exercise 5.5. (1) Let $(x_n) \subset (0,1)$ be a decreasing sequence converging to 0. We denote by δ_{x_n} the Dirac mass at x_n. Prove that the series $G = \sum_{n\geq 1}(-1)^n \delta_{x_n}$ converges in $H^{-1}(0,1)$. Check that $G = \frac{d}{dx}(\chi_\omega)$, where ω is an open subset of $(0,1)$ to be described.

(2) Let $\theta \in W^{1,\infty}(\mathbb{R},\mathbb{R})$ with θ' continuous on $\mathbb{R}\backslash\{0\}$ and, for all $p \geq 1$,

$$\theta'(1/2p) = t, \quad \theta'(1/(2p+1)) = 0 \quad (t > 0 \text{ given, small}).$$

Prove that the series $\sum_{n\geq 1}(-1)^n \delta_{1/n}/(1 + \theta'(1/n))$ does not converge in $H^{-1}(0,1)$. Compare it with $G \circ (I + \theta)$ when we choose $x_n = (I + \theta)(1/n)$ for all $n \geq 1$.

Exercise 5.6. Let $\Omega := (0,1)\times(0,1)$ be the unit square in $\mathbb{R}^2$. Prove that $\theta \in C^{1,\infty} \to p(\theta) = P(\Omega_\theta)$ is of class C^∞ and that

$$p'(0).\xi = \int_\Gamma \operatorname{div}\xi - (D\xi.n).n \quad \forall \xi \in C^{1,\infty}.$$

Prove that differentiability holds even on $W^{1,\infty}$. Check that, if $\xi = (\xi_1, \xi_2)$ is supported in a neighborhood of 0, we have $p'(0).\xi = -(\xi_1(0,0) + \xi_2(0,0))$.

Exercise 5.7. Prove by an explicit one-dimensional computation that one cannot drop the assumption $v_0 \in H^2_{\mathrm{loc}}(\mathbb{R}^N)$ to obtain the differentiability of the functional J defined in (5.27) (Hint: check that if v_0 is only in H^1, J is not necessarily differentiable.)

Exercise 5.8. Compute the mean curvature at each point of an ellipsoid in $\mathbb{R}^3$.

Exercise 5.9. State a differentiation result for the functional J defined by

$$J(\Omega) = \int_\Omega |\nabla u_\Omega|^2\,dx + \int_\Omega (u_\Omega - u_0)^2\,dx,$$

where u_Ω is the solution of the boundary value problem

$$\begin{cases} -\Delta u = f & \text{in } \Omega,\\ \frac{\partial u}{\partial n} + \alpha u = g & \text{on } \partial\Omega. \end{cases}$$

Here u_0, α are given, with $\alpha \in (0,\infty)$ and $u_0 \in L^2_{\mathrm{loc}}(\mathbb{R}^N)$.

Exercise 5.10. Find the adjoint problem in Exercise 5.9 and deduce a new expression for the derivative of J.

Exercise 5.11. We choose the assumptions of Corollary 5.2.3. The goal is to show that, if there exist two extensions $t \longrightarrow \tilde{f}(t), \overline{f}(t) \in L^1(\mathbb{R}^N)$ of f that are differentiable at 0, then

$$\frac{\partial \tilde{f}}{\partial t}(0, x) = \frac{\partial \overline{f}}{\partial t}(0, x) \quad \forall\, x \in \Omega. \tag{5.118}$$

(1) Prove that, for all $\varphi \in C_0^\infty(\mathbb{R}^N)$, $\int_{\Omega_t} \varphi$ converges to $\int_\Omega \varphi$ as $t \longrightarrow 0$. Deduce that χ_{Ω_t} converges a.e. to χ_Ω and that $|\Omega_t \cap \Omega| \rightarrow |\Omega|$ when $t \longrightarrow 0$.

(2) Check that there exists a sequence (t_n) decreasing to 0 such that $t_n^{-1}[\tilde{f}(t_n) - \tilde{f}(0)]$ (resp. $t_n^{-1}[\overline{f}(t_n) - \overline{f}(0)]$) converges a.e. to $\frac{\partial \tilde{f}}{\partial t}(0)$ (resp. $\frac{\partial \overline{f}}{\partial t}(0)$) and verifying $|\Omega \backslash \Omega_{t_n}| \leq 2^{-n}$. We introduce

$$E_n = \cap_{k \geq n} \big(\Omega_{t_k} \cap \Omega\big).$$

Show that $|\Omega \backslash E_n| \longrightarrow 0$ as $n \longrightarrow +\infty$. We set $E = \cup_{n \geq 1} E_n$.

(3) Prove that a.e. $z \in E$, there exists $n_z > 0$ such that

$$f(t_k, z) = \tilde{f}(t_k, z) = \overline{f}(t_k, z) \quad \forall\, k \geq n_z.$$

Deduce (5.118).

Chapter 6

Geometric properties of the optimum

This chapter is devoted to the study of the possible geometric properties of any solution of a shape optimization problem (or even in some cases of any critical shape). In the sequel, we assume that we have proved existence (and sometimes uniqueness) of a solution Ω^* for the problem. Now, we want to determine whether this solution is *symmetric*, *star shaped*, or *convex*, or even simply *connected*. These questions are actually difficult and the favorable situations where we are able to give a positive answer are not so frequent.

In this chapter, we will mainly use two different methods. The first consists in working directly on the shape functional to prove, for example, that the symmetric sets are "better" than the nonsymmetric (see below). The second consists in working on the equations satisfied by the optimum, for example Euler–Lagrange equations obtained thanks to the derivative with respect to the domain.

6.1 Symmetry

6.1.1 Introduction

In this section, four methods are given to prove that the solution of a shape optimization problem has some symmetry. Of course, it is necessary that the data themselves possess this symmetry, at least intrinsically. Sometimes this is not enough, and the minimizer may not have the conjectured symmetry. We will illustrate this with some examples and give methods to prove nonsymmetry results of this kind.

The two first methods we describe here are inspired by similar questions in the classical field of calculus of variations. Indeed, let us consider a function u that solves a minimization problem of the form

$$J(u) = \min_{v \in V} J(v).$$

If we want to prove that this function u has some symmetry property, two strategies can be considered:

- We introduce a *symmetrized* or *rearranged* function u^* of u that has the desired symmetry property. Then we prove that $J(u^*) \le J(u)$. This already proves that u^* is also solution. If we want to prove that all solutions are symmetric, we must use either a uniqueness result or prove that $J(u^*) < J(u)$ if $u^* \neq u$.

- We consider the optimality conditions (or Euler–Lagrange equations) of the problem. Then we work on this system of (partial) differential equations, often by using the maximum principle, to prove that the solution u has some symmetry.

In shape optimization problems, these two ideas can be adapted. In the next section, we introduce two kinds of rearrangements: Schwarz symmetrization (or decreasing radial rearrangement) and the Steiner symmetrization. The latter is especially useful in shape optimization problems since, among all possible rearrangements (see [204], [212], [30], [239] for a comprehensive study of the rearrangements) it provides a proof of the symmetry with respect to a given hyperplane. Schwarz symmetrization is useful when we expect the solution of the problem to be a ball. The ideas developed in this section are mainly due to G. Pólya and G. Szegő; cf. [260] and [258].

In the third section, we investigate the overdetermined boundary value problem obtained when we make the optimality conditions explicit thanks to differentiation with respect to the domain. To prove symmetry for such problems, one very efficient tool is the method of moving planes introduced by Aleksandrov[1] ([10]) and popularized by Gidas–Ni–Nirenberg ([150]) and, in this kind of situation, by J. Serrin ([275]).

In the fourth section, another method, based on a reflection process together with a variational principle, is presented. This method is less known than the previous ones, but can be very efficient too, particularly in some situations not covered by the other methods.

In the fifth section, we present an original method. It consists in introducing a new shape optimization problem whose minima are precisely the solutions of the overdetermined problem obtained in Section 6.1.3. Then, by making the optimality conditions for this new problem explicit, we are able to get enough information to characterize the minima.

Finally, in Section 6.1.6, we show how perturbation techniques can be helpful to prove some nonsymmetry results. This method is presented in the context of Newton's aerodynamic problem. We show that the radial minimizer is not the global minimizer of the problem.

In the present section let us concentrate on the model problem (4.30). In Theorem 4.5.2, we proved existence of a minimizer. Let us recall that for any $f \in L^2_{\mathrm{loc}}(\mathbb{R}^N)$ we define

$$J(\Omega) = \min\{J(\omega);\ \omega \text{ a quasi-open set, } |\omega| = m\}, \tag{6.1}$$

where $J(\omega) = j(u_\omega)$ with $j(v) = \int_{\mathbb{R}^N} \frac{1}{2}|\nabla v|^2 - fv$ and $u_\omega = u_\omega^f$ is the solution of the

[1]Aleksandr Danilovic ALEKSANDROV, 1912–1999, Russian, was one of the great geometers of the 20th century.

Dirichlet problem on ω with a right-hand side f:

$$\begin{cases} -\Delta u = f & \text{in } \Omega, \\ u = 0 & \text{on } \partial\Omega. \end{cases} \tag{6.2}$$

Let us also recall that $j(u_\omega) = -\frac{1}{2}\int_\omega |\nabla u_\omega|^2$.

We assume here that f is symmetric with respect to some hyperplane. As is shown below with some examples, this does not necessarily imply that any optimal shape $\widetilde{\Omega}$ is symmetric. It is even possible that no symmetric optimal shape exists. But if we assume that f is *Steiner symmetric*, it can be proved that at least one optimal shape has to be symmetric. We will provide four different methods to prove this.

6.1.2 The method of Steiner symmetrization

We present here the method of Steiner symmetrization. For further results and properties of different kinds of symmetrization and rearrangement, and for proofs, we refer to [30], [204], [239], [212].

Let us begin with definitions of the Steiner symmetrization of sets and functions. We recall that $|M|$ denotes the Lebesgue measure of a (measurable) set M. We will not specify the underlying dimension if no misunderstanding is possible (we consider here sets of dimension 1 and of dimension N as well).

Let $N \geq 2$ and $\Omega \subset \mathbb{R}^N$ be a measurable set. We denote by Ω' the projection of Ω on $\mathbb{R}^{N-1}$, namely,

$$\Omega' := \left\{x' \in \mathbb{R}^{N-1} \text{ such that there exists } x_N \text{ with } (x', x_N) \in \Omega\right\},$$

and for $x' \in \mathbb{R}^{N-1}$, we denote by $\Omega(x')$ the intersection of Ω with $\{x'\} \times \mathbb{R}$:

$$\Omega(x') := \{x_N \in \mathbb{R} \text{ such that } (x', x_N) \in \Omega\}, \quad x' \in \Omega'.$$

Observe that if Ω is open, the sets $\Omega(x')$ are also open and $x' \to |\Omega(x')|$ is lower semicontinuous.

Definition 6.1.1. Let $\Omega \subset \mathbb{R}^N$ be a measurable set. Then the set

$$\Omega^\star := \left\{x = (x', x_N) \text{ such that } -\tfrac{1}{2}|\Omega(x')| < x_N < \tfrac{1}{2}|\Omega(x')|,\ x' \in \Omega'\right\}$$

is the Steiner symmetrization of Ω with respect to the hyperplane $x_N = 0$.

It is easy to check that $\Omega^\star$ is open if Ω is an open set.

We point out that, even if Ω is itself symmetric with respect to $x_N = 0$, it does not necessarily coincide with its Steiner symmetrization. Indeed, the latter has to be,

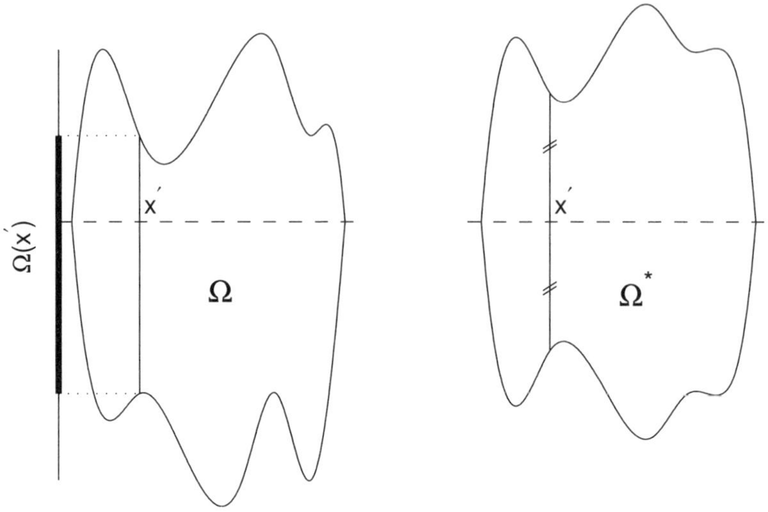

Figure 6.1. The Steiner symmetrization: on the left is the open set Ω, on the right is its symmetrization Ω^*.

by construction, convex in the direction x_N. More precisely, we have

$$\Omega = \Omega^\star \iff \begin{cases} \Omega \text{ is symmetric with respect to } x_N = 0, \\ \Omega \text{ is convex in the direction } x_N. \end{cases}$$

Let us now consider a nonnegative measurable function u, defined on Ω, with the following property:

$$\begin{cases} \text{for any } c > 0, \text{ the level sets } \{x_N \in \mathbb{R},\ u(x', x_N) > c\} \\ \text{have a finite Lebesgue measure.} \end{cases} \tag{6.3}$$

One can define for almost any $x' \in \mathbb{R}^{N-1}$, the distribution function of u by

$$m_u(x', c) := |\{x_N \in \mathbb{R};\ u(x', x_N) > c\}|, \quad c > 0. \tag{6.4}$$

This is a nonincreasing function that is right continuous with respect to c.

Definition 6.1.2. Let u satisfy (6.3) and $m_u(x', c)$ be defined by (6.4). Let us consider the function $y = Y(x', c) := \frac{1}{2} m_u(x', c)$. Then its right-continuous inverse function, denoted by $u^\star(x', \cdot)$, satisfies

$$c = u^\star(x', y) = u^\star(x', -y),$$

is defined on $\Omega^\star$, and is called the Steiner symmetrization of u with respect to the hyperplane $x_N = 0$.

Remark 6.1.3. An equivalent definition is the following. Let us set

$$\forall\, x \in R^N, \quad u^\star(x) := \sup\{c \in \mathbb{R} : \ x \in \Omega_c^\star\},$$

where Ω_c is the level set $\{u > c\}$ and $\Omega_c^\star$ is its Steiner symmetrization. One can visualize Definition 6.1.2 by seeing the graph of the function u as a hill above the hyperplane $x_N = 0$. The graph of $u^\star$ is obtained by rearranging each level set $\{u > c\}$ by means of Definition 6.1.1.

One can remark that, as for $\Omega^\star$, the level sets of $u^\star$ are convex in the direction x_N and symmetric. The following property also holds: if f is a nonnegative function defined on $\mathbb{R}^N$ and satisfying (6.3), then

$$f = f^\star \iff \begin{cases} \forall\, x' \in \mathbb{R}^{N-1}, \forall\, x_N \in \mathbb{R}, & f(x', x_N) = f(x', -x_N); \\ \forall\, c > 0, & \{f > c\} \text{ is convex in the direction } x_N. \end{cases} \tag{6.5}$$

Let us now gather in one theorem the main properties of Steiner symmetrization. For the proofs, we refer for example to [204], [211], [260].

Theorem 6.1.4. *Let $\Omega \subset \mathbb{R}^N$ be a measurable set, and let u, v be two nonnegative functions defined on Ω and satisfying* (6.3). *Let $\Omega^\star$, $u^\star$, and $v^\star$ denote their respective Steiner symmetrizations. Then*

(i) $|\Omega| = |\Omega^\star|$;

(ii) if F is continuous from $\mathbb{R}_+^\star$ into $\mathbb{R}_+$, then

$$\int_\Omega F(u)(x)\, dx = \int_{\Omega^\star} F(u^\star)(x)\, dx \qquad \text{(equimeasurability)};$$

(iii) (Hardy[2]–Littlewood[3] inequality)

$$\int_\Omega uv(x)\, dx \le \int_{\Omega^\star} u^\star v^\star(x)\, dx;$$

(iv) if u belongs to the Sobolev space $W_0^{1,p}(\Omega)$ with $p \ge 1$, then $u^\star \in W_0^{1,p}(\Omega^\star)$ and

$$\int_\Omega |\nabla u(x)|^p\, dx \ge \int_{\Omega^\star} |\nabla u^\star(x)|^p\, dx \qquad \text{(Pólya inequality)}.$$

[2]Godfrey Harold HARDY, 1877–1947, English, professor at Cambridge and Oxford. He was very creative in several fields including diophantine analysis, distribution of prime numbers, the Riemann zeta function, and the summation of divergent series.

[3]John Edensor LITTLEWOOD, 1885–1977, English, student and collaborator of G. H. Hardy. He shared with him a passion and interest for number theory and its analytic aspects.

Let us go back to problem (6.1), where Ω minimizes the functional $J(\omega) = j(u_\omega)$ among the quasi-open sets of given measure m, with $j(v) := \int_{\mathbb{R}^N} \frac{1}{2}|\nabla v|^2 - fv$.

Theorem 6.1.5. *Let $f \in L^2_{\mathrm{loc}}(\mathbb{R}^N)$ such that $f \geq 0$ and $f = f^\star$. Then for every solution Ω of* (6.1), *$J(\Omega^\star) = J(\Omega)$. Therefore, among all optimal sets, at least one is symmetric with respect to the hyperplane $\{x_N = 0\}$.*

Proof. Let Ω be a minimizing set for problem (6.1) and $\Omega^\star$ its Steiner symmetrization. Let $u_{\Omega^\star}$ be the solution of the Dirichlet problem on $\Omega^\star$ with the data $f^\star$, and let $u^\star$ be the Steiner symmetrization of u_Ω. According to Theorem 6.1.4, $|\Omega^\star| = |\Omega|$ and

$$J(\Omega) = \int_\Omega \frac{1}{2}|\nabla u_\Omega|^2 - f u_\Omega \geq \int_{\Omega^\star} \frac{1}{2}|\nabla u^\star|^2 - f^\star u^\star.$$

Since $f = f^\star$, the second integral above is also equal to $j(u^\star)$. By definition of $u_{\Omega^\star}$, it is not less than $j(u_{\Omega^\star}) = J(\Omega^\star)$. □

Remark 6.1.6. We do not claim that all the minimizing domains are symmetric. An elementary counterexample is given by the following one-dimensional situation: we choose

$$f(x) = \begin{cases} 1 & \text{if } x \in [-1, 1], \\ 1/|x| & \text{if } x \notin [-1, 1], \end{cases}$$

with a volume constraint equal to 1. It is easy to check that the solutions are all intervals of the kind $(a, a + 1)$ contained in $[-1, 1]$.[4]

Nevertheless, one can wonder whether it is possible, in some cases, to prove that all the optimal shapes are symmetric. Except for the case where we know that the solution is unique, one can obtain such results by strengthening the assumptions on the data f. For example, if f is analytic, it is proved in [204, Sect. II.7] that equality $\int_{\Omega^\star} |\nabla u^\star(x)|^2 = \int_\Omega |\nabla u_\Omega(x)|^2$ implies $u_\Omega(z + \cdot) = u^*$, where z is a vector orthogonal to the hyperplane $x_N = 0$, and then $\Omega = z + \Omega^*$.

Remark 6.1.7. The assumption (6.5) on the function f may appear purely technical, but it is quite essential. Actually, the symmetry result can even be false if f is symmetric but does not satisfy (6.5). Let us give a one-dimensional example: let f be the (symmetric) function defined by

$$f(x) = \begin{cases} 1 & \text{if } x \in (-\infty, -4] \cup [-2, 2] \cup [4, +\infty), \\ -\frac{1}{2}x^2 - 3x - 3 & \text{if } x \in (-4, -2), \\ -\frac{1}{2}x^2 + 3x - 3 & \text{if } x \in (2, 4). \end{cases}$$

Then for $m = 1$ in (6.1), the optimal domain Ω of J is the interval $(5/2, 7/2)$ (or its reflection $(-7/2, -5/2)$).[5]

[4] Hint: according to Theorem 6.1.5, we know that the interval $(-1/2, 1/2)$ is optimal. Now it is clear that the energy remains constant for any other interval as above.

[5]

Remark 6.1.8. If the function f is radially decreasing, it is possible, with the same argument, to prove that the ball is a solution. For that purpose, we use the *Schwarz decreasing radial rearrangement*. It is defined in the following way. For any measurable set ω, we define its symmetrization ω^* as the ball with the same volume. The functions are rearranged with the same principle: each level set is rearranged in the ball with the same volume. If f is radially decreasing, then $f = f^*$. All the results of Theorem 6.1.4 are still valid (see [30], [204], [212], [239]).

Working in exactly the same way as in the proof of Theorem 6.1.5, we have $J(\Omega^*) \le J(\Omega)$, with Ω^* a ball.

As a corollary we find that, when $f \equiv 1$, the ball is the optimal domain. This was an old conjecture of Saint-Venant[6] who was looking for the cross-sectional shape of a beam that would maximize its torsional rigidity. G. Pólya solved this conjecture (with the above-mentioned method) in [258]. We will see later that the ball is the only maximizer, at least among regular domains.

We can prove the famous Rayleigh–Faber–Krahn inequality with the same method. This inequality, conjectured by Lord Rayleigh in his book *The Theory of Sound*, corresponds to the minimization of the first eigenvalue of the Laplacian with Dirichlet boundary condition, with a volume constraint (see Chapter 4):

Theorem 6.1.9. *Let $\lambda_1(\Omega)$ denote the first eigenvalue of the Laplacian with Dirichlet boundary condition on Ω, and let B be a ball of the same volume as Ω. Then*

$$\lambda_1(B) = \min \{\lambda_1(\Omega),\ \Omega \textit{ measurable},\ |\Omega| = |B|\} .$$

Moreover, if Ω is a quasi-open set not equal quasi-everywhere to a ball with $|\Omega| = |B|$, then $\lambda_1(\Omega) > \lambda_1(B)$.

To prove this result, let us use the variational characterization of $\lambda_1(\Omega)$ as a minimum:

$$\lambda_1(\Omega) = \min_{v \in H_0^1(\Omega), v \neq 0} \frac{\int_\Omega |\nabla v(x)|^2\,dx}{\int_\Omega v^2(x)\,dx}.$$

Sketch of the proof. Let I be an optimal domain and let I^+ be the union of its connected components, which are closer to $+3$ than to -3. Let $\tilde{f}$ be the function equal to f on $[2,4]$ and to 1 elsewhere. By symmetrization around the point $+3$ and applying Theorem 6.1.5 to $\tilde{f}$, we can prove that we get a lower energy on replacing I^+ by an interval of the same length centered at $+3$. We make the same construction around -3 with the union of the connected components closer to -3. It remains to prove that it is better to replace the union of two intervals centered at $+3$ and -3 by one interval centered at $+3$. For that purpose, we remark that the energy of the interval $[3-a, 3+a]$ is given by $\varphi(2a) = -2\int_0^a g(x)$ where $g(x) = \frac{x^2}{4}(1-\frac{x^2}{3})^2$. Since g is nondecreasing on $[0,1]$, it is easy to check that for $\alpha \in [0,1]$, we have $\varphi(1) \le \varphi(\alpha) + \varphi(1-\alpha)$. □

[6] Adhémar Jean Claude Barré de SAINT-VENANT, 1797–1886, French, student at École Polytechnique, then engineer of Ponts et Chaussées. He worked in mechanics, elasticity, hydrostatics, and hydrodynamics. Along with Stokes, he is credited with the correct statement of the Navier–Stokes equations.

Then we introduce u^*, the Schwarz rearrangement of the first eigenfunction u on Ω, and we prove the result exactly as above. For an analysis of the case of equality, we refer to the discussion in [204] or to the more recent paper [103]. In particular, it follows that the ball is the unique minimizer of λ_1 among open sets (of given volume) up to sets of zero capacity (which do not affect the value of λ_1). Let us remark that it is proved in [57] that a quasi-open optimal set is indeed open.

We will present some results on the next two eigenvalues λ_2 and λ_3 in Section 6.4.1.

6.1.3 Using optimality conditions together with a maximum principle

In this section, we assume that Ω is a bounded regular (at least C^2) open set that is a (local) minimum for problem (6.1). According to the optimality conditions obtained in Chapter 5 and, in particular, following (5.105), we have (here c is a Lagrange multiplier)

Proposition 6.1.10. *Let Ω be an open set of class C^2 solving* (6.1). *Then there exists a constant c such that $|\nabla u_\Omega| = |\frac{\partial u_\Omega}{\partial n}| = c$ on $\partial\Omega$.*

We are going to show that, if $f = f^\star$ (in the sense of (6.5)), any optimal shape of (6.1) (and even any critical shape in the sense of the above proposition) has a plane of symmetry. According to Remark 6.1.6, we know that it is not necessarily the plane $x_N = 0$, but we will see that it is always a plane parallel to this. For this purpose, we will use the classical method of moving planes introduced by Aleksandrov (see [10]), where he was able to prove that any surface with constant mean curvature is actually a ball), and developed in this context by J. Serrin in [275].

Theorem 6.1.11. *Let us assume that f is a nonnegative, continuous function that satisfies $f = f^\star$. Let us assume that Ω is an open set of class C^2 with $-\frac{\partial u_\Omega}{\partial n} = c$ (constant) on $\partial\Omega$, where u_Ω is the solution of the Dirichlet problem* (6.2). *Then for every connected component ω of Ω, there exists a real number λ such that ω is symmetric with respect to the hyperplane $x_N = \lambda$.*

Proof. Let us first remark that if $\widetilde{\Omega}$ is the set symmetric to Ω with respect to the plane $x_N = 0$, since f is symmetric, $u_{\widetilde{\Omega}}$ is symmetric to u_Ω and therefore $\widetilde{\Omega}$ is also a solution. Of course, since we are not able to prove uniqueness of the optimal solution, this is not sufficient to conclude the symmetry of Ω.

Let ω be an (open) connected component of Ω. Let us denote by T_λ the hyperplane $x_N = \lambda$. When λ is large enough, T_λ does not intersect ω because the latter is bounded. As λ decreases, T_λ will begin to intersect ω. Then T_λ cuts a "cap" $\Sigma(T_\lambda) = \{x \in \omega;\ x_N > \lambda\}$ from ω. We denote by $\Sigma'(T_\lambda)$ the reflection of $\Sigma(T_\lambda)$ with respect to T_λ. At the beginning of this process, according to the regularity of Ω (and hence of ω), $\Sigma'(T_\lambda)$ lies entirely inside ω, until either

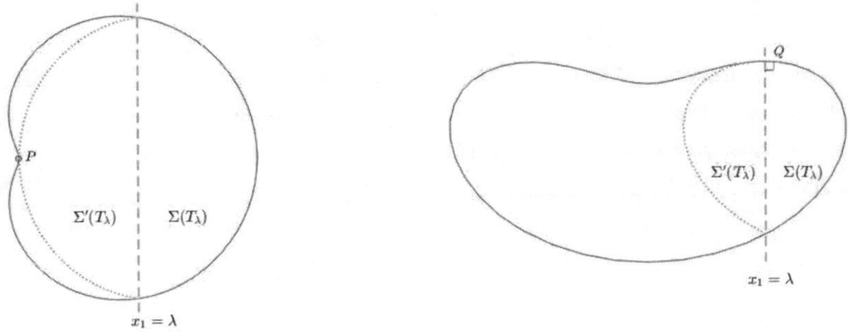

Figure 6.2. The moving plane method: (left) case (1) and (right) case (2).

(1) $\Sigma'(T_\lambda)$ becomes internally tangent to the boundary of ω at one point P which does not belong to T_λ,

or

(2) T_λ reaches a position where it is orthogonal to the boundary of ω at some point Q.

We still denote by T_λ the hyperplane when one of the two above positions is reached. To simplify notation, we introduce $\Sigma \equiv \Sigma(T_\lambda)$ and $\Sigma' \equiv \Sigma'(T_\lambda)$. Let us now consider a new function v on Σ' by reflection:

$$v(x', x_N) := u_\Omega(x', 2\lambda - x_N).$$

By properties of the Laplacian operator, v satisfies

$$\begin{cases} -\Delta v(x', x_N) = f(x', 2\lambda - x_N) & \text{in } \Sigma', \\ v = u & \text{on } \partial\Sigma' \cap T_\lambda, \\ v = 0, \ -\frac{\partial v}{\partial n} = \mathrm{c} & \text{on } \partial\Sigma' \setminus T_\lambda. \end{cases} \tag{6.6}$$

Since Σ' is contained in ω by construction, one can consider the function $w = u - v$ in Σ'. It satisfies

$$-\Delta w(x', x_N) = f(x', x_N) - f(x', 2\lambda - x_N). \tag{6.7}$$

Let us prove that

$$\forall\, (x', x_N) \in \Sigma', \quad f(x', x_N) - f(x', 2\lambda - x_N) \geq 0. \tag{6.8}$$

This is obviously true for $\lambda = 0$ by symmetry of f. Now, we can always assume $\lambda > 0$. Indeed, if it were not the case, according to the claim at the beginning of this proof, we could work with the symmetric domain $\widetilde{\Omega}$ for which we would have $\lambda > 0$.

Let us recall that $x_N \in [0, +\infty) \to f(x', x_N)$ is nonincreasing. On Σ', we have $x_N < \lambda$, thus $x_N < 2\lambda - x_N$, and then if $\lambda > x_N \geq 0$, the fact that $f(x', x_N) - f(x', 2\lambda - x_N) \geq 0$ is a consequence of the monotonicity of f. Now if we assume that $-(2k+2)\lambda \leq x_N \leq -2k\lambda \leq 0$, with $k \geq 0$, we have $0 \leq (2k+2)\lambda \leq 2\lambda - x_N \leq (2k+4)\lambda$ and then, in the same way using the monotonicity of f, we have

$$f(x', 2\lambda - x_N) \leq f(x', (2k+2)\lambda) = f(x', -(2k+2)\lambda) \leq f(x', x_N).$$

This proves (6.8). Thus, it follows from (6.7) that w is superharmonic on Σ'. Consequently w has its minimum on the boundary of Σ'. Now, by construction, $w \geq 0$ on $\partial\Sigma'$, and then, according to the strong maximum principle, either

$$u - v > 0 \quad \text{in } \Sigma', \tag{6.9}$$

or else $u \equiv v$ on at least one connected component C of Σ'. In the latter case, since u cannot vanish in ω and $u = v = 0$ on $\partial C \cap [x_N < \lambda]$, C must coincide with the part of ω that is on the left of T_λ. This implies that ω is symmetric with respect to T_λ and proves the theorem.

It remains to prove that (6.9) cannot occur. Let us assume first that we are in case (1): Σ' is internally tangent to the boundary of ω at some point P. Then $u - v = 0$ at P and consequently, according to (6.9) and using the Hopf[7] maximum principle at a boundary point, we should have

$$\frac{\partial(u - v)}{\partial n} > 0 \quad \text{at } P.$$

But this would contradict the fact that $\frac{\partial u}{\partial n} = \frac{\partial v}{\partial n} = -c$ at P, and hence (6.9) is impossible in case (1).

In case (2), the situation is more complicated since we cannot use the Hopf maximum principle at the boundary point Q, because Σ' has a right angle at this point and therefore does not possess the required regularity (an interior ball condition is needed). Nevertheless, we can prove that the function $u - v$ has a zero of order at least 2 at point Q. Now the contradiction will come thanks to the following version of the Hopf maximum principle due to J. Serrin (we refer to [275] for more details):

Lemma 6.1.12 (Serrin). *Let $D^\star$ be an open set of class C^2 and T a hyperplane containing the normal vector to $\partial D^\star$ at a point Q. We denote by D one of the parts of $D^\star$ that is located on one side of T.*

---

[7]Heinz HOPF, 1894–1971, German and later Swiss. After an outstanding thesis in Riemannian geometry, he made deep contributions to algebraic geometry.

Let w be a superharmonic function of class C^2 in the closure of D, such that $w > 0$ in D and $w = 0$ at Q. Let s be a unit vector that nontangentially points into D at Q. Then either

$$\frac{\partial w}{\partial s} > 0 \quad \text{or} \quad \frac{\partial^2 w}{\partial s^2} > 0 \quad \text{at } Q.$$

To finish the proof of the theorem, we apply this lemma to our situation: if $w = u - v > 0$ in Σ', since $w = 0$ at point Q, we have

$$\frac{\partial(u-v)}{\partial s} > 0 \quad \text{or} \quad \frac{\partial^2(u-v)}{\partial s^2} > 0 \quad \text{at } Q.$$

But this contradicts the fact that u and v have the same first and second partial derivatives at Q. Indeed, since $u = v$ and $\partial u/\partial n = \partial v/\partial n$ on $\partial\Sigma'$, the tangential derivatives coincide out of Q, and also in Q by continuity; on the other hand, the normal derivatives are equal because $u = v$ on T_λ, which is normal to $\partial\omega$ at Q. Thus we obtain equality of all first and second derivatives. □

6.1.4 Variational methods

When the solution of the state equation is obtained, via a *variational principle* as the minimum of some functional, another method can be used: "cut and reflect". In this section, we need to assume that we have been able to prove *existence* of an open set that is a minimizer for the functional. For that purpose, one can use any of the methods described in Chapter 4.

Let us explain the method of cut and reflect on the model problem (6.1), where we choose $f \equiv 1$ in (6.2) for the sake of simplicity. Let us denote by Ω a minimizer of J among open sets of given volume. We recall that

$$J(\Omega) = -\frac{1}{2}\int_\Omega |\nabla u_\Omega|^2,$$

where the solution u_Ω of (6.2) is also a minimizer on $H_0^1(\Omega)$ of the functional

$$j(v) := \frac{1}{2}\int_\Omega |\nabla v|^2 - \int_\Omega v. \tag{6.10}$$

Moreover, $J(\Omega) = j(u_\Omega)$. Now let us choose any unit vector δ and cut Ω in two parts of equal volume by a hyperplane H_δ orthogonal to δ. Let us denote these two parts by Ω^+ (resp. Ω^-). Now, one of the two quantities $\frac{1}{2}\int_{\Omega^+} |\nabla u_\Omega|^2 - \int_{\Omega^+} u_\Omega$ and $\frac{1}{2}\int_{\Omega^-} |\nabla u_\Omega|^2 - \int_{\Omega^-} u_\Omega$ is certainly greater than or equal to the other. Without loss of generality, let us assume that

$$\frac{1}{2}\int_{\Omega^-} |\nabla u_\Omega|^2 - \int_{\Omega^-} u_\Omega \le \frac{1}{2}\int_{\Omega^+} |\nabla u_\Omega|^2 - \int_{\Omega^+} u_\Omega.$$

Let us now make a reflection and introduce $\widehat{\Omega} = \Omega^- \cup \sigma(\Omega^-)$, where σ is the orthogonal reflection through the hyperplane H_δ. We also introduce

$$\widehat{u}(x) = \begin{cases} u(x) & \text{if } x \in \Omega^-, \\ u(\sigma(x)) & \text{if } x \in \sigma(\Omega^-). \end{cases} \tag{6.11}$$

Let us compute the energy of $\widehat{\Omega}$:

$$J(\widehat{\Omega}) = -\frac{1}{2}\int_{\widehat{\Omega}} |\nabla u_{\widehat{\Omega}}|^2.$$

Now, the function $\hat{u}$ defined in (6.11) is admissible in the variational formulation (6.10), and therefore

$$\begin{aligned} J(\widehat{\Omega}) = \frac{1}{2}\int_{\widehat{\Omega}} |\nabla u_{\widehat{\Omega}}|^2 - \int_{\widehat{\Omega}} u_{\widehat{\Omega}} &\leq \frac{1}{2}\int_{\widehat{\Omega}} |\nabla \widehat{u}|^2 - \int_{\widehat{\Omega}} \widehat{u} \\ &= 2\left(\frac{1}{2}\int_{\Omega^-} |\nabla u|^2 - \int_{\Omega^-} u\right) \\ &\leq \frac{1}{2}\int_{\Omega} |\nabla u_\Omega|^2 - \int_\Omega u_\Omega = J(\Omega). \end{aligned}$$

Since Ω is a minimizer of the functional J, we must have equality in the above chain of inequalities. This implies, in particular, that $\widehat{u} = u_{\widehat{\Omega}}$ is the solution of the PDE (6.2) on $\widehat{\Omega}$. Now, since $\Delta(u - \widehat{u}) = 0$ in $\Omega \cap \widehat{\Omega}$ and $u - \widehat{u} = 0$ in $\Omega^- \subset \Omega \cap \widehat{\Omega}$, we have by analyticity, $u - \widehat{u} = 0$ in $\Omega \cap \widehat{\Omega}$. It follows that $\Omega^+ = \Omega^-$, and the domain is symmetric with respect to H_δ, because if this were not the case, we would have $\widehat{u} = 0$ inside $\widehat{\Omega}$, contradicting the strong maximum principle.

6.1.5 Using another shape optimization problem

In this section, we also consider the case $f \equiv 1$ and we assume that Ω is a regular critical shape for the minimization problem. We recall that this means

$$\begin{cases} -\Delta u = 1 & \text{in } \Omega, \\ u = 0 & \text{on } \partial\Omega, \\ \frac{\partial u}{\partial n} = \text{constant} = -c & \text{on } \partial\Omega. \end{cases} \tag{6.12}$$

The idea is the following. We introduce another functional $J_1 = J_1(\omega)$ defined for any bounded open set ω of $\mathbb{R}^N$ such that the solutions of problem (6.12) are exactly the minima of the functional J_1. Writing the optimality conditions for the functional J_1, it is possible to prove that the mean curvature of the boundary is constant for any minimum (or even any critical point) of J_1. In other words, the method consists in introducing another shape optimization problem to prove the desired symmetry result. This method was explained in [102], which gives another proof of the following result.

Theorem 6.1.13 (Serrin). *Let Ω be a connected and bounded open set of class C^2 in $\mathbb{R}^N$. Then problem* (6.12) *has a solution with $c > 0$ if and only if Ω is a ball.*

Let us remark that the condition $c > 0$ follows from the Hopf maximum principle.

Let us denote by $\mathcal{O}$ the class of all connected and bounded open sets of class C^2 in $\mathbb{R}^N$. With each $\omega \in \mathcal{O}$, we associate $u_\omega = u^1_\omega$, the solution of the Dirichlet problem for the Laplacian with a right-hand side $f \equiv 1$. According to classical regularity results (see, e.g., [7], [151], [54]), $u_\omega \in C^1(\overline{\omega}) \cap H^2(\Omega)$. The new functional J_1 that we consider here is

$$J_1(\omega) = N \int_{\partial\omega} |\nabla u_\omega|^3 - (N+2) \int_\omega |\nabla u_\omega|^2, \quad \omega \in \mathcal{O}.$$

Lemma 6.1.14. *We have $J_1(\omega) \geq 0$ for any ω in $\mathcal{O}$ and $J_1(\Omega) = 0$ if Ω is a solution of problem* (6.12).

Proof. Let us begin with a simple inequality (used in [291]): for any $\omega \in \mathcal{O}$,

$$1 = (\Delta u_\omega)^2 \leq N \sum_{i=1}^N \left(\frac{\partial^2 u_\omega}{\partial x_i^2}\right)^2 \leq N \sum_{i,j=1}^N \left(\frac{\partial^2 u_\omega}{\partial x_i \partial x_j}\right)^2. \tag{6.13}$$

But

$$2 \sum_{i,j=1}^N \left(\frac{\partial^2 u_\omega}{\partial x_i \partial x_j}\right)^2 = \Delta\big(|\nabla u_\omega|^2\big) - 2\nabla(\Delta u_\omega).\nabla u_\omega = \Delta\big(|\nabla u_\omega|^2\big).$$

Multiplying (6.13) by $u_\omega \geq 0$ and integrating over ω yield

$$\int_\omega u_\omega \leq \frac{N}{2} \int_\omega u_\omega \Delta\big(|\nabla u_\omega|^2\big).$$

We integrate by parts and use the positivity of u_Ω, which implies $\frac{\partial u_\omega}{\partial n} = -|\nabla u_\omega|$ on $\partial\omega$, to get

$$\int_\omega u_\omega \leq \frac{N}{2} \left[\int_{\partial\omega} |\nabla u_\omega|^3 + \int_\omega |\nabla u_\omega|^2 \Delta u_\omega\right].$$

Since $-\Delta u_\omega = 1$ and $\int_\omega |\nabla u_\omega|^2\,dx = \int_\omega u_\omega\,dx$, we have $J_1(\omega) \geq 0$.

To prove that $J_1(\Omega) = 0$, let us first remark that

$$J_1(\Omega) = Nc^3|\partial\Omega| - (N+2) \int_\Omega |\nabla u_\Omega|^2. \tag{6.14}$$

Indeed, to compute the last integral, we use the classical Rellich formula, valid for any $v \in C^1(\overline{\Omega}) \cap H^2(\Omega)$ (see, e.g., [264]):

$$2\int_{\partial\Omega}(x.\nabla v)\frac{\partial v}{\partial n} - \int_{\partial\Omega}(x.n)|\nabla v|^2 = 2\int_{\Omega}(x.\nabla v)\Delta v + (2-N)\int_{\Omega}|\nabla v|^2.$$

We apply this formula to $v = u_\Omega$, and using $\nabla u_\Omega = -|\nabla u_\Omega| n = -cn$ on $\partial\Omega$, we obtain

$$c^2\int_{\partial\Omega}(x.n) = -2\int_{\Omega}(x.\nabla u_\Omega) + (2-N)\int_{\Omega}|\nabla u_\Omega|^2.$$

From a simple integration by parts, we get

$$\int_{\Omega} x.\nabla u_\Omega = -N\int_{\Omega} u_\Omega = -N\int_{\Omega}|\nabla u_\Omega|^2, \qquad \int_{\partial\Omega}(x.n) = \int_{\Omega}\operatorname{div} x = N|\Omega|,$$

and therefore $c^2 N|\Omega| = c^2\int_{\partial\Omega}(x.n) = (2+N)\int_{\Omega}|\nabla u_\Omega|^2$. On the other hand, since

$$|\Omega| = \int_{\Omega} dx = -\int_{\Omega}\Delta u_\Omega = -\int_{\partial\Omega}\frac{\partial u_\Omega}{\partial n} = c|\partial\Omega|, \tag{6.15}$$

using (6.14), we get $J_1(\Omega) = 0$. □

Now we use the shape derivative to write the optimality condition. This will give us enough information on the minima of J_1 to be able to conclude. Let $V \in C^1(\mathbb{R}^N, \mathbb{R}^N)$, $\omega \in \mathcal{O}$ be given and $\omega_t = (I + tV)(\omega)$. Introducing the mean curvature H of $\partial\omega$ we get, as explained in Chapter 5,

Lemma 6.1.15. *The derivative of $t \to j(t) = J_1(\omega_t)$ is given by*

$$j'(0) = \int_{\partial\omega}\left(\left[(2N-2)|\nabla u_\omega|^2 - 2NH|\nabla u_\omega|^3\right]V.n - 3N|\nabla u_\omega|^2\frac{\partial u'_\omega}{\partial n}\right), \tag{6.16}$$

where u'_ω is the solution of

$$\begin{cases}\Delta u'_\omega = 0 & \text{in } \omega,\\ u'_\omega = -\frac{\partial u_\omega}{\partial n}V.n & \text{on } \partial\omega.\end{cases} \tag{6.17}$$

Proof. We have already given the expression for the derivative of $t \to \int_{\omega_t}|\nabla u_{\omega_t}|^2$ (see (5.103)) and it is equal to $\int_{\partial\omega}|\nabla u_\omega|^2(V.n)$. Proposition 5.4.18 gives an expression of the derivative of $t \to h(t) = \int_{\partial\omega_t}|\nabla u_{\omega_t}|^3$, namely,

$$h'(0) = \int_{\partial\omega} 3|\nabla u_\omega|\nabla u_\omega.\nabla u'_\omega + (V.n)\left[\frac{\partial|\nabla u_\omega|^3}{\partial n} + H|\nabla u_\omega|^3\right].$$

Using $\nabla u_\omega = -|\nabla u_\omega| n$, the first term above is also given by

$$\int_{\partial\omega} |\nabla u_\omega| \nabla u_\omega . \nabla u'_\omega = -\int_{\partial\omega} |\nabla u_\omega|^2 n . \nabla u'_\omega = -\int_{\partial\omega} |\nabla u_\omega|^2 \frac{\partial u'_\omega}{\partial n}.$$

For the second term, we use the fact that $H = \operatorname{div}\left(\frac{\nabla u_\omega}{|\nabla u_\omega|}\right)$ according to Proposition 5.4.8, and therefore

$$H = \frac{1}{|\nabla u_\omega|} + \frac{1}{|\nabla u_\omega|^2} \nabla(|\nabla u_\omega|) . \nabla u_\omega.$$

Thus

$$\begin{aligned} \frac{\partial |\nabla u_\omega|^3}{\partial n} &= 3|\nabla u_\omega|^2 \nabla(|\nabla u_\omega|) . n = -3|\nabla u_\omega| \nabla(|\nabla u_\omega|) . \nabla u_\omega \\ &= -3|\nabla u_\omega|^3 \left(H - \frac{1}{|\nabla u_\omega|}\right). \end{aligned}$$

We obtain formula (6.16) by collecting the above expressions. □

Proof of Theorem 6.1.13. Let Ω be a solution of problem (6.12). According to Lemma 6.1.14, Ω is also a minimum for the functional J_1. Thus the expression for $-j'(0)$ in Lemma 6.1.15 is equal to zero. Let us replace $|\nabla u_\Omega|$ here by $c = |\Omega|/|\partial\Omega|$ to get

$$0 = 2c^2 \int_{\partial\Omega} [N - 1 - NHc](V.n) - 3Nc^2 \int_{\partial\Omega} \frac{\partial u'_\Omega}{\partial n}.$$

But $\int_{\partial\Omega} \frac{\partial u'_\Omega}{\partial n} = \int_\Omega \Delta u'_\Omega = 0$, and therefore

$$0 = \int_{\partial\Omega} [N - 1 - NHc] V.n = 0, \quad \text{for any } V \in C^2(\mathbb{R}^N, \mathbb{R}^N).$$

This proves that the mean curvature of $\partial\Omega$ is constant and consequently Ω is a ball, according to a classical result due to Aleksandrov (see [42] or [282]).

Conversely, if Ω is a ball, it is easy to construct a radial solution of (6.12): indeed, if u_1 is a solution of the problem in the unit ball with $c = 1$, then $u(x) = c^2 u_1(x/c)$ is a solution of the problem with the data c on the ball of radius c. □

Remark 6.1.16. The motivation for the previous approach was to find an alternative proof of Serrin's theorem that would not use the maximum principle. Indeed there are several overdetermined problems similar to (6.12) for which Serrin's method does not apply. Classical examples are the conjectures of Schiffer[8], which are still open.

[8]Menahem Max SCHIFFER, 1911–1997, born in Berlin. He went to Jerusalem University in 1938. His contributions concerned the calculus of variations and its application to geometrical questions in complex analysis.

The problem is to prove that the only domains for which there exists an eigenfunction of the Dirichlet–Laplacian solution of the overdetermined problem

$$\begin{cases} -\Delta u = \lambda u & \text{in } \Omega, \\ u = 0 & \text{on } \partial\Omega, \\ |\nabla u| = \text{constant} & \text{on } \partial\Omega \end{cases} \tag{6.18}$$

are the balls. The same problem holds for the eigenfunctions of the Neumann–Laplacian: prove that if a domain Ω is such that there exists u solving

$$\begin{cases} -\Delta u = \lambda u & \text{in } \Omega, \\ \frac{\partial u}{\partial n} = 0 & \text{on } \partial\Omega, \\ u = \text{constant} & \text{on } \partial\Omega \end{cases} \tag{6.19}$$

(with $\lambda > 0$), then Ω must be a ball. Actually problem (6.19) is equivalent to the famous Pompeiu[9] problem, which has yielded many works in different fields of analysis. See [297] for an extended bibliography.

6.1.6 A case of nonsymmetry: The Newton problem

In the *Principia Mathematica*, Newton[10] studied the aerodynamics (or the resistance) of a body moving in a weakly dense fluid. If we define this body as the graph of a positive and concave function u defined on a plane domain D, a simplified model of the problem consists in defining the resistance of this body as the quantity

$$R(u) := \int_D \frac{1}{1 + |\nabla u(x)|^2}\, dx. \tag{6.20}$$

Now a natural question, raised by I. Newton, is to find for a given height the most aerodynamic body, that is, the one of minimal resistance. Mathematically, this corresponds to

> find a function $u \in W^{1,\infty}(D)$, u concave, $0 \le u \le M$ that minimizes the functional R defined in (6.20).

This problem enters the framework of the calculus of variations. In this general context, the existence of a solution has been proved in [81]; see also [82], [219], [218], [256], or the survey [78] for more details and results on this problem.

[9]Dimitri POMPEIU, 1873–1954, Romanian mathematician. He defended his thesis in Paris in 1905 and was a professor in Iasi, Bucarest, and Cluj. He made contributions to the theory of functions of complex variables and to rational mechanics.

[10]Sir Isaac NEWTON, 1643–1727, English. He was not only a mathematician, but certainly one of the greatest scientists of all time. He founded the calculus and modern analysis.

In his work, I. Newton considered the case where D is the unit disk and was able to compute the radial minimum explicitly. Indeed, since in the radial case the problem is one-dimensional, the minimal function $u = u(r)$ must make the derivative at $t = 0$ of

$$t \to \int_0^1 r\left[1 + \{u'(r) + tv'(r)\}^2\right]^{-1}$$

vanish for all $v = v(r)$. This leads to the Euler–Lagrange equation:

$$\frac{ru'(r)}{\left(1 + (u'(r))^2\right)^2} = \text{constant}. \tag{6.21}$$

Equation (6.21) has a unique solution u^* that is shown in Figure 6.3 (in the case $M = 1$) and that is given by $u^*(r) = M$ for $r \in [0, r_0]$, and for $r \in [r_0, 1]$, parameterized by

$$\forall\, t \in [1, T], \quad r(t) = \frac{r_0}{4}\frac{\left(1 + t^2\right)^2}{t}, \qquad u^*(t) = M - r(t)f(t), \tag{6.22}$$

$$f(t) = \frac{t}{(1 + t^2)^2}\left(-\frac{7}{4} + \frac{3}{4}t^4 + t^2 - \log t\right), \qquad T = f^{-1}(M), \qquad r_0 = 4T/(1 + T^2)^2.$$

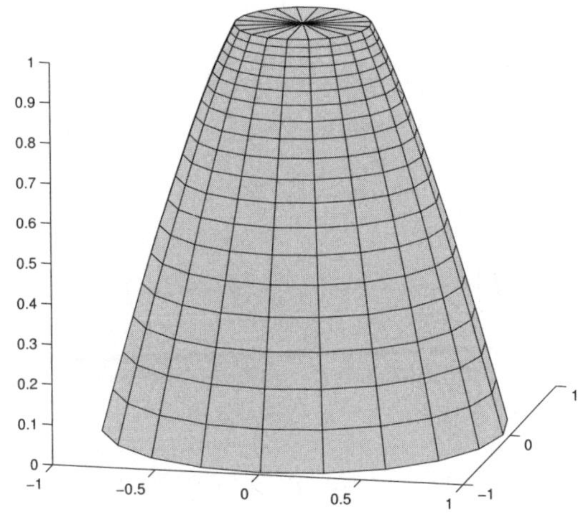

Figure 6.3. The optimal solution for the Newton problem among radial functions.

The existence of a "plateau" may seem not intuitive: for a very aerodynamic body we would expect a sharp point. Actually, here the body travels in a fluid with a few particles, so this intuition is not valid.

There is a very natural symmetry question that appears now: if D is a unit disk in $\mathbb{R}^2$, is the minimum of the functional R radial? In other words, is the minimum of R in the class $\mathcal{C}_M = \{u \in W^{1,\infty}(\Omega),\ u \text{ concave}, 0 \le u \le M\}$ the function u^* defined in (6.22)?

Strangely enough, this elementary question remained open for a long time, probably because everyone was thinking that the answer should be positive. But the answer is NO. It was proved in [60], and we give this proof below.

More generally, we can wonder what could be the different strategies for proving that the optimum of some functional *is not symmetric*. We refer to the beautiful paper [206] for several examples of nonsymmetry and an overview of different strategies for the proof. Let us give here some simple ones. Let us assume, for example, that we want to prove that the optimum does not have radial symmetry. Let us denote by $\widetilde{\Omega}$ the minimum *among* radially symmetric sets. Then we can

- find necessary optimality conditions (of first or second order) that are not satisfied by $\widetilde{\Omega}$;
- prove that we can improve the functional thanks to nonradial perturbations of $\widetilde{\Omega}$;
- find an explicit (nonradial) domain better than $\widetilde{\Omega}$.

For the Newton problem, we will prove the following.

Theorem 6.1.17. *The function u^* defined in* (6.22) *does not minimize the functional R in the class $\mathcal{C}_M$. Therefore, the minimum of R in the class $\mathcal{C}_M$ is not radially symmetric.*

Proof. We follow the proof given by Brock–Ferone–Kawohl in [60]. The first and second derivatives of the functional R at the point u^* are given by

$$R'(u^*)\varphi = -2\int_D \frac{\nabla u^*.\nabla\varphi}{\left(1+|\nabla u^*|^2\right)^2}, \tag{6.23}$$

$$R''(u^*)(\varphi,\varphi) = \int_D \frac{2}{\left(1+|\nabla u^*|^2\right)^3}\left[-(1+|\nabla u^*|^2)|\nabla\varphi|^2 + 4(\nabla u^*.\nabla\varphi)^2\right]. \tag{6.24}$$

If u^* were a local minimum of R on $\mathcal{C}_M$, we would have $R''(u^*)(\varphi,\varphi) \ge 0$ for any regular φ such that for all $r \in [0, r_0]$, $\varphi(r) = 0 = \varphi(1)$ (here we use the strict concavity of u^* on $[r_0, 1]$, which ensures that $u^* + \epsilon\varphi \in \mathcal{C}_M$ for ϵ small enough). Now let us choose $\varphi(r,\theta) = \eta(r)\sin(k\theta)$ with $\operatorname{supp}\eta \subset (r_0, 1)$. Since $|\nabla\varphi|^2 = (\eta'(r)\sin(k\theta))^2 + \frac{k^2}{r^2}(\eta(r)\cos(k\theta))^2$, the expression between the brackets in (6.24) becomes

$$[\,] = -(1+|u_r^*|^2)\,(\eta'(r)\sin(k\theta))^2 + \frac{k^2}{r^2}(\eta(r)\cos(k\theta))^2 + 4(u_r^*.\varphi_r)^2. \tag{6.25}$$

Thus we observe that the quantity (6.24) becomes negative for k large enough. □

6.2 Convexity

6.2.1 Introduction

Another geometrical question that is natural to consider is whether an optimal shape is convex. It can appear to be more difficult to prove or even to suspect convexity than symmetry, since it is often clear that the data of a problem contain some symmetry, while it can be unclear whether some convexity is hidden in the data. Actually, in the problems we are going to consider here, we will always consider, among the data, a fixed convex domain K and the optimal domain, depending on K will inherit its convexity property.

To prove convexity, it seems difficult to use a rearrangement procedure as in the previous section. Actually, a natural idea could consist in replacing any set ω by its convex hull, say $\widetilde{\omega}$, and for a function u defined on ω, we could introduce a new function $\tilde{u}$ whose level sets are the convex hull of the level sets of u. Unfortunately, it seems difficult, even impossible, to find a general relation between the integral of $\tilde{u}$ and the corresponding integral of u. And it seems even worse for the integral involving the gradient of u. This is why such a method seems hopeless.

Therefore, we will always work with the overdetermined problem that is obtained from the optimality conditions. First, we will show the convexity of the solution by proving that the solution cannot be strictly included in its convex hull, using the maximum principle in a clever way. Later on, we will give two other methods, specific to the two-dimensional case, which consist in proving that the curvature of the solution is nonnegative. We refer to [192] and [205] for the original proofs and more details.

For the sake of simplicity, let us consider in this section a model problem that consists in minimizing the capacity with a volume constraint. More precisely, let K be a given compact convex set K in $\mathbb{R}^N$ and choose a positive constant $m > |K|$. We seek the domain Ω containing K, with measure m, that minimizes $\mathrm{cap}_\Omega(K)$, the capacity of K relative to Ω (see (3.47) for the definition of relative capacity). The existence of a minimizer was proved in Exercise 4.12. Let us denote by u_Ω the capacitary potential of K relative to Ω. If Ω is open, this potential is the solution of (see Chapter 2)

$$\begin{cases} \Delta u_\Omega = 0 & \text{in } \Omega\backslash K, \\ u_\Omega = 0 & \text{on } \partial\Omega, \\ u_\Omega = 1 & \text{on } K, \end{cases} \tag{6.26}$$

$$\text{and } \mathrm{cap}_\Omega K := J(\Omega) = \int_{\Omega\backslash K} |\nabla u_\Omega|^2. \tag{6.27}$$

To get the optimality conditions, let us introduce $j(t) = J((I + tV)(\Omega))$, where

$V \in C^1(\mathbb{R}^N, \mathbb{R}^N)$. According to the results of Chapter 5, if the minimum Ω is regular enough, there exists a constant c such that

$$c\int_{\partial\Omega} V.n = j'(0) = 2\int_{\Omega\setminus K} \nabla u_\Omega.\nabla u' + \int_{\partial\Omega} |\nabla u_\Omega|^2 V.n,$$

where u' is the solution of the problem

$$\begin{cases} \Delta u' = 0 & \text{in } \Omega\setminus K, \\ u' = -\frac{\partial u_\Omega}{\partial n} V.n & \text{on } \partial\Omega, \\ u' = 0 & \text{on } K. \end{cases} \tag{6.28}$$

It follows that

$$c\int_{\partial\Omega} V.n = 2\int_{\partial\Omega} u' \frac{\partial u_\Omega}{\partial n} V.n + \int_{\partial\Omega} |\nabla u_\Omega|^2 V.n = -\int_{\partial\Omega} |\nabla u_\Omega|^2 V.n,$$

and therefore

Proposition 6.2.1. *The regular open sets Ω that contain K and minimize the capacity of K with a volume constraint are such that the potential u_Ω of K relative to Ω satisfies the overdetermined condition*

$$|\nabla u_\Omega| = \textit{constant} = c \quad \textit{on } \partial\Omega. \tag{6.29}$$

6.2.2 Comparison with the convex hull

We are going to prove the following result related both to the geometry of the optimum and to the uniqueness of the solution for the overdetermined problem (6.26), (6.29) (and also for the associated shape optimization problem).

Theorem 6.2.2. *Let K be a compact convex set and (u, Ω) a solution of* (6.26)+(6.29). *Then Ω is convex and there is no other solution.*

Proof. Let $\Omega^\star$ be the convex hull of Ω and $u^\star$ the capacitary potential of K relative to $\Omega^\star$, solving (6.26) with $\Omega^\star$ instead of Ω. We have to prove that Ω cannot be strictly included in $\Omega^\star$. For that purpose, we will use the following result.

Lemma 6.2.3.

$$|\nabla u^\star(x)| \geq c \quad \forall\, x \in \Omega^\star\setminus K.$$

Let us grant this lemma and proceed with the proof. The argument will make use of what is known as Lavrentiev's[11] principle.

[11] Mikhail Alekseevich LAVRENTIEV, 1900–1980, Moscow, Kiev, Novosibirsk, then Moscow. He worked in complex analysis (conformal maps and their geometric use for partial differential equations).

Without loss of generality, we can assume that K contains the origin. In the convex domain $\Omega_t^\star := \{x : x/t \in \Omega^\star\}$, we define the function

$$v_t(x) = u^\star\left(\frac{x}{t}\right), \quad \text{where } 0 < t < 1.$$

Let us assume, for the purpose of contradiction, that $\Omega^\star \backslash \Omega \neq \emptyset$. In this case,

$$t_0 = \sup\{t : \Omega_t^\star \subset \Omega\} < 1. \tag{6.30}$$

Since K is convex, we have $K_{t_0} = \{x : x/t_0 \in K\} \subset K$. Since $v_{t_0} - u_\Omega$ is harmonic in $\Omega_{t_0}^\star \backslash K$ and takes negative values on $\partial(\Omega_{t_0}^\star \backslash K)$, the maximum principle implies that $v_{t_0} \leq u_\Omega$ in $\Omega_{t_0}^\star \backslash K$.

Now let $x_0 \in \partial\Omega_{t_0}^\star \cap \partial\Omega$ ($\neq \emptyset$ by assumption). Using Lemma 6.2.3 and (6.29), we have

$$\frac{c}{t_0} \leq \frac{1}{t_0} \lim_{\substack{y/t_0 \to x_0 \\ y/t_0 \in \Omega^\star \backslash K}} |\nabla u^\star(y/t_0)| = \lim_{\substack{y \to x_0 \\ y \in \Omega_{t_0}^\star \backslash K}} |\nabla v_{t_0}(y)| \leq \lim_{\substack{y \to x_0 \\ y \in \Omega_{t_0}^\star \backslash K}} |\nabla u_\Omega(y)| = c,$$

which contradicts (6.30) and finishes the proof of the first part of the theorem.

For the second part, let Ω_1, Ω_2 be two solutions. The same argument as above (Lavrentiev's principle) can be applied to both Ω_1 and Ω_2 to prove that $\Omega_1 \subset \Omega_2 \subset \Omega_1$. □

Proof of Lemma 6.2.3. For $0 < s < 1$, let us denote by $\mathcal{L}_s$ the level sets of $u^\star$, that is,

$$\mathcal{L}_s = \{x \in \Omega^\star : u^\star(x) > s\}.$$

By a result due to Lewis [227] (see also [86]), these level sets are convex. Let y be a point of $\partial\mathcal{L}_s$. Without loss of generality, we can assume that y is the origin and that $e_1 = (1, 0, \ldots, 0)$ is the exterior normal vector to $\partial\mathcal{L}_s$ at the point y. Then $\{x_1 = 0\}$ is a support hyperplane of $\mathcal{L}_s$ at the point y and $K \subset [x_1 < 0]$.

Let us now choose a point $z \in \partial\Omega^\star \cap \{x_1 > 0\}$ that is as far as possible from the hyperplane $\{x_1 = 0\}$ (see Figure 6.4).

By the geometric properties of the convex hull, one can assume that z belongs to $\partial\Omega \cap \partial\Omega^\star$. Let us now consider the harmonic function defined by

$$v(x) = u^\star(x) + ax_1,$$

where $0 < a < c$, on the open set $\Omega^\star \cap \{x_1 > 0\}$. Since v is harmonic in $\Omega^\star \cap \{x_1 > 0\}$, its maximum is attained on the boundary, either at z or at the origin.

Let us assume, for a moment, that the maximum is attained at z. By the maximum principle, since $\Omega \subset \Omega^\star$, we have $u^\star \geq u$ on $\Omega \backslash K$, and hence

$$|\nabla u^*(z)| = -\frac{\partial u^*}{\partial n}(z) \geq -\frac{\partial u}{\partial n}(z) = |\nabla u(z)|.$$

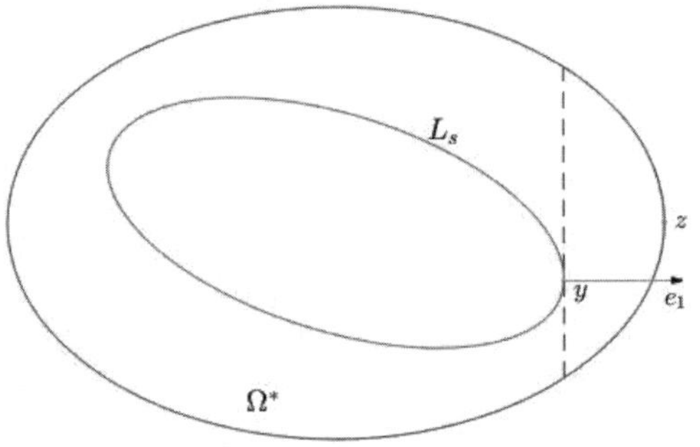

Figure 6.4. Proof of Lemma 6.2.3.

Therefore, if z were a maximum point for v, we would have

$$0 \le \frac{\partial v}{\partial n}(z) = \frac{\partial u^*}{\partial n}(z) + a \le \frac{\partial u}{\partial n}(z) + a = -c + a, \tag{6.31}$$

which would contradict the fact that $a < c$. Thus, the maximum of v is attained at the origin, which implies

$$-\frac{\partial v}{\partial x_1}(0) \ge 0$$

or

$$|\nabla u^\star(0)| \ge a.$$

Since $a \in (0, c)$ is arbitrary, Lemma 6.2.3 follows. □

We now present, without too many details (we refer to [204] and [86] for more), another way to get convexity by proving directly that the minimal domain coincides with its convex hull. Let us explain it for a problem similar to the previous one, though more general.

Let us consider a function F of class $C^{2,\alpha}$ on the interval $[0, 1]$. We assume F to be convex, nonnegative, and nondecreasing with $F'(0) = 0$. We denote by f the derivative of F, which we extend by 0 for negative t. We also consider a convex set K, and we want to minimize the functional

$$J(\Omega) = \frac{1}{2}\int_\Omega |\nabla u|^2\,dx + \int_\Omega F(u)\,dx,$$

among all open sets Ω containing K, where u is the solution of the following semilinear boundary value problem:

$$\begin{cases} \Delta u = f(u) & \text{in } \Omega \backslash K, \\ u = 0 & \text{on } \partial\Omega, \\ u = 1 & \text{on } K. \end{cases} \tag{6.32}$$

According to the maximum principle and the fact that f is nondecreasing, we have $0 < u \leq 1$ in Ω. We extend the function u by 0 outside Ω.

Theorem 6.2.4. *Let Ω be a regular open set that minimizes the functional J. If K is convex, then Ω is also convex.*

Idea of the proof. If Ω^* denotes the convex hull of the minimum Ω, the key is to prove the following inequality:

$$Q(x_1, x_2) := u\left(\tfrac{1}{2}(x_1 + x_2)\right) - \min(u(x_1), u(x_2)) \geq 0 \quad \forall\, x_1, x_2 \in \mathbb{R}^N. \tag{6.33}$$

Indeed, this inequality implies, by induction and the density of numbers of the kind $k/2^n$ in the interval $[0, 1]$, that for any $x_1, x_2 \in \mathbb{R}^N$ and for any $t \in [0, 1]$,

$$u(tx_1 + (1 - t)x_2) \geq \min(u(x_1), u(x_2)),$$

which exactly means that the level sets $\{x \in \mathbb{R}^N,\ u(x) > c\}$ are convex. This implies, in particular, that Ω is convex, since it corresponds to the level set $c = 0$.

To prove (6.33), we argue by contradiction. We assume that there exist two points x_1, x_2 such that Q attains a negative minimum:

$$u\left(\tfrac{1}{2}(x_1 + x_2)\right) < \min(u(x_1), u(x_2)). \tag{6.34}$$

Since $u \geq 0$ in Ω^* with $u = 0$ on the complement of Ω, we see that (6.34) implies that the points x_1, x_2 are both in Ω. Therefore we have $\frac{1}{2}(x_1 + x_2) \in \Omega^*$.

Then we successively prove (by contradiction, using the minimality of the pair (x_1, x_2)),

- $u(x_1) = u(x_2) > 0$;
- $x_1 \in \Omega \backslash K$, $x_2 \in \Omega \backslash K$, and $\frac{1}{2}(x_1 + x_2) \in \Omega^* \backslash K$;
- $\nabla u(x_1)$ is collinear with $\nabla u(x_2)$ (in the same direction);
- $\nabla u(\frac{1}{2}(x_1 + x_2))$ is not zero, collinear with (in the same direction as) $\nabla u(x_1)$ or $\nabla u(x_2)$;
- $\frac{1}{2}(x_1 + x_2) \in \Omega \backslash K$;

- we introduce three positive numbers $a := \left|\nabla u(\frac{1}{2}(x_1+x_2))\right|$, $b := |\nabla u(x_1)|$, and $c := |\nabla u(x_2)|$, and we let $\mu = \frac{c}{b+c}$; then

$$\begin{aligned} &\frac{1}{a} = \frac{1}{2}\left(\frac{1}{b}+\frac{1}{c}\right), \\ &\frac{1}{a^2}\Delta u(\frac{1}{2}(x_1+x_2)) \geq \frac{\mu}{b^2}\Delta u(x_1) + \frac{1-\mu}{c^2}\Delta u(x_2). \end{aligned} \tag{6.35}$$

The conclusion comes using (6.35), the relation $\Delta u = f(u)$, and the monotonicity of f, which lead to the desired contradiction. □

6.2.3 Nonnegative curvature

In this section, we discuss another interesting idea consisting in proving that the boundary of the domain Ω solution of (6.26), (6.29) necessarily has a nonnegative mean curvature (here we assume Ω to be of class C^2). Of course this implies convexity *in dimension* 2. We present this result with strong assumptions to avoid technicalities and to concentrate on the new ideas, which are quite general.

Let us restrict ourselves here to dimension 2. The definition of a star-shaped open set is recalled at the beginning of the next section (Definition 6.3.1). We want to prove

Proposition 6.2.5. *Let $K \subset \Omega \subset \mathbb{R}^2$ with K a compact, convex set with boundary of class C^2 and Ω an open set of class C^2. We assume that Ω is star shaped with respect to some point of K and we assume that the capacitary potential u_Ω of K relative to Ω is a solution of* (6.26)+(6.29) *with $c \neq 0$. Then $\partial\Omega$ has a nonnegative curvature and therefore Ω is convex.*

We use the following elementary lemma.

Lemma 6.2.6. *Let u be a harmonic function on an open set ω of $\mathbb{R}^2$, and assume that $|\nabla u(x)|^2$ does not vanish on ω. Then the function* $\log(|\nabla u(x)|^2)$ *is harmonic on ω.*

This lemma is easily proved by a direct computation or thanks to complex analysis by writing u (locally) as the real part of a holomorphic function f, in which case $\log(|\nabla u(x)|^2)$ is twice the real part of $\log(f'(z))$.

Proof of Proposition 6.2.5. Let us first prove that $v(x) = |\nabla u_\Omega(x)|^2$ does not vanish on $\Omega\backslash K$. Without loss of generality, we can assume that K contains the origin O and that Ω is star shaped with respect to O. The function $w(z) = w(x,y) = \nabla u_\Omega . z = x(u_\Omega)_x + y(u_\Omega)_y$ is harmonic in $\Omega\backslash K$. Since u is constant on the boundary of this open set ($= \partial K \cup \partial\Omega$) and because K and Ω are star shaped with respect to O, we have $w = (\nabla u_\Omega . n) n.z < 0$ (see Proposition 6.3.2). By the maximum principle, it follows that $w < 0$ on $\Omega\backslash K$, which implies that v cannot vanish on this open set (nor on its boundary).

According to Lemma 6.2.6, the function $\log(|\nabla u_\Omega(x)|^2)$ is harmonic on $\Omega\backslash K$ and continuous on its closure. Thus, it attains its minimum on the boundary, either on ∂K or on $\partial\Omega$. Let us prove that it cannot be attained on ∂K. Indeed, if it was the case, the Hopf maximum principle would imply (we denote by ν the "exterior" normal vector to ∂K, directed to $\Omega\backslash K$)

$$-\frac{\partial}{\partial\nu}\left(|\nabla u_\Omega|^2\right) = -2\frac{\partial u_\Omega}{\partial\nu}\frac{\partial^2 u_\Omega}{\partial\nu^2} < 0 \quad \text{at a boundary point of } \partial K,$$

which is incompatible with the relation on ∂K given by Proposition 5.4.12:

$$\Delta u_\Omega = H\frac{\partial u_\Omega}{\partial\nu} + \frac{\partial^2 u_\Omega}{\partial\nu^2} = 0 \tag{6.36}$$

(where $H \geq 0$ is the curvature of ∂K). It follows that the minimum of $|\nabla u_\Omega(x)|^2$ is attained on the exterior boundary, and hence at each point of this boundary since $|\nabla u_\Omega|$ is constant equal to c. Therefore

$$2\frac{\partial u_\Omega}{\partial n}\frac{\partial^2 u_\Omega}{\partial n^2} < 0 \quad \text{on } \partial\Omega.$$

Now we apply (6.36) on $\partial\Omega$, which yields

$$\left(\frac{\partial u_\Omega}{\partial n}\right)^2 H(x) \geq 0 \quad \text{on } \partial\Omega,$$

and the positivity of the curvature of $\partial\Omega$ follows. □

6.3 Star-shapedness

Let us first recall the definition of a star-shaped set (see also Definition 2.4.3):

Definition 6.3.1. A set Ω is star shaped with respect to one of its points x_0 if for any x in Ω, the segment $[x_0, x]$ is completely contained in Ω.

If $x_0 = 0$ (which we will often assume), this is equivalent to saying that $t\Omega \subset \Omega$ for any $t \in [0, 1]$ (where $t\Omega$ denotes the image of Ω by the homothety of center 0 and ratio t). We will also use the following characterization, which is easy to check:

Proposition 6.3.2. *If Ω is a Lipschitz open set, then Ω is star shaped with respect to the origin if and only if $x.n \geq 0$ for almost every $x \in \partial\Omega$ (n denotes the exterior normal vector to $\partial\Omega$ at point x).*

6.3.1 Use of subsolutions and supersolutions

Let us consider again the problem of minimization of the capacity with a volume constraint as in the previous section. Here a compact set K that is star shaped with respect to the origin is given. We saw in (6.29) that any regular open set Ω that minimizes $\mathrm{cap}_\Omega(K)$ with a given volume is such that the associated capacitary potential solves the overdetermined problem

$$\begin{cases} \Delta u = 0 & \text{in } \Omega\backslash K, \\ u = 0 & \text{on } \partial\Omega, \\ u = 1 & \text{on } K, \\ |\nabla u| = c > 0 & \text{on } \partial\Omega. \end{cases} \tag{6.37}$$

To prove the existence of a solution for this overdetermined problem, the variational method of minimizing the capacity with a volume constraint is probably the most natural. Yet there exists another method, inspired by the technique of subsolutions and supersolutions in partial differential equations, introduced in this context by A. Beurling in [43]. For any open set ω containing K, we denote by u_ω the capacitary potential of K relative to ω. We use the term *subsolution* (resp. *supersolution*) of problem (6.37) for any open set ω such that, for any $x \in \partial\omega$,

$$\liminf_{y\to x, y\in\omega} |\nabla u_\omega(y)| \geq c \qquad \Big(\text{resp. } \limsup_{y\to x, y\in\omega} |\nabla u_\omega(y)| \leq c\Big).$$

We have the following result, established by A. Beurling [43] in dimension 2, and generalized in [191] for dimension N:

Theorem 6.3.3. *Let us assume that there exist a subsolution Ω_0 and a supersolution Ω_1 of problem* (6.37) *with $\Omega_0 \subset \Omega_1$. Then, there exists a solution (Ω, u_Ω) to* (6.37) *with $\Omega_0 \subset \Omega \subset \Omega_1$.*

Remark 6.3.4. In practice it is easy to exhibit subsolutions and supersolutions. For a supersolution, we can choose a ball with a large radius R. We prove that it is a supersolution by comparing its capacitary potential to that of an annulus, which can be computed explicitly. For the subsolution, we choose a domain that is closed to K such as, for example, a level set $\{x;\ u(x) > 1 - \varepsilon\}$, where u is a fixed capacitary potential. Let us apply these techniques to the case where K is star shaped. Uniqueness of the solution is also true here as in the convex case.

Theorem 6.3.5. *If K is star shaped, there exists a unique solution Ω to the overdetermined problem* (6.37). *Thus the functional $\omega \to \mathrm{cap}_\omega(K)$ has only one regular minimizer (regular in the sense that the derivative with respect to the domain is well defined; see Chapter 5). Moreover Ω is star shaped.*

Proof. The existence follows from Theorem 6.3.3 and Remark 6.3.4. Let us prove uniqueness. If there exist two solutions Ω_1 and Ω_2, let us denote by u_1 and u_2 their respective capacitary potentials. According to the maximum principle, $\Omega_1 \cap \Omega_2$ is a supersolution. Indeed the capacitary potential $u_{1,2}$ of $\Omega_1 \cap \Omega_2$ satisfies $u_{1,2} \leq \min(u_1, u_2)$ and hence, for example on the common parts of the boundary of Ω_1 and $\Omega_1 \cap \Omega_2$, $u_{1,2} = u_1 = 0$, which implies $|\nabla u_{1,2}| \leq |\nabla u_1| = c$. Since, it is always possible to find a subsolution close to K (see Remark 6.3.4), according to Theorem 6.3.3, there exists a solution Ω_3 included in $\Omega_1 \cap \Omega_2$. In other words, we can always assume that the two solutions Ω_1 and Ω_2 are such that $\Omega_1 \subset \Omega_2$.

Let us apply the Lavrentiev principle. Letting t be the largest real number (less than 1) such that $t\Omega_2 \subset \Omega_1$, we have $t > 0$. Let u_t be the capacitary potential of the domain whose interior boundary is $\partial(tK)$ (which is included in K since K is star shaped) and whose exterior boundary is $\partial(t\Omega_2)$. Obviously, $u_t(x) = u_2(x/t)$. According to the maximum principle, we have $u_t \leq u_1$ in $t\Omega_2$. Now, by the definition of t, there exists a point, say x_0, common to $\partial\Omega_1$ and $\partial(t\Omega_2)$. At this point, by comparison of the gradients,

$$|\nabla u_t(x_0)| = \frac{|\nabla u_2(x_0/t)|}{t} \leq |\nabla u_1(x_0)| = c. \tag{6.38}$$

But since the point x_0/t belongs to $\partial\Omega_2$, the left side of (6.38) must be equal to c/t, which shows that t is necessarily equal to 1, and consequently $\Omega_1 = \Omega_2$.

Let us now prove that Ω is star shaped (this proof comes from [288]). Let us denote by Ω the unique solution of the problem. Let t be a fixed real number, $0 < t < 1$. Let us introduce the domain whose interior boundary is $\partial(tK)$ ($\subset K$) and whose exterior boundary is $\partial(t\Omega)$. Let us denote by u_t the corresponding capacitary potential: $u_t(x) = u(x/t)$. For t in a neighborhood of 1, in order that $K \subset t\Omega_2$, we also introduce the domain whose interior boundary is ∂K and exterior boundary is $\partial(t\Omega)$, with u' denoting the corresponding capacitary potential. By the maximum principle, we have $u_t \leq u'$ in $t\Omega$, and hence, on the exterior boundary, where both functions vanish, we have

$$|\nabla u_t(x)| = \frac{|\nabla u(x/t)|}{t} = \frac{c}{t} \leq |\nabla u'(x)|. \tag{6.39}$$

Therefore $|\nabla u'| \geq c$ on $\partial(t\Omega)$, which means that $t\Omega$ is a subsolution. According to Remark 6.3.4, which ensures the existence of a "large" supersolution, and applying Theorem 6.3.3 together with the uniqueness result, we get $t\Omega \subset \Omega$ for any $t < 1$, which proves the desired result thanks to Proposition 6.3.2. □

6.3.2 Use of a star-shaped rearrangement

As for the case of symmetry, we can also use a rearrangement technique to prove that the solution of a shape optimization problem is star shaped.

For the sake of simplicity, we work in 2 dimensions with polar coordinates. The extension to 3 dimensions is straightforward. We follow the presentation proposed by B. Kawohl in [204], and we refer to this same book for the proofs and more details.

Let Ω be an open set containing the origin. For the rearrangement of Ω into an open set, star shaped with respect to 0, a natural idea would consist in placing in any direction going through the origin, a segment whose length is the "quantity of matter" contained by Ω in this direction:

$$\Omega^* = \{(r,\theta);\ 0 \leq r \leq r(\theta),\ \theta \in [0,2\pi)\},$$

where $r(\theta)$ represents the length or one-dimensional Lebesgue measure of the intersection of Ω with the half-line of angle θ. However, when we use this definition to rearrange the level sets of functions, the new rearranged function does not enjoy the expected good properties as in cases of Steiner or Schwarz symmetrization. This is why we need to change the definition by introducing a weight g along the radius.

In the remainder of this section, we assume that all the open sets contain a ball B_ε of center 0 and with a fixed radius ε. The reason for this technical assumption is to avoid the singularity of the polar coordinates at the origin. Let us also fix a function $g : (0,+\infty) \to (0,+\infty)$ and denote by G one of its antiderivatives. For $\theta \in [0,2\pi)$ fixed, we define the different quantities

$$\Omega(\theta) := \{r \geq \varepsilon \text{ such that } (r,\theta) \in \Omega\},$$

$$l(\theta) := \int_{\Omega(\theta)} g(r)\,dr, \qquad h(\theta) := l(\theta) + G(\varepsilon), \qquad R(\theta) := G^{-1}(h(\theta)).$$

By construction, we can see that $R(\theta)$ does not depend on ε (if $\epsilon_1 < \epsilon$, we have $l_1(\theta) = G(\epsilon) - G(\epsilon_1) + l(\theta)$). We now define the star-shaped rearrangement of Ω as the open set

$$\Omega^* := \{x = (|x|\cos\theta, |x|\sin\theta) \in \mathbb{R}^2,\ 0 \leq |x| < R(\theta)\}, \quad \theta \in [0,2\pi).$$

Similarly, we can define the rearrangement of a function u by rearranging its level sets in the same way. For that purpose, we need a hypothesis ensuring that all level sets contain the fixed ball B_ε:

$$\begin{cases} u : \Omega \to [0,+\infty) \text{ Lipschitz, } u = 0 \text{ on } \partial\Omega, \\ u \text{ attains its maximum at every point of } B(0,\epsilon). \end{cases} \tag{6.40}$$

The star-shaped rearrangement u^* is then defined on Ω^* by

$$u^*(x) := \sup\{c \in \mathbb{R}^+ : x \in \Omega_c^*\},$$

where Ω_c denotes the level set $\Omega_c=\{x : u(x)> c\}$, and Ω_c^* is its star-shaped rearrangement.

In practice, we are going to work with functions g defined by

$$g(r) := r^{1-p} \quad \text{where } p \text{ is a real number greater or equal to } 0.$$

(In dimension 3, we would choose the functions $g(r) := r^{2-p}$). The case $p = 1$ is the "natural" case described above. When we use the function $g(r) := r^{1-p}$, we will denote by $\Omega^{*(p)}$ (resp. $u^{*(p)}$) the star-shaped rearrangement of Ω (resp. u). The first question that comes now is, are these rearrangements equimeasurable? Actually, due to the formula for change of variables in polar coordinates, only the rearrangement corresponding to $p = 0$ is equimeasurable:

Proposition 6.3.6. *For $p = 0$, the area of Ω is equal to the area of its rearrangement $\Omega^{*(0)}$.*

Proof. By definition of $l(\theta)$,

$$|\Omega| = \pi\varepsilon^2 + \int_0^{2\pi} l(\theta)\, d\theta,$$

while

$$|\Omega^{*(0)}| = \pi\varepsilon^2 + \int_0^{2\pi}\int_\varepsilon^{R(\theta)} r\, dr\, d\theta = \pi\varepsilon^2 + \int_0^{2\pi} \frac{R^2(\theta) - \varepsilon^2}{2}\, d\theta.$$

Now by definition of $R(\theta)$, we have $\frac{R^2(\theta)-\varepsilon^2}{2} = h(\theta) - \varepsilon^2/2 = l(\theta)$, and the result follows. □

For $p \neq 0$, the previous result does not hold, but nevertheless we have an inequality (see [204]):

Proposition 6.3.7. *Let Ω be a Lipschitz open set. Then for any $p \geq 1$ we have*

$$\Omega^{*(p)} \subset \Omega^{*(0)}, \quad \textit{and therefore} \quad |\Omega^{*(p)}| \leq |\Omega| = |\Omega^{*(0)}|.$$

Let us now state a theorem, analogous to Theorem 6.1.4, that summarizes the main properties linking u to its star-shaped rearrangement $u^{*(p)}$:

Theorem 6.3.8. *Let $u \in W^{1,p}(\Omega)$ satisfy (6.40) for $p > 1$, and let F be a nondecreasing, continuous function from $\mathbb{R}_+^*$ to $\mathbb{R}$. Then the following inequalities hold:*

- $u^{*(0)}(x) \geq u^{*(p)}(x)$ *for any* $x \in \Omega^{*(p)}$;
- $\int_\Omega F(u) \geq \int_{\Omega^{*(p)}} F(u^{*(p)})$;

- $\int_\Omega |\nabla u|^p \geq \int_{\Omega^{*(p)}} |\nabla u^{*(p)}|^p$.

Let us now give an application of these properties, similar to the one presented in the previous section. We consider a compact set $K \subset \mathbb{R}^N$, $N \geq 2$, *star shaped with respect to every point in the ball B_ε centered at the origin and of radius ε.*

Let J be the functional defined by

$$J(\Omega) := \int_\Omega |\nabla u_\Omega|^2 + c^2|\Omega|, \tag{6.41}$$

where u_Ω is the capacitary potential of K in Ω. We prove a first result about the minima of J, that uses the star-shaped property of K. This result can easily be generalized to other functionals of the same kind. This theorem can also be a first step to prove a uniqueness result for a shape optimization problem.

Theorem 6.3.9. *Let Ω_1 and Ω_2 be two open sets that minimize the functional J defined by* (6.41). *Then either $\Omega_1 \subset \Omega_2$ or $\Omega_2 \subset \Omega_1$.*

To prove this theorem, we adapt an idea of Friedman and Phillips described in [145]. We first need the following lemma, which is also interesting in itself.

Lemma 6.3.10. *If Ω_1 and Ω_2 are two minimizers of the functional J, then $\Omega_1 \cap \Omega_2$ and $\Omega_1 \cup \Omega_2$ are also minimizers.*

Proof. Let us denote by u_1, u_2, v, and w the capacitary potentials of K respectively associated to Ω_1, Ω_2, $\Omega_1 \cap \Omega_2$, and $\Omega_1 \cup \Omega_2$. Let us introduce the functions $v_1 = \min(u_1, u_2) \in H_0^1(\Omega_1 \cap \Omega_2)$ and $w_1 = \max(u_1, u_2) \in H_0^1(\Omega_1 \cup \Omega_2)$. Since Ω_1 and Ω_2 are minimizers, we already know that

$$J(\Omega_1) = J(\Omega_2) \leq J(\Omega_1 \cap \Omega_2) \tag{6.42}$$

and

$$J(\Omega_1) = J(\Omega_2) \leq J(\Omega_1 \cup \Omega_2). \tag{6.43}$$

By the definition of capacity,

$$\int_{\Omega_1\cap\Omega_2} |\nabla v|^2 \leq \int_{\Omega_1\cap\Omega_2} |\nabla v_1|^2, \qquad \int_{\Omega_1\cup\Omega_2} |\nabla w|^2 \leq \int_{\Omega_1\cup\Omega_2} |\nabla w_1|^2. \tag{6.44}$$

Now, it is classical to check (by decomposing $\Omega_1 \cup \Omega_2$ into the union of subsets where $u_1 < u_2$, $u_1 > u_2$, and $u_1 = u_2$) that

$$\int_{\Omega_1} |\nabla u_1|^2 + \int_{\Omega_2} |\nabla u_2|^2 = \int_{\Omega_1\cup\Omega_2} |\nabla w_1|^2 + \int_{\Omega_1\cap\Omega_2} |\nabla v_1|^2. \tag{6.45}$$

Therefore, summing the inequalities (6.44) yields

$$\int_{\Omega_1\cap\Omega_2} |\nabla v|^2 + \int_{\Omega_1\cup\Omega_2} |\nabla w|^2 \le \int_{\Omega_1} |\nabla u_1|^2 + \int_{\Omega_2} |\nabla u_2|^2. \tag{6.46}$$

Moreover, since

$$c^2\left(|\Omega_1 \cup \Omega_2| + |\Omega_1 \cap \Omega_2|\right) = c^2\left(|\Omega_1| + |\Omega_2|\right), \tag{6.47}$$

summing (6.46) and (6.47) produces

$$J(\Omega_1 \cup \Omega_2) + J(\Omega_1 \cap \Omega_2) \le J(\Omega_1) + J(\Omega_2). \tag{6.48}$$

However, (6.48), together with (6.42) and (6.43), proves that equality holds:

$$J(\Omega_1 \cup \Omega_2) = J(\Omega_1 \cap \Omega_2) = J(\Omega_1) = J(\Omega_2). \tag{6.49}$$

The lemma is proved. We can also deduce that we have equality in the inequalities (6.44), namely $v = v_1$ and $w = w_1$, by the uniqueness of the capacitary potential. □

Proof of Theorem 6.3.9. Since K is star shaped with respect to every point of the ball B_ε, each ray coming from the origin meets the boundary of K at only one point, and we can construct a homeomorphism of ∂K onto the sphere of center O and of radius ε (as in [41, Sect. 11.3]). This proves, in particular, that ∂K is connected in dimension $N \ge 2$.

Let us now consider the open set $\Omega_1 \cap \Omega_2 \backslash K$. Since ∂K is connected, this open set has only one connected component that has ∂K on its boundary. Let us prove that there are no other connected components. If ω was another such connected component, its boundary would be contained in $\partial\Omega_1 \cup \partial\Omega_2$. Thus the function $v = \min(u_1, u_2)$, which is harmonic according to the proof of the previous lemma, and which is zero on $\partial\omega$, would be identically zero. This is impossible because u_1 and u_2 are positive on $\Omega_1 \cap \Omega_2$.

Therefore the open set $\Omega = \Omega_1 \cap \Omega_2 \backslash K$ is connected. The function $w = \max(u_1, u_2)$ is also harmonic on Ω, by the proof of the previous lemma. Consequently, the harmonic function $u_1 - w$, which is nonpositive on $\partial\Omega$, is either negative in Ω or identically zero. The first case means that $u_1 < u_2$ in Ω, which is only possible if $\Omega_1 \subset \Omega_2$. Otherwise, if Ω_1 contains a part of the boundary of Ω_2, the inequality $u_1 < u_2$ cannot hold. The second case implies that $u_2 \le u_1$, but using the same reasoning with $u_2 - w$, we prove that either $\Omega_2 \subset \Omega_1$ or $u_1 = u_2$, which means that $\Omega_1 = \Omega_2$. □

Let us come back to the initial question: we wanted to prove that the minima of J are star shaped. More precisely, we can state the following.

Theorem 6.3.11. *Let us assume that K is star shaped with respect to every point of a ball B_ϵ centered at the origin. If there exists a Lipschitz open set Ω that minimizes the functional J, then it is unique and star shaped with respect to the origin.*

Proof. Let us denote by u the capacitary potential of the open set Ω that minimizes of the functional J. Let $u^{*(2)}$ denote the star-shaped rearrangement of u. According to Proposition 6.3.7 and Theorem 6.3.8 we have

$$J(\Omega^{*(2)}) = \int_{\Omega^{*(2)}} |\nabla u^{*(2)}(x)|^2 + c^2|\Omega^{*(2)}| \leq \int_\Omega |\nabla u(x)|^2 + c^2|\Omega| = J(\Omega). \quad (6.50)$$

This means that $\Omega^{*(2)}$ is a minimum of J because it also contains K. In particular $|\Omega^{*(2)}| = |\Omega|$, as otherwise the above inequality would be strict. Moreover, according to Theorem 6.3.9, one of the two minima is contained in the other. Consequently, because both are Lipschitz, they coincide. □

6.4 Some other topological and geometrical properties

6.4.1 Connectedness

Another natural question is whether the solution of a shape optimization problem is connected or not. It is a priori a difficult question. In some sense, it is related to topological optimization, which we will discuss in the next chapter. In this section, we are simply going to give two examples for eigenvalues of the Dirichlet–Laplacian. In the first case (that of λ_2), we prove that the minimum is necessarily disconnected and, more precisely, it is the union of two identical balls. In the second case (that of λ_3), we can prove that the minimum is connected but we are not able to identify it. We also give an example of a situation where one can prove the nonconnectedness of the solution by passing to the limit.

6.4.1.1 Eigenvalues of the Laplacian. Let us begin with the case of the second eigenvalue of the Dirichlet–Laplacian. We are going to use the Rayleigh–Faber–Krahn inequality proved in Theorem 6.1.9.

Theorem 6.4.1. *The minimum of $\lambda_2(\Omega)$ (the second eigenvalue of the Dirichlet–Laplacian) among quasi-open sets of given volume m, is attained for the union of two identical balls of volume $m/2$, and only in this case.*

Proof. Let Ω be any quasi-open set. Let $\varphi \in H^1_0(\Omega)$ be an eigenfunction associated to $\lambda_2(\Omega)$ satisfying

$$\forall\, v \in H^1_0(\Omega), \quad \int_\Omega \nabla\varphi\nabla v = \lambda_2(\Omega)\int_\Omega \varphi v, \qquad \int_\Omega \varphi^2 = 1, \ \int_\Omega \varphi\varphi_1 = 0, \quad (6.51)$$

where φ_1 is an eigenfunction associated to the first eigenvalue of Ω. According to the variational characterization (we can replace φ_1 by $|\varphi_1|$), we know that $\varphi_1 \geq 0$ on Ω.

Let us define $\omega_1 = [\varphi > 0]$. We can assume that $|\omega_1| > 0$ (otherwise, we change φ into $-\varphi$). Then $\psi := \varphi_{|\omega_1} = \varphi^+ \in H_0^1(\omega_1)$. Thus applying (6.51) to the functions $v \in H_0^1(\omega_1)$ shows that ψ is an eigenfunction on ω_1 associated to the eigenvalue $\lambda_2(\Omega)$. In particular, $\lambda_2(\Omega) \geq \lambda_1(\omega_1)$.

Now let $\omega_2 = \Omega \backslash \omega_1$ and $\zeta := \varphi_{|\omega_2} = -\varphi^- \in H_0^1(\omega_2)$. If $\zeta \equiv 0$, since $0 = \int_\Omega \varphi\varphi_1 = \int_{\omega_1} \psi\varphi_1$, this implies that $\varphi_1 \equiv 0$ on ω_1 and hence $\varphi_1 \in H_0^1(\omega_2)$, $\varphi_1 \not\equiv 0$. Therefore $\lambda_1(\Omega)$ is also an eigenvalue of ω_2 and thus $\lambda_2(\Omega) \geq \lambda_1(\Omega) \geq \lambda_1(\omega_2)$. If $\zeta \not\equiv 0$, by applying (6.51) to any $v \in H_0^1(\omega_2)$, we see that $\lambda_2(\Omega)$ is also an eigenvalue of ω_2 and, in particular, $\lambda_2(\Omega) \geq \lambda_1(\omega_2)$. We have proved that, in any case,

$$\lambda_2(\Omega) \geq \max\{\lambda_1(\omega_1), \lambda_1(\omega_2)\}.$$

Let us now introduce the domain Ω^* defined as the union of two disjoint balls ω_1^* and ω_2^* with the same volume as ω_1 and ω_2 respectively. Any eigenvalue of Ω^* is also an eigenvalue of ω_1^*, ω_2^*. Conversely, if u is an eigenfunction of ω_1^*, the function $\tilde{u}$ equal to u on ω_1^* and zero on ω_2^*, is an eigenfunction of Ω for the same eigenvalue (and symmetrically if we exchange the roles of ω_1^* and ω_2^*). The monotonicity of the eigenvalues with respect to inclusion yields that, for example, if $|\omega_1^*| \geq |\omega_2^*|$, then $\lambda_1(\Omega^*) = \lambda_1(\omega_1^*)$ and $\lambda_2(\Omega^*) \leq \lambda_1(\omega_2^*) = \max(\lambda_1(\omega_1^*), \lambda_1(\omega_2^*))$. Therefore

$$\lambda_2(\Omega^*) \leq \max(\lambda_1(\omega_1^*), \lambda_1(\omega_2^*)) \leq \max(\lambda_1(\omega_1), \lambda_1(\omega_2)) \leq \lambda_2(\Omega), \tag{6.52}$$

the second inequality above coming from Theorem 6.1.9 (and it is strict if Ω is not a ball). The relation (6.52) proves that the minimum of λ_2 has to be sought among the unions of two balls, and we easily check that the optimal solution corresponds to two equal balls, and only to this case. □

The case of the other eigenvalues of the Dirichlet–Laplacian is much more difficult and remains essentially open. Let us nevertheless give a result due to Wolf and Keller [293] which, in some sense, generalizes the previous theorem. It is also useful to reduce the number of possibilities when looking for a minimizer. In particular, this result will imply that the minimum of λ_3 is attained by a connected set in dimension 2.

Let us recall that we saw in Chapter 4 (Corollary 4.7.12) that there exists a (quasi-open) set that minimizes λ_n among all quasi-open sets of $\mathbb{R}^N$ of volume 1 contained within some fixed domain, and existence holds also in $\mathbb{R}^N$; see (4.58) and the comments around it in Section 4.8. Let us denote by D_n^* such a set that realizes the minimum in $\mathbb{R}^N$ and by $\lambda_n^* = \lambda_n(D_n^*)$ the minimum value of λ_n. As usual, we denote by $t\Omega$ the image of the set Ω by a homothety of ratio t. We have

Theorem 6.4.2 (Wolf–Keller). *Let us assume that D_n^* is the union of (at least) two disjoint sets, each of them with a positive measure. Then*

$$\left(\lambda_n^*\right)^{N/2} = \min_{1\le j\le (n-1)/2} \left(\left(\lambda_j^*\right)^{N/2} + \left(\lambda_{n-j}^*\right)^{N/2}\right). \tag{6.53}$$

If i is a value of $j \le (n-1)/2$ for which the above minimum is achieved, then we can choose

$$D_n^* = \left[\left(\frac{\lambda_i^*}{\lambda_n^*}\right)^{1/2} D_i^*\right] \bigcup \left[\left(\frac{\lambda_{n-i}^*}{\lambda_n^*}\right)^{1/2} D_{n-i}^*\right]. \tag{6.54}$$

Proof. Let us write $D_n^* = D_1 \cup D_2$ (disjoint union) with $|D_1| > 0$, $|D_2| > 0$, and $|D_1| + |D_2| = 1$. Let u_n^* be an eigenfunction of the Laplacian on D_n^*, corresponding to the eigenvalue λ_n^*. Then u_n^* cannot be identically zero on one of the components of D_n^*, say for example D_1. In particular, λ_n^* is an eigenvalue of D_1: $\lambda_n^* = \lambda_i(D_1)$ for an integer $i \le n$ and we denote by i precisely the largest of those integers. We cannot have $i = n$, since otherwise we could decrease λ_n^* by dilating D_1 up to the volume of D_n^*, which would contradict the minimality of λ_n^*. Thus $i \le n-1$. Since λ_n^* is the nth eigenvalue of D_n^*, there must exist at least $n-i$ eigenvalues of D_2 smaller than λ_n^* in such a way that $\lambda_{n-i}(D_2) \le \lambda_n^*$. If $\lambda_{n-i}(D_2) < \lambda_n^*$, we could decrease $\lambda_n^* = \max\{\lambda_i(D_1), \lambda_{n-i}(D_2)\}$ by dilating D_1 and shrinking D_2, keeping the total volume equal to 1. This would also contradict the minimality of λ_n^*. Therefore finally, $\lambda_{n-i}(D_2) = \lambda_i(D_1) = \lambda_n^*$.

Now we still get a minimum for λ_n^* by replacing D_1 by $|D_1|^{1/N} D_i^*$ (which has the same volume and a better possible λ_i) and by replacing D_2 by $|D_2|^{1/N} D_{n-i}^*$. Consequently we have

$$\lambda_i(D_1) = |D_1|^{-2/N}\lambda_i^* = \lambda_n^* = |D_2|^{-2/N}\lambda_{n-i}^* = \lambda_{n-i}(D_2).$$

Finally the condition $|D_1| + |D_2| = 1$ yields $\left(\lambda_n^*\right)^{N/2} = \left(\lambda_i^*\right)^{N/2} + \left(\lambda_{n-i}^*\right)^{N/2}$.

Let us now consider the set $\widetilde{D}_j$ defined for $j = 1, \dots, n-1$ by

$$\widetilde{D}_j = \left[\left(\frac{\left(\lambda_j^*\right)^{N/2}}{\left(\lambda_j^*\right)^{N/2} + \left(\lambda_{n-j}^*\right)^{N/2}}\right)^{1/N} D_j^*\right] \bigcup \left[\left(\frac{\left(\lambda_{n-j}^*\right)^{N/2}}{\left(\lambda_j^*\right)^{N/2} + \left(\lambda_{n-j}^*\right)^{N/2}}\right)^{1/N} D_{n-j}^*\right]. \tag{6.55}$$

Each $\widetilde{D}_j$ has volume 1 and the same jth eigenvalue as its first component, together with the same $(n-j)$th eigenvalue of its second component, which are equal to

$$\left(\left(\lambda_j^*\right)^{N/2} + \left(\lambda_{n-j}^*\right)^{N/2}\right)^{2/N}.$$

It follows that $\lambda_n(\widetilde{D}_j)$ is also given by the same value. Since $\lambda_n^* \le \lambda_n(\widetilde{D}_j)$ and $\lambda_n^* = \lambda_n(\widetilde{D}_i)$ for some index i, λ_n^* is the minimal value of $\lambda_n(\widetilde{D}_j)$. Moreover $\widetilde{D}_i$

is optimal for any index i that realizes the minimum in (6.53) and the theorem is proved. □

Remark 6.4.3. The previous proof can be extended to quasi-open sets D_1, D_2 such that $\mathrm{cap}(D_1 \cap D_2) = 0$. For that purpose, we can use the property that $H^1_0(D_1 \cup D_2) = H^1_0(D_1) \cup H^1_0(D_2)$ (see [66]).

Generally the value of λ^*_n is not known except for $n = 1$ and 2 (see Chapter 1). Nevertheless it can be approximated numerically. The previous theorem is also an interesting tool to prove that the optimal set is connected:

Corollary 6.4.4. *The open set of the plane D^*_3, of measure 1, that minimizes the third eigenvalue of the Dirichlet–Laplacian, is connected.*

Proof. Indeed, let us assume that the optimal set is not connected. According to the Wolf–Keller theorem, we have $\lambda^*_3 = \lambda^*_1 + \lambda^*_2$ ($i = 1$ is the only possible value). Now, we know that $\lambda^*_1 = \pi j^2_{01} = 18.168\ldots$ (where j_{01} is the first zero of the Bessel function J_0), while, according to Theorem 6.4.1, $\lambda^*_2 = 2\lambda^*_1 = 36.336\ldots$. Thus $\lambda^*_1 + \lambda^*_2 = 54.504\ldots$. However, by definition λ^*_3 is less than or equal to the third eigenvalue of the disk of area 1, which is $\lambda_3(D_1) = \pi j^2_{11} = 46.125\ldots$. This shows that λ^*_3 cannot be equal to $\lambda^*_1 + \lambda^*_2$. □

Remark 6.4.5. It is also possible to prove that D^*_3 is connected in dimension 3 (see [69]). The conjecture states that D^*_3 is a ball in dimension 2 and the union of three identical balls in dimension greater than or equal to 4. Numerical simulations show that it is a "pinched ball" in dimension 3 (see the discussion in Chapter 1).

In the paper [293], Wolf and Keller give some credit to the conjecture in the plane by proving that the disk is a local minimizer for λ_3. Since λ_3 is a double eigenvalue of the disk, we cannot use simply the shape derivative (a double eigenvalue is not differentiable, so we would need to use subgradients for example). Wolf and Keller use sophisticated perturbation arguments, in the spirit of Lord Rayleigh, to prove that the third eigenvalue of a domain Ω_ε, given in polar coordinates by $r = R(\theta, \varepsilon)$ where R has the expansion

$$R(\theta, \varepsilon) = 1 + \varepsilon \sum_{n=-\infty}^{\infty} a_n e^{in\theta} + \varepsilon^2 \sum_{n=-\infty}^{\infty} b_n e^{in\theta} + O(\varepsilon^3), \tag{6.56}$$

satisfies

$$\lambda_3(\Omega_\varepsilon) = \pi j^2_{11}(1 + 2|\varepsilon|\,|a_2|) + O(\varepsilon^2).$$

In the case where $a_2 \neq 0$, we immediately get the desired result. When $a_2 = 0$, we must consider the expansion a little further and we reach the same conclusion.

6.4.1.2 Case of nonconnectedness. Let us now give an example of nonconnectedness for the problem of minimization of capacity with a given volume. This example comes from [139]. Let an open set Ω be given in $\mathbb{R}^2$, for which we look for $K \subset \Omega$, with a given area, that minimizes the functional $J(K) := \operatorname{cap}_\Omega K$ (see Exercise 4.11). Let us consider the case when the open set Ω is made of two adjacent disks of radius 1 connected with a thin channel, as in Figure 6.5.

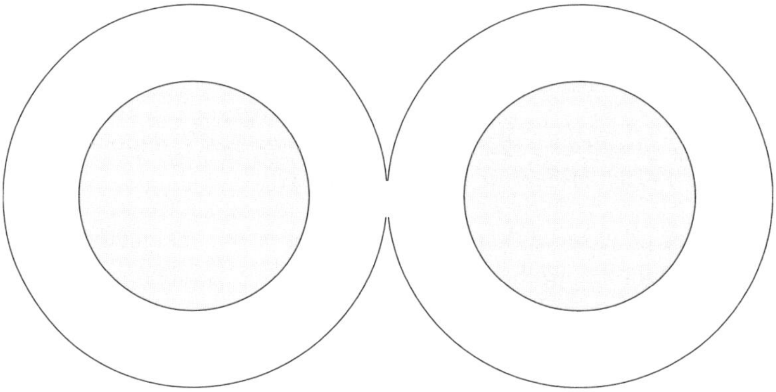

Figure 6.5. An example of nonconnectedness for the minimization of capacity.

When the width of the channel tends to 0, Ω converges to $\widetilde{\Omega} = B((-1,0);1) \cup B((1,0);1)$. Now, for the open set $\widetilde{\Omega}$, it is easy (using Schwarz symmetrization) to check that the minima of the capacity are to be chosen among

$$A_1 = B((-1,0);\rho), \quad A_2 = B((1,0);\rho), \quad \text{or} \quad A_3 = B((-1,0);r) \cup B((1,0);r),$$

with $2r^2 = \rho^2$. Now (see (3.56)),

$$\operatorname{cap}_{\widetilde{\Omega}}(A_1) = \operatorname{cap}_{\widetilde{\Omega}}(A_2) = -2\pi/\log\rho, \qquad \operatorname{cap}_{\widetilde{\Omega}}(A_3) = -4\pi/\log r.$$

Thus for $r > \frac{1}{2}$, we have $\operatorname{cap}_{\widetilde{\Omega}}(A_3) < \operatorname{cap}_{\widetilde{\Omega}}(A_1) = \operatorname{cap}_{\widetilde{\Omega}}(A_2)$. Therefore the solution for $\widetilde{\Omega}$ is not connected. By continuity, it follows that the solution for Ω is not connected when the channel is thin enough.

Let us remark that for $r = \frac{1}{2}$, the problem on $\widetilde{\Omega}$ has 3 distinct minima, 2 of which are connected.

6.4.2 A geometric property of the normal

The moving plane method explained in Section 6.1.3 can be used to prove some properties different from symmetry. Let us consider for example the following shape optimization problem.

Let a function $f \in L^2(\mathbb{R}^N)$ be given with a compact support K, and let us denote by K^* the convex hull of K. For any open set ω containing K, we introduce the function $u_\omega := u_\omega^f$ solving the usual Dirichlet problem on ω, and we consider the functional

$$J(\omega) := \int_\omega |\nabla u_\omega(x)|^2 - f u_\omega(x)\, dx + a^2|\omega|.$$

Theorem 6.4.6 (Shahgholian). *Any regular open set Ω that minimizes the functional J satisfies what is known as the* geometric property of the normal: *at any point $x \in \partial\Omega$, the interior normal vector (when it exists) intersects K^*, the convex hull of K.*

Proof. We give just the main arguments of the proof (see [276] for the details). Let us assume, for the purpose of contradiction, that there exists a point $x \in \partial\Omega \setminus K^*$ whose interior normal vector $\nu(x)$ does not intersect K^*. Then there exists a plane T containing $\nu(x)$ and such that $T \cap K^* = \emptyset$. Let us denote by η the unit vector orthogonal to T and opposite to K^*. We can write the plane T as

$$T = \{y \in \mathbb{R}^N;\ \eta \cdot y = t_0\}$$

for some t_0, and we denote more generally by T_t the plane $\{y \in \mathbb{R}^N;\ \eta \cdot y = t\}$ (and thus $T = T_{t_0}$). We set

$$\Omega_t = \{y \in \Omega;\ \eta \cdot y > t\}$$

and let $\mathrm{Ref}(\Omega_t)$ be the reflection of Ω_t with respect to T_t (see Figure 6.6).

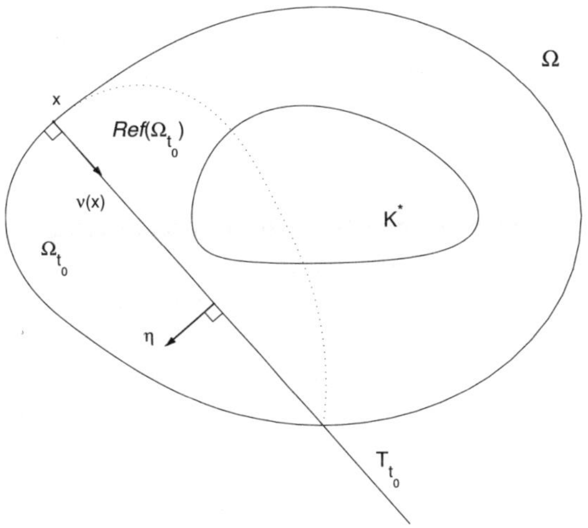

Figure 6.6. The geometric property of the normal (the plane T_{t_0} should meet K^*).

Here T_{t_0} is orthogonal to $\partial\Omega$ at the point x since it contains the normal vector to $\partial\Omega$ at x. We can always assume that t_0 is the largest real number t for which this situation occurs (otherwise, we could translate in the direction of η until the situation where we meet another plane T_t orthogonal to $\partial\Omega$).

Let us first prove the following.

Lemma 6.4.7. $\mathrm{Ref}(\Omega_{t_0}) \subset \Omega$.

Proof. Let us assume that $\mathrm{Ref}(\Omega_{t_0}) \not\subset \Omega$. Let

$$t' = \sup\{t;\ \mathrm{Ref}(\Omega_t)\backslash\Omega \neq \emptyset\}.$$

Then $\mathrm{Ref}(\Omega_{t'})$ is in Ω and its boundary is internally tangent to $\partial\Omega$ at some point $x^0 \notin T_{t'}$.

Let us introduce $\tilde{u}$ such that $\tilde{u}(y) = u(y')$ (y' being the reflection of y with respect to $T_{t'}$). Then $v = u - \tilde{u}$ is superharmonic in $\mathrm{Ref}(\Omega_{t'})$ and positive on $\partial(\mathrm{Ref}(\Omega_{t'}))$. Consequently either $v = 0$ on one of the connected components of $\mathrm{Ref}(\Omega_{t'})$ (and we prove that is not possible), or $v > 0$ on $\mathrm{Ref}(\Omega_{t'})$ and attains its minimum at x^0. But then, according to the maximum principle of Hopf,

$$0 > \frac{\partial v}{\partial \nu}(x^0) = \frac{\partial u}{\partial \nu}(x^0) - \frac{\partial \tilde{u}}{\partial \nu}(x^0) = -1 + 1 = 0,$$

where ν is the interior normal vector to Ω at the point x^0. This gives the desired contradiction. □

Now we prove, following Serrin in [275], that x is a double zero of the function $v = u - \tilde{u}$ ($\tilde{u}$ being the function defined by $\tilde{u}(y) = u(y')$, where y' is the reflection of y with respect to T_{t_0}), and we finish the proof by using Lemma 6.1.12 by Serrin, stated in Section 6.1.3. □

Remark 6.4.8. This geometric property can be used a posteriori to prove the existence of a minimum, as is shown in [32]. Indeed, since we know that the minimizer must possess this geometric property of the normal, it is possible to minimize the functional J in the class of Lipschitz open sets that have this property. This geometric constraint gives enough compactness and also continuity to prove existence. Moreover, we can prove that this geometric property is preserved by Hausdorff convergence.

6.4.3 Some other geometric properties

Let us now give, without proof, a result showing how some geometric properties of the data can be transferred to the solution. For this purpose, let us again consider the problem of minimizing the capacity of a given compact set K (see (6.26), (6.29)), penalized by the volume: here the functional is

$$J(\omega) = \int_{\omega} |\nabla u_{\omega}|^2 + c^2|\omega|,$$

where u is the solution of (6.26).

Let us denote by $\Omega \subset \mathbb{R}^2$ a minimizer of J, which we assume to be a regular open set. The following theorem is due to Acker; see [2], [3] and also see [223] for similar results. Roughly speaking, it states that Ω cannot have a shape more complicated than K.

Theorem 6.4.9 (Acker). *Let e be a fixed unit vector. We assume that the boundary of K is a curve of class C^2 and that it contains only a finite number p of segments σ (possibly degenerated to a point), where the normal vector to ∂K is $\pm e$. Let $p_1, p_2, \dots, p_n$ be the set of points of $\partial\Omega$ having e as a normal vector (interior or exterior) numbered in order on the curve $\partial\Omega$.*

Then $n \le p$, and to every point p_i we can associate a segment σ_i of ∂K such that

- *the segments $\sigma_1, \sigma_2, \dots, \sigma_n$ are distinct and numbered in increasing order on ∂K;*
- *if p_i is a local maximum (resp. minimum) in the direction e, then σ_i is a local maximum (resp. minimum) on ∂K in the direction e.*

Moreover if K is convex in one direction, Ω is also convex in the same direction.

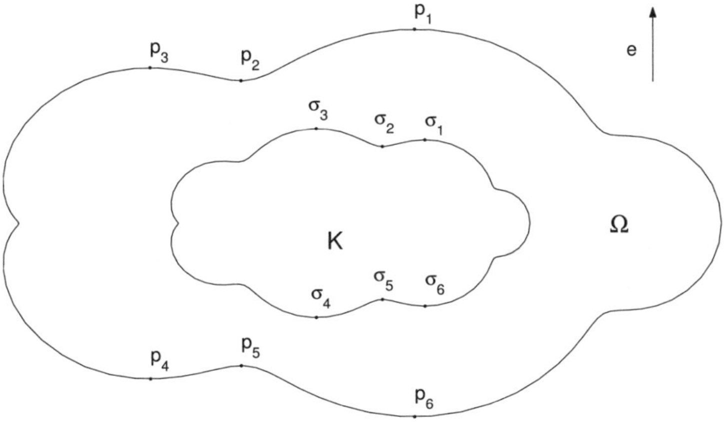

Figure 6.7. The exterior domain Ω^* is simpler than the data K.

The proof of this theorem heavily relies on conformal maps and is therefore restricted to dimension 2 (see the above references).

Chapter 7

Relaxation and homogenization

7.1 Introduction

In this chapter, we present the techniques of relaxation and homogenization that are often used in shape optimization problems and we describe the link between them. We saw in Chapter 4 that many problems do not have optimal shapes. It is therefore useful to look for a solution outside the class of domains of $\mathbb{R}^N$.

The extension of the notion of solution and the need to use a wider class of objects in which we try to minimize some functional is the principle of *relaxation*. It is a general principle that generally allows us to find an optimal object that replaces an optimal shape when the latter does not exist. But this technique can also be employed to prove existence of an optimal shape in some cases: after proving existence of a solution in the relaxed framework, we write the optimality conditions and we are sometimes able to prove that the optimal *relaxed* object is indeed a shape. We will give two such examples in Section 7.2.3. Another idea is to use the optimal relaxed object to find a quasi-optimal solution. This is done, for example, in structural mechanics (see, e.g., [12], [35], [37] and the references therein), where the quasi-optimal shape is obtained from the homogenization solution through a penalization technique. We also refer, for these questions, to the lecture notes by L. Tartar on homogenization [287].

Before introducing relaxation, let us make the link between the different notions of γ-convergence, Γ-convergence, G-convergence, and convergence in the sense of Mosco. These notions were introduced in recent decades independently, but are closely related. Let us recall that γ-convergence was introduced in Chapter 3.

Once again, we choose to work in the framework of homogeneous Dirichlet boundary conditions, because the situation is much simpler than for other boundary conditions. Nevertheless, we will also explain what can be adapted to other cases.

7.1.1 Γ-convergence.

Γ-convergence was introduced by E. de Giorgi in 1975 to answer the following question. Let us consider a sequence of minimization problems associated to functionals J_n. What is the topology we have to consider on the underlying metric space in order that the convergence of J_n to J implies the convergence of the minima of J_n to the minima of J?

Definition 7.1.1. Let $(J, (J_n))$ be a sequence of functionals defined on a metric space X. We say that J_n Γ-converges to J (denoted by $J_n \xrightarrow{\Gamma} J$) if for any $x \in X$,

(i) $\forall\, x_n \to x,\ J(x) \le \liminf J_n(x_n)$;

(ii) $\exists\, x_n \to x,\ J(x) \ge \limsup J_n(x_n)$.

In the mathematical literature, this notion is sometimes named differently (see, e.g., [28] for epi-convergence). Let us now give a fundamental result that is actually stronger than simple convergence of minima (see, e.g., [116, Cor. 7.20]).

Theorem 7.1.2 (de Giorgi). *Let J_n be a sequence of functionals that Γ-converges to a functional J and that is bounded from below independently of n. Assume there exists a sequence ε_n of positive real numbers converging to 0, and a sequence x_n relatively compact, such that*

$$J_n(x_n) \le \inf J_n + \varepsilon_n.$$

Then

(1) J has a minimum and $\min J = \lim_{n\to\infty} \inf J_n$*;*

(2) any accumulation point of the sequence x_n achieves the minimum of J.

For the proof of this theorem and the others stated below, we refer to the literature on this topic, like [116] or [28]. Let us remark that, by construction, any Γ-limit is lower semicontinuous.

Another important result is the compactness of Γ-convergence:

Theorem 7.1.3. *If X is separable, then Γ-convergence is sequentially compact: from any sequence of functionals J_n defined on X with values in $\overline{\mathbb{R}}$, one can extract a subsequence that Γ-converges.*

Let us finally quote a useful "stability" result: if J_n Γ-converges to J and if K is a continuous functional, then $J_n + K$ Γ-converges to $J + K$.

7.1.1.1 Link with γ-convergence. Let us come back to the example of the Dirichlet problem on a bounded open set $\Omega \subset D$ with $f \in L^2(D)$:

$$\begin{cases} -\Delta u_\Omega = f \text{ in } \Omega, \\ u_\Omega \in H_0^1(\Omega). \end{cases} \tag{7.1}$$

We know that u_Ω can be characterized as the unique minimum of the functional $\tilde{J}_\Omega$ defined on $H_0^1(\Omega)$ by

$$\tilde{J}_\Omega(v) = \frac{1}{2}\int_\Omega |\nabla v(x)|^2\,dx - \int_\Omega f v(x)\,dx.$$

Now if we consider a sequence of domains Ω_n, one cannot directly consider the Γ-convergence of the sequence of functionals $\tilde{J}_{\Omega_n}$ since these are not defined on the same space. It is not enough to consider the functional $\tilde{J}_\Omega$ on $H_0^1(D)$ since then we lose the information that the minimum is to be searched for in $H_0^1(\Omega)$. Therefore the idea consists in introducing this information by "penalizing" the functions that are not in $H_0^1(\Omega)$ via the indicator of $H_0^1(\Omega)$: we introduce

$$\forall\, v \in H_0^1(D), \quad I_\Omega(v) = \begin{cases} 0 & \text{if } v \in H_0^1(\Omega), \\ +\infty & \text{otherwise.} \end{cases}$$

Then we consider the new functional J_Ω defined on $H_0^1(D)$ by

$$J_\Omega(v) = \frac{1}{2}\int_D |\nabla v(x)|^2\, dx - \int_D f v(x)\, dx + I_\Omega(v). \tag{7.2}$$

By construction, u_Ω a solution of (7.1) is the unique minimum of J_Ω on $H_0^1(D)$ (because $\tilde{J}_\Omega(v) = J_\Omega(v)$ if $v \in H_0^1(\Omega)$ and $+\infty$ otherwise).

The link between γ-convergence and Γ-convergence is described by

Proposition 7.1.4. *A sequence of open sets Ω_n γ-converges to Ω if and only if, for any $f \in L^2(D)$, J_{Ω_n} Γ-converges to J_Ω.*

Proof. The sufficient condition is a simple application of de Giorgi's Theorem 7.1.2 together with the characterization of $u_{\Omega,f}$ as the minimum of the functional J_Ω.

To prove the necessary condition, we use the characterization of γ-convergence in terms of the Mosco convergence of corresponding Sobolev spaces (see Proposition 3.5.5).

We want to prove part (ii) of the definition of Γ-convergence. Let $v \in H_0^1(D)$. Either $v \notin H_0^1(\Omega)$ and then the result is obvious since the left-hand side of the inequality is $+\infty$, or $v \in H_0^1(\Omega)$ and then, according to (M1) of Mosco convergence, there exists a sequence $v_n \in H_0^1(\Omega_n)$ that converges strongly to v in $H_0^1(D)$. But for v_n and v, we have $J_\Omega(v) = \tilde{J}_\Omega(v)$ and $J_{\Omega_n}(v_n) = \tilde{J}_{\Omega_n}(v_n)$. Passing to the limit, thanks to the strong convergence of v_n to v, it follows that $\lim J_{\Omega_n}(v_n) = J_\Omega(v)$ and part (ii) is proved.

Let us now prove part (i). Let v_n be a sequence converging to v (strongly in $H_0^1(D)$). If $v_n \notin H_0^1(\Omega_n)$, except for a finite number of integers n, the right-hand side of the inequality in part (i) is $+\infty$ and the result is obvious. Otherwise, there exists a subsequence $v_{n_k} \in H_0^1(\Omega_{n_k})$ that converges to v. Now we use (M2) of Mosco convergence: $v \in H_0^1(\Omega)$ and we conclude as above by passing to the limit in the expression of $J_{\Omega_{n_k}}(v_{n_k}) = \tilde{J}_{\Omega_{n_k}}(v_{n_k})$. □

We saw that the compactness of Γ-convergence always allows us to extract from the sequence J_{Ω_n} a subsequence that is Γ-convergent. The main questions of the relaxation process are, is it possible to identify its limit and what happens if this limit is not of the kind J_Ω? We will discuss these questions in Section 7.2.2.

7.1.2 *G*-convergence

The similar notion of G-convergence is defined in homogenization theory. Let us give an example in thermal engineering. Let us assume that the media Ω is obtained as a mixture of different components and let us denote by $a_\varepsilon(x)$ its conductivity (as a function of $x \in \Omega$). We assume that this conductivity depends on a (small) parameter ε. If Ω is heated with a source f (with its boundary maintained in the ice), the temperature into the media can be obtained by solving the following stationary heat equation:

$$\begin{cases} -\operatorname{div}(a_\varepsilon(x)\nabla u_\varepsilon) = f & \text{in } \Omega, \\ u_\varepsilon = 0 & \text{on } \partial\Omega. \end{cases} \tag{7.3}$$

As a definition of G-convergence, we introduce

Definition 7.1.5. The sequence of conductivities $a_{\varepsilon_n}(x)$ is said to G-converge as ε_n tends to 0 to a homogenized conductivity A^* (which is a matrix in this case), if for any $f \in H^{-1}(\Omega)$, the solution u_{ε_n} of (7.3) converges weakly in $H_0^1(\Omega)$ to u^* a solution of

$$\begin{cases} -\operatorname{div}(A^*\nabla u^*) = f & \text{in } \Omega, \\ u^* = 0 & \text{on } \partial\Omega. \end{cases} \tag{7.4}$$

In the above definition, it is possible to replace the conductivities $a_\varepsilon(x)$ by matrices A_ε assumed to be uniformly elliptic. It is useful, for example, when this notion of G-convergence is used in other contexts like linear elasticity. Let us mention that there exists a stronger notion introduced by Murat–Tartar (see [244] or [245]) called H-convergence. In this case, it is assumed that u_{ε_n} weakly converges to u^* but also that $A_{\varepsilon_n}\nabla u_{\varepsilon_n}$ weakly converges to $A^*\nabla u^*$. Let us remark that in the case of symmetric matrices A_ε and A^*, G-convergence and H-convergence are equivalent. We will not further consider this notion of H-convergence but instead refer to the literature.

7.1.2.1 Link with Γ-convergence. Let us now explain the link between G-convergence and Γ-convergence. The domain Ω is now fixed, thus the choice of functional is simpler since it suffices to consider the energy functional associated to problem (7.3) or (7.4). Let us denote by J_ε^f the functional defined on $H_0^1(\Omega)$ by

$$J_\varepsilon^f(v) = \frac{1}{2}\int_\Omega a_\varepsilon(x)|\nabla v(x)|^2\,dx - \int_\Omega f v(x)\,dx, \tag{7.5}$$

while J_*^f denotes the functional defined on $H_0^1(\Omega)$ by

$$J_*^f(v) = \frac{1}{2}\int_\Omega (A^*\nabla v(x)).\nabla v(x)\,dx - \int_\Omega fv(x)\,dx. \tag{7.6}$$

(Of course, if $f \in H^{-1}(\Omega)$, we must replace $\int_\Omega fv(x)\,dx$ by the duality product $\langle f, v\rangle$). We have

Proposition 7.1.6. *The conductivity $a_{\varepsilon_n}(x)$ G-converges to A^* if and only if, for any $f \in H^{-1}(\Omega)$, the functional $J_{\varepsilon_n}^f$ (7.5) Γ-converges to J_*^f (7.6) for the weak topology on $H_0^1(\Omega)$.*

Proof. Applying de Giorgi's theorem and the characterization of u_ε and u^* as (unique) minima of J_ε^f and J_*^f respectively implies the result in the sense $\Longleftarrow$.

Let us now prove the necessary condition. First let us notice that the result does not depend on the function (or distribution) f. Indeed since

$$J_\varepsilon^f(v) - J_\varepsilon^g(v) = \langle g - f, v\rangle$$

is independent of ε, it is easy to check that Γ-convergence for a particular f implies Γ-convergence for any f.

The key point is the following observation: if $a_{\varepsilon_n}(x)$ G-converges to A^*, then we have convergence of energies:

$$\int_\Omega a_{\varepsilon_n}(x)|\nabla u_{\varepsilon_n}(x)|^2\,dx \longrightarrow \int_\Omega (A^*\nabla u^*(x)).\nabla u^*(x)\,dx. \tag{7.7}$$

This is obtained through an integration by parts of equations (7.3) and (7.4) multiplied by u_{ε_n} and u^* respectively.

We mimic the proof of Γ-convergence. Let $v \in H_0^1(\Omega)$ be fixed for all the sequence. Let us introduce $f = -\operatorname{div}(A^*\nabla v)$ (in such a way that $v = u^*$ for this particular f) and the sequence of functions u_{ε_n}, solutions of (7.3) with the right-hand side f. By G-convergence, u_{ε_n} weakly converges to u^* in $H_0^1(\Omega)$ and then we have both convergence of $\langle f, u_{\varepsilon_n}\rangle$ to $\langle f, u^*\rangle$ and of $\int_\Omega a_{\varepsilon_n}(x)|\nabla u_{\varepsilon_n}(x)|^2\,dx$ to $\int_\Omega (A^*\nabla u^*(x)).\nabla u^*(x)\,dx$. Thus we have been able to find a sequence of functions u_{ε_n} converging to v and such that $J_{\varepsilon_n}^f(u_{\varepsilon_n}) \longrightarrow J_*^f(v)$, which is part (ii) of Γ-convergence.

To prove part (i), let us consider any sequence v_{ε_n} weakly converging to v. We still use the sequence u_{ε_n} introduced above. Since u_{ε_n} is characterized as the minimum of $J_{\varepsilon_n}^f$, we have

$$J_{\varepsilon_n}^f(u_{\varepsilon_n}) \le J_{\varepsilon_n}^f(v_{\varepsilon_n}).$$

Now since $J_{\varepsilon_n}^f(u_{\varepsilon_n}) \to J_*^f(v)$, we get the result passing to the limit inf in the above inequality. □

It is possible to prove that any Γ-limit of J_ϵ^f is of the kind J_*^f (see [116, Thm. 22.1]). Therefore, thanks to compactness of the Γ-convergence (Theorem 7.1.3), it is possible to deduce from the previous proposition (see, e.g., [116, Thm. 22.2])

Corollary 7.1.7. *From any family of conductivities a_ε satisfying $0 < \alpha \le a_\epsilon \le M$, one can extract a subsequence that G-converges to a matrix A^*.*

7.1.2.2 Other properties of G-convergence. In the G-convergence framework, the key point is to identify the possible G-limits for the conductivities a_ε. In other words, we try to identify the G-closure of the set of admissible a_ε. It can be done in some cases as we will see in Section 7.3. Nevertheless, there are lots of problems (e.g., in elasticity) for which this identification is not yet done. Let us state a final result about G-convergence that will be useful in Section 7.3 and is interesting in itself.

Proposition 7.1.8. *Let A_{ε_n} be a sequence of symmetric matrices that G-converge to a matrix A^*. Let us assume, moreover,*

$$\begin{cases} A_{\varepsilon_n} \rightharpoonup A_+ \textit{in } (L^\infty(\Omega))^{N^2} \textit{weak-}*, \textit{ coefficient by coefficient;} \\ A_{\varepsilon_n}^{-1} \rightharpoonup A_-^{-1} \textit{in } (L^\infty(\Omega))^{N^2} \textit{weak-}*, \textit{ coefficient by coefficient.} \end{cases} \tag{7.8}$$

Then

$$A_-(x) \le A^*(x) \le A_+(x) \quad \textit{almost everywhere on } \Omega.$$

Let us recall that if A and B are two symmetric matrices, $A \le B$ means that $B - A$ is positive, that is, $(Bx, x) \ge (Ax, x)$ for any vector x. For the proof of this proposition, we refer to [244]. There is an important consequence for the eigenvalues of A^*, the G-limit of a sequence of conductivities $a_\varepsilon(x)\,\mathrm{Id}$. Indeed, let us consider a body built of two composite materials with respective conductivities α and β. We refer to Section 7.3 for extended comments on this topic, but let us state here the following.

Corollary 7.1.9. *Let ω_n be a sequence of subsets of Ω and let us consider some material with a component ω_n of conductivity $\alpha > 0$ and another component $\Omega\backslash\omega_n$ of conductivity $\beta > 0$. Let us denote by $a_n = \alpha\chi_{\omega_n} + \beta(1 - \chi_{\omega_n})$ the sequence of corresponding conductivities. We assume that the sequence of characteristic functions χ_{ω_n} weakly-$*$ converge in $L^\infty(\Omega)$ to a function θ and that the conductivities a_n G-converge to a matrix A^*.*

Then the eigenvalues of the matrix A^* lie between the harmonic mean and the arithmetic mean of α and β:

$$\frac{1}{\frac{\theta}{\alpha} + \frac{1-\theta}{\beta}} \le \lambda(A^*) \le \theta\alpha + (1-\theta)\beta.$$

Proof. We want to apply Proposition 7.1.8 to the sequence of matrices a_n Id. Since $a_n \rightharpoonup (\alpha\theta + \beta(1-\theta))$ in $L^\infty(\Omega)$ weak-$*$, we have $A_+ = (\alpha\theta + \beta(1-\theta))\,\mathrm{Id}$. In the same way, since

$$\frac{1}{a_n} = \frac{1}{\alpha}\chi_{\omega_n} + \frac{1}{\beta}(1-\chi_{\omega_n}) \rightharpoonup \frac{1}{\alpha}\theta + \frac{1}{\beta}(1-\theta),$$

we have $A_-^{-1} = \frac{1}{\alpha}\theta + \frac{1}{\beta}(1-\theta)\,\mathrm{Id}$. Now since the eigenvalues of the (positive) matrices $A_+ - A^*$ and $A^* - A_-$ are respectively $\alpha\theta + \beta(1-\theta) - \lambda(A^*)$ and $\lambda(A^*) - 1/\big(\frac{\theta}{\alpha} + \frac{1-\theta}{\beta}\big)$, the result follows. □

7.2 Relaxation for the Dirichlet problem

7.2.1 Introduction

The principle of relaxation is the following. Let us suppose that we want to solve an optimization problem like

$$\min_{x\in X} J(x). \tag{7.9}$$

Let us assume that we are not able to prove existence of a solution (possibly because there *is no* solution!). Then we try to embed X in a larger space X^* and we introduce a new functional J^* defined on X^* that extends J (in the sense that J and J^* coincide on X) in such a way that the following conditions are satisfied:

(1) the *relaxed* problem $\min_{x^*\in X^*} J^*(x^*)$ has a solution;

(2) $\inf_X J = \min_{X^*} J^*$;

(3) the minimizing sequences of problem (7.9) converge (up to a subsequence) to a solution of the relaxed problem;

(4) if the initial problem (7.9) has a solution, then it is also a solution of the relaxed problem.

Usually, the relaxed space X^* is built as a completion of X in such a way that X is dense in X^*. Since it is better to have a space X^* that is *as small as possible*, the important thing is to understand what the behavior of the minimizing sequences of problem (7.9) is. Therefore, the main difficulty is to properly identify the space X^*. In this chapter, we will see some examples where this identification is possible.

In order to get the first condition above, it is natural to choose a space X^* that is compact and a functional J^* that is lower semicontinuous. Then it is possible to get conditions (2), (3), and (4) by imposing the following properties:

$$\text{if } (x_n) \subset X \text{ converge to } x^* \in X^*, \text{ then } J^*(x^*) \le \liminf J(x_n); \tag{7.10}$$

$$\begin{aligned}&\text{for any } x^* \in X^*, \text{ there exists a sequence } x_n \text{ of } X \text{ that converge to } x^* \\ &\qquad \text{and such that } J^*(x^*) = \liminf J(x_n).\end{aligned} \tag{7.11}$$

Remark 7.2.1. In practice it is not necessary to impose that J^* coincides with J on X, though it is natural. Consequently, one can choose as the relaxed functional J^* the lower semicontinuous envelope of J defined by

$$J^*(x^*) := \inf\{\liminf J(x_n),\ x_n \to x^*\}. \tag{7.12}$$

In this case, it is clear that conditions (7.10) and (7.11) are fulfilled.

7.2.2 Completion for γ-convergence

7.2.2.1 Statement of the problem. Let us consider again the Dirichlet problem for the Laplacian. We fix a regular domain D that contains all the open sets we will consider here and we let $f \in H^{-1}(D)$. For any quasi-open set $\omega \subset D$, let us denote by $u_\omega = u_\omega^f$ the solution of the Dirichlet problem

$$\begin{cases} u_\omega \in H_0^1(\omega), \\ \forall\, \varphi \in H_0^1(\omega), \quad \int_\omega \nabla u_\omega \nabla \varphi = \int_\omega \varphi f. \end{cases} \tag{7.13}$$

Here, X is the set of all quasi-open sets ω contained in D. If needed, it is possible to add some constraints, like a volume constraint for example (without modification, also see below).

In this section, we want to minimize two different kinds of functionals. The first kind consists of an integration on the whole domain D, while for the second kind we integrate over ω. More precisely, let us consider a function

$$\begin{cases} j : D \times \mathbb{R} \longrightarrow \mathbb{R}, \\ j \text{ is measurable, continuous in } s \in \mathbb{R}, \text{ a.e. } x, \\ \forall\, (x, s) \in D \times \mathbb{R}, \quad |j(x, s)| \le a(x) + b|s|^q, \end{cases} \tag{7.14}$$

where $a \in L^1(D)$, $b \in [0, +\infty)$, $q \in [1, 2N/(N-2))$. Also let p be a fixed vector in $\mathbb{R}^N$. Then

Lemma 7.2.2. *For any measurable set $\omega \subset D$, the functionals defined on $H_0^1(D)$ by*

$$v \mapsto \int_\omega j(x, v(x))\, dx, \qquad v \mapsto \int_\omega |\nabla v - p|^2$$

are respectively continuous and lower semicontinuous for the weak convergence of $H_0^1(D)$.

This is a classical result that we have used previously. To prove continuity of the first functional, we use the compact embedding of $H_0^1(D)$ into $L^q(D)$ and the dominated convergence theorem. For the second one, we use the fact that the norm is l.s.c. for weak convergence.

Let us now introduce the two functionals

$$J_1(\omega) := \int_D j(x, u_\omega(x))\, dx + \alpha \int_D |\nabla u_\omega(x) - p|^2\, dx, \tag{7.15}$$

$$J_2(\omega) := \int_\omega j(x, u_\omega(x))\, dx + \alpha \int_\omega |\nabla u_\omega(x) - p|^2\, dx, \tag{7.16}$$

where u_ω is the solution of (7.13). We want to describe the relaxed problem associated to the minimization of the functionals J_1 or J_2 on the set $X = \mathcal{A}(D)$ of quasi-open sets contained in D. Let us begin with the case without volume constraint; we will consider a volume constraint on $|\omega|$ later.

7.2.2.2 Completion of the set $\mathcal{A}(D)$. The characterization of the completion of the set $\mathcal{A}(D)$ will lead us to consider partial differential equations like

$$\begin{cases} -\Delta u + u\mu = f \text{ in } D, \\ u \in H_0^1(D), \end{cases} \tag{7.17}$$

with a measure $\mu \in \mathcal{M}_0(D)$. We recall that this set of measures is defined by (see Chapter 3)

Definition 7.2.3. We define $\mathcal{M}_0(D)$ to be the set of nonnegative Borel measures vanishing on sets of zero H^1-capacity.

First of all, let us make precise the meaning of problem (7.17) and its solution. By similarity with the classical situation, we will say that $u_\mu \in H_0^1(D)$ is the solution of (7.17) if this is the unique minimizer of the (strictly convex) functional

$$F(v) = \frac{1}{2}\int_D |\nabla v|^2 + \frac{1}{2}\int_D v^2 d\mu - \int_D fv \le +\infty.$$

Let us remark that the expression of the functional F shows that we have to look for the minimizer in the subspace $H_0^1(D) \cap L^2_\mu(D)$, where $L^2_\mu(D)$ denotes the space of measurable functions that are square integrable for the measure μ. Here, and in the sequel, when we write $\int_D v^2\,d\mu$, in fact we consider $\int_D \tilde{v}^2\,d\mu$, where $\tilde{v}$ is the quasi-continuous representative of v, and this is meaningful when $\mu \in \mathcal{M}_0(D)$.

We can also characterize the solution u_μ of problem (7.17) by a variational formulation:

$$u_\mu \in H_0^1(D),\ \forall\, v \in H_0^1(D) \cap L_\mu^2(D), \quad \int_D \nabla u_\mu . \nabla v + \int_D u_\mu v\, d\mu = \int_D fv. \tag{7.18}$$

Let us note that the usual maximum and comparison principles hold true for a problem like (7.17): if $f \geq 0$ in D, then $u \geq 0$ in D and if $f_1 \leq f_2$, then $u_1 \leq u_2$.

For reasons that will be made clear later, we will identify any open set ω with the nonnegative Borel measure μ_ω defined by

$$\mu_\omega(A) = \begin{cases} 0 & \text{if } \operatorname{cap}_D(A \backslash \omega) = 0, \\ +\infty & \text{if } \operatorname{cap}_D(A \backslash \omega) > 0. \end{cases} \tag{7.19}$$

Taking into account the characterization of $H_0^1(\omega)$ given in Theorem 3.3.42, we have

$$\begin{cases} v \in H_0^1(\omega) & \Longrightarrow \int_D v^2\, d\mu_\omega = 0, \\ v \in H_0^1(D) \backslash H_0^1(\omega) & \Longrightarrow \int_D v^2\, d\mu_\omega = +\infty. \end{cases} \tag{7.20}$$

Indeed, in the second case, $\operatorname{cap}_D([v \neq 0] \backslash \omega) > 0$.

It is easy to check that $u_{\mu_\omega} = u_\omega$: indeed, if $v \notin H_0^1(\omega)$, $F(v) = +\infty$ and if $v \in H_0^1(\omega)$, the functional F coincides with that of problem (7.13)). This property allows us to embed $\mathcal{A}(D)$ in $\mathcal{M}_0(D)$ by the map $\omega \to \mu_\omega$ and we will see that $\mathcal{M}_0(D)$ is actually the completion of $\mathcal{A}(D)$ for γ-convergence. Let us first remark that this notion of γ-convergence naturally extends to $\mathcal{M}_0(D)$:

Definition 7.2.4. We say that a sequence of measures μ_n of $\mathcal{M}_0(D)$ γ-converges to a measure $\mu \in \mathcal{M}_0(D)$ if, for any $f \in H^{-1}(D)$, the sequence of solutions u_{μ_n} of problem (7.17) associated to μ_n weakly converges in $H_0^1(D)$ to u_μ.

It is possible to prove (cf. [116]) that the topology of γ-convergence is metrizable. Thus it is possible to introduce the completion of $\mathcal{A}(D)$ (or more precisely of the set of measures μ_ω) for γ-convergence.

We are now in a position to state the main result of this section:

Theorem 7.2.5 (Dal Maso–Mosco). *The set $\mathcal{M}_0(D)$ is a realization of the completion of $\mathcal{A}(D)$ for γ-convergence. Moreover $\mathcal{M}_0(D)$ is compact for γ-convergence.*

We are not going to prove this theorem here. Let us just remark that the compactness is a consequence of the equivalence between γ-convergence and Γ-convergence proved above, together with Theorem 7.1.3. Let us also explain how, from a sequence of open sets ω_n, one builds a measure μ that is an accumulation point of ω_n (or more precisely of μ_{ω_n}) for γ-convergence.

As in the case of Proposition 3.2.5, we strongly use the right-hand side $f \equiv 1$ in problem (7.13). Let us denote by $w_n = w^1_{\omega_n}$ the solution of

$$w_n \in H^1_0(\omega_n),\ \forall\, \varphi \in H^1_0(\omega_n), \qquad \int_{\omega_n} \nabla w_n \nabla \varphi = \int_{\omega} \varphi. \tag{7.21}$$

Let us remark that

Lemma 7.2.6. $1 + \Delta w_n \in \mathcal{M}_0(D)$.

Proof. We note $w_n = w$ and we introduce $p_k : \mathbb{R} \to \mathbb{R}$ defined by

$$p_k(r) = \begin{cases} 0, & r \le 0, \\ kr, & r \in [0, 1/k], \\ 1, & r \ge 1. \end{cases}$$

We have $p_k(w) \in H^1_0(\omega)$. According to (7.21), for any $\psi \in C_0^\infty(D)^+$, we have

$$\int_D \psi p_k(w) = \int_D \nabla[\psi p_k(w)] \nabla w \ge \int_D p_k(w) \nabla \psi \nabla w.$$

Letting k going to $+\infty$ yields

$$\int_D \psi \ge \int_{[w>0]} \psi \ge \int_D \nabla \psi \nabla w.$$

This says that $1 + \Delta w$ is a nonnegative distribution. Thus it is a nonnegative measure which, according to Proposition 3.3.35, belongs to $\mathcal{M}_0(D)$. Indeed, $1 + \Delta w = -\Delta(z - w) \ge 0$, where we choose $z := u^1_{\mathcal{O}}$ with $\mathcal{O}$ a bounded open set containing $\overline{D}$, on which the function z is C^∞. □

Let us consider again the sequence w_n defined in (7.21). Using a Poincaré inequality, we see that this sequence is bounded in $H^1_0(D)$. Therefore, we can extract a subsequence (still denoted w_n) that converges weakly in $H^1_0(D)$ to some function w. According to the previous lemma, we also know that $\nu := 1 + \Delta w$ is a nonnegative measure on D. Therefore, if we expect that the sequence ω_n γ-converge to a measure μ, we should have at the limit,

$$-\Delta w + w\mu = 1.$$

In other words, formally, the measure μ can be defined as ν / w at least where w is strictly positive. More precisely, one can prove

Proposition 7.2.7. *Following the previous notation, the sequence of measures μ_{ω_n} (or the sequence of open sets ω_n) γ-converge to the measure $\mu \in \mathcal{M}_0(D)$ defined by*

$$\mu(B) = \begin{cases} \int_B \frac{d\nu}{w} & \text{if } \operatorname{cap}_D(B \cap \{w = 0\}) = 0, \\ +\infty & \text{if } \operatorname{cap}_D(B \cap \{w = 0\}) > 0. \end{cases} \tag{7.22}$$

Let us prove here just that if μ is defined by (7.22), then w is a solution of (7.17) (with the right-hand side $f = 1$). For more details, we refer to the literature on this topic, for example, to [118], [28], [29]; see also [119] for the nonlinear case.

Let us first remark that $w\, d\mu = d\nu$. Thus

$$\int_D w^2 d\mu = \int_D w\, d\nu = \langle 1 + \Delta w, w\rangle_{H^{-1}\times H_0^1} < +\infty,$$

this last equality coming from Proposition 3.3.35. Consequently $w \in L^2_\mu(D)$.

Then let $v \in H_0^1(D) \cap L^2_\mu(D)$. Using Proposition 3.3.35, we have

$$\int_D \nabla w \nabla v + \int_D wv\, d\mu = \langle -\Delta w, v\rangle + \int_D v\, d\nu = \int_D v.$$

This shows that w is a variational solution of $-\Delta w + w\mu = 1$.

7.2.2.3 Relaxation of the functional. Let us begin with the simpler case, where the functional we want to minimize is given by

$$J_1(\omega) := \int_D j(x, u_\omega(x))\, dx + \alpha \int_D |\nabla u_\omega - p|^2. \tag{7.23}$$

We would like to define $J_1^*(\mu)$, where μ is a measure in $\mathcal{M}_0(D)$, by

$$J_1^*(\mu) := \int_D j(x, u_\mu(x))\, dx + \alpha \int_D |\nabla u_\mu - p|^2.$$

But it cannot work because of the gradient term. Thus we need to use a trick that consists in writing

$$J_1(\omega) = J_1(\omega) + \alpha \int_D u_\omega^2\, d\mu_\omega,$$

since this last integral vanishes by definition of the measure μ_ω. Therefore, we introduce the relaxed functional

$$J_1^*(\mu) := \int_D j(x, u_\mu(x))\, dx + \alpha \left(\int_D |\nabla u_\mu - p|^2 + \int_D u_\mu^2\, d\mu\right). \tag{7.24}$$

Proposition 7.2.8. *The functional J_1^* defined by* (7.24) *is the relaxation of J_1 for γ-convergence: if ω_n γ-converges to μ, then $J_1(\omega_n) \longrightarrow J_1^*(\mu)$. In particular, conditions* (7.10) *and* (7.11) *are fulfilled.*

Proof. Let ω_n be a sequence of open sets that γ-converges to a measure μ. It means that u_{ω_n} a solution of (7.13) weakly converges in $H_0^1(D)$ to u_μ a solution of (7.17). According to Lemma 7.2.2, we infer that

$$\int_D j(x, u_{\omega_n}(x))\,dx \longrightarrow \int_D j(x, u_\mu(x))\,dx.$$

Let us now look at the other integral. We have

$$\int_D |\nabla u_{\omega_n} - p|^2 = \int_D |\nabla u_{\omega_n}|^2 - 2\int_D \nabla u_{\omega_n}.p + |p|^2|D|.$$

This implies, using equation (7.13),

$$\int_D |\nabla u_{\omega_n} - p|^2 = \int_D f u_{\omega_n} - 2\int_D \nabla u_{\omega_n}.p + |p|^2|D|.$$

By weak convergence of u_{ω_n} to u_μ in $H_0^1(D)$, it follows that

$$\int_D |\nabla u_{\omega_n} - p|^2 \longrightarrow \int_D f u_\mu - 2\int_D \nabla u_\mu.p + |p|^2|D|.$$

Finally using equation (7.17), we prove that

$$\int_D |\nabla u_\mu|^2 + \int_D u_\mu^2\,d\mu = \int_D f u_\mu,$$

and then

$$\int_D |\nabla u_{\omega_n} - p|^2 \longrightarrow \int_D |\nabla u_\mu|^2 - 2\int_D \nabla u_\mu.p + |p|^2|D| + \int_D u_\mu^2\,d\mu,$$

which proves the expected result. □

Let us now consider the case of the functional J_2 defined by

$$J_2(\omega) := \int_\omega j(x, u_\omega(x))\,dx + \alpha \int_\omega |\nabla u_\omega - p|^2. \tag{7.25}$$

Here, we do not consider any volume constraint. Let us note that we can write

$$\begin{aligned} J_2(\omega) &= \int_D j(x, u_\omega(x))\,dx - \int_{D\setminus\omega} j(x, u_\omega(x))\,dx \\ &\quad + \alpha\Big(\int_D |\nabla u_\omega - p|^2 - \int_{D\setminus\omega} |\nabla u_\omega - p|^2\Big) \\ &= \int_D j(x, u_\omega(x))\,dx - \int_{D\setminus\omega} j(x, 0)\,dx + \alpha\Big(\int_D |\nabla u_\omega - p|^2 - \int_{D\setminus\omega} |p|^2\Big) \end{aligned}$$

or

$$J_2(\omega) = \int_D j(x, u_\omega(x))\, dx + \alpha \int_D |\nabla u_\omega - p|^2$$
$$- \int_D \left(j(x,0) + \alpha|p|^2\right) dx + \int_\omega \left(j(x,0) + \alpha|p|^2\right) dx.$$

In the above expression, we can see that for the first three integrals we will be able to use previous results because the domain of integration is D. Thus it remains to consider only the case of the last integral, which is of the kind $J_0(\omega) = \int_\omega g(x)\, dx$ with g an integrable function on D. To make a relaxation of this integral, we choose to consider its lower semicontinuous envelope, which is defined by

$$J_0^*(\mu) = \inf \{\liminf J_0(\omega_n),\ \omega_n\ \gamma\text{-converges to } \mu\}\,.$$

To give a different expression for this relaxed integral J_0^*, we need another definition. Let us denote by w_μ the solution of the problem

$$\begin{cases} -\Delta w_\mu + w_\mu \mu = 1 \text{ in } D, \\ w_\mu \in H_0^1(D), \end{cases} \tag{7.26}$$

and let us introduce the positivity set of w_μ:,

$$A_\mu = \{x \in D,\ w_\mu(x) > 0\}. \tag{7.27}$$

Since w_μ is quasi-continuous (see Chapter 3), A_μ is a quasi-open set. Let us also remark that if u_μ is the solution of problem (7.17) with any right-hand side f, then $u_\mu = 0$ on $D \backslash A_\mu$. To prove it, we begin with the case when f is nonnegative and bounded: $0 \le f \le M$. In this case, by maximum principle we have $0 \le u_\mu \le M w_\mu$ and then u_μ vanishes on the set where $w_\mu = 0$. The case $f \in L^\infty(D)$ follows by using $f = f^+ - f^-$ and we conclude by a classical truncation argument.

We can now state a proposition that gives another expression of the relaxed functional when J_0 is of the previous kind.

Proposition 7.2.9. *Let g be a function in $L^1(D)$ and J_0 the functional defined by*

$$J_0(\omega) := \int_\omega g(x)\, dx. \tag{7.28}$$

Then the lower semicontinuous envelope for γ-convergence of the functional J_0 is

$$J_0^*(\mu) := \inf \left\{\int_B g(x)\, dx,\ B \text{ a Borel set},\ A_\mu \subset B\right\}. \tag{7.29}$$

For the proof, which is based on arguments from measure theory, we refer to [80].

If g is a nonnegative function, it is clear that the infimum in (7.29) is achieved for $B = A_\mu$. More generally, if we decompose g into $g = g^+ - g^-$, we have, for any Borel set B that contains A_μ,

$$\int_B g(x)\,dx = \int_B g^+(x) - g^-(x)\,dx \geq \int_{A_\mu} g^+(x)\,dx - \int_B g^-(x)\,dx.$$

Therefore, taking the infimum,

$$\inf_{B, A_\mu \subset B} \int_B g(x)\,dx \geq \int_{A_\mu} g^+(x)\,dx - \sup_{B, A_\mu \subset B} \int_B g^-(x)\,dx.$$

Now if we choose in particular the Borel set $B = A_\mu \cup \{x \in D,\ g(x) \leq 0\}$, we can see that we have equality in the above inequality. If we have no constraint on the volume, the above supremum is obviously achieved when we choose $B = D$. Thus we have proved

Corollary 7.2.10. *The relaxed functional J_0^* is also given by*

$$J_0^*(\mu) := \int_{A_\mu} g^+(x)\,dx - \int_D g^-(x)\,dx = \int_{A_\mu} g(x)\,dx - \int_{D \setminus A_\mu} g^-(x)\,dx, \tag{7.30}$$

where g^+ and g^- denote, respectively, the positive and negative parts of g.

Let us come back to the initial problem. We can summarize the previous developments in the following theorem (in the final expression of J_2^* we gather the integrals on D and on A_μ to simplify).

Theorem 7.2.11. *With the previous notation, the relaxed functional J_2^* is defined on $\mathcal{M}_0(D)$ by*

$$J_2^*(\mu) = \int_{A_\mu} j(\cdot, u_\mu) + \alpha|\nabla u_\mu - p|^2 + \alpha \int_D u_\mu^2\,d\mu - \int_{D \setminus A_\mu} j^-(\cdot, 0). \tag{7.31}$$

In fact, J_2^* is the lower semicontinuous envelope of J_2 for γ-convergence. In particular, conditions (7.10) *and* (7.11) *are fulfilled.*

7.2.2.4 The case with a volume constraint. To end this section, let us now give several analogous results in the case where we put some volume constraints on the class of admissible open sets. To be able to deal with all kind of constraints at the same time, let us use, as in [80] the following notation: we denote by T a closed subinterval of the interval $[0, |D|]$ and we write the constraint as $|\omega| \in T$. The more frequent cases will be, for example,

- $T = \{c\}$: we look for a set with measure equal to c;
- $T = [0, c]$: we look for a set with measure less than or equal to c;
- $T = [c, |D|]$: we look for a set with measure greater than or equal to c.

The first question is, what is the behavior of the constraint with respect to γ-convergence? More precisely, we have seen that in the case without constraint, the closure of $\mathcal{A}(D)$ is $\mathcal{M}_0(D)$. What happens here?

Proposition 7.2.12. *Let $T = [m, M] \subset [0, |D|]$ and $\mathcal{A}_T$ be the set of quasi-open subsets of D whose volume is in T. Then the closure of $\mathcal{A}_T$ for γ-convergence is the subset of $\mathcal{M}_0(D)$,*

$$X_T^* = \{\mu \in \mathcal{M}_0(D),\ |A_\mu| \leq M\}.$$

Proof. We prove here only one inclusion and we refer to [80] for the reverse inclusion.

Let $\mu \in \mathcal{M}_0(D)$ and let ω_n be a sequence of quasi-open sets that γ-converges to μ. Let us consider the function f defined on $D \times \mathbb{R}$ by

$$f(x, s) = \begin{cases} 1 & \text{if } s > 0, \\ 0 & \text{if } s \leq 0. \end{cases}$$

This function is clearly lower semicontinuous on $D \times \mathbb{R}$. By definition of A_μ, we have then $|A_\mu| = \int_D f(x, w_\mu(x))\, dx$ (where w_μ is the solution of (7.26)).

By definition of γ-convergence, we know that w_n (a solution of (7.21)) weakly converges in $H_0^1(D)$ to w_μ and strongly in $L^2(D)$ by compactness of the embedding of $H_0^1(D)$ into $L^2(D)$. Up to extracting a subsequence, we can also assume that w_n converges almost everywhere to w_μ on D. By using Fatou lemma and the l.s.c. of f, we get

$$|A_\mu| = \int_D f(x, w_\mu(x))\, dx \leq \liminf \int_D f(x, w_n(x))\, dx = \liminf |\omega_n| \leq M.$$

This proves that any γ-limit of sequences of $\mathcal{A}_T$ is in X_T^*. □

The relaxation of the functional J_1 in the case with constraint is exactly the same as in the case without constraint. Thus the relaxed problem is, for $X = \mathcal{A}_T$ and $J = J_1$,

$$\begin{cases} \text{minimize } J_1^*(\mu) := \int_D j(\cdot, u_\mu) + \alpha|\nabla u_\mu - p|^2 + \alpha \int_D u_\mu^2\, d\mu, \\ \text{with } \mu \in \mathcal{M}_0(D),\ |A_\mu| \leq \sup T. \end{cases}$$

For J_2, it is a little bit more complicated. First of all, we must consider the volume constraint when we take the infimum on the Borel sets B that contain A_μ. Now, we cannot choose $B = D$ to have the infimum. Let us recall below the general expression of J_2^* (an expression that may often be simplified).

Proposition 7.2.13. *We have*

$$J_2^*(\mu) := \int_{A_\mu} j(\cdot, u_\mu) + \alpha|\nabla u_\mu - p|^2 + \alpha \int_D u_\mu^2 \, d\mu \\ + \inf\left\{\int_{B\setminus A_\mu} j(x,0)\,dx,\ B \text{ a Borel set},\ A_\mu \subset B,\ |B| \in T\right\}. \tag{7.32}$$

Remark 7.2.14. If the function j is positive and if the interval T is $T = [0, M]$, it is clear that the inf in the above expression is attained for $B = A_\mu$ and then it is equal to 0.

7.2.3 Another example

7.2.3.1 Introduction. We now consider another example of relaxation for the Dirichlet problem. In this example, we will show how we can prove existence of a classical solution through the use of the relaxed formulation.

The problem we consider here can be seen as the stationary (and simplified) version of an optimal control problem where we want to stabilize a vibrating body using actuators distributed into the domain. Let us denote by Ω the body we want to stabilize, by ω the subdomain, $\omega \subset \Omega$, the support of the actuators (χ_ω as usual denotes the characteristic function of ω), and by u the state of the system (modeling the vibrations). We consider a force f acting on the system, and we assume f to be nonnegative. The state u is the solution of the system

$$\begin{cases} -\Delta u + \chi_\omega u = f & \text{in } \Omega, \\ u = 0 & \text{on } \partial\Omega. \end{cases} \tag{7.33}$$

The functional we want to minimize is the sum of the internal energy and of the cost:

$$J(\omega) = \int_\Omega |\nabla u|^2 + \int_\Omega \chi_\omega u^2. \tag{7.34}$$

Let us observe that, multiplying equation (7.33) by u and integrating over Ω yields another expression of the functional: $J(\omega) = \int_\Omega fu$.

Of course, we need to put some constraints on the domain ω we are looking for, otherwise it is clear that the solution would be Ω (if we want to control on some subdomain we may as well control on the whole domain). The more natural constraint here is to fix the volume of ω, which is proportional to the cost of the actuators.

For the physical motivation of this model and, in particular, to understand why the optimal control with support ω should be u, we refer to the paper [183].

To summarize, the shape optimization problem we want to solve here consists in looking for a domain ω of volume $|\omega| \leq c$ that minimizes the functional (7.34) with a state function u, a solution of (7.33). Let us remark that the unknown ω appears only in the term without derivative of the state equation (which makes the situation simpler).

Remark 7.2.15. Two domains ω_1 and ω_2 that coincide almost everywhere (in the sense that their characteristic functions χ_{ω_1} and χ_{ω_2} are equal almost everywhere) define the same state function u and the same energy. The class of admissible domains is to be understood modulo the equivalence

$$\omega_1 \simeq \omega_2 \quad \Longleftrightarrow \quad \chi_{\omega_1} = \chi_{\omega_2} \ a.e.$$

In the sequel, we will not distinguish between a domain and its equivalence class.

Instead of choosing the constraint $|\omega| \leq C$, one can actually saturate it and choose $|\omega| = C$, because the functional J is nonincreasing with respect to inclusion:

Proposition 7.2.16. *The functional J defined in* (7.34) *is nonincreasing for the inclusion "if $\omega_1 \subset \omega_2$ then $J(\omega_1) \geq J(\omega_2)$". Consequently,*

$$\inf\left\{J(\omega),\ |\omega| \leq C\right\} = \inf\left\{J(\omega),\ |\omega| = C\right\}.$$

Proof. Let $u_i, i = 1, 2$ be the solution of (7.33) relative to ω_i. By difference,

$$\begin{cases} -\Delta(u_1 - u_2) + \chi_{\omega_1}(u_1 - u_2) = \chi_{\omega_2 - \omega_1} u_2 & \text{in } \Omega, \\ u_1 - u_2 = 0 & \text{on } \partial\Omega. \end{cases} \tag{7.35}$$

Now, since f is nonnegative, u_2 is also nonnegative by the maximum principle, and then the same comparison principle ensures $u_1 - u_2 \geq 0$ in Ω. The result follows immediately from the fact that

$$J(\omega_1) - J(\omega_2) = \int_\Omega f(u_1 - u_2)\, dx. \tag{7.36}$$

□

We have already considered the case of a functional nonincreasing with respect to inclusion. It was the content of the Buttazzo–Dal Maso theorem (Theorem 4.7.6 in Chapter 4). We then saw that the extra assumption of lower semicontinuity of the functional for γ-convergence guarantees the existence of a solution for the shape optimization problem. Can we apply this theorem here? Is the functional J defined by (7.33) and (7.34) lower semicontinuous for γ-convergence? Unfortunately no! We can see this in two different ways:

- First through an indirect proof showing that the Buttazzo–Dal Maso theorem does not apply. For that purpose, we can use the analysis given in the section "Relaxation" below and give an explicit example of a minimum of the functional that is really relaxed and better than every domain. In the radial case, or in dimension 1, where we can do explicit computations, this is indeed possible.

- Next by a direct proof providing a sequence of domains ω_n that γ-converges to a domain ω, but such that $J(\omega) > \liminf J(\omega_n)$. We will show this just after Theorem 7.2.18, by choosing for ω_n the (planar) domain that is under the graph of the function $x \mapsto 2 + \sin(nx)$:

$$\omega_n = \left\{(x, y) \in \mathbb{R}^2;\ 0 < x < \pi,\ 0 < y < 2 + \sin(nx)\right\}$$

(this is a counterexample that we have already used in Chapter 2).

7.2.3.2 Relaxation. The set X on which we want to minimize J can be completely described here, using characteristic functions, as the following subset of $L^\infty(\Omega)$:

$$X = \left\{l \in L^\infty(\Omega),\ 0 \le l \le 1,\ l(l-1) = 0\ a.e.,\ \int_\Omega l(x)\,dx = C\right\}, \tag{7.37}$$

where we assume $C \in (0, |\Omega|)$. This set is not closed for the topology that is naturally associated with the problem, namely the weak-$*$ topology in $L^\infty(\Omega)$. Its closure is in fact its closed convex hull:

Proposition 7.2.17. *The closure of the set X of characteristic functions, for the weak-$*$ topology of L^∞ is*

$$X^* = \overline{\operatorname{conv}(X)} = \left\{l \in L^\infty(\Omega),\ 0 \le l \le 1,\ \int_\Omega l(x)\,dx = C\right\}. \tag{7.38}$$

In other words X^* is the weak-$*$ closed convex hull of X in $L^\infty(\Omega)$ and X is exactly the set of extreme points of the convex set X^*.

Proof. Let $X^* = \{l \in L^\infty(\Omega),\ 0 \le l \le 1,\ \int_\Omega l(x)\,dx = C\}$. Then it is easy to prove the following:

- The set X is included into X^* and since this one is convex, the convex hull of X is also included into X^*.

- X^* is closed for the weak-$*$ topology. Here we cannot use a convexity argument (a set that is convex and strongly closed is not necessarily closed for the weak-$*$ topology; see [54]). In fact we use only that, if $l \in L^\infty(\Omega)$ is such that $\int_\Omega l\varphi(x)\,dx \ge 0$ for any nonnegative function φ of $L^1(\Omega)$, then $l \ge 0$ almost everywhere. Therefore $\overline{X} \subset \overline{\operatorname{conv}(X)} \subset X^*$.

- Any point of X is an extreme point of the convex set X^* and there are no other extreme points.

In order to prove the equality $\overline{X} = \overline{\text{conv}(X)} = X^*$, we must now show that $X^* \subset \overline{X}$. Thus, we have to prove that any function l such that $0 \le l(x) \le 1$, $\int_\Omega l(x)\,dx = C$ is the weak-$*$ limit of a sequence of characteristic functions χ_{ω_n} such that $\int_\Omega \chi_{\omega_n}(x)\,dx = C$. We will prove it according to the following four steps:

First step: Ω is a cube, $\Omega = Q = [0, L]^N$, and l is a constant function $l(x) = \alpha \in (0, 1)$ (the cases $\alpha = 0$ or 1 are trivial).

Second step: Ω is any open set, l is still a constant function $l(x) = \alpha \in (0, 1)$, and we do not assume any constraint on $\int_\Omega l(x)\,dx$.

Third step: Ω is any open set, l is now any function such that $0 \le l(x) \le 1$, and we do not assume any constraint on $\int_\Omega l(x)\,dx$.

Fourth step: the general case is Ω is any open set, l is any function such that $0 \le l(x) \le 1$, and we do assume the constraint $\int_\Omega l(x)\,dx = C$.

First step: Let $l(x) = \alpha$ be a constant, $\alpha \in (0, 1)$. We split the cube $Q = [0, L]^N$ into small cubic cells and on any of these cells, we consider a small cube whose volume is a proportion α of the total volume of the cell. More precisely, let us introduce $a = (\alpha)^{1/N}$ and let us set

$$\omega_n^Q = \left[\bigcup_{k=0}^{n-1} \left[L\frac{k}{n}, L\frac{k+a}{n}\right]\right]^N.$$

The volume of ω_n^Q is exactly αL^N. Thus it is classical to prove that for any function φ of $L^1(Q)$, we have

$$\int_Q \chi_{\omega_n^Q}\varphi(x)\,dx = \int_{\omega_n^Q} \varphi(x)\,dx \longrightarrow \alpha \int_Q \varphi(x)\,dx.$$

(We start by proving it for a continuous function for which it is the classical theory of integration, then we conclude by density). Therefore, we have

$$\chi_{\omega_n^Q} \rightharpoonup \alpha \quad \text{in } L^\infty(Q) \text{ weakly-}*. \tag{7.39}$$

Second step: For a bounded open set Ω, we consider a cube Q that contains Ω and we set $\omega_n = \omega_n^Q \cap \Omega$ in such a way that $\chi_{\omega_n} = \chi_{\omega_n^Q} \cdot \chi_\Omega$. Thus for any function $\varphi \in L^1(\Omega)$, we have

$$\int_{\omega_n} \varphi(x)\,dx = \int_Q \chi_{\omega_n^Q}(\chi_\Omega\varphi(x))\,dx \longrightarrow \alpha \int_Q (\chi_\Omega\varphi(x))\,dx = \alpha \int_\Omega \varphi(x)\,dx.$$

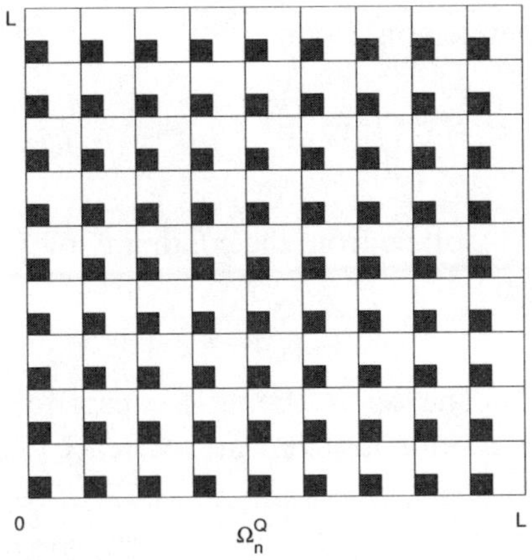

Figure 7.1. A sequence of open sets whose characteristic function weakly converges to α.

Third step: We begin with the approximation of step functions (functions that are piecewise constant). Let $s = \sum_{i \in I} \alpha_i \chi_{A_i}$ be such a function (with disjoints A_i), and for each fixed index i, we choose a sequence $\chi_{\omega_n^i}$ that converges weakly-$*$ to α_i. Then

$$\chi_{\{\cup_{i\in I}(\omega_n^i \cap A_i)\}} = \sum_{i\in I} \chi_{\omega_n^i \cap A_i} \stackrel{*}{\rightharpoonup} \sum_{i \in I} \alpha_i \chi_{A_i} = s.$$

(In the above expression we use the fact that $\chi_{A\cap B} = \chi_A \chi_B$ and that $\chi_{A\cup B} = \chi_A + \chi_B$ if A and B are disjoint.)

Thus any step function in X^* is in $\overline{X}$. It follows that any function f in X^* is in $\overline{X}$, because there always exists a sequence of step functions converging to f.

Fourth step: Let us now take into account the constraint $\int_\Omega l(x)\,dx = C$. For any sequence of characteristic functions χ_{ω_n} that we consider, we have $\int_\Omega \chi_{\omega_n}(x)\,dx \longrightarrow C$. Let us set $\varepsilon_n = C - \int_\Omega \chi_{\omega_n}(x)\,dx$. If $\epsilon_{n_k} < 0$ for some subsequence, we remove from the set ω_{n_k} a subset o_k of measure $-\epsilon_{n_k}$ and we have $\chi_{\omega_{n_k}\setminus o_k} \stackrel{*}{\rightharpoonup} l$ because χ_{o_k} converges to 0 in L^1. Otherwise, we add to ω_n a set o_n disjoint of ω_n of measure ϵ_n and $\chi_{\omega_n \cup o_n} = \chi_{\omega_n} + \chi_{o_n} \stackrel{*}{\rightharpoonup} l$, which concludes the proof. □

The definition of the relaxed functional J^* now comes in a natural way. With each element $l \in X^*$, we associate the function u_l a solution of the problem

$$\begin{cases} -\Delta u_l + l u_l = f & \text{in } \Omega, \\ u_l = 0 & \text{on } \partial\Omega, \end{cases} \tag{7.40}$$

and the functional J^* is defined on X^* by

$$J^*(l) = \int_\Omega |\nabla u_l|^2 + l u_l^2 = \int_\Omega f u_l. \tag{7.41}$$

Properties (7.10) and (7.11) follow from the continuity of J^* for the weak-$*$ topology (see Theorem 7.2.18 below).

7.2.3.3 Properties of J^*.

The set X^* being compact for the weak-$*$ topology, it remains to prove that J^* is continuous to ensure existence of a solution for the relaxed minimization problem.

Theorem 7.2.18. *The functional J^* is continuous on $L^\infty(\Omega)$ for the weak-$*$ topology. Thus there exists $l^* \in X^*$ that minimizes J^* on X^*. Moreover,*

$$\inf_{\omega \in X} J(\omega) = \min_{l \in X^*} J^*(l) = J^*(l^*). \tag{7.42}$$

Proof. Let l_n be a sequence in $L^\infty(\Omega)$ that weakly-$*$ converges to a function l. Let us denote by u_n the solution of (7.40) associated with l_n and by u the solution associated with l. By difference, $v_n = u_n - u$ satisfies

$$\begin{cases} -\Delta v_n + l_n v_n = (l - l_n) u & \text{in } \Omega, \\ v_n = 0 & \text{on } \partial\Omega. \end{cases} \tag{7.43}$$

In particular, the variational formulation of (7.43) implies

$$\int_\Omega |\nabla v_n|^2 + \int_\Omega l_n v_n^2 = \int_\Omega (l - l_n) u v_n. \tag{7.44}$$

Since l_n is bounded in $L^\infty(\Omega)$, it follows, according to (7.44), that v_n is bounded in $H_0^1(\Omega)$. Therefore, by the Rellich theorem, we can extract a subsequence v_{n_k} that converges strongly in $L^2(\Omega)$ to a function $v \in H_0^1(\Omega)$. Consequently, $u v_{n_k}$ converges strongly to uv in $L^1(\Omega)$ and since $l - l_n \overset{*}{\rightharpoonup} 0$ in $L^\infty(\Omega)$, the right-hand side of (7.44) tends to 0. Now, since $\int_\Omega l_n v_n^2 \geq 0$, this implies that v_{n_k} converges (strongly) to 0 in $H_0^1(\Omega)$. Finally, since 0 is the only accumulation point of the sequence v_n, the whole sequence tends to 0 in $H_0^1(\Omega)$. The continuity of J^* follows.

The existence of a minimizer is a consequence of the compactness of X^* for the weak-$*$ topology. Finally, equality between the infimum on X and the minimum on X^* comes from the fact that X^* is the closure of X and that J^* coincides with J on X. □

Remark 7.2.19. As already mentioned, we are going to deduce from this theorem that the functional J is not lower semicontinuous for γ-convergence. Let ω_n be the set in the plane defined by

$$\omega_n = \left\{(x, y) \in \mathbb{R}^2;\ 0 < x < \pi,\ 0 < y < 2 + \sin(nx)\right\}.$$

Then, ω_n converges for the Hausdorff distance to the rectangle $\omega = (0, \pi) \times (0, 1)$. Moreover, since ω_n is a sequence of simply connected planar domains, the Šverak Theorem 3.4.14) applies: ω_n γ-converges to ω. Now it is classical to verify that the sequence of characteristic functions χ_{ω_n} converges in L^∞ weak-$*$ to the function l defined by

$$l(x, y) = \begin{cases} 1 & \text{if } (x, y) \in \omega = (0, \pi) \times (0, 1), \\ \frac{1}{2} & \text{if } (x, y) \in (0, \pi) \times (1, 3), \\ 0 & \text{if } (x, y) \notin (0, \pi) \times (0, 3). \end{cases}$$

According to Theorem 7.2.18, it follows that u_{ω_n}, a solution of (7.33), converges strongly in $H_0^1(\Omega)$ to u_l a solution of (7.40) for the function l defined above. Thus

$$J(\omega_n) = \int_\Omega f u_{\omega_n} \longrightarrow \int_\Omega f u_l.$$

Now since $l > \chi_\omega$, the maximum principle yields (by subtraction) that $u_\omega > u_l$ in Ω. Therefore

$$J(\omega) = \int_\Omega f u_\omega > J^*(l) = \int_\Omega f u_l = \lim J(\omega_n).$$

This proves that J is not lower semicontinuous for γ-convergence.

Let us now emphasize another property of the functional J^*: the convexity. It is not so frequent to be able to use this important property in shape optimization!

Proposition 7.2.20. *The functional J^* is convex on $L^\infty(\Omega)$.*

Proof. It follows immediately from the variational formulation of problem (7.40). Since u_l achieves the minimum on $H_0^1(\Omega)$ of $v \mapsto \frac{1}{2}\int_\Omega |\nabla v|^2 + l v^2 - \int_\Omega f v$, which is an affine function in l, we have

$$-\frac{1}{2} J^*(l) = -\frac{1}{2}\int_\Omega |\nabla u_l|^2 + l u_l^2 = \min_{v \in H_0^1(\Omega)} \left\{\frac{1}{2}\int_\Omega |\nabla v|^2 + l v^2 - \int_\Omega f v\right\},$$

which is concave as an infimum of affine functions. Therefore J^* is convex. □

Among the consequences of this convexity, we will use later

- the fact that the first-order optimality condition is necessary and sufficient;

- the fact that if J^* has two minima l_1 and l_2, then all the points on the segment $[l_1, l_2]$ are also minima.

Remark 7.2.21. Let us suppose that, instead of minimizing the functional J, we want to *maximize* it (which is not really relevant from a physical point of view, but this is not the point). In the same way, using the relaxed formulation, we can prove existence of a maximizer in X^* for the functional J^* (which is continuous on the convex compact set X^*). Now it is well known that a convex function on a convex compact set achieves its maximum *at an extreme point*. Since the extreme points of X^* are precisely the characteristic functions (see Proposition 7.2.17), we prove in this way that J has a classical maximizer. It is a first example where going through the relaxed formulation allows us to prove the existence of a classical solution. For another example in a similar situation, we refer to [113].

7.2.3.4 Optimality conditions. Let us now try to characterize the minima, by using the optimality conditions. We first give the expression of the derivative of the functional J^*.

Proposition 7.2.22. *The functional J^* is Fréchet-differentiable at any point $l \in L^\infty(\Omega)$ and*

$$\langle J^{*\prime}(l), h\rangle = -\int_\Omega h u_l^2, \tag{7.45}$$

where u_l is the solution of (7.40).

Proof. Let h be fixed in $L^\infty(\Omega)$. By subtracting equations (7.40) for $l + h$ and l, we get

$$-\Delta(u_{l+h} - u_l) + l(u_{l+h} - u_l) = -h u_{l+h}. \tag{7.46}$$

Multiplying (7.46) by $(u_{l+h} + u_l)$ and integrating on Ω yields

$$\int_\Omega |\nabla u_{l+h}|^2 - |\nabla u_l|^2 + l u_{l+h}^2 - l u_l^2 = -\int_\Omega h u_{l+h}(u_{l+h} + u_l). \tag{7.47}$$

Thus we have

$$J^*(l + h) - J^*(l) + \int_\Omega h u_l^2 = -\int_\Omega h u_l(u_{l+h} - u_l), \tag{7.48}$$

which implies

$$\left| J^*(l + h) - J^*(l) + \int_\Omega h u_l^2 \, dx \right| \le \|h\|_\infty \|u_{l+h} - u_l\|_2 \|u_l\|_2. \tag{7.49}$$

Now thanks to (7.46) and Poincaré inequality, we have

$$\|u_{l+h} - u_l\|_2 \le C\|h u_{l+h}\|_2 \le 2C\|h\|_\infty \|u_l\|_2,$$

and the result follows from (7.49). □

Let us now consider a minimum of J^* that we denote by l^*. Let us introduce three subsets of Ω that will play a role in the following discussion:

Definition 7.2.23. Let l^* be a point in X^*. We define the following three sets, which depend only on l^*:

$$\begin{cases} \Omega_0 = \{x \in \Omega,\ l^*(x) = 0\}, \\ \Omega^* = \{x \in \Omega,\ 0 < l^*(x) < 1\}, \\ \Omega_1 = \{x \in \Omega,\ l^*(x) = 1\}. \end{cases}$$

Of course, these sets are defined only up to a set of zero measure because l^* is only in $L^\infty(\Omega)$. The above equalities and inequalities have to be understood almost everywhere. Since we want to write the optimality conditions satisfied by l^*, we have to characterize the tangent cone $T'(l^*)$ to X^* at the point l^* in $L^\infty(\Omega)$. It is usually defined as follows: an element h belongs to $T'(l^*)$ if, for any sequence t_n decreasing to 0, there exists a sequence $h_n \in L^\infty(\Omega)$ that converges uniformly to h such that, for any n, $l^* + t_n h_n \in X^*$.

Lemma 7.2.24. *The tangent cone $T'(l^*)$ to the convex set X^* at the point l^* is the set of functions h of $L^\infty(\Omega)$ such that*

(i) $\int_\Omega h(x)\,dx = 0$;

(ii) $\|\chi_{Q_n^0} h^-\|_\infty \to 0$ *when* $n \to \infty$, *where* $Q_n^0 = \{x \in \Omega,\ l^*(x) \le 1/n\}$;

(iii) $\|\chi_{Q_n^1} h^+\|_\infty \to 0$ *when* $n \to \infty$, *where* $Q_n^1 = \{x \in \Omega,\ l^*(x) \ge 1 - 1/n\}$.

Remark 7.2.25. For the proof, we refer to [33] or [106]. The condition

$$h(x) \ge 0 \text{ in } \Omega_0 \text{ and } h(x) \le 1 \text{ in } \Omega_1, \quad \int_\Omega h = 0, \tag{7.50}$$

is clearly necessary for a point h to be in the tangent cone $T'(l^*)$, but not sufficient as shown by the following counterexample. We choose $\Omega = (-1, 1)$ and

$$l^*(x) = \begin{cases} 0 & \text{on } (-1, 0], \\ x & \text{on } (0, 1) \end{cases} \quad \text{and} \quad h(x) = \begin{cases} 0 & \text{on } (-1, 0], \\ (-1)^n & \text{on } (\frac{1}{n+1}, \frac{1}{n}),\ n \ge 2, \\ \theta & \text{on } (\frac{1}{2}, 1), \end{cases}$$

where θ is such that $\int_\Omega h = 0$. Then, it is easy to check that the condition $x + t_n h_n(x) \ge 0$ on $[0, 1]$ prevents any sequence h_n converging uniformly to h and then h cannot be in the tangent cone to l^* although (7.50) is satisfied.

The first-order optimality condition is written

$$\forall\, h \in T'(l^*), \quad \langle J'(l), h\rangle \geq 0. \tag{7.51}$$

Since J^* and X^* are convex, the necessary condition is also sufficient; therefore (7.51) is a characterization of a minimum of J^*.

We are now going to rewrite this condition by using the sets Ω_0, Ω^*, and Ω_1, which were introduced above.

Theorem 7.2.26. *Let l^* be an element of X^* and u^* the solution of problem* (7.40) *corresponding to l^*. Let Ω_0, Ω_1, Ω^* be defined as above. Then l^* is a minimum of the functional J^* if and only if*

(i) u^* is constant on Ω^*;

(ii) for all $(x_0, x^*, x_1) \in \Omega_0 \times \Omega^* \times \Omega_1$, we have $u^*(x_0) \leq u^*(x^*) \leq u^*(x_1)$.

In the following, we will denote by c^* the value taken by u^* on Ω^*.

Proof. Let us assume that l^* is a minimizer of J^* and let us introduce

$$\Omega_n^* = \{x \in \Omega,\ 1/n \leq l^* \leq 1 - 1/n\}\,.$$

We are going to prove that u^* is constant on Ω_n^*. Since $\Omega^* = \bigcup_{n>0} \Omega_n^*$ (increasing union), this will prove the first point. For the purpose of contradiction, let us assume that u^* is not constant on Ω_n^*. Then, it is possible to find two measurable subsets ω_1 and ω_2 in Ω_n^* such that

$$|\omega_1| = |\omega_2| \quad \text{and} \quad \int_{\omega_1} {u^*}^2 < \int_{\omega_2} {u^*}^2. \tag{7.52}$$

Let us choose h defined by

$$h(x) = \begin{cases} -1 & \text{in } \omega_1, \\ +1 & \text{in } \omega_2, \\ 0 & \text{elsewhere.} \end{cases}$$

By using Lemma 7.2.24, we see that h belongs to the tangent cone $T'(l^*)$. But according to (7.52), we also have

$$\langle J'(l^*), h\rangle = -\int_\Omega h{u^*}^2 = -\int_{\omega_2} {u^*}^2 + \int_{\omega_1} {u^*}^2 < 0,$$

which would contradict the optimality condition (7.51).

We can prove the second point in an analogous way by assuming that there exists a set of positive measure ω_0 in Ω_0 such that

$$u^*_{/\omega_0} > u^*_{/\Omega^*} = \text{constant}.$$

Then we choose ω^* in Ω^*_n, with $|\omega_0| = |\omega^*|$, and we conclude using a function h that satisfies $h = 1$ in ω_0 and $h = -1$ in ω^*.

Conversely, let us assume that the couple (l^*, u^*) satisfies points (i) and (ii) of the theorem. Let h be a point of the tangent cone $T'(l^*)$. According to the remark that follows Lemma 7.2.24, h is positive on Ω_0 and negative on Ω_1; thus

$$\begin{aligned} -\int_\Omega hu^{*^2} &= -\int_{\Omega_0} hu^{*^2} - \int_{\Omega^*} hu^{*^2} - \int_{\Omega_1} hu^{*^2} \\ &\geq -\int_{\Omega_0} hc^{*^2} - \int_{\Omega^*} hc^{*^2} - \int_{\Omega_1} hc^{*^2} \\ &= -c^{*^2} \int_\Omega h = 0. \end{aligned}$$

Consequently, l^* satisfies the optimality condition (7.51) and, taking into account the convexity of J^*, l^* is indeed a minimizer of J^*. □

Remark 7.2.27. Using the classical regularity results for an elliptic PDE like (7.40) ($f \in L^2(\Omega)$), the function u^* a solution of (7.40) belongs to the Sobolev space $H^2(\Omega)$, and then u^* is continuous in dimensions $N = 1, 2, 3$ (by the Sobolev embedding $H^2 \hookrightarrow C^0$). By the homogeneous Dirichlet boundary condition, the set Ω_0 is nonempty and it contains a neighborhood of the boundary of Ω: indeed, otherwise $c^* = 0$. But since u^* cannot vanish on a set of positive measure ($f \not\equiv 0$), it would follow that $|\Omega^*| = 0$, namely that l is a characteristic function. Since $C \in (0, |\Omega|)$, necessarily $|\Omega_0| > 0$.

We also have $c^* = \sup_{x\in\Omega_0} u^*(x)$ and, if Ω_1 is nonempty, $c^* = \inf_{x\in\Omega_1} u^*(x)$. Finally the open set $[u^* < c^*] \subset \Omega_0$ is connected if $\partial\Omega$ is itself connected. Indeed, if ω were a connected component of this open set that does not meet a neighborhood of $\partial\Omega$, we would have $u^* = c^*$ on $\partial\omega$; but $-\Delta u^* = f \geq 0$ in ω (since $l^* = 0$ in Ω_0) and by the maximum principle, this would imply $u^* \geq c^*$ in ω, which is not possible.

7.2.3.5 Back to a classical solution. We are now going to explain how we can, in some situations, use the optimality condition described in Theorem 7.2.26 to prove that the minimizer l^*, which is a priori relaxed, is in fact a classical minimizer. Under some conditions that we are going to describe, we can indeed prove that l^* is necessarily a characteristic function. From the definitions of the sets Ω_0, Ω^*, and Ω_1, it means that the set Ω^* is empty (or of zero measure, which is the same in our case). Then, by definition we will have $l^* = \chi_{\Omega_1}$. Let us state such a result. It is only a sufficient condition; we could write others in the same spirit.

Theorem 7.2.28. *Let $u_0 = u_\Omega^f$ be the solution of the Dirichlet problem in Ω. Then, there exists a unique solution to the minimization problem* $\min_{X^*} J^*$ *and this solution is a classical one (l^* is a characteristic function) as soon as any of the three following conditions is satisfied:*

(i) $u_0 \le f$ in Ω;

(ii) $f \le -\Delta f$ in Ω;

(iii) $C > |\{x \in \Omega,\ u_0(x) > \alpha\}|$, $\alpha = \inf\{f(x),\ x$ such that $u_0(x) > f(x)\}$.

Remark 7.2.29. Let us recall that $C = \int_\Omega l^*$ is the constraint of the problem. These conditions are only sufficient. The second one is very easy to check. The other two require a simple PDE to be solved on Ω. Of course, condition (iii) can be seen as a generalization of condition (i).

For example, in the case of the disk $\Omega = D(0, R)$ of $\mathbb{R}^2$, if we choose the constant function $f \equiv m$, we have in polar coordinates, $u_0(r, \theta) = m\frac{R^2-r^2}{4}$. Thus,

- if $R \le 2$, condition (i) is verified for any volume constraint;
- while if $R > 2$, condition (i) is no longer verified, but condition (iii) is verified as soon as $C > \pi(R^2 - 4)$.

In the radial case, the computations can actually be done almost explicitly and one can prove (see Theorem 7.2.31 below) that the minimum l^* is a characteristic function when f is a decreasing function of r.

Proof of Theorem 7.2.28. We have to prove that, under the above assumptions, the set Ω^* associated to the minimizer l^* is necessarily empty. This will prove that all minimizers are characteristic functions. Now since J^* is convex, we know that if l_1^* and l_2^* are two minimizers then $tl_1^* + (1-t)l_2^*$ is also a minimizer of J^* for any $t \in (0,1)$. Since the characteristic functions are extremal points of the set X^*, the uniqueness of the solution follows.

Let l^* be a minimizer and let us assume that the corresponding set Ω^* is nonempty. Therefore we have $u^* = c^* =$ constant on Ω^*. Since u^* is in $H^2(\Omega)$, this implies that $u^*_{x_i} = 0$ a.e. on the set $\{u^* = c^*\}$ (see Lemma 3.1.8 and Remark 3.1.10). Using the same argument, we can see that $u^*_{x_i x_i}$ vanishes a.e. on the set $\{u^*_{x_i} = 0\}$ and then also on the set $\{u^* = c^*\}$. It follows that $-\Delta u^* = 0$ a.e. in Ω^*. But from the equation satisfied by u^*, there comes

$$l^* u^* = l^* c^* = f \text{ in } \Omega^*. \tag{7.53}$$

Now thanks to the maximum principle, we know that $u^* \le u_0$ in Ω, and then (7.53) gives

$$l^* = \frac{f}{c^*} \ge \frac{f}{u_0} \text{ in } \Omega^*.$$

If we assume condition (i), we have $l^* \ge 1$ in Ω^*, which contradicts the definition of Ω^*.

Now if we assume condition (ii), let us denote by v the function $f - u_0$. We have

$$\begin{aligned} -\Delta v &= -\Delta f + \Delta u_0 = -\Delta f - f \ge 0 \text{ in } \Omega, \\ v &= f \ge 0 \text{ on } \partial\Omega. \end{aligned}$$

Then since v is superharmonic, $v = f - u_0 \ge 0$ in Ω and we can apply condition (i).

Finally, if Ω^* is nonempty, we saw in (7.53) that $l^* = \frac{f}{c^*}$ in Ω^*. But since, by definition, $l^* < 1$ in Ω^*, we should have $f < c^*$. Thus, Ω^* is necessarily included in the set $\{x \in \Omega,\ f(x) < u_0\}$ because $u^* \le u_0$ by the maximum principle. It follows that, on Ω^*,

$$1 > l^* = \frac{f}{c^*} \ge \frac{\inf\{f(x),\ x \text{ such that } u_0(x) > f(x)\}}{c^*} = \frac{\alpha}{c^*},$$

which means $c^* > \alpha$. Consequently, $\Omega^* \cup \Omega_1$, which is included in the set $\{x \in \Omega,\ u^* \ge c^*\}$ must satisfy

$$\Omega^* \cup \Omega_1 \subset [u^* \ge c^*] \subset [u^* > \alpha] \subset [u_0(x) > \alpha]. \tag{7.54}$$

Now, since $C = \int_\Omega l^* = |\Omega_1| + \int_{\Omega^*} l^* \le |\Omega^* \cup \Omega_1|$, condition (iii) is incompatible with the inclusion (7.54). This proves that Ω^* must be empty as soon as condition (iii) is satisfied. □

Remark 7.2.30. The optimality conditions given in Theorem 7.2.26 can also be used in other situations:

- The numerical computation of the solution: an algorithm based on direct research of the sets Ω_0, Ω^*, and Ω_1 satisfying the requested conditions is shown in [183] and happens to be really efficient.

- In the radial case, one can also use them to look explicitly for the solution by solving a simple system of ordinary differential equations. In this spirit, we can prove the following theorem, for which we also refer to [183].

Theorem 7.2.31. *We assume that Ω is the unit ball and $f = f(r)$ is a radial, positive, and decreasing function in $L^2(\Omega)$:*

(i) There exists a classical solution $l^* = \chi_{\Omega_1}$.

(ii) Any classical solution is radially symmetric (Ω_1 is the ball of volume C centered at the origin).

(iii) There are no other radially symmetric solutions.

7.3 Relaxation by homogenization

7.3.1 Presentation of the problem

Let us now consider a model problem that is very classical in heat conduction. It appears in different places in the literature, see, for example, [244], [287], [11], [83].

Suppose we have two different materials, of respective conductivities α and β with $0 < \alpha < \beta < +\infty$. We want to fill a given domain Ω with a mixture of these two materials. Let us denote by A (resp. $\Omega\backslash A$) the part of the body occupied by the material of conductivity α (resp. β). The conductivity of Ω can be written

$$a(x) = \alpha\chi_A(x) + \beta(1 - \chi_A(x)). \tag{7.55}$$

Imposing the amount of each material corresponds to writing the constraint

$$|A| = \int_\Omega \chi_A(x)\,dx = c. \tag{7.56}$$

Let us now assume that the body Ω is heated by some source f and its boundary maintained at temperature 0. Then, the temperature inside Ω is obtained by solving the heat equation

$$\begin{cases} -\operatorname{div}(a(x)\nabla u) = f & \text{in } \Omega, \\ u = 0 & \text{on } \partial\Omega. \end{cases} \tag{7.57}$$

Then we can be interested in minimizing a functional like

$$J(A) := \int_A g(x, u(x))\,dx + \int_{\Omega\backslash A} h(x, u(x))\,dx, \tag{7.58}$$

where we can assume that the functions g and h are of Carathéodory type (see (7.14)) and satisfy the usual conditions,

$$|g(x,s)| + |h(x,s)| \le a_0(x) + b_0|s|^2 \quad \text{with } a_0 \in L^1(\Omega),\ b_0 \in [0, +\infty). \tag{7.59}$$

As in the previous example, a minimizing sequence χ_{A_n} converges weakly-$*$ in $L^\infty(\Omega)$ to some function θ, with values between 0 and 1, that satisfies $\int_\Omega \theta(x)\,dx = c$. In the situation of the previous section, the convergence of u_n to the u limit was easily deduced (because the perturbation in the PDE was only in the zero-order term of the equation). Here this is not the case and we have to use the notion of G-convergence introduced in Section 7.1.2.

Thus, if $\chi_{A_n} \overset{*}{\rightharpoonup} \theta$, we have

$$a_n(x) = \alpha\chi_{A_n}(x) + \beta(1 - \chi_{A_n}(x)) \overset{*}{\rightharpoonup} \alpha\theta(x) + \beta(1 - \theta(x)) \quad \text{in } L^\infty(\Omega).$$

By the compactness property of G-convergence (see Corollary 7.1.7), it follows that there exists a subsequence, still denoted a_n, such that $a_n \operatorname{Id}$ G-converges to some matrix A^*. By definition this means that u_n weakly converges to u^* a solution of

$$\begin{cases} -\operatorname{div}(A^*\nabla u^*) = f & \text{in } \Omega, \\ u^* = 0 & \text{on } \partial\Omega. \end{cases} \tag{7.60}$$

Moreover, by using the assumptions verified by g and h,

Proposition 7.3.1. *If χ_{A_n} converges weakly-$*$ in $L^\infty(\Omega)$ to θ, then $J(A_n)$ (defined in* (7.58)*) converges in $\mathbb{R}$ to $J^*(\theta)$ defined by*

$$J^*(\theta) := \int_\Omega [\theta(x)g(x, u^*(x)) + (1 - \theta(x))h(x, u^*(x))]\, dx, \tag{7.61}$$

where u^* is a solution of (7.60).

Proof. Using the assumptions on the functions g and h, we deduce that $g(x, u_n(x))$ and $h(x, u_n(x))$ strongly converge in $L^1(\Omega)$ to $g(x, u^*(x))$ and $h(x, u^*(x))$, respectively.

Now $J(A_n)$ can be written

$$J(A_n) = \int_\Omega \chi_{A_n}(x)g(x, u_n(x)) + (1 - \chi_{A_n}(x))h(x, u_n(x))\, dx.$$

Since χ_{A_n} converges weakly-$*$ in $L^\infty(\Omega)$ to θ, the result follows. □

7.3.2 Relaxation

The relaxed problem consists in looking for some θ in

$$\mathcal{X} = \left\{\theta \in L^\infty(\Omega),\ 0 \le \theta(x) \le 1 \text{ a.e.},\ \textstyle\int_\Omega \theta(x)\, dx = c\right\}$$

that minimizes the functional J^* defined in (7.61). Now, the problem stated as such is not really satisfactory. In particular, we would like to describe more precisely what the G-limit of matrices $a_n \operatorname{Id}$ could be, as we did in Section 7.2.2 for the Dirichlet problem. This is why we are led to introduce the set

$$M_\theta := \{G\text{-limits of } a_n \operatorname{Id}, \text{ with } \chi_{A_n} \overset{*}{\rightharpoonup} \theta\}. \tag{7.62}$$

The relaxed problem is now posed on the set X^* defined by

$$X^* = \{(\theta, A^*);\ \theta \in L^\infty(\Omega),\ A^* \in M_\theta,\ 0 \le \theta(x) \le 1 \text{ a.e.}, \int_\Omega \theta(x)\,dx = c\}$$

on which we have to minimize the functional

$$J^*(\theta, A^*) = \int_\Omega [\theta(x) g(x, u^*(x)) + (1-\theta(x)) h(x, u^*(x))]\,dx,$$

where $u^*(x)$ is the function of $H_0^1(\Omega)$ that is a solution of $-\operatorname{div}(A^* \nabla u^*) = f$.

The set X^* is naturally endowed with the weak-$*$ topology for the first component in θ and the topology of G-convergence for the second component. As explained above, with this topology X^* is sequentially compact and J^* is continuous. Thus the minimization problem has a solution.

Let us now describe more precisely the set M_θ. The following result is due to Lurie–Cherkaev in dimension 2 and Murat–Tartar for the general case.

Theorem 7.3.2. *Let us introduce the arithmetic and harmonic means of α and β with respect to θ:*

$$a^+(\theta) = \theta\alpha + (1-\theta)\beta, \qquad a^-(\theta) = \left(\frac{\theta}{\alpha} + \frac{(1-\theta)}{\beta}\right)^{-1}.$$

Then M_θ is composed of the symmetric matrices $A^*(x)$ whose eigenvalues on each point $(\lambda_1(x), \lambda_2(x), \dots, \lambda_N(x))$ satisfy

$$a^-(\theta) \le \lambda_i(x) \le a^+(\theta), \quad i = 1 \dots N, \tag{7.63}$$

$$\begin{cases} \displaystyle\sum_{i=1}^N \frac{1}{\lambda_i - \alpha} \le \frac{1}{a^-(\theta) - \alpha} + \frac{N-1}{a^+(\theta) - \alpha}, \\ \displaystyle\sum_{i=1}^N \frac{1}{\beta - \lambda_i} \le \frac{1}{\beta - a^-(\theta)} + \frac{N-1}{\beta - a^+(\theta)}. \end{cases} \tag{7.64}$$

We are not going to prove this theorem here. Let us recall that inequalities (7.63) were proved in Corollary 7.1.9. We observe that such a precise characterization of the set M_θ of all G-limits is quite exceptional. In general, what is difficult to get is the converse, namely to prove that a matrix A^* whose eigenvalues satisfy (7.63) and (7.64) is the G-limit of a sequence of matrices like $a_n \operatorname{Id}$. This can only be obtained by complex computations, but these are interesting since they can give an explicit construction of an approximating sequence (see the main contributions by Murat–Tartar).

Let us plot in the plane (see Figure 7.2) the possible set for $\lambda_1(x), \lambda_2(x)$.

We observe that this set is convex. This property will in fact imply that M_θ is itself convex. We now follow the presentation of Murat–Tartar [244].

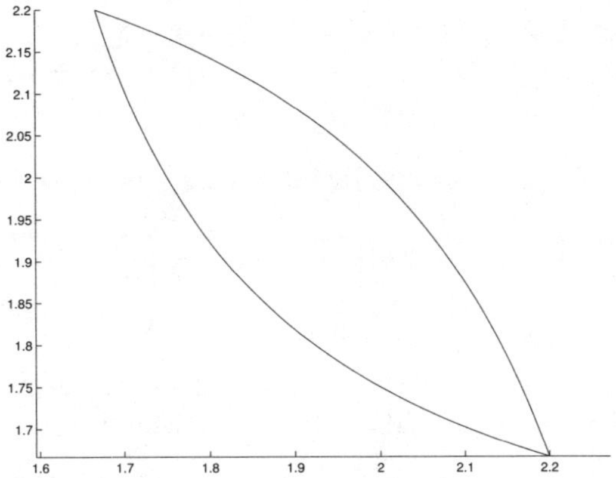

Figure 7.2. The convex set in the plane of all possible couples of eigenvalues $(\lambda_1(x), \lambda_2(x))$.

Proposition 7.3.3. *For any fixed θ in $[0, 1]$, the set M_θ is a convex set of symmetric matrices.*

We will use the following algebraic lemma for the proof.

Lemma 7.3.4. *Let φ be a convex function from $\mathbb{R}$ to $\overline{\mathbb{R}}$ and c a real number. The set of $N \times N$ symmetric matrices satisfying $\sum_{i=1}^{N} \varphi(\mu_i) \leq c$, where $\mu_1, \mu_2, \ldots, \mu_N$ are the eigenvalues of the matrices, is a convex set of $\mathcal{M}_N(\mathbb{R})$.*

Proof. Indeed, let us identify any matrix with its linear map (through the canonical basis in $\mathbb{R}^N$). The orthonormal basis of eigenvectors is obtained through an orthogonal passage matrix P. Therefore, in any orthonormal basis, if A_{ii} denotes the ith diagonal term of the matrix of the corresponding linear map, we have

$$A_{ii} = \sum_{k=1}^{N} p_{ik} \mu_k p_{ik} = \sum_{k=1}^{N} p_{ik}^2 \mu_k,$$

which is a convex combination of the μ_k since P is an orthogonal matrix. Consequently $\varphi(A_{ii}) \leq \sum_{k=1}^{N} p_{ik}^2 \varphi(\mu_k)$ and after summing,

$$\sum_{i=1}^{N} \varphi(A_{ii}) \leq \sum_{k=1}^{N} \sum_{i=1}^{N} p_{ik}^2 \varphi(\mu_k) = \sum_{k=1}^{N} \varphi(\mu_k). \tag{7.65}$$

Inequality (7.65) being valid for any matrix A in any orthonormal basis, we deduce the equivalence

$$\left\{\sum_{k=1}^{N} \varphi(\mu_k) \leq c\right\} \iff \left\{\sum_{i=1}^{N} \varphi(A_{ii}) \leq c \text{ in any orthonormal basis}\right\}.$$

Now, since the function defined on $\mathcal{M}_N(\mathbb{R})$ by $A \mapsto \sum_{i=1}^N \varphi(A_{ii})$ is clearly convex and since the inverse image of $(-\infty, c]$ by this map is a convex subset of $\mathcal{M}_N(\mathbb{R})$, the lemma is proved. □

Proof of Proposition 7.3.3. It follows by introducing the two convex functions

$$\varphi_1(t) = \begin{cases} \frac{1}{t-\alpha} & \text{if } t \in (\alpha, a^+(\theta)], \\ +\infty & \text{otherwise}, \end{cases} \qquad \varphi_2(t) = \begin{cases} \frac{1}{\beta-t} & \text{if } t \in [a^-(\theta), \beta), \\ +\infty & \text{otherwise}. \end{cases}$$

(In such a way, if one of the eigenvalues μ_i goes out of the interval $[a^-(\theta), a^+(\theta)]$, we have $\varphi_1(\mu_i)$ or $\varphi_2(\mu_i)$, which takes the value $+\infty$.) Then we set

$$c_1 = \frac{1}{a^-(\theta) - \alpha} + \frac{N-1}{a^+(\theta) - \alpha}, \qquad c_2 = \frac{1}{\beta - a^-(\theta)} + \frac{N-1}{\beta - a^+(\theta)},$$

and we have just to observe that M_θ can be defined as the intersection of the two convex sets $\left\{\sum_{k=1}^N \varphi_1(\mu_k) \le c_1\right\}$ and $\left\{\sum_{k=1}^N \varphi_2(\mu_k) \le c_2\right\}$. □

7.3.3 Optimality conditions

Let us now write the optimality conditions satisfied by the optimal couple (θ^*, A^*). We will see that the analysis is quite similar to that of Section 7.2.3.4. We are first going to write it in the general case, then we will see an interesting and representative example where it can be used to get qualitative information on the solution. In particular, we are going to see an example where we can prove that no classical solution exists.

In order to be able to differentiate the functional, we must make more assumptions on the data g and h. We will assume that $u \mapsto g(x, u)$ and $u \mapsto h(x, u)$ are Gâteaux[1]-differentiable on $H_0^1(\Omega)$, this is in particular satisfied if we assume (see [203])

$$\begin{cases} \frac{\partial g}{\partial t} \text{ and } \frac{\partial h}{\partial t} \text{ are Carathéodory functions (see (7.14))}, \\ \left|\frac{\partial g}{\partial t}(x,t)\right| + \left|\frac{\partial h}{\partial t}(x,t)\right| \le a_1(x) + b_1|t| \text{ with } a_1 \in L^1(\Omega). \end{cases}$$

Thus we can state

Proposition 7.3.5. *The functional J^* is Gâteaux-differentiable at any point (θ, A) and its derivative is given by*

$$dJ^*_{(\theta,A)}(\sigma, B) = \int_\Omega (\sigma[g(\cdot,u) - h(\cdot,u)] - (B\nabla u, \nabla p)), \tag{7.66}$$

[1]René Eugène GÂTEAUX, 1889–1914, French, was among the first “dead for the country” during the First World War. In spite of his too short career, his name remains important in differential calculus.

where u is the state associated to A, a solution of

$$- \operatorname{div}(A\nabla u) = f, \quad u \in H_0^1(\Omega), \tag{7.67}$$

and p is the adjoint state, a solution of

$$- \operatorname{div}(A\nabla p) = \theta \frac{\partial g}{\partial t}(\cdot, u) + (1-\theta)\frac{\partial h}{\partial t}(\cdot, u), \quad p \in H_0^1(\Omega). \tag{7.68}$$

Proof. This is classical, let us just do a formal computation. An expansion at order 1 of $J^*(\theta + t\sigma, A + tB)$ gives

$$\begin{aligned} dJ^*_{(\theta,A)}(\sigma, B) &= \lim_{t\to 0} \frac{J^*(\theta + t\sigma, A + tB) - J^*(\theta, A)}{t} \\ &= \int_\Omega \sigma[g(\cdot,u) - h(\cdot,u)] + \left[\theta \frac{\partial g}{\partial t}(\cdot,u) + (1-\theta)\frac{\partial h}{\partial t}(\cdot,u)\right] v, \end{aligned}$$

where v is the solution of the problem

$$- \operatorname{div}(A\nabla v) = \operatorname{div} B\nabla u, \quad v \in H_0^1(\Omega). \tag{7.69}$$

To eliminate v in the expression of dJ^*, we multiply equation (7.69) by p and equation (7.68) by v. We then get (7.66) by identification. □

The optimality conditions can now be obtained by writing that $dJ^* \geq 0$ at point (θ^*, A^*), for any variation (σ, B) in the cone of admissible directions. Let us first fix $\sigma = 0$ and let only B vary. The convexity of the set M_θ (Proposition 7.3.3) shows that $B = C - A^*$ is an admissible variation for any matrix $C \in M_\theta$. Replacing in (7.66), we get

$$\forall\, C \in M_{\theta(x)}, \quad \int_\Omega (C(x)\nabla u, \nabla p)\, dx \leq \int_\Omega (A^*(x)\nabla u, \nabla p)\, dx. \tag{7.70}$$

Let ω be any open subdomain of Ω. If we replace the matrix C by the matrix $\widetilde{C}$ defined by

$$\widetilde{C}(x) = \begin{cases} C(x) & \text{on } \omega, \\ A^*(x) & \text{on } \Omega\backslash\omega, \end{cases}$$

which is still in $M_{\theta(x)}$, then relation (7.70) remains valid. This shows that relation (7.70) can be localized, namely,

$$\text{for a.e. } x \in \Omega,\ \forall\, C \in M_{\theta(x)}, \quad (C(x)\nabla u, \nabla p) \leq (A^*(x)\nabla u, \nabla p). \tag{7.71}$$

In other words, $A^*(x)$ is a solution of a pointwise maximization problem that involves the two vectors $\nabla u(x)$ and $\nabla p(x)$. This problem can easily be solved thanks to the following geometric lemma:

Lemma 7.3.6. *Let C_θ be the set of $N \times N$ symmetric matrices whose eigenvalues are between the two numbers $a^-(\theta)$ and $a^+(\theta)$ and let e and e' be two unit vectors of $\mathbb{R}^N$. Then the following three properties are equivalent:*

$$(i) \quad A \in C_\theta \text{ and, for all } C \in C_\theta,\ (Ce, e') \le (Ae, e'). \tag{7.72}$$

$$(ii) \quad Ae = \tfrac{1}{2}(a^+(\theta) + a^-(\theta))e + \tfrac{1}{2}(a^+(\theta) - a^-(\theta))e'. \tag{7.73}$$

$$(iii) \quad A(e + e') = a^+(\theta)(e + e') \text{ and } A(e - e') = a^-(\theta)(e - e'). \tag{7.74}$$

Proof. Let us have a look at Figure 7.3.

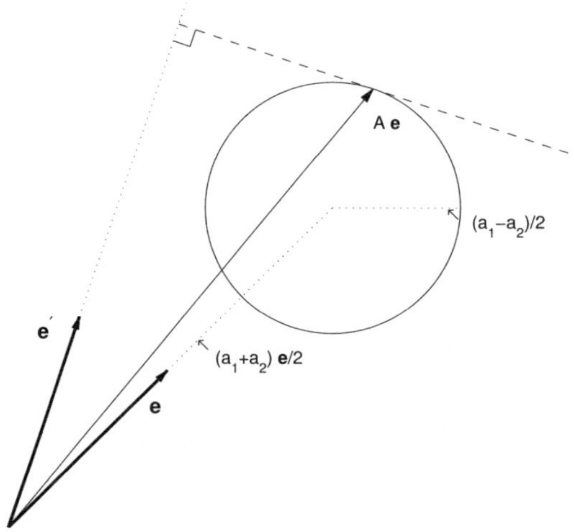

Figure 7.3. Proof of Lemma 7.3.6.

For any matrix $C \in C_\theta$, the extremity of the vector Ce lies inside the circle of center $\frac{1}{2}(a^+(\theta) + a^-(\theta))e$ and of radius $\frac{1}{2}(a^+(\theta) - a^-(\theta))$. Indeed, we have

$$\left|Ce - \tfrac{1}{2}(a^+(\theta) + a^-(\theta))e\right| \le \left\|C - \tfrac{1}{2}(a^+(\theta) + a^-(\theta))\,\mathrm{Id}\right\|_2 .$$

But the matrix $C - \frac{1}{2}(a^+(\theta) + a^-(\theta))\,\mathrm{Id}$ being symmetric, its norm (related to the Euclidean norm) is given by its spectral radius; see, for example, [104]. Now, taking into account the assumption on the eigenvalues of C, this is estimated from above by

$$\max\left\{\left|a^-(\theta) - \tfrac{1}{2}(a^+(\theta) + a^-(\theta))\right|, \left|a^+(\theta) - \tfrac{1}{2}(a^+(\theta) + a^-(\theta))\right|\right\}$$

or

$$\left\|C - \tfrac{1}{2}(a^+(\theta) + a^-(\theta))\,\mathrm{Id}\right\|_2 \le \tfrac{1}{2}(a^+(\theta) - a^-(\theta)),$$

which proves the assertion.

Now, among all possible matrices $C \in C_\theta$, it is clear that those that will maximize the scalar product (Ae, e') are those for which the extremity of the vector Ae takes the position shown in the figure (the tangent to the circle is here orthogonal to the vector e'). Thus the expression of Ae is given by (7.73). Since A is symmetric, it is also a solution of the maximization problem where we exchange the roles of e and e'. In other words we get Ae' by exchanging the roles of e and e' in formula (7.73). Formula (7.74) follows easily and Lemma 7.3.6 is proved. □

Let us remark that the above lemma gives a characterization of matrices that satisfy the maximum property, not in the set M_θ, but in the bigger set C_θ. To be able to apply it to our situation, we need to prove the existence of a matrix $A \in M_\theta$ that satisfies condition (7.73) or condition (7.74).

Let us first consider the case where $e \neq e'$. Relations (7.74) show that $e + e'$ and $e - e'$ (which are orthogonal) must be eigenvectors of A associated respectively to the eigenvalues $a^+(\theta)$ and $a^-(\theta)$. By completing the basis of eigenvectors in the orthogonal subspace in such a way that $a^+(\theta)$ is the unique eigenvalue for all these eigenvectors, we get a matrix that satisfies conditions (7.63) and (7.64). Therefore it is in M_θ. Let us observe that the matrix we just defined in fact corresponds to a material that is laminated in a direction orthogonal to a vector $\tilde{e}$, where $\tilde{e}$ is collinear to $e - e'$ and orthogonal to $e + e'$.

In the case where $e = e'$, we make an analogous construction by simply imposing $Ae = a^+(\theta)e$, since the second relation of (7.74) is meaningless.

Remark 7.3.7. We may notice that the previous analysis uses only the following properties of the set M_θ:

- The matrices in M_θ have eigenvalues between $a^-(\theta)$ and $a^+(\theta)$.
- There exists a matrix in M_θ whose first two eigenvalues are $a^+(\theta)$ and $a^-(\theta)$.

This remark can be useful for more complex problems for which the knowledge of the set M_θ could be much less precise.

Let us come back to our optimality conditions satisfied by the couple (θ^*, A^*). Lemma 7.3.6 will apply when the two vectors ∇u (which, once normalized, corresponds to e) and ∇p (which, once normalized, corresponds to e') are nonzero. When this is the case, let us introduce their angle φ defined by $\nabla u . \nabla p = |\nabla u|\,|\nabla p| \cos\varphi$. We can summarize the previous results as follows.

Proposition 7.3.8. *If (θ^*, A^*) is a minimum of the functional J^* then, outside the set where $|\nabla u|\,|\nabla p| = 0$, we have*

$$\begin{cases} A^* \frac{\nabla u}{|\nabla u|} = \frac{1}{2}(a^+(\theta) + a^-(\theta))\frac{\nabla u}{|\nabla u|} + \frac{1}{2}(a^+(\theta) - a^-(\theta))\frac{\nabla p}{|\nabla p|}, \\ A^* \frac{\nabla p}{|\nabla p|} = \frac{1}{2}(a^+(\theta) + a^-(\theta))\frac{\nabla p}{|\nabla p|} + \frac{1}{2}(a^+(\theta) - a^-(\theta))\frac{\nabla u}{|\nabla u|}, \\ (A^*\nabla u, \nabla p) = |\nabla u||\nabla p|(a^+(\theta)\cos^2\frac{\varphi}{2} - a^-(\theta)\sin^2\frac{\varphi}{2}). \end{cases} \tag{7.75}$$

Remark 7.3.9. The previous proposition does not give any information in the case where ∇u or ∇p vanishes. In fact, as explained in [244] (see also [261]), one can always find an optimal solution that uses only laminated materials. This means that we can consider the case where the matrix A^* has $a^-(\theta)$ as a simple eigenvalue and $a^+(\theta)$ as an eigenvalue of order $N-1$.

Now let us make θ vary. For that purpose, let us introduce $(\theta(x,t); A(x,t))$, defined for $x \in \Omega$ and $t \in [0,1]$ such that $\theta(x,0) = \theta^*$, and $A(x,t)$ that satisfy

$$\begin{cases} A(x,t) = A^*(x) & \text{if } |\nabla u(x)|\,|\nabla p(x)| = 0, \\ (A(x,t)\nabla u(x), \nabla p(x)) = |\nabla u(x)|\,|\nabla p(x)| & \\ \qquad \times (a^+(\theta(x,t)) \cos^2 \frac{\varphi}{2} - a^-(\theta(x,t)) \sin^2 \frac{\varphi}{2}) & \text{otherwise.} \end{cases}$$

(One can build such a matrix $A(x,t)$ by following the same process as described in the proof of Lemma 7.3.6.) In the above expression, u and p denote respectively the state and the adjoint state associated to the optimal couple (θ^*, A^*).

Let us now differentiate $J^*(\theta(t), A(t))$ with respect to t at $t = 0$. Taking into account (7.66) and from the expression of $A(x,t)$, we get

$$\frac{d}{dt}_{|t=0} J^*(\theta(t), A(t)) = \int_\Omega \theta'(0)[g(\cdot,u) - h(\cdot,u)] - (A'(0)\nabla u, \nabla p)$$

or

$$\begin{aligned} \frac{d}{dt}_{|t=0} J^*(\theta(t), A(t)) = {} & \int_\Omega \theta'(0)[g(\cdot,u) - h(\cdot,u)] \\ & - \int_\Omega |\nabla u||\nabla p| \left(\frac{d}{dt}_{|t=0} a^+(\theta(t)) \cos^2 \frac{\varphi}{2} - \frac{d}{dt}_{|t=0} a^-(\theta(t)) \sin^2 \frac{\varphi}{2} \right). \end{aligned} \tag{7.76}$$

By using the expression of the harmonic mean and of the arithmetic mean and introducing the function $Q = Q(x)$ defined by

$$Q = g(\cdot,u) - h(\cdot,u) + \frac{\beta - \alpha}{\alpha\beta} |\nabla u|\,|\nabla p| \left(\alpha\beta \cos^2 \frac{\varphi}{2} - (a^-(\theta))^2 \sin^2 \frac{\varphi}{2} \right), \tag{7.77}$$

one can state

Proposition 7.3.10. *For any admissible variation* $\delta\theta = \frac{d\theta}{dt}\,|_{t=0}$ *(i.e., such that* $\int_\Omega \delta\theta = 0$ *and that the curve stays in the convex set* $0 \le \theta \le 1$*), we have*

$$\int_\Omega \delta\theta\, Q(x)\, dx \ge 0, \tag{7.78}$$

where $Q(x)$ *is defined by* (7.77).

Using the same discussion as in Section 7.2.3.4 for Theorem 7.2.26, we deduce from (7.78) the existence of a Lagrange multiplier c^* such that

Corollary 7.3.11. *If* (θ^*, A^*) *is a minimizer of* J^**, there exists a constant* c^* *such that*

$$\begin{cases} \theta^*(x) = 0 & \Longrightarrow \quad Q(x) \geq c^*, \\ 0 < \theta^*(x) < 1 & \Longrightarrow \quad Q(x) = c^*, \\ \theta^*(x) = 1 & \Longrightarrow \quad Q(x) \leq c^*, \end{cases} \tag{7.79}$$

where $Q(x)$ *is defined by* (7.77).

7.3.4 An example of an application

To show how we can use the optimality conditions, we will consider the case where we want to minimize the following functional J defined for any measurable subset ω of $\Omega \subset \mathbb{R}^2$ with a given measure c by

$$J(\omega) = \int_\Omega f(x)u(x)\,dx, \tag{7.80}$$

where u is the solution of

$$\begin{cases} -\operatorname{div}\left(\alpha\chi_\omega(x) + \beta(1-\chi_\omega(x))\nabla u\right) = f & \text{in } \Omega, \\ u = 0 & \text{on } \partial\Omega. \end{cases} \tag{7.81}$$

In other words, we choose here $g(x,s) = h(x,s) = f(x)s$.

According to the previous analysis, the relaxed problem can be written here as follows: we look for an optimal couple (θ^*, A^*) in

$$X^* = \left\{(\theta, A);\ \theta \in L^\infty(\Omega),\ A \in M_\theta,\ 0 \leq \theta(x) \leq 1 \text{ a.e.}, \textstyle\int_\Omega \theta(x)\,dx = c\right\},$$

where M_θ is defined in (7.62), that minimizes the functional

$$J^*(\theta, A) = \int_\Omega f(x)u(x)\,dx,$$

with u the solution in $H_0^1(\Omega)$ of

$$\begin{cases} -\operatorname{div}(A\nabla u) = f & \text{in } \Omega, \\ u = 0 & \text{on } \partial\Omega. \end{cases} \tag{7.82}$$

The adjoint state is given by solving equation (7.68), whose right-hand side is f in this case. We find the same equation (7.82), so that we have here $u = p$.

Let us look for the conditions satisfied by an optimal couple (θ^*, A^*). According to Proposition 7.3.8, at points where $\nabla u^* \neq 0$, we must have

$$A^*\nabla u^* = a^+(\theta^*)\nabla u^*. \tag{7.83}$$

Of course this relation (7.83) is still true when $\nabla u(x) = 0$. Thus equation (7.82) can be written more simply as

$$\begin{cases} -\operatorname{div}\left(a^+(\theta^*)\nabla u^*\right) = f & \text{in } \Omega, \\ u = 0 & \text{on } \partial\Omega. \end{cases} \tag{7.84}$$

In formula (7.77), we must take the angle $\varphi = 0$. Thus $Q(x)$ is given by

$$Q(x) = (\beta - \alpha)|\nabla u^*(x)|^2, \tag{7.85}$$

and Corollary 7.3.11 yields the existence of a constant c_1^* such that

$$\begin{cases} \theta^*(x) = 0 & \Longrightarrow & |\nabla u^*(x)| \geq c_1^*, \\ 0 < \theta^*(x) < 1 & \Longrightarrow & |\nabla u^*(x)| = c_1^*, \\ \theta^*(x) = 1 & \Longrightarrow & |\nabla u^*(x)| \leq c_1^*. \end{cases} \tag{7.86}$$

Let us now prove that if we add the extra assumption that

$$\text{for a.e. } x, \quad f(x) \neq 0, \tag{7.87}$$

then there cannot exist a classical solution (i.e., θ^* cannot be a characteristic function like χ_{ω^*}, at least with a regular ω^*).

Let us assume, for contradiction, that there exists a classical solution $\theta^* = \chi_{\omega^*}$ with a regular ω^* (e.g., of class C^2). Let us denote by ω a connected component of ω^* and by γ the part of its boundary contained in Ω. From each side of the interface γ, there are two regions:

- In one region (ω), we have $\theta^* = 1$. We denote by u_1 the solution of the equation in this region. Moreover we have here $a^+(\theta^*) = \alpha$.
- In the other region, we have $\theta^* = 0$. We denote by u_0 the solution of the equation in this region. Moreover we have here $a^+(\theta^*) = \beta$.

Since equation (7.84) is satisfied in Ω, we have continuity of u^* through the interface:

$$u_0 = u_1 \quad \text{on } \gamma, \tag{7.88}$$

and continuity of the jump of the normal derivative,

$$\beta\frac{\partial u_0}{\partial n} = \alpha\frac{\partial u_1}{\partial n} \quad \text{on } \gamma. \tag{7.89}$$

The relation (7.88) implies equality of the tangential derivatives on γ: $\partial_\gamma u_0 = \partial_\gamma u_1$. Now the optimality conditions (7.86) show that $|\nabla u_0| \geq c_1^* \geq |\nabla u_1|$ approaching the interface. Therefore, from the equality of the tangential components, we must have on γ that $\frac{\partial u_0}{\partial n} \geq \frac{\partial u_1}{\partial n}$. But since $\beta > \alpha$, the conjunction of the previous inequality with equality (7.89) implies

$$\frac{\partial u_0}{\partial n} = \frac{\partial u_1}{\partial n} = 0 \quad \text{and} \quad \partial_\gamma u_0 = \partial_\gamma u_1 = c_1^* \quad \text{on } \gamma. \tag{7.90}$$

Let us remark that c_1^* cannot vanish: otherwise, from (7.86), we would have $|\nabla u_1| = 0$ in ω and, coming back to (7.84), $f = 0$ in ω, which would contradict the assumption (7.87).

Now two different situations may occur:

- Either ω is strictly contained in Ω, which implies that γ is a closed curve, and then $\partial_\gamma u_1 = c_1^*$ on the whole $\partial\omega$;
- or the boundary of ω has a common part with the boundary of Ω, which means that γ reaches the boundary of Ω. In this case, $\partial_\gamma u_1 = c_1^*$ on γ and $\partial_\gamma u_1 = 0$ on $\partial\omega\backslash\gamma$ (because $u_1 = 0$ on $\partial\omega\backslash\gamma$).

In both cases, we can deduce that

$$0 = \int_{\partial\omega} \partial_\gamma u_1 = c_1^*|\gamma|,$$

which implies that γ has a length that is zero. This gives the expected contradiction.

Remark 7.3.12. In the case where we want to maximize the functional J defined in (7.80) (or minimize $-J$), we need to make the following modifications in the previous analysis. Now we have

- $p = -u^*$;
- $\varphi = \pi$;
- $Q(x) = -\frac{(\beta-\alpha)}{\alpha\beta}(a^-(\theta^*))^2|\nabla u^*(x)|^2$;

and the optimality conditions are now written as follows: there exists a constant c_1^* such that

$$\begin{cases} \theta^*(x) = 0 & \Longrightarrow \quad \beta|\nabla u^*(x)| \leq c_1^*, \\ 0 < \theta^*(x) < 1 & \Longrightarrow \quad a^-(\theta^*)|\nabla u^*(x)| = c_1^*, \\ \theta^*(x) = 1 & \Longrightarrow \quad \alpha|\nabla u^*(x)| \geq c_1^*. \end{cases} \tag{7.91}$$

We no longer get a contradiction assuming existence of a classical solution, as $\theta^* = \chi_{\omega^*}$. But now we know that the tangential derivative must vanish on the interface γ between the zone $\theta^* = 0$ and the zone $\theta^* = 1$. In particular, this implies that u_1 and the normal derivative $\frac{\partial u_1}{\partial n}$ must be constant on γ.

In the particular case $f = 1$, this has an interesting and funny consequence: if we look for a connected component of the optimal solution $\omega \subset\subset \Omega$ completely contained in Ω, the previous relations, together with Serrin's Theorem 6.1.11, show that ω is necessarily a ball. Consequently, by the unique continuation principle for analytic functions, we must have $u^* = C(A - r^2)$ on the whole Ω. This implies that Ω itself is necessarily a ball!

Bibliography

[1] E. Acerbi, G. Buttazzo, Reinforcement problems in the calculus of variations, *Ann. Inst. H. Poincaré Anal. Non Linéaire* **3**(4) (1986), 273–284. Zbl 0607.73018

[2] A. Acker, On the geometric form of free boundaries satisfying a Bernoulli condition, *Math. Meth. Appl. Sci.* **6** (1984), 449–456. Zbl 0558.35081

[3] A. Acker, On the geometric form of free boundaries satisfying a Bernoulli condition II, *Math. Meth. Appl. Sci.* **8** (1986), 387–404. Zbl 0616.35084

[4] D. R. Adams, L. I. Hedberg, *Function Spaces and Potential Theory*, Springer, 1995. Zbl 0834.46021

[5] R. A. Adams, *Sobolev Spaces*, Academic Press, New York, London, 1975. Zbl 0314.46030

[6] S. Agmon, *Lectures on Elliptic Boundary Value Problems*, Van Nostrand Math Studies, 1965. Zbl 0142.37401

[7] S. Agmon, A. Douglis, L. Nirenberg, Estimates near the boundary for solutions of elliptic partial differential equations satisfying general boundary conditions I, *Comm. Pure Appl. Math.* **12** (1959), 623–727. Zbl 0093.10401

[8] G. Alberti, G. Bouchitté, G. Dal Maso, The calibration method for the Mumford-Shah functional, *C. R. Acad. Sci. Paris Sér. I Math.* **329**(3) (1999), 249–254. Zbl 0948.49005

[9] G. Alessandrini, E. Rosset, The inverse conductivity problem with one measurement: Bounds on the size of the unknown object, *SIAM J. Appl. Math.* **58**(4) (1998), 1060–1071. Zbl 0953.35141

[10] A. D. Alexandroff, A characteristic property of spheres, *Ann. Mat. Pura Appl.* **58** (1962), 303–315. Zbl 0107.15603

[11] G. Allaire, An introduction to homogenization and bounds on effective properties applied to optimal shape design, *School on Homogenization, ICTP Trieste*, 1993. https://www.mat.uniroma2.it/~braides/ICTP93/ICTP93Allaire2.pdf.

[12] G. Allaire, *Shape Optimization by the Homogenization Method*, Applied Mathematical Sciences **146**, Springer, New York, 2002. Zbl 0990.35001

[13] G. Allaire, *Conception optimale de structures*, Collection Mathématiques et Applications **58**, Springer, 2007. Zbl 1132.49033

[14] G. Allaire, E. Bonnetier, G. Francfort, F. Jouve, Shape optimization by the homogenization method, *Numer. Math.* **76** (1997), 27–68. Zbl 0889.73051

[15] G. Allaire, R. V. Kohn, Optimal design for minimum weight and compliance in plane stress using extremal microstructures, *Europ. J. Mech. A/Solids* **12**(6) (1993), 839–878. Zbl 0794.73044

[16] F. Almgren, J. Taylor, The geometry of soap films and soap bubbles, *Sci. Am.* **235** (1976), 82.

[17] H. W. Alt, L. A. Caffarelli, Existence and regularity for a minimum problem with free boundary, *J. reine angew. Math.* **325** (1981), 105–144. Zbl 0449.35105

[18] L. Ambrosio, G. Buttazzo, An optimal design problem with perimeter penalization, *Calc. Var. Partial Differential Equations* **1**(1) (1993), 55–69. Zbl 0794.49040

[19] A. Ancona, *Théorie du potentiel dans les espaces fonctionnels à forme coercive*, *Cours de 3è cycle*, Université de Paris 6, 1973.

[20] B. Andrews, J. Clutterbuck, Proof of the fundamental gap conjecture, *J. Amer. Math. Soc.* **24** (2011), 899–916. Zbl 1222.35130

[21] S. Andrieux, A. Ben Abda, M. Jaoua, On the inverse emergent plane crack problem, *Math. Methods Appl. Sci.* **21**(10) (1998), 895–906. Zbl 0916.35129

[22] P. Antunes, P. Freitas, Numerical optimization of low eigenvalues of the Dirichlet and Neumann Laplacians, *J. Optim. Theory Appl.* **154**(1) (2012), 235–257. Zbl 1252.90076

[23] M. S. Ashbaugh, Open problems on eigenvalues of the Laplacian, in *Analytic and Geometric Inequalities and Their Applications* **4787**, T. M. Rassias, H. M. Srivastava (eds.), Kluwer, 1999. Zbl 1158.35396

[24] M. S. Ashbaugh, R. Benguria, Proof of the Payne-Pólya-Weinberger conjecture, *Bull. Amer. Math. Soc.* **25**(1) (1991), 19–29. Zbl 0736.35075

[25] M. S. Ashbaugh, R. Benguria, On Rayleigh's conjecture for the clamped plate and its generalization to three dimensions, *Duke Math. J.* **78** (1995), 1–17. Zbl 0935.74041

[26] M. S. Ashbaugh, R. Benguria, Isoperimetric inequalities for eigenvalues of the Laplacian, *Spectral Theory and Mathematical Physics: A Festschrift in Honor of Barry Simon's 60th Birthday,* Proc. Sympos. Pure Math. **76**, Part 1, Amer. Math. Soc., Providence, RI, 2007, 105–139. Zbl 1221.35261

[27] M. S. Ashbaugh, D. Bucur, On the isoperimetric inequality for the buckling of a clamped plate, *Z. angew. Math. Phys.* **54**(5) (2003), 756–770. Special issue dedicated to Lawrence E. Payne. Zbl 1030.49043

[28] H. Attouch, *Variational Convergence for Functions and Operators*, Applicable Mathematics Series, Pitman, London, 1984. Zbl 0561.49012

[29] H. Attouch, C. Picard, Variational inequalities with varying obstacles: the general form of the limit problem, *J. Funct. Anal.* **50**(3) (1983), 329–386.

[30] C. Bandle, *Isoperimetric Inequalities and Applications*, Pitman, London, 1980. Zbl 0436.35063

[31] N. Banichuk, *Introduction to Optimization of Structures*, Springer, New York, 1990. Zbl 0708.73051

[32] M. Barkatou, A. Henrot, Un résultat d'existence en optimisation de forme en utilisant une propriété géométrique de la normale, *ESAIM Control Optim. Calc. Var.* **2** (1997), 105–123. Zbl 0920.49027

[33] E. Bednarczuk, M. Pierre, E. Rouy, J. Sokolowski, Tangent sets in some functional spaces, *Nonlinear Anal. Ser. A: Theory Methods* **42**(5) (2000), 871–886. Zbl 0995.90097

[34] Z. Belhachmi, D. Bucur, G. Buttazzo, J. M. Sac-Epée, Shape optimization problems for eigenvalues of elliptic operators, *ZAMM Z. angew. Math. Mech.* **86**(3) (2006), 171–184. Zbl 1087.49032

[35] M. Bendsoe, *Optimization of Structural Topology, Shape, and Material*, Springer, 1995. Zbl 0822.73001

[36] M. Bendsoe, N. Kikuchi, Generating optimal topologies in structural design using a homogenization method, *Comp. Meth. Appl. Mech. Eng.* **71** (1988), 197–224. Zbl 0671.73065

[37] M. Bendsoe, C. Mota Soares, *Topology Optimization of Structures*, Kluwer Academic Press, Dordrecht, 1993. Zbl 0789.73004

[38] M. Bendsoe, O. Sigmund, *Topology Optimization, Theory, Methods and Applications*, Springer, 2004. Zbl 1059.74001

[39] A. Berger, The eigenvalues of the Laplacian with Dirichlet boundary conditions in $\mathbb{R}^2$ are almost never minimized by disks, *Ann. Glob. Anal. Geom.* **47**(3) (2015), 285–304. Zbl 1346.35133

[40] A. Berger, *Optimisation du spectre du Laplacien avec conditions de Dirichlet et de Neumann dans* $\mathbb{R}^2$ *et* $\mathbb{R}^3$. Thesis, Universities of Neuchâtel and Grenoble, 2015.

[41] M. Berger, *Convexes et polytopes, polyèdres réguliers, aires et volumes*, CEDIC, Paris; Nathan Information, Paris, 1978. Zbl 0423.51001

[42] M. Berger, B. Gostiaux, *Differential Geometry: Manifolds, Curves and Surfaces*, Graduate Texts in Mathematics **115**, Springer, 1988. Zbl 0629.53001

[43] A. Beurling, *On Free Boundary Problems for the Laplace Equation*, Seminars on Analytic Functions I, Institute Advance Studies Seminars (1958), Princeton, pp. 248–263. Zbl 0099.08302

[44] C. Bianchini, A. Henrot, P. Salani, An overdetermined problem with non-constant boundary condition, *Interfaces Free Bound.* **16**(2) (2014), 215–241. Zbl 1297.35153

[45] W. Blaschke, *Kreis und Kugel*, Leipzig, 1916. Zbl 46.1109.01

[46] A. Bonnet, Caractérisation des minima globaux de la fonctionnelle de Mumford-Shah en segmentation d'images, *C. R. Acad. Sci. Paris Sér. I Math.* **321**(8) (1995), 1121–1126. Zbl 0920.49022

[47] M. Bonnet, Shape identification using acoustic measurements: a numerical investigation using BIE and shape differentiation, *Inverse Problems in Engineering Mechanics (Tokyo, 1992)*, Springer, Berlin, 1992, 191–200. Zbl 0829.73029

[48] M. H. Bossel, Membranes élastiquement liées: extension du théorème de Rayleigh-Faber-Krahn et de l'inégalité de Cheeger, *C. R. Acad. Sci. Paris Sér. I Math.* **302**(1) (1986), 47–50. Zbl 0606.73018

[49] G. Bouchitté, I. Fragalà, I. Lucardesi, Shape derivatives for minima of integral functionals, *Math. Program., Ser. B* **148**(1) (2014), 111–142. Zbl 1304.49086

[50] G. Bouchitté, I. Fragalà, I. Lucardesi, A variational method for second order shape derivatives, *SIAM J. Control Optim.* **54**(2) (2016), 1056–1084. Zbl 1338.49085

[51] B. Bourdin, A. Chambolle, Implementation of an adaptive finite-element approximation of the Mumford-Shah functional, *Numer. Math.* **85**(4) (2000), 609–646. Zbl 0961.65062

[52] M. Boutkrida, J. Mossino, G. Moussa, On nonhomogeneous reinforcements of varying shape and different exponents, *Boll. Unione Mat. Ital. Sez. B Artic. Ric. Mat.* **8**(2) (1999), 517–536. Zbl 0935.49005

[53] A. Braides, G. Dal Maso, Non-local approximation of the Mumford-Shah functional, *Calc. Var. Partial Differential Equations* **5**(4) (1997), 293–322. Zbl 0873.49009

[54] H. Brezis, *Analyse fonctionnelle*, Masson, Paris, 1983. Zbl 0511.46001

[55] H. Brezis, F. Browder, Sur une propriété des espaces de Sobolev, *C. R. Acad. Sci.* **287** (1978), 113–115. Zbl 0381.46019

[56] T. Briançon, Regularity of optimal shapes for the Dirichlet's energy with volume constraint, *ESAIM: Control Optim. Calc. Var.* **10** (2004), 99–122. Zbl 1118.35078

[57] T. Briançon, M. Hayouni, M. Pierre, Lipschitz continuity of state functions in some optimal shaping, *Calc. Var. Partial Differential Equations* **23**(1) (2005), 13–32. Zbl 1062.49035

[58] T. Briançon, J. Lamboley, Regularity of the optimal shape for the first eigenvalue of the Laplacian with volume and inclusion constraints, *Ann. Inst. H. Poincaré Anal. Non Linéaire* **26**(4) (2009), 1149–1163. Zbl 1194.49059

[59] F. Brock, An isoperimetric inequality for eigenvalues of the Stekloff problem, *ZAMM Z. angew. Math. Mech.* **81**(1) (2001), 69–71. Zbl 0971.35055

[60] F. Brock, V. Ferone, B. Kawohl, A symmetry problem in the calculus of variations, *Calc. Var. Partial Differential Equations* **4**(6) (1996), 593–599. Zbl 0856.49018

[61] K. Bryan, L. F. Caudill, Stability and reconstruction for an inverse problem for the heat equation, *Inverse Problems* **14**(6) (1998), 1429–1453. Zbl 0914.35152

[62] D. Bucur, Characterization for the Kuratowski limits of a sequence of Sobolev spaces, *J. Differential Equations* **151** (1999), 1–19. Zbl 0924.46028

[63] D. Bucur, Uniform concentration-compactness for Sobolev spaces on variable domains, *J. Differential Equations* **162**(2) (2000), 427–450. Zbl 0957.49027

[64] D. Bucur, Minimization of the k-th eigenvalue of the Dirichlet Laplacian, *Arch. Ration. Mech. Anal.* **206**(3) (2012), 1073-1083. Zbl 1254.35165

[65] D. Bucur, G. Buttazzo, *Variational Methods in Shape Optimization Problems*, Progress in Nonlinear Differential Equations and Their Applications **65**, Birkhäuser Boston, MA, 2005. Zbl 1117.49001

[66] D. Bucur, G. Buttazzo, I. Figueiredo, On the attainable eigenvalues of the Laplace operator, *SIAM J. Math. Anal.* **30** (1999), 527–536. Zbl 0920.35099

[67] D. Bucur, G. Buttazzo, A. Henrot, Existence results for some optimal partition problems, *Adv. Math. Sci. Appl.* **8**(2) (1998), 571–579. Zbl 0915.49006

[68] D. Bucur, G. Buttazzo, C. Nitsch, Symmetry breaking for a problem in optimal insulation, *J. Math. Pures Appl. (9)* **107**(4) (2017), 451–463.

[69] D. Bucur, A. Henrot, Minimization of the third eigenvalue of the Dirichlet Laplacian, *Proc. Roy. Soc. London* **456** (2000), 985–996. Zbl 0974.35082

[70] D. Bucur, A. Henrot, A new isoperimetric inequality for elasticae, *J. Eur. Math. Soc. (JEMS)* **19**(11) (2017), 3355–3376. Zbl 06802927

[71] D. Bucur, A. Henrot, J. Sokolowski, A. Zochowski, Continuity of the elasticity system solutions with respect to the geometrical domain variations, *Adv. Math. Sci. Appl.* **11**(1) (2001), 57–73. Zbl 1001.74008

[72] D. Bucur, D. Mazzoleni, A surgery result for the spectrum of the Dirichlet-Laplacian, *SIAM J. Math. Anal.* **47**(6) (2015), 4451–4466. Zbl 1331.49062

[73] D. Bucur, N. Varchon, Boundary variation for a Neumann problem, *Ann. Scuola Norm. Sup. Pisa Cl. Sci.* **4**(29) (2000), 807–821. Zbl 1072.35063

[74] D. Bucur, N. Varchon, A duality approach for the boundary variation of Neumann problems, *SIAM J. Math. Anal.* **34** (2002), 460–477. Zbl 1055.35037

[75] D. Bucur, J. P. Zolésio, N-dimensional shape optimization under capacitary constraints, *J. Differential Equations* **123**(2) (1995), 504–522. Zbl 0847.49029

[76] D. Bucur, J. P. Zolésio, Anatomy of the shape Hessian via Lie brackets, *Ann. Mat. Pura Appl. (IV)* **173** (1997), 127–143. Zbl 0939.49023

[77] G. Buttazzo, Thin insulating layers: The optimization point of view, *Material Instabilities in Continuum Mechanics (Edinburgh, 1985–1986)*, pp. 11–19, Oxford Science Publications, Oxford University Press, New York, 1988. Zbl 0645.73015

[78] G. Buttazzo, A survey on the Newton problem of optimal profiles, in *Variational Analysis and Aerospace Engineering*, Springer Optimization and Its Applications **33** (2009), pp. 33–48. Zbl 1182.49041

[79] G. Buttazzo, G. Dal Maso, An existence result for a class of shape optimization problems, *Arch. Rational Mech. Anal.* **122** (1993), 183–195. Zbl 0811.49028

[80] G. Buttazzo, G. Dal Maso, A. Garroni, A. Malusa, On the relaxed formulation of some shape optimization problems, *Adv. Math. Sci. Appl.* **7**(1) (1997), 1–24. Zbl 0971.49024

[81] G. Buttazzo, V. Ferone, B. Kawohl, Minimum problems over sets of concave functions and related questions, *Math. Nachr.* **173** (1995), 71–89. Zbl 0835.49001

[82] G. Buttazzo, B. Kawohl, On Newton's problem of minimal resistance, *Math. Intelligencer* **15**(4) (1993), 7–12. Zbl 0800.49038

[83] E. Cabib, G. Dal Maso, On a class of optimum problems in structural design, *J. Optim. Theory Appl.* **56** (1988), 39–65. Zbl 0617.49005

[84] X. Cabré, X. Ros-Oton, Sobolev and isoperimetric inequalities with monomial weights, *J. Differential Equations* **255**(11) (2013), 4312–4336. Zbl 1293.46018

[85] X. Cabré, X. Ros-Oton, J. Serra, Sharp isoperimetric inequalities via the ABP method, *J. Eur. Math. Soc. (JEMS)* **18**(12) (2016), 2971–2998. Zbl 1357.28007

[86] L. A. Caffarelli, J. Spruck, Convexity properties of solutions to some classical variational problems, *Comm. Partial Differential Equations* **7** (1982), 1337–1379. Zbl 0508.49013

[87] J. Céa, *Optimisation, théorie et algorithmes*, Dunod, Paris, 1971. Zbl 0211.17402

[88] J. Céa, Problems of shape optimal design, in *Optimization of Distributed Parameters Structures, Part II*, Haug, Céa (eds.), Sijthoff& Noordhoff, Alphen aan den Rijn, 1981, 1005–1048. Zbl 0517.73096

[89] J. Céa, S. Garreau, Ph. Guillaume, M. Masmoudi, The shape and topological optimizations connection, *Comput. Methods Appl. Mech. Engrg.* **188**(4) (2000), 713–726. Zbl 0972.74057

[90] J. Céa, A. J. Gioan, J. Michel, Quelques résultats sur l'identification de domaines, *Calcolo* **10** (1974), 207–232. Zbl 0303.93023

[91] A. Chambolle, G. Dal Maso, Discrete approximation of the Mumford-Shah functional in dimension two, *M2AN Math. Model. Numer. Anal.* **33**(4) (1999), 651–672. Zbl 0943.49011

[92] A. Chambolle, F. Doveri, Continuity of Neumann linear elliptic problems on varying two-dimensional bounded open sets, *Comm. Partial Differential Equations* **22** (1997), 811–840. Zbl 0901.35019

[93] A. Chambolle, C. Larsen, C^∞ regularity of the free boundary for a two-dimensional optimal compliance problem, *Calc. Var. Partial Differential Equations* **18**(1) (2003), 77–94. Zbl 1026.49031

[94] T. Chatelain, M. Choulli, Clarke generalized gradient for eigenvalues, *Comm. Appl. Anal.* **1**(4) (1997), 443–454. Zbl 0896.47016

[95] D. Chenais, On the existence of a solution in a domain identification problem, *J. Math. Anal. Appl.* **52** (1975), 189–289. Zbl 0317.49005

[96] D. Chenais, Sur une famille de variétés à bord lipschitziennes, application à un problème d'identification de domaine, *Ann. Inst. Fourier* **4**(27) (1977), 201–231. Zbl 0333.46020

[97] A. Cherkaev, R. V. Kohn (eds.), *Topics in the Mathematical Modelling of Composite Materials*, Progress in Nonlinear Differential Equations and Their Applications **31**, Birkhäuser, Boston, 1997. Zbl 0870.00018

[98] A. Cherkaev, K. Lurie, Effective characteristics of composite materials and the optimal design of structural elements, *Uspekhi Mekhaniki* **9** (1986), 3–81.

[99] J. Choe, M. Ghomi, M. Ritoré, The relative isoperimetric inequality outside convex domains in $\mathbb{R}^n$, *Calc. Var. Partial Differential Equations* **29**(4) (2007), 421–429. Zbl 1116.58016

[100] G. Choquet, Theory of capacities, *Ann. Inst. Fourier* **5** (1953/54), 131–295. Zbl 0064.35101

[101] G. Choquet, *Lectures on Analysis, Vol. I*, W. A. Benjamin, 1969. Zbl 0181.39602

[102] M. Choulli, A. Henrot, Use of the domain derivative to prove symmetry results in partial differential equations, *Math. Nachr.* **192** (1998), 91–104. Zbl 0912.35114

[103] A. Cianchi, N. Fusco, Functions of bounded variation and rearrangements, *Arch. Ration. Mech. Anal.* **165**(1) (2002), 1–40. Zbl 1028.49035

[104] P. G. Ciarlet, *Mathematical Elasticity: Three Dimensional Elasticity*, Elsevier Science Publishers, 1988. Zbl 0648.73014

[105] D. Cioranescu, F. Murat, A strange term from outer space, *Proceedings of Seminars of Collège de France: Nonlinear PDEs and Their Applications*, vol. 2, Research Notes in Math. **60**, H. Brezis, J. L. Lions (eds.), Pitman, 1982. Zbl 0496.35030

[106] R. Cominetti, J. P. Penot, Tangent sets to unilateral sets, *C. R. Acad. Sci. Paris, Sér. I* **321** (1995), 1631–1636. Zbl 0866.49026

[107] G. Cortesani, Asymptotic behaviour of a sequence of Neumann problems, *Comm. Partial Differential Equations* **22** (1997), 1691–1729. Zbl 0895.35011

[108] O. Coulaud, A. Henrot, Numerical approximation of a free boundary problem arising in electromagnetic shaping, *SIAM J. Num. Anal.* **31** (1994), 1109–1127. Zbl 0804.65129

[109] S. J. Cox, The generalized gradient at a multiple eigenvalue, *J. Funct. Anal.* **133** (1995), 30–40. Zbl 0884.47008

[110] S. J. Cox, B. Kawohl, P. X. Uhlig, On the optimal insulation of conductors, *J. Optim. Theory Appl.* **100**(2) (1999), 253–263. Zbl 0915.90264

[111] S. J. Cox, M. Ross, Extremal eigenvalue problems for starlike planar domains, *J. Differential Equations* **120** (1995), 174–197. Zbl 0837.35101

[112] S. J. Cox, M. Ross, The maximization of Neumann eigenvalues on convex domains. Preprint available from the authors.

[113] S. J. Cox, P. X. Uhlig, Where best to hold a drum fast, *SIAM J. Optim.* **9**(4) (1999), 948–964. Zbl 0966.35092

[114] R. Courant, D. Hilbert, *Methods of Mathematical Physics*, Vols. 1 and 2, Wiley, New York, 1953 and 1962. Zbl 0051.28802, Zbl 0099.29504

[115] M. Crouzeix, Variational approach of a magnetic shaping problem, *Eur. J. Mech. B Fluids* **10** (1991), 527–536. Zbl 0741.76089

[116] G. Dal Maso, *An Introduction to* Γ*-Convergence*, Birkhäuser, Boston, 1993. Zbl 0816.49001

[117] G. Dal Maso, A. Defranceschi, Limits of nonlinear Dirichlet problems in varying domains, *Manuscripta Math.* **61** (1988), 251–278. Zbl 0653.49017

[118] G. Dal Maso, U. Mosco, Wiener's criterion and Γ-convergence, *Appl. Math. Optim.* **15** (1987), 15–63. Zbl 0644.35033

[119] G. Dal Maso, F. Murat, Asymptotic behaviour and correctors for Dirichlet problems in perforated domains with homogeneous monotone operators, *Ann. Scuola Norm. Sup. Pisa, Cl. Sci.* **24**(4) (1997), 239–290. Zbl 0899.35007

[120] M. Dambrine, On variations of the shape hessian and sufficient conditions for the stability of critical shapes, *Rev. R. Acad. Cien., Serie A Math.* **96**(1), (2002), 95–121. Zbl 1025.49025

[121] M. Dambrine, M. Pierre, About stability of equilibrium shapes, *M2AN Math. Model. Numer. Anal.* **34**(4), (2000), 811–834. Zbl 0966.49023

[122] D. Daners, A Faber-Krahn inequality for Robin problems in any space dimension, *Math. Ann.* **335** (2006), 767–785. Zbl 1220.35103

[123] R. Dautray and J. L. Lions (eds.), *Analyse mathématique et calcul numérique*, Vols. I and II, Masson, Paris, 1984. Zbl 0683.35001 Zbl 0664.47001

[124] G. David, C^1-arcs for minimizers of the Mumford-Shah functional, *SIAM J. Appl. Math.* **56**(3) (1996), 783–888. Zbl 0870.49020

[125] G. David, Global minimizers of the Mumford-Shah functional, in *Current Developments in Mathematics, 1997, Cambridge, MA*, International Press, Boston, MA, 1997, 219–224. Zbl 0927.00019

[126] G. David, Singular sets of minimizers for the Mumford-Shah functional, *Progress in Mathematics* **233**, Birkhäuser, Boston, 2005. Zbl 1086.49030

[127] G. David, S. Semmes, On the singular sets of minimizers of the Mumford-Shah functional, *J. Math. Pures Appl.* **9**(75) (1996), 299–342. Zbl 0853.49010

[128] E. de Giorgi, Sulla proprieta isoperimetrica dell'ipersfera, nelle classi degli insiemi aventi frontiera orientata di misura nita, *Memorie Accad. Naz. Lincei, serie 8* **5**, 1958. Zbl 0116.07901

[129] M. Delfour, J. P. Zolésio, Velocity method and Lagrangian formulation for the computation of the shape Hessian, *SIAM Control and Optimization* **29**(6) (1991), 1414–1442. Zbl 0747.49007

[130] M. Delfour, J. P. Zolésio, Shape analysis via oriented distance functions, *J. Funct. Anal.* **123** (1994), 120–201. Zbl 0814.49032

[131] M. Delfour, J. P. Zolésio, *Shapes and Geometries. Analysis, Differential Calculus, and Optimization*, second edition. Advances in Design and Control **22**. Society for Industrial and Applied Mathematics (SIAM), Philadelphia, PA, 2011. Zbl 1251.49001

[132] C. Dellacherie, *Ensembles analytiques, capacités, mesures de Hausdorff*, Lecture Notes in Mathematics **295**. Springer, Berlin-New York, 1972. Zbl 0259.31001

[133] N. Dunford, J. T. Schwartz, *Linear Operators*, Pure and Applied Mathematics **7**, Interscience Publishers, New York, London 1958. Zbl 0084.10402

[134] K. Eppler, Fréchet-differentiability and sufficient optimality conditions for shape functionals, in *Optimal Control of P.D.E.* **133**, K. H. Hoffmann, G. Leugering, F. Tröltzsch (eds.), Birkhäuser (1999), 133–143. Zbl 0935.49019

[135] L. C. Evans, R. F. Gariepy, *Measure Theory and Fine Properties of Functions*, Studies in Advanced Math., CRS Press, 1992. Zbl 0804.28001

[136] G. Faber, Beweis, dass unter allen homogenen Membranen von gleicher Fläche und gleicher Spannung die kreisförmige den tiefsten Grundton gibt, *Sitz. Ber. Bayer. Akad. Wiss.* (1923), 169–172. Zbl 49.0342.03

[137] H. Federer, *Geometric Measure Theory*, Die Grundlehren der Mathematischen Wissenschaften [Fundamental Principles of Mathematical Sciences] **153**, Springer New York, New York 1969. Zbl 0176.00801

[138] R. Finn, *Equilibrium Capillary Surfaces*, Grundlehren der Mathematischen Wissenschaften [Fundamental Principles of Mathematical Sciences] **284**. Springer, New York, 1986. Zbl 0583.35002

[139] M. Flucher, M. Rumpf, Bernoulli's free-boundary problem, qualitative theory and numerical approximation, *J. reine angew. Math.* **486** (1997), 165–204. Zbl 0909.35154

[140] J. Frehse, Capacity methods in the theory of partial differential equations, *Jahresber. Dtsch. Math.-Ver.* **84** (1982), 1–44. Zbl 0486.35002

[141] P. Freitas, D. Krejčiřík, The first Robin eigenvalue with negative boundary parameter, *Adv. Math.* **280** (2015), 322–339. Zbl 1317.35151

[142] G. Frémiot, Eulerian semiderivatives of the eigenvalues for Laplacian in domains with cracks, *Adv. Math. Sci. Appl.* **12**(1) (2002), 115–134. Zbl 1030.49044

[143] A. Friedman, Reinforcement of the principal eigenvalue of an elliptic operator, *Arch. Rat. Mech. Anal.* **73** (1980), 1–17. Zbl 0439.35052

[144] A. Friedman, B. Gustafsson, Identification of the conductivity coefficient in an elliptic equation, *SIAM J. Math. Anal.* **18**(3) (1987), 777–787. Zbl 0644.35091

[145] A. Friedman, D. Phillips, The free boundary of a semilinear elliptic equation, *Trans. Amer. Math. Soc.* **282** (1984), 153–182. Zbl 0552.35079

[146] A. Friedman, M. Vogelius, Determining cracks by boundary measurements, *Indiana Univ. Math. J.* **38**(3) (1989), 527–556. Zbl 0697.35165

[147] N. Fujii, Second order necessary conditions in a domain optimization problem, J. Optim. Theory Appl. **65**(2) (1990), 223–245. Zbl 0675.49021

[148] P. R. Garabedian, M. Schiffer, Convexity of domain functionals, J. Anal. Math. **2** (1953), 281–368. Zbl 0052.33203

[149] S. Garreau, Ph. Guillaume, M. Masmoudi, The topological asymptotic for PDE systems: the elasticity case, *SIAM J. Control Optim.* **39**(6) (2001), 1756–1778. Zbl 0990.49028

[150] B. Gidas, W. M. Ni, L. Nirenberg, Symmetry and related properties via the maximum principle, *Comm. Math. Phys.* **68** (1979), 209–243. Zbl 0425.35020

[151] D. Gilbarg, N. S. Trudinger, *Elliptic Partial Differential Equations of Second Order*, Reprint of the 1998 edition, Classics in Mathematics, Springer, Berlin, 2001. Zbl 1042.35002

[152] A. Girouard, N. Nadirashvili, I. Polterovich, Maximization of the second positive Neumann eigenvalue for planar domains, *J. Differential Geom.* **83**(6) (2009), 637–661. Zbl 1186.35120

[153] A. Girouard, I. Polterovich, Spectral geometry of the Steklov problem (survey article), *J. Spectr. Theory*, **7**(2) (2017), 321–359. Zbl 06737069

[154] E. Giusti, *Minimal Surfaces and Functions of Bounded Variations*, Monograph in Math. **80**, Birkhäuser, 1984. Zbl 0545.49018

[155] R. D. Grigorieff, Diskret kompakte Einbettungen in Sobolewschen Räumen, *Math. Ann.* **197** (1972), 71–85. Zbl 0222.35002

[156] P. Grisvard, *Elliptic Problems in Nonsmooth Domains*, Monographs and Studies in Mathematics **24**, Pitman, Boston, MA, 1985. Zbl 0695.35060

[157] M. Grun-Rehomme, Caractérisation du sous-différentiel d'intégrandes convexes dans les espaces de Sobolev, *J. Math. Pures Appl.* **56** (1977), 149–156. Zbl 0314.35001

[158] Ph. Guillaume, Intrinsic expression of the derivatives in domain optimization problems, *Numer. Funct. Anal. Optim.* **17**(1–2) (1996), 93–112. Zbl 0846.49020

[159] Ph. Guillaume, M. Masmoudi, Dérivées d'ordre supérieur en optimisation de domaines, *C. R. Acad. Sci. Paris* **315** Série I (1992), 859–862. Zbl 0758.49014

[160] Ph. Guillaume, M. Masmoudi, Calcul numérique des dérivées d'ordre supérieur en conception optimale de formes, *C. R. Acad. Sci. Paris* **316**, Série I (1992), 1091–1096. Zbl 0758.49014

[161] Ph. Guillaume, M. Masmoudi, Computation of high order derivatives in optimal shape design, *Numer. Math.* **67** (1994), 231–250. Zbl 0792.65044

[162] J. Hadamard, *Mémoire sur le problème d'analyse relatif à l'équilibre des plaques élastiques encastrées*, (1908), dans Oeuvres de J. Hadamard, CNRS Paris 1968. Zbl 39.1022.01

[163] E. M. Harrell, P. Kröger, K. Kurata, On the placement of an obstacle or a well so as to optimize the fundamental eigenvalue, *SIAM J. Math. Anal.* **33**(1) (2001), 240–259. Zbl 0994.47015

[164] J. Haslinger, R. A. E. Mäkinen, *Introduction to Shape Optimization. Theory, Approximation, and Computation*, Advances in Design and Control, SIAM, Philadelphia, PA, 2003. Zbl 1020.74001

[165] J. Haslinger, P. Neitaanmäki, *Finite Element Approximation for Optimal Shape, Material and Topology Design*, second edition, John Wiley & Sons, Chichester, 1996. Zbl 0845.73001

[166] J. Hass, M. Hutchings, R. Schlafly, The double bubble conjecture, *Electron. Res. Announc. Math. Soc.* **1**(3) (1995), 98–102. Zbl 0864.53007

[167] J. Hass, R. Schlafly, Double bubbles minimize, *Ann. of Math.* **151**(2) (2000), 459–515. Zbl 0970.53008

[168] B. Hauchecorne, D. Suratteau, *Des mathématiciens de A à Z*, Ellipses, 1996. Zbl 0901.01025

[169] F. Hausdorff, *Set Theorie*, Chelsea, 1962. Zbl 0060.12401

[170] V. P. Havin, V. G. Maz'ja, Nonlinear potential theory, *Uspehi Mat. Nauk* **27** (1972), 67–138. Zbl 0247.31010

[171] M. Hayouni, *Existence et régularité pour des problèmes d'optimisation de formes*, Thèse de l'Université Henri Poincaré, 1997.

[172] M. Hayouni, Lipschitz continuity of the state function in a shape optimization problem, *J. Conv. Anal.* **6**(1) (1999), 71–90. Zbl 0948.49021

[173] M. Hayouni, Sur la minimisation de la première valeur propre du laplacien, *C. R. Acad. Sci. Paris Sér. I Math.* **330**(7) (2000), 551–556. Zbl 0960.49028

[174] M. Hayouni, A. Novruzi, Sufficient condition for the existence of solutions of a free boundary problem, *Quart. Appl. Math.* **60**(3) (2002), 425–435. Zbl 1054.35146

[175] M. Hayouni, M. Pierre, Domain continuity for an elliptic operator of fourth order, *Commun. Contemp. Math.* **4**(1) (2002), 1–14. Zbl 1008.35019

[176] L. I. Hedberg, Spectral synthesis and stability in Sobolev spaces, *Euclidean Harmonic Analysis*, Lecture Notes in Mathematics **779**, J. J. Benedetto (ed.), Springer, Berlin 1980. Zbl 0469.31003

[177] J. Heinonen, T. Kilpeläinen, O. Martio, *Nonlinear Potential Theory of Degenerate Elliptic Equations*, Oxford Science Publications, 1993. Zbl 0780.31001

[178] L. L. Helms, *Introduction to Potential Theory*, R. Krieger, 1975. Zbl 0188.17203

[179] A. Henrot, Subsolutions and supersolutions in a free boundary problem, *Ark. Mat.* **32**(1) (1994), 79–98. Zbl 0809.35172

[180] A. Henrot, Minimization problems for eigenvalues of the Laplacian, *J. Evol. Equ.* **3** (2003), 443–461. Zbl 1049.49029

[181] A. Henrot, *Extremum Problems for Eigenvalues of Elliptic Operators*, Frontiers in Mathematics, Birkhäuser, Basel, 2006. Zbl 1109.35081

[182] A. Henrot (ed.), *Shape Optimization and Spectral Theory*, de Gruyter, Berlin, 2017 (open access). Zbl 06705915

[183] A. Henrot, H. Maillot, Optimization of the shape and the location of the actuators in an internal control problem, *Boll. Unione Mat. Ital. Sez. B Artic. Ric. Mat. 8* **4**(3) (2001), 737–757. Zbl 1177.49061

[184] A. Henrot, E. Oudet, Minimizing the second eigenvalue of the Laplace operator with Dirichlet boundary conditions, *Arch. Ration. Mech. Anal.* **169** (2003), 73–87. Zbl 1055.35080

[185] A. Henrot, M. Pierre, Un problème inverse en formage des métaux liquides, *(*M2AN Math. Model. Numer. Anal.) **23** (1989), 155–177. Zbl 0672.65101

[186] A. Henrot, M. Pierre, About existence of equilibria in electromagnetic casting, Quart. Appl. Math. **49** (1991), 563–575. Zbl 0731.76092

[187] A. Henrot and M. Pierre, *About Critical Points of the Energy in an Electromagnetic Shaping Problem*, Lecture Notes in Control and Information Sciences **178**, J. P. Zolézio (ed.), 1992, 238–253. Zbl 0820.35139

[188] A. Henrot, M. Pierre, *Variation et optimisation de formes. Une analyse géométrique*, Mathématiques et Applications **48**, Springer, Berlin, Heidelberg, 2005. Zbl 1098.49001

[189] A. Henrot, M. Pierre, M. Rihani, Finite dimensional reduction for the positivity of some second shape derivatives, *Methods Appl. Anal.* **10**(3) (2003), 1–20. Zbl 1068.49031

[190] A. Henrot, M. Pierre, M. Rihani, Positivity of the shape Hessian and instability of some equilibrium shapes, *Med. J. Math.* **1** (2004), 195–214. Zbl 1072.49028

[191] A. Henrot, D. Seck, Retour à une ancienne approche pour un problème à frontière libre classique, *Publications Mathématiques de Besançon - Analyse non linéaire* **15** (1997), 29–40. Zbl 0996.35088

[192] A. Henrot, H. Shahgholian, Convexity of free boundaries with Bernoulli type boundary condition, *Nonlinear Anal.* **28** (1997), 815–823. Zbl 0863.35117

[193] A. Henrot, G. Villemin, An optimum design problem in magnetostatics, *M2AN Math. Model. Numer. Anal.* **36**(2) (2002), 223–239. Zbl 1054.49030

[194] A. Henrot, D. Zucco, Optimizing the first Dirichlet eigenvalue of the Laplacian with an obstacle, to appear, see arxiv.org/abs/1702.01307

[195] J. Hersch, The method of interior parallels applied to polygonal or multiply connected membranes, *Pacific J. Math.* **13** (1963), 1229–1238. Zbl 0127.40603

[196] F. Hirsch, G. Lacombe, *Éléments d'analyse fonctionnelle*, Masson, Paris, 1997. Zbl 0878.46001

[197] L. Holzleitner, Hausdorff convergence of domains and boundaries for shape optimum design, *Control Cybernet.* **30**(1) (2001), 23–44. Zbl 1011.49029

[198] I. Hong, On an inequality concerning the eigenvalue problem of membrane, *Kodai Math. Sem. Rep.* **6** (1954), 113–114. Zbl 0057.08805

[199] A. Jameson, L. Martinelli, Aerodynamic shape optimization techniques based on control theory, *Computational Mathematics Driven by Industrial Problems (Martina Franca, 1999)*, Lecture Notes in Mathematics **1739**, Springer, Berlin, 2000, 151–221. Zbl 0954.00021

[200] D. D. Joseph, Parameter and domain dependence of eigenvalues of elliptic partial differential equations, *Arch. Rat. Mech. Analysis* **24** (1967), 325–401. Zbl 0152.11302

[201] J.-P. Kahane, P.-G. Lemarié-Rieusset, *Séries de Fourier et ondelettes*, Cassini, 1998. Zbl 0966.42001

[202] T. Kato, *Perturbation Theory for Linear Operators*, Springer, 1966. Zbl 0148.12601

[203] O. Kavian, *Introduction à la théorie des points critiques*, Mathématiques et Applications **13**, Springer, 1993. Zbl 0797.58005

[204] B. Kawohl, *Rearrangements and Convexity of Level Sets in PDE*, Lecture Notes in Mathematics **1150**, Springer, 1985. Zbl 0593.35002

[205] B. Kawohl, Some qualitative properties of nonlinear partial differential equations, *Nonlinear Diffusion Equations and Their Equilibrium States, II (Berkeley, CA, 1986)*, Mathematical Sciences Research Institute Publications **13**, Springer, New York-Berlin, 1988, 19–31. Zbl 0666.35035

[206] B. Kawohl, Symmetry or not?, *Math. Intelligencer* **20**(2) (1998), 16–22. Zbl 0947.00004

[207] B. Kawohl, O. Pironneau, L. Tartar, J. P. Zolésio, *Optimal Shape Design. Lectures given at the Joint C.I.M./C.I.M.E. Summer School held in Tróia, June 1–6, 1998*, A. Cellina, A. Ornelas (eds.) Lecture Notes in Mathematics **1740**, 2000. Zbl 0954.00031

[208] M. V. Keldyš, On the solvability and the stability of the Dirichlet problem, *Amer. Math. Soc. Trans.* **2**(51) (1966), 1–73. Zbl 0179.43901

[209] O. D. Kellog, *Foundations of Potential Theory*, Frederick Ungar, New York, 1929. Zbl 55.0282.01

[210] J. Kennedy, An isoperimetric inequality for the second eigenvalue of the Laplacian with Robin boundary conditions, *Proc. Amer. Math. Soc.* **137**(2) (2009), 627–633. Zbl 1181.35156

[211] S. Kesavan, On two functionals connected to the Laplacian in a class of doubly connected domains, *Proc. Roy. Soc. Edinburgh Sect. A* **133**(3) (2003), 617–624. Zbl 1046.35021

[212] S. Kesavan, *Symmetrization and Applications*, Series in Analysis **3**, World Scientific, 2006. Zbl 1110.35002

[213] E. Ya. Khruslov, The method of orthogonal projections and the Dirichlet problem in domains with a fine-grained boundary, (Russian) *Mat. Sb. (N.S.)* **88** (130) (1973), 38–60. Zbl 0251.35032

[214] E. Ya. Khruslov, The first boundary value problem in domains with a complex boundary for higher order equations, (Russian) *Mat. Sb. (N.S.)* **103(145)**(2(6)) (1977), 614–629, 632. Zbl 0396.35039

[215] R. V. Kohn, G. Strang, Optimal design and relaxation of variational problems I-II-III, *Comm. Pure Appl. Math.* **39** (1986), 113–137, 139–182, 353–377. Zbl 0609.49008, Zbl 0621.49008, Zbl 0694.49004

[216] E. Krahn, Über eine von Rayleigh formulierte Minimaleigenschaft des Kreises, *Math. Ann.* **94** (1924), 97–100. Zbl 51.0356.05

[217] E. Krahn, Über Minimaleigenschaften der Kugel in drei und mehr Dimensionen, *Acta Comm. Univ. Dorpat.* **A9** (1926), 1–44. Zbl 52.0510.03

[218] T. Lachand-Robert, E. Oudet, Minimizing within convex bodies using a convex hull method, *SIAM J. Optim.* **16** (2005)(2), 368–379. Zbl 1104.65056

[219] T. Lachand-Robert, M. Peletier, Newton's problem of the body of minimal resistance in the class of convex developable functions, *Math. Nachr.* **226** (2001), 153–176. Zbl 1048.49011

[220] J. Lamboley, About Hölder-regularity of the convex shape minimizing λ_2, *Appl. Anal.***90**(2) (2011), 263–278. Zbl 1209.49059

[221] J. Lamboley, A. Novruzi, M. Pierre, Estimates of first and second order shape derivatives in nonsmooth multidimensional domains and applications, *J. Funct. Anal.* **270**(7) (2016), 2616–2652. Zbl 1345.49050

[222] J. Lamboley, M. Pierre, Structure of shape derivatives around irregular domains and applications, *J. Convex Anal.* **14**(4) (2007), 807–822. Zbl 1123.49034

[223] K. E. Lancaster, Qualitative behavior of solutions of elliptic free boundary problems, Pacific J. Math. **154** (1992), 297–316. Zbl 0769.35066

[224] N. S. Landkof, *Foundations of Modern Potential Theory*, Springer, Berlin 1972. Zbl 0253.31001

[225] E. Laporte, P. Le Tallec, *Numerical Methods in Sensitivity Analysis and Shape Optimization*, Birkhäuser, 2003. Zbl 1013.74002

[226] J. C. Leger, Flatness and finiteness in the Mumford-Shah problem, *J. Math. Pures Appl. 9* **78** (1999), 431–459. Zbl 0942.49030

[227] J. L. Lewis, Capacitary functions in convex rings, *Arch. Rat. Mech. Anal.* **66** (1977), 201–224. Zbl 0393.46028

[228] P. L. Lions, The concentration-compactness principle in the calculus of variations. The locally compact case. I, II, *Ann. Inst. H. Poincaré Anal. Non Linéaire* **1**(2, 4) (1984), 109–145, 223–283. Zbl 0541.49009, Zbl 0704.49004

[229] V. A. Marchenko, E. Ya. Hruslov, *Boundary Value Problems in Domains with a Fine-Grained Boundary*, Izdat. Naukova Dumka, Kiev, 1974. Zbl 0289.35002

[230] V. G. Maz'ja, *Sobolev Spaces*, Springer, 1985. Zbl 0692.46023

[231] D. Mazzoleni, A. Pratelli, Existence of minimizers for spectral problems, *J. Math. Pures Appl.* **100**(3) (2013), 433–453. Zbl 1296.35100

[232] N. G. Meyers, A theory of capacities for potentials of functions in Lebesgue classes, *Math. Scand.* **26** (1970), 255–292. Zbl 0242.31006

[233] A. M. Micheletti, Metrica par famiglie di domini limitati e proprietà generiche degli autovalori, *Ann. Sc. Norm. Pisa Ser.* **26**(3) (1972), 683–694. Zbl 0255.35028

[234] F. Mignot, F. Murat, J. P. Puel, Variation d'un point de retournement par rapport au domaine, *Comm. Partial Differential Equations* **4** (1979), 1263–1297. Zbl 0422.35039

[235] B. Mohammadi, Mesh adaption and automatic differentiation for optimal shape design, *Int. J. Comput. Fluid Dyn.* **10**(3) (1998), 199–211. Zbl 0924.76088

[236] B. Mohammadi, O. Pironneau, *Applied Shape Optimization for Fluids*, second edition. Numerical Mathematics and Scientific Computation, Oxford University Press, Oxford, 2009. Zbl 1179.65002

[237] J. M. Morel, The Mumford-Shah conjecture in image processing, Séminaire Bourbaki, vol. 1995/96; *Astérisque* **241** (1997), Exp. No. 813, 4, 221–242. Zbl 0879.49002

[238] F. Morgan, *Geometric Measure Theory, A Beginner's Guide*, Academic Press, 1995. Zbl 0819.49024

[239] J. Mossino, *Inégalités isopérimétriques et applications en physique*, Travaux en Cours, Hermann, Paris, 1984. Zbl 0537.35002

[240] A. Munnier, Thèse de l'Université de Franche-Comté, Besançon, 2000.

[241] F. Murat, The Neumann sieve, *Nonlinear Variational Problems (Isola d'Elba, 1983)*, Research Notes in Mathematics **127**, Pitman, Boston MA, London, 1985, 24–32. Zbl 0586.35037

[242] F. Murat, J. Simon, *Quelques résultats sur le contrôle par un domaine géométrique*, Publication du Laboratoire d'Analyse Numérique de l'Université Paris 6 (1974), 1–46.

[243] F. Murat, J. Simon, *Sur le contrôle par un domaine géométrique*, Publication du Laboratoire d'Analyse Numérique de l'Université Paris 6 **189** (1976).

[244] F. Murat, L. Tartar, *Calcul des variations et homogénéisation*, Homogenization methods: theory and applications in physics, Collect. Dir. Études Rech. Elec. France **57**, Eyrolles, Paris, 1985, 319–369.

[245] F. Murat, L. Tartar, Calculus of variations and homogenization, *Topics in the Mathematical Modelling of Composite Materials*, A. Cherkaev et al. (eds.), Progress in Nonlinear Differential Equations and Their Applications **31**, Birkhäuser Boston, Boston, MA, 1997, 139–173. Zbl 0939.35019

[246] N. S. Nadirashvili, Rayleigh's conjecture on the principal frequency of the clamped plate, *Arch. Rational Mech. Anal.* **129** (1995), 1–10. Zbl 0826.73035

[247] A. Novotny, J. Sokołowski, *Topological Derivatives in Shape Optimization.* Interaction of Mechanics and Mathematics. Springer, Heidelberg, 2013. Zbl 1276.35002

[248] A. Novruzi, M. Pierre, Structure of shape derivatives, *J. Evol. Equ.* **2**(3) (2002), 365–382. Zbl 1357.49150

[249] Y. S. Osipov, A. P. Suetov, A problem of J.-L. Lions, *Dokl. Akad. Nauk SSSR* **276**(2) (1984), 288–291. Zbl 0595.35043

[250] R. Osserman, The isoperimetric inequality, *Bull. Amer. Math. Soc.* **84** (1978), 1182–1238. Zbl 0411.52006

[251] E. Oudet, Numerical minimization of eigenmodes of a membrane with respect to the domain, *ESAIM Control Optim. Calc. Var.* **10** (2004), 315–335. Zbl 1076.74045

[252] L. E. Payne, Isoperimetric inequalities and their applications, *SIAM Rev.* **9** (1967), 453–488. Zbl 0154.12602

[253] L. E. Payne, Some comments on the past fifty years of isoperimetric inequalities, *Inequalities (Birmingham, 1987)*, Lecture Notes in Pure and Applied Mathematics **129**, Dekker, New York, 1991, 143–161. Zbl 0723.52003

[254] M. Pierre, J. R. Roche, Numerical simulation of tridimensional electromagnetic shaping of liquid metals, *Numer. Math.* **65**(1) (1993), 203–217. Zbl 0792.65096

[255] O. Pironneau, *Optimal Shape Design for Elliptic Systems*, Springer Series in Computational Physics, Springer, New York, 1984. Zbl 0534.49001

[256] A. Plakhov, Newton's problem of minimal resistance under the single-impact assumption, *Nonlinearity* **29**(2) (2016), 465–488. Zbl 1339.49041

[257] P. Plotnikov, J. Sokołowski, *Compressible Navier-Stokes Equations. Theory and Shape Optimization*, Monografie Matematyczne. Instytut Matematyczny PAN **73**, Birkhäuser-Springer Basel AG, Basel, 2012. Zbl 1260.35002

[258] G. Polya, Torsional rigidity, principal frequency, electrostatic capacity and symmetrization, *Quart. Appl. Math.* **6** (1948), 267–277. Zbl 0037.25301

[259] G. Polya, On the characteristic frequencies of a symmetric membrane, *Math. Z.* **63** (1955), 331–337. Zbl 0065.08703

[260] G. Polya, G. Szegö, *Isoperimetric Inequalities in Mathematical Physics*, Annals of Mathematics Studies **27**, Princeton University Press, 1951. Zbl 0044.38301

[261] U. E. Raitum, Extension of extremal problems connected with linear elliptic equations, *Dokl. Akad. Nauk SSSR* **243** (1978), 281–283.

[262] J. Rauch, M. Taylor, Potential and scattering on wildly perturbed domain, *J. Funct. Anal.* **18** (1975), 27–59. Zbl 0293.35056

[263] P. A. Raviart, J. M. Thomas, *Introduction à l'analyse numèrique des équations aux dérivées partielles*, Masson, 1983. Zbl 0561.65069

[264] F. Rellich, Darstellung der eigenwerte $\Delta u + \lambda u$ durch ein randintegral, *Math. Z.* **46** (1940), 635–646. Zbl 0023.04204

[265] B. Rousselet, Shape design sensitivity of a membrane, *J. Optim. Theory Appl.* **40** (1983), 595–623. Zbl 0497.73097

[266] W. Rudin, *Real and Complex Analysis*, McGraw-Hill, 1974. Zbl 0278.26001

[267] G. Salinetti, R. J.-B. Wets, On the convergence of sequences of convex sets in finite dimensions, *SIAM Rev.* **21** (1979), 18–33. Zbl 0421.52003

[268] E. Sanchez-Palencia, *Nonhomogeneous Media and Vibration Theory*, Lecture Notes in Physics **127**, Springer, Berlin-New York, 1980. Zbl 0432.70002

[269] G. Savaré, G. Schimperna, Domain perturbations and estimates for the solutions of second order elliptic equations, *J. Math. Pures Appl.* **81** (2002), 1071–1112 . Zbl 1112.35312

[270] M. Schiffer, Hadamard's formula and variations of domain-functions, *Amer. J. Math.* **68** (1946), 417–448. Zbl 0060.23706

[271] E. Schmidt, Uber das isoperimetrische Problem in Raum von n dimensionen, *Math Z.* **44** (1939), 689–788. Zbl 65.0833.02

[272] R. Schoen, S.-T. Yau, *Lectures on Differential Geometry*, Conference Proceedings and Lecture Notes in Geometry and Topology **1**, International Press, Cambridge, MA, 1994. Zbl 0830.53001

[273] H. A. Schwarz, *Gesammelte mathematische Abhandlungen* **2**, Springer, Berlin 1890. Zbl 22.0031.04

[274] L. Schwartz, *Théorie des distributions*, Hermann, Paris, 1966. Zbl 0149.09501

[275] J. Serrin, A symmetry problem in potential theory, *Arch. Rational Mech. Anal.* **43** (1971), 304–318. Zbl 0222.31007

[276] H. Shahgholian, Quadrature surfaces as free boundaries, *Ark. Mat.* **32** (1994), 475–492. Zbl 0827.31004

[277] J. Simon, Differentiation with respect to the domain in boundary value problems, *Num. Funct. Anal. Optimiz.* **2** (1980), 649–687. Zbl 0471.35077

[278] J. Simon, Variations with respect to domain for Neumann condition, *Control of Distributed Parameter Systems, Proceedings of the 1986 IFAC Congress at Pasadena*, Pergamon Press, 1987, 395–400. Zbl 0693.93004

[279] I. M. Singer, B. Wong, S.-T. Yau, An estimate of the gap of the first two eigenvalues in the Schrödinger operator, *Ann. Scuola Norm. Sup. Pisa Cl. Sci. (4)* **12**(2) (1985), 319–333. Zbl 0603.35070

[280] J. Sokołowski, J. P. Zolesio, *Introduction to Shape Optimization. Shape Sensitivity Analysis*, Springer Series in Computational Mathematics **16**, Springer, Berlin, 1992. Zbl 0761.73003

[281] J. Sokołowski, A. Zochowski, On the topological derivative in shape optimization, *SIAM J. Control Optim.* **37** (1999), 1251–1272. Zbl 0940.49026

[282] M. Spivak, *Differential Geometry*, vol. 4, 1975. Zbl 0306.53002

[283] J. Steiner, *Gesammelte Werke*, vol. 2, Berlin 1882. Zbl 14.0018.04

[284] F. Stummel, Perturbation theory for Sobolew spaces, *Proc. Roy. Soc. Edinburgh Sect. A* **1** (1974/75), 5–49. Zbl 0296.47014

[285] V. Šverak, On optimal shape design, *J. Math. Pures Appl.* **72**(6) (1993), 537–551. Zbl 0849.49021

[286] G. Szegö, Inequalities for certain eigenvalues of a membrane of given area, *J. Rational Mech. Anal.* **3** (1954), 343–356. Zbl 0055.08802

[287] L. Tartar, An introduction to the homogenization method in optimal design, in *Optimal Shape Design (Tróia, 1998)*, A. Cellina, A. Ornelas (eds.), Lecture Notes in Mathematics **1740**, Springer, Berlin, 2000, 47–156. Zbl 1040.49022

[288] D. E. Tepper, Free boundary problem - the starlike case, *SIAM J. Math. Anal.* **6** (1975), 503–505. Zbl 0299.31004

[289] M. van den Berg, On condensation in the free-boson gas and the spectrum of the Laplacian, *J. Stat. Phys.* **31**(3) (1983), 623–637.

[290] H. F. Weinberger, An isoperimetric inequality for the N-dimensional free membrane problem, *J. Rational Mech. Anal.* **5** (1956), 633–636. Zbl 0071.09902

[291] H. F. Weinberger, Remark on the preceding paper of Serrin, *Arch. Rational Mech. Anal.* **43** (1971), 319–320. Zbl 0222.31008

[292] R. Weinstock, Inequalities for a classical eigenvalue problem, *J. Rational Mech. Anal.* **3** (1954), 745–753. Zbl 0056.09801

[293] S. A. Wolf, J. B. Keller, Range of the first two eigenvalues of the Laplacian, *Proc. R. Soc. London A* **447** (1994), 397–412. Zbl 0816.35097

[294] S.-T. Yau, Problem section, *Seminar on Differential Geometry*, Annals of Mathematics Studies **102**, Princeton University Press, Princeton, NJ, 1982, 669–706. Zbl 0479.53001

[295] S.-T. Yau, Open problems in geometry, *Differential Geometry. Part 1: Partial Differential Equations on Manifolds, (Los Angeles, CA, 1990)*, Proc. Symp. Pure Math. **54**, AMS, Providence, RI, 1993, 1–28. Zbl 0801.53001

[296] Q. Yu, J. Q. Zhong, Lower bounds of the gap between the first and second eigenvalues of the Schrödinger operator, *Trans. Amer. Math. Soc.* **294**(1) (1986), 341–349. Zbl 0593.53030

[297] L. Zalcman, A bibliographical survey of the Pompeiu problem, *Approximation by Solutions of Partial Differential Equations*, B. Fuglede et al. (eds.), Kluwer, 1992, 185–194. Zbl 0830.26005

[298] W. P. Ziemer, *Weakly Differentiable Functions*, Graduate Texts in Mathematics, Springer, Berlin, 1989. Zbl 0692.46022

General Index

Index of Mathematicians